Springer Aerospace Technology

The *Springer Aerospace Technology* series is devoted to the technology of aircraft and spacecraft including design, construction, control and the science. The books present the fundamentals and applications in all fields related to aerospace engineering. The topics include aircraft, missiles, space vehicles, aircraft engines, propulsion units and related subjects.

More information about this series at http://www.springer.com/series/8613

Kozlov A. I. · Logvin A. I. · Sarychev V. A. ·
Shatrakov Y. G. · Zavalishin O. I.

Introduction to the Theory of Radiopolarimetric Navigation Systems

Springer

Kozlov A. I.
Moscow, Russia

Logvin A. I.
Moscow, Russia

Sarychev V. A.
Saint-Petersburg, Russia

Shatrakov Y. G.
Saint-Petersburg, Russia

Zavalishin O. I.
Moscow, Russia

ISSN 1869-1730 ISSN 1869-1749 (electronic)
Springer Aerospace Technology
ISBN 978-981-13-8394-6 ISBN 978-981-13-8395-3 (eBook)
https://doi.org/10.1007/978-981-13-8395-3

This Springer imprint is published by the registered company Springer Nature Singapore Pte Ltd.
The registered company address is: 152 Beach Road, #21-01/04 Gateway East, Singapore 189721, Singapore

Foreword

Now you are holding a book in your hand which occupies a specific place in this series of books devoted to the radio navigation systems. Therefore, it shall arouse special interest among its potential readers. The other books in this series we have offered to your attention so far have examined the navigation technologies supported by the corresponding radio electronic systems. Its technological effectiveness is due to the fact that the transport systems function in a specially created navigation environment (field) where all actions are quite strictly defined. Thus, all the processes "take place inside the intellect" of the vehicle, and we should only "insert" the current parameters of the transport means and take the instructions received on board the plane into account, according to some predetermined rules, if we have to make some route adjustments. You may come across some surprises in the navigational sighting, if "there are some things nobody has ever thought of before" or if there is an unpredicted failure situation, which might also be specially created by some "dark forces," especially since there is an increase in their number and equipment status, and they become more skillful, inventive and perfect year by year. For example, it has become necessary to take some measures to ensure the security of the information the civil aviation authorities have started thinking about quite recently.

Besides, the unmanned and individual (private) flights are making their way in the world self-confidently, no matter what kind of restrictions they may have on the state level. Of course, there can be more failure situations in this field, which is conditioned by their individual nature. According to Gödel's incompleteness theorem widely-known today, which can be applied to any technological system, we may handle the unplanned situations only due to the functional capabilities of such systems. In this case, we should be able to fend off the challenges we may come across because of the damages or destruction of the navigation space (field) created before. In this case, we should be prepared to get along without its services as a last resort, and provide reliable navigation definitions "by our own means." In such cases the book we suggest you to read suggests you to declare about the autonomous navigation, if the location of the vehicle and the intentions of the people participating in the transportation of goods are estimated only by using the means of information available on board the vehicle or those available by using ground-based navigation systems.

The necessity to provide navigation in any weather and at any time of the day automatically makes it possible to use autonomous navigation systems and on-board radar systems which are quite powerful tools for providing information. By the way, it should be noted that such autonomous systems have appeared on board the aircraft for a long time. These are, first of all, the on-board meteorological radars and collision avoidance systems which typically use, for example, the traffic alert and collision avoidance system (TCAS); radar systems created for preventing the collision of the aircraft with the ground; autonomous landing systems, etc.

So, there are always some radar modules on board the aircraft which should be "taught to observe more things and to do it more thoroughly" in order to make it possible to take the necessary actions to provide safe flights (although these can be only local flights) or to complete the aerial work. Thus, we can say that the autonomous navigation is closely connected with the necessity to increase the information capabilities and characteristics of the on-board radar systems. The authors of this book suggested to take this path, throwing away the achievements obtained in the field of radio physics, or, to be more precise, one of its powerful branches called "radio polarimetry," especially since the radio-polarimetric information is based mainly on the peculiarities of the radar observation object (radar target), as the changes in the polarization state of the transceiver signal are caused only by the geometrical and electrophysical characteristics of the radar target. It highlights the additional information about the observation objects which is so necessary for making much more grounded decisions when dealing with autonomous navigation in the field of the on-board radar system.

Perhaps that's why the authors of this book who have focused their attention mainly on the radiophysical aspects of expanding the capabilities of the autonomous radio navigation system have called it "Introduction." It's up to the readers of this book to decide how to use the possibilities provided by radio polarimetry for expanding the functional capacities of both on-board and ground radio navigation systems. Of course, we may come to this conclusion when we speak about the information systems used for the provision of other means of transport. The structure and the logic of the composition of the book can be easily restored according to its table of contents.

Therefore, we should not discuss them right now, especially because each of the authors has worked in the field of civil aviation for some time and is widely-known for his/her numerous publications on different issues relating to radio polarimetry and avionics not only in our country, but also abroad.

I am sure that the book presented to your attention will be interesting, first of all, for the founders of avionics. It will also be interesting for the users of avionics. It will certainly take its worthy place in the training of specialists in relevant spheres of activity.

Moscow, Russia

Lukin D. S.
Honored Worker of Sciences of the Russian Federation
Laureate of State Prize of the USSR
Professor, Doctor of Physical and Mathematical Sciences

Preface: Autonomous Radio Navigation Systems State and Prospects

Despite the “all-transport” title of this book, almost all attention is given to problems associated with the navigation support of air transport. This is due to the fact that since air transport operates in 4D-space (time + spatial coordinates), and even with the most saturated dynamics, all that is used in aviation transport can be used by other modes of transport. In its turn, high dynamics will lead to the fact that the management of transport makes it necessary to take into account also all the accelerations in the movement of aircraft, which are also manifested in the vectorial hypostasis. So the phase space, where the appropriate control actions for aviation transport are formed, has a dimension much more than 4D, which, incidentally, is observed in other modes of transport, but for aviation transport this dimension is always more. Today, air transportation is carried out within the framework of a specially created and maintained navigation information field (now global), the actions in which are strictly regulated by relevant regulatory enactments. Generally speaking, actions in such a field significantly simplify the procedures of navigational definitions, including due to restrictions on the dynamics of aircraft.

However, despite the marked regulation of both the navigation field itself and the actions in it, the multidimensional dynamics for aircraft and the continuous growth of air transportation still lead to a powerful intellectualization of air transport, where the main target (imperative) is the desire to transform information confidence and security into functional the possibilities of means of transport and, first of all, to ensure the safety of transportation and the compaction of flights, especially in transport saturated regions. Moreover, this information confidence allows “putting” additional air flights into the current traffic regulations. Today even more often the thesis is formulated that information resources are a kind of “alternative fuel” of the twenty-first century.

This allows us to believe that now as a measure of the level of development of aviation transport, the assessment of the level of its equipping with information systems, which in critical situations is able to select the effective tactics of action (first of all from the point of view of ensuring security) for the crew of an aircraft, is becoming an increasingly important measure. Such situations can occur in the event of a violation of the navigational field created for the provision of air transportation, with the actions of pilots of private air transport, in the process of performing aviation work.

The aforementioned possibility of densifying air traffic based on the corresponding information equipment of the navigation field and the increasing rate of issuance of navigational definitions has led to the emergence of a new mode of transport—multimodal, which can be called "synthetic" because it uses technology involving several "classical" modes of transport during transporting goods and passengers. The fact of the existence of multimodal (intermodal) transport is not particularly disputed today. Moreover, its operation is trying to implement in a single navigation field, the basis of which is the means of information support of precisely air transportation. Navigation fields supporting other modes of transport interact with the "aviation field" with network technologies with appropriate mutual adaptation.

The functioning of multimodal transport requires the deployment of a branched and powerful navigation field, actively using satellite and network technologies for its maintenance. As a rule, the functioning of multimodal transport is associated with the creation of multimodal corridors and the setting of the priority of cargo transportation.

It is obvious that the factors such as

- the failure situations caused, among other things, by the actions of various criminal and terrorist groups;
- carrying out of aerial works;
- organization of multimodal transportations;
- actions of private aviation pilots which are often unpredictable;
- increasing penetration of unmanned aerial systems;
- inhomogeneity (heterogeneity) of the provision of information and maintenance of the navigation field (for example, because of its being far from the relevant centres or because of the influence of the navigation restrictions caused by the conditions of observation);
- possible non-regulated actions (collisions) of the other participants taking part in air transportation of goods;
- the need for joint effective use of the navigation space by civilian aircraft and by military aviation;
- and flying in regions with rapidly changing weather conditions or unexpected natural or anthropogenic influences may easily lead to the destruction of the global navigation field for some field or to its local disturbances causing the crew of the aircraft to take control by themselves to some degree of completeness. In this sense, a local navigation field around the aircraft is created, where the crew requires much more detailed navigation information about the current situation that has arisen than it is able to provide a global navigational field, especially if it is violated. Actions in such a local navigation field in the book are called autonomous navigation. The source of the local navigation field is mainly the information means of the aircraft, primarily radar, as today only

they are able to implement the appropriate information support of the aircraft irrespective of the weather conditions and observation time, in the contaminated atmosphere at the distances demanded for making the required corrections of the flight route.

The autonomous radio navigation acquires particular importance during transport works, such as the elimination of the consequences of natural and technological disasters, environmental monitoring, actions taken in conflict areas, as well as construction and assembly work. In this case the navigation field is highlighted, formed, shaped and analyzed, and its state is carefully estimated.

Of course, the knowledge of the autonomous navigation and corresponding "local" navigation field are used in any concept related to the development of air transport to some extent. Thus, the concept of the CNS/ATM (communication, navigation, surveillance/air traffic management) system proposed by the ICAO (International Civil Aviation Organization) and its services are aimed at maintaining safe intervals between the aircraft and the obstacles within the air space, except the "canonical" ATC (air traffic control) services. The ATM services include: the air traffic flow management (ATFM) used for the optimal traffic flow management during certain periods of time when the needs exceed the carry-on capacity of the ATC in the dispatch area); the airspace management (ASM) carried out by the dynamic distribution of the temporary windows and, in some cases, through the distribution of the parts of the airspace among different categories of consumers, based on their short-term needs; the flight information service (FIS) aimed at collecting, processing and disseminating the flight information for the purpose of helping the pilots to carry out safe and effective flights; and the warning about the danger of the alert level (AL) aimed at ensuring the starting speed of searching and rescuing the aircraft in distress. It is required to create a corresponding local navigation field in which autonomous navigation is implemented for the realization of all these functions. The air transportation procedures which are the essence of the ATM are provided by on-board and ground-based information systems which carry out telecommunication, navigation and surveillance functions.

Special interest in the implementation of autonomous navigation arouses the realization of the CNS/ATM concept of the monitoring function, which represents the process of information support for air traffic participants, to some extent responsible for the state, safety and regularity of air transportation, as well as for aviation work and aircraft testing , which provides appropriately structured information on the coordinates of the aircraft and its state necessary for making decisions on aircraft management and airspace organization.

Especially it is necessary to allocate monitoring systems dedicated to correct flight trajectories and aircraft actions based on the observation of structures and formations dangerous for flight. At the same time, the problem is not only put on obstacle detection to the flight of an aircraft, but also on recognition and identification, and also on judging the aircraft safety. (Here we use the triad of concepts: detection, recognition, DRI identification which is widely used in foreign literature. In this context detection is associated with target selection against the background of a variety of types of interference, recognition with distinction of some target observed in the group with another class targets, as for the identification, it is connected with the target differentiation from same class targets).

This group of observing systems includes weather observation systems located both on Earth and on aircraft, the purpose of which is to provide air traffic participants with the meteorological information obtained as a result of a variety of methods of observation sufficient for a safe flight of the aircraft. Currently, the main means of remote monitoring of meteorological conditions in civil aviation are radar, more precisely, on-board and ground-based primary radar systems PSR. There is a consistent trend in providing any on-board radar facilities with functions of detection and assessment of meteorological hazards. Particular attention has recently been drawn to the problem of estimating the W/S (windshear) by radar means.

The entire meteorological information received from various sources is concentrated and prepared to provide the crew with integrated ITWS (Integrated Terminal Weather Systems). These systems display both current meteorological information and a short-term weather forecast with the identification of phenomena that are especially important for the aircraft in the state of the atmosphere.

By the implementation of surveillance functions, airborne collision avoidance systems play an important role, for example, the TCAS (Airborne) Traffic Alert Collision Avoidance System, which fixes the dangerous approach of another aircraft. The aforementioned systems provide the pilot with information on the separation. These systems can ensure the creation of warning and protection zones around the aircraft, as well as the issuance of information allowing the pilot to control and resolve possible collisions. The CNS/ATM concept was based on the recognition that TCAS-type on-board equipment will be able to provide the necessary information about the surrounding airspace and objects therein, especially in areas where it is difficult to communicate with the Earth, and there is no opportunity to help to prevent the threat of an accident in which case—Either errors in the communication channel or from the controller. The on-board TCAS system operates independently of ground equipment, providing the pilot with advisory information on potential conflict situations with aircraft (more specifically, on the parameters of the conflict situation).

To prevent the risk of collision with the Earth, especially in the absence of visibility, the aircraft must be equipped with appropriate "observational" means of warning of a dangerous approach to the Earth. These systems allow the aircraft to increase the time for maneuver in the presence of an appropriate dangerous situation. They are intended to provide a controlled flight with a rounding into the terrain CFIT (Controlled Flight Into Terrain). To this class should include surveillance systems that serves individual stages of vehicle movement—automatic parking systems, security alarm systems, automatic landing systems for aircraft, information means of robotic vehicles, etc.

With the development of surveillance systems, the individuality of those of other air transportations (especially private ones and with the help of unmanned vehicles) is increasingly ensured—transportation technologies are increasingly approaching the use of information on current traffic. Such information for the management of vehicles and the state of transport space is intended to deliver just a variety of surveillance means combined into information networks. The intellectualization of transport greatly facilitated the growth of the role of the semantic information (about the nature and danger of the objects of observation) and the pragmatic (about the intentions of objects in the zone of the aircraft's actions) information in information flows organized in transport complexes for their operation. In the theory of artificial intelligence, it is precisely the increase in the specific gravity of the semantic and pragmatic information used by the system that is considered as a necessary condition for the formation of databases and knowledge that are the main object of research for this theory.

A fundamentally new type of navigation services, directly stemming from the main target of the CNS/ATM concept, is the transition to Free Flights on the most profitable trajectories for users, instead of the current order of aircraft movement along strictly regulated corridors. In its turn, the management of the movement of mass transport flows within the transport technologies "for the client" assumes the mandatory deployment of automated traffic control systems, both of single vehicles and their aggregate. Such management requires equipping the navigation field with a set of information sensors capable of obtaining semantic and pragmatic information. The most important and most significant sensors should be placed on the vehicles themselves and it is necessary to organize the appropriate exchange of current information between the participants in the transport process (in the CNS/ATM concept this task is realized by the monitoring function).

The most problematic issues of the development of intelligent radio electronic information support for aircraft operations are in state (military) aviation. Obviously, this thesis simply does not require any special comments. The tasks that are being solved today by the aviation complexes of military aircraft are accompanied by huge information loads (and overloads) on the radio electronic systems used. And here the only way out can be offered—to move within the framework of autonomous navigation to modern intellectual technologies, specially dedicated for "increase" of bottlenecks in information flows.

In order to streamline and structure information flows about the environment, they must contain information redundancy, that is, the vehicle crew makes decisions based on a variety of information about a complex and dynamically changing situation. Hence the need to include in the airborne radio electronic complex a variety of information sensors, most of which work as part of providing autonomous radionavigation at the limit of its sensitivity or even catching only "traces" of such information.

After we have substantiated the relevance and direction of efforts to create autonomous radio navigation systems for civil aviation and the transport complex in general, we will focus on one very interesting and, in our opinion, very promising direction in the creation of such systems—radio polarimetric. In turn, the radio polarimetric trend develops within the framework of radiophysics, a science whose ancestor is considered in our country by Corresponding Member of the USSR Academy of Sciences Sergei Mikhailovich Rytov. He is the author of the remarkable two-volume "Introduction to Statistical Radiophysics" (1976 and 1978), actively used in scientific research on radio polarimetry and in training relevant specialists. In 1961, unfortunately, in a hard-to-get edition he published a very interesting article "Influence of the Faraday effect in the ionosphere on the backscattering diameter of radar targets" where the main research paths in radar polarimetry were anticipated. In this article, such issues were thoroughly considered as the scattering matrix with divergent input and output polarization bases, the own polarization of scatterers having a scattering plane, the total polarization reception, the possibility of virtual polarization selection, the propagation channel influence on the scattering matrix measurement results, the normalization of the scattering matrix elements, methods for measuring the scattering matrix. C. M. Rytov studied in detail the coherent structure of the signals, and he also substantiated the prospects for using the phase-dissipative theorem for processing signals from radiometric systems. In 1938 he formulated a theorem, now called the S. M. Rytov theorem, about the polarization structure of the rays of electromagnetic waves.

Our choice of radio polarimetric direction for increasing the functionality of radio polarimetric autonomous navigation systems is due to the fact that scattered (reflected) radio waves carry much more information about radar targets and scattering objects than can be obtained by classical radar, which estimates only the coordinate characteristics of the object of observation (availability and number of objects in the resolution element, range, speed, angular coordinates). That is why the interest in radio-polarimetric methods is so stable today, which made it possible in principle to use practically all the information contained in the received radio wave. Radio polarimetry allows, in principle, to obtain non-coordinate information about the characteristics of objects of observation, which we have already linked above with semantic and pragmatic information about the region in which the aircraft is autonomously forced to operate. Under non-coordinate (non-kinematic) parameters (characteristics) of targets, we mean information about the type, class of the object, its dimensions, electrophysical properties of the surface, structural properties, as well as the intentions of the object and its significance. Obviously, for autonomous navigation, and even in critical conditions, any information about the non-coordinate characteristics of objects surrounding the aircraft of the situation is truly priceless. Today, in civil aviation, such information is recorded in sufficient volume and with good quality only with the radio-meteorological support of flights.

The polarization structure of signals is also actively used in telecommunications, radio navigation, television, technical diagnostics, medicine, scientific experiment, by ensuring information security. A practically immense area of applications of methods and achievements of radio polarimetry consists of systems and complexes of weapons and military equipment.

In the area of radio polarimetry, the studies of Russian (before the collapse of the Soviet Union—Soviet) scientists have been and are proceeding in a common direction with foreign research in radiophysics. Among foreign scientists working in the radio polarimetric direction, we would like to mention here Huynen I. R., Boerner W. M., Ligthart L. P., Saillard J. and Pottier E., Yamaguchi, Krogager E., Czyz Z. H., Luneburg E., Ke J., Poelman A. J., Mott H., Carpenter D.,

Beckmann P. and many others, whose works are actively used by our scientists and engineers. Of the numerous Russian scientists who left their noticeable mark in radio polarimetry, we will mention here our colleagues: Potekhin V. A., Kanareikin D. B., Tatarinov V. N., Rodimov A. P., Popovsky V. V., Melititsky V. A., Pozdnyak S. I., Gusev K. G., Bogorodsky V. V., Kutuza B. G., Bakhrakh L. D., Dulevich V. E., Akinshin N. V., Pavlov N. F., Khlusov V. A., Tuchkov L. T., Korostelev A. A.., Schestopalov Yu., Stepanenko V. D., Shchukin G. G., Shevchenko V. V., Rusinov V. R., Zhivotovskij L. N., Mosionzchic A. Andreev G. N., Starich A. N., Markelov V. N., Grishenko A. A., Mironov O. G., Polyansky V. A., Pozdyshev V. E., Ostrovityanov R. V., Monakov A. A., Maslov V. Yu., Lartsov S. V., Bad Length N. N., Demidov Y. M., Kolyadov V. A., Lutin E. A., Kijan O. N., Gorbachev A. A., Korabljev A. Ju., Vagapov R. Ch., Isaeva O. V., Egorova E. S., Sulaev S. A., and others, most of which were quite seriously "noted" in civil and military aviation, as well as their numerous publications in foreign publications. Here it is pertinent to mention the most fundamental scientific publication on issues and problems of radio polarimetric—a three-volume monograph, "Polarization of radio waves", written by the authors of this book by A. I. Kozlov, A. I. Logvin and A. I., Sarychev V. A. ("Polarization of radio waves", Moscow, 2005, 2007, 2008, publishing house "Radiotekhnika", total volume—1900 pages).

The purpose of this book is to expand the penetration of radio polarimetric technologies into autonomous radio navigation systems, primarily those that support air transportation. This circumstance immediately deduces the proposed book from the stream of books about radio navigation systems. The point is that the emphasis on radio polarimetric technologies forces us to address not to the onboard and ground-based radioelectronic systems forming the navigation field, but to the remote observation objects themselves and to the processes responsible for the formation of the corresponding signals with imprinting in them non-coordinate information about objects where confidently the semantic and pragmatic components prevail. In other words, interest in such autonomous navigation systems immediately brings to the fore the radiophysics with its specific research discourse and mathematical apparatus. No existing computer-aided design technologies, such as CAD/CAM/CAE, are able to bypass this difficulty, because every significant result in radiophysics, as a rule, radically disturbs the information technologies used in the creation of radio navigation and radar systems. Besides, such technologies in radiophysics provide a powerful imprint of the individuality of the formation of signals by specific objects, moreover so that the main efforts for the radiophysical provision of radio navigation systems are aimed at identifying this individuality (i.e., non-coordinate characteristics).

Thus, the book to your attention is largely radiophysical rather than radio navigational, but, in our opinion, without adequate radio-physical support, any noticeable progress in autonomous radionavigation systems is simply impossible. This circumstance is intended to form an expected circle of readers, which should be quite wide, because the book considers the origins of signal generation for such a very important class of radio navigation systems as autonomous, and the structure and characteristics of such signals undoubtedly "pull" for themselves effective algorithms for their processing, directly involved in the formation of technologies for managing vehicles in those new situations and realities that we are constantly exposed to the rapidly changing modern world where one of the factors and mechanisms affecting it is the transport complex, primarily the aviation one.

Moscow, Russia Kozlov A. I.
Moscow, Russia Logvin A. I.
Saint-Petersburg, Russia Sarychev V. A.
Saint-Petersburg, Russia Shatrakov Y. G.
Moscow, Russia Zavalishin O. I.

Contents

Part I
Polarization Structure of the Autonomous Radionavigation Systems Signals

Chapter 1
Radiophysical Provision of Radio Polarimetric Navigation Systems

1.1 Basic Concepts and Definitions of Radio Polarimetry

The electromagnetic field, as a carrier of information about the objects under study, is a field of interrelated vector values—the electric (electric vector $\vec{E}$) and magnetic (magnetic vector $\vec{H}$) field intensity. Any spatial or temporal variation in one of them changes the spatial–temporal distribution of the other in accordance with Maxwell's equation:

$$\left.\begin{aligned} &\operatorname{rot}\vec{H} - \frac{\partial \vec{D}}{\partial t} = \vec{j} \\ &\operatorname{rot}\vec{E} + \frac{\partial \vec{B}}{\partial t} = 0 \\ &\operatorname{div}\vec{D} = \rho \\ &\operatorname{div}\vec{B} = 0, \end{aligned}\right\} \tag{1.1}$$

where $\vec{D}$ is the electrical vector, $\vec{B}$ is the vector of magnetic induction, $\vec{j}$ is the current density vector, and ρ is the free charge density.

The relation of field vectors in some linear isotropic environment is usually characterized by equations:

$$\vec{B} = \mu\vec{H};\ \vec{D} = \varepsilon\vec{E};\ \vec{j} = \sigma\vec{E} + \vec{j}_{\text{CT}}, \tag{1.2}$$

where μ is the magnetic capacity, ε dielectric capacity, σ electrical conductivity, and $\vec{j}_{\text{CT}}$ extraneous current density.

In case of anisotropic environment, the relation between $\vec{D}$ and $\vec{E}$ vectors, as well as between $\vec{B}$ and $\vec{H}$, is described by the relations: $\vec{D} = \widehat{\varepsilon}\vec{E},\ \vec{B} = \widehat{\mu}\vec{H}$, in which $\widehat{\varepsilon}$ and $\widehat{\mu}$ represent some tensors. For non-uniform environments, values ε, μ, and σ are coordinate functions, and in case of properties of the environment changes in time–time functions.

Kozlov A. I. et al., *Introduction to the Theory of Radiopolarimetric Navigation Systems*, Springer Aerospace Technology, https://doi.org/10.1007/978-981-13-8395-3_1

For some environments, for example, to the ionosphere, the presence of ε and σ dependence from the direction of electromagnetic wave propagation is also significant.

The term of the electromagnetic waves polarization is directly related to Maxwell's equation vector nature. Due to the fact that relationship between the vectors $\vec{E}$ and $\vec{H}$ the knowledge of at least one of them, in principle, allows to determine the second, so the concept of polarization is associated directly only with the $\vec{E}$ electric vector.

Over time, the end of $\vec{E}$ vector defined at some point in space will describe some curve, which can either be in a plane or be in some spatial curve. Since the direction of the electromagnetic wave propagation is always perpendicular to the $\vec{E}$ vector, the first case relates to the situation where this direction does not change during the whole observation time, and the second to the situation where such a change takes place. In the future, in the whole monograph, with the exception of specially specified cases, this will only concern flat curves.

It is said that the electromagnetic wave has one or another polarization at a given point of space depending on kind of $\vec{E}$ vector the hodograph has at this point.

Before making any classification of electromagnetic waves, let us return to Maxwell's equations. As you can see from Formula (1.1), the source of the electromagnetic field is some extraneous currents $\vec{j}_{\mathrm{CT}}$ or the alternating electric or magnetic fields generated by them, as well as alternating currents.

As it follows from the electrodynamics, it is possible to determine the $\vec{E}$ and $\vec{H}$ values at any point in space, if the distribution of currents on a closed surface surrounding this point is known. These currents should be considered as a source of the required electromagnetic field for such a task. From the point of view of Maxwell's equation solution by virtue of their linearity, the above means that the $\vec{E}$ field at each point will be a certain linear combination of these currents linear functional. If the dependence of the current on time is determined, then the $\vec{E}$ law of variation will also be determined; if $\vec{j}$ is random, then the change in time of $\vec{E}$ vector will be random.

In geophysics tasks, the electromagnetic field source is either current flowing in the transmitting antenna, which is typical for remote reconnaissance problems by active radiolocation methods, or current generated by the thermal chaotic motion of electrons and charges inside and on the surface of geophysical objects, which is typical for microwave radiometry methods. In the case where the field scattered by objects is analyzed, the currents arising in the object under the action of the irradiating wave can be considered as the source of the received electromagnetic wave.

If under study of the characteristics of the electromagnetic wave, the transmitting and receiving antennas, as well as the object of study, remain motionless, the field $\vec{E}$ at any point will repeat the nature of the $\vec{j}$ law of variation.

Let us consider the simplest case when the transmitter emits a monochromatic frequency wave ω_0. Due to the periodicity of the $\vec{j}$ current changes in time, the vector $\vec{E}$ will also change periodically; i.e., its hodograph will be a closed line. Moreover, since Maxwell's equations are actually written for the components of vectors $\vec{E}$ and $\vec{H}$ (E_x, E_y, E_z, H_x, etc.), then under the sinusoidal law of variation of $\vec{j}$, the same law will take place for E_x, E_y (the Z-axis coincides with the wave propagation direction). This means that in the plane E_x, E_y the end of the vector $\vec{E}$ will describe the ellipse.

Let us consider the second case when the current law of variation is modulated by a signal $A(t)$ slowly changing in comparison with the frequency period ω_0; i.e., it is a formula $A(t)\cos(\omega_0 t + \varphi)$ (quasi-monochromatic wave). In this situation, ellipse, keeping its shape, will only change its size obviously.

Electromagnetic waves, the $\vec{E}$ vector hodograph of which describes unaltering in shape (but not in size) ellipse, called fully polarized (FPW), or polarized waves. It should not be thought that FPW can only be monochromatic or quasi-monochromatic, and a common case of their occurrence will be discussed below.

In another extremum situation, where the $\vec{j}_{\text{CT}}$ current is a random time function, the field $\vec{E}$ will also be a random time function.

Let us proceed to the analysis of orthogonal components of the $\vec{E}(E_x, E_y)$ field. If it turns out that the average power per each of them will be the same, and there will be no correlation between E_x and E_y, then such a wave is called a completely non-polarized wave (NPW) or non-polarized waves.

All other waves not covered by the mentioned extremum cases belong to the class of partially polarized waves (PPWs). There are a lot of reasons leading to the appearance of NPW. First of all, as it follows from the above, their occurrence takes place when the current $\vec{j}_{\text{CT}}$ is a mixture of, for example, quasi-monochromatic $\vec{j}_{\text{KM}}$ and random $\vec{j}_{\text{CL}}$ components. PPW can also arise in the scattering of FPW on the object, the reflective properties of which change over time. Finally, the PPW can be obtained by forming a superposition of completely polarized and non-polarized waves. It will be shown below that any PPW can always be presented in an unambiguous way as a sum of FPW and NPW. The ratio of PPW power to the total power of the wave is called the m polarization coefficient.

Let's move to a more detailed description of the main parameters and characteristics of different wave types.

The independent solution of Maxwell's equations in an arbitrary Cartesian coordinate system, the axes of which do not coincide with the electric vector $\vec{E}$ of the monochromatic wave with frequency ω_0 propagating in the direction $\vec{r}$, can be given by three complex projections on the axis of this system, the origin of which is placed in the considered point:

$E_j(t) = E_{jm}\exp\left(\varphi_j - \vec{k}\vec{r}\right)\exp(i\omega_0 t)$, $(j = x, y, z)$ where E_{jm} are amplitudes of relevant field projections, $\vec{k}$ wave vector, $k = 2\pi/\lambda$ wave number, λ wavelength, φ_j initial phases, and t time.

Complex numbers $E_j = E_{jm}\exp\varphi_j$ are called complex oscillation amplitudes along the axes X, Y, Z, the directions of which are determined by the basic unit vectors $\vec{e}_x, \vec{e}_y, \vec{e}_z$. These values can be considered as projections of the complex field strength vector amplitude of the: $\vec{E} = E_x\vec{e}_x + E_y\vec{e}_y + E_z\vec{e}_z$.

As mentioned above, to determine the polarization of the wave electric field at the point of space means to find the law of variation of the vector $\vec{E}$ at this point in time. By rotating the coordinate system in space, one of the components of a plane wave can be always turned to zero.

To be specific, we assume that the wave propagates along the z-axis, then $E_j(t) = E_{jm}\exp(\varphi_j - kz)\exp(j\omega_0 t)$, and $E_z = 0$, $(j = x, y)$, and therefore the wave electric vector can be represented as:

$$\vec{E}(t) = \left(E_{xm}\mathrm{e}^{\mathrm{i}\varphi_x}\vec{e}_x + E_{ym}\mathrm{e}^{\mathrm{i}\varphi_y}\vec{e}_y\right)\mathrm{e}^{\mathrm{i}(\omega_0 t - kz)}.$$

Making obvious transformations, we have:

$$\vec{E}(t) = \left(E_{xm}\vec{e}_x + E_{ym}\mathrm{e}^{2\mathrm{i}\delta}\,\vec{e}_y\right)\mathrm{e}^{\mathrm{i}\varphi_0}\mathrm{e}^{\mathrm{i}(\omega_0 t - kz)}, \tag{1.3}$$

where $2\delta = \varphi_y - \varphi_x$ is orthogonal component phase difference $(0 \le \delta < 2\pi)$ and $\varphi_0 \equiv \varphi_x$ acts as the initial phase of the x-component of the electric vector.

The formula for $\vec{E}(t)$ is a parametric setting of the trajectory of the end of the vector $\vec{E}(t)$ at a random point in space and determines the wave polarization. Excluding time, it is possible to find the formula of the trajectory of the end of the vector (hodograph) in the *XOY* plane:

$$\left(\frac{X}{E_{xm}}\right)^2 + \left(\frac{Y}{E_{ym}}\right)^2 - \frac{2XY}{E_{xm}E_{ym}}\cos 2\delta = \sin^2 2\delta, \tag{1.4}$$

where the $X \equiv \mathrm{Re}E_x$, $Y \equiv \mathrm{Re}E_y$ notations are input, while $E_{xm} = |E_x|$, $E_{ym} = |E_y|$.

As it is seen, the $\vec{E}(t)$ vector hodograph is a quadratic curve and (due to the limitations of E_{xm} and E_{ym}) corresponds to the equation of the ellipse (polarization ellipse). Such radio wave is called an *elliptically polarized electromagnetic wave*. In coordinate form, it can be represented as follows:

$$\left.\begin{aligned} E_x &= E_{xm}\mathrm{e}^{\mathrm{i}\varphi_0}\mathrm{e}^{\mathrm{i}(\omega_0 t - kz)} \\ E_y &= E_{ym}\mathrm{e}^{\mathrm{i}\varphi_0}\mathrm{e}^{2\mathrm{i}\delta}\mathrm{e}^{\mathrm{i}(\omega_0 t - kz)} \end{aligned}\right\}, \tag{1.5}$$

where E_{xm} and E_{ym} are real numbers (in this case, one or both of them can be negative numbers).

Subsequent records drop the inappreciable multiplier $\mathrm{e}^{\mathrm{i}(\omega_0 t - kz)}$.

Waves which characteristics satisfy the Formula (1.5) belong to the class of the so-called fully polarized electromagnetic waves (FPW), which are called such electromagnetic waves, the form (but not the size!) of the polarization ellipse of which remains constant over time.

The power transferred by the electromagnetic wave is obviously proportional to the $\left(|E_x|^2 + |E_y|^2\right) = \left(E_{xm}^2 + E_{ym}^2\right) = P_0$ sum, which gives the possibility to enter the notations:

$$\left.\begin{aligned} E_{xm} &= \sqrt{P_0}\cos\gamma \\ E_{ym} &= \sqrt{P_0}\sin\gamma \end{aligned}\right\}, \tag{1.6}$$

where $0 \le \gamma < 2\pi$.

Using the entered notations, it is possible to write the following formulae for the so-called complex amplitudes instead of the Formula (1.5):

$$\left.\begin{aligned} E_x &= \sqrt{P_0}\mathrm{e}^{\mathrm{i}\varphi_0}\cos\gamma \\ E_y &= \sqrt{P_0}\mathrm{e}^{\mathrm{i}\varphi_0}\mathrm{e}^{2\mathrm{i}\delta}\sin\gamma \end{aligned}\right\}. \tag{1.7}$$

In this case, instead of the Formula (1.4) we obtain:

$$\left(\frac{X}{\cos\gamma}\right)^2 + \left(\frac{Y}{\sin\gamma}\right)^2 - \frac{4XY}{\sin 2\gamma}\cos 2\delta = P_0 \sin^2 2\delta. \tag{1.8}$$

So, as can be seen from Eqs. (1.4) and (1.8), the wave polarization state is characterized by three parameters E_{xm}, E_{ym}, δ or P_0, γ, δ.

An important concept in radio polarimetry is the concept of orthogonally polarized electromagnetic waves. Two waves are orthogonally polarized if their polarization ellipses have the same shape, while the large axes of these ellipses are mutually perpendicular, and the directions of rotation of the electric vectors are opposite.

From the foregoing, the second of the analytical record of the expression for the wave, orthogonal to the wave defined by the expression (1.7), will have the following form:

$$\left.\begin{aligned} E_x &= -\sqrt{P_0}\mathrm{e}^{\mathrm{i}\varphi_1}\mathrm{e}^{-2\mathrm{i}\delta}\sin\gamma \\ E_y &= \sqrt{P_0}\mathrm{e}^{\mathrm{i}\varphi_1}\cos\gamma \end{aligned}\right\}. \tag{1.9}$$

In the Formula (1.9), the change in the direction of rotation of the electric vector is taken into account by the change in the sign of the exponent. It should be noted that orthogonally polarized waves are not required to have the same initial phase, which is taken into account by changing the phase φ_0 to phase φ_1.

It should not be supposed that fully polarized waves can only be monochromatic. Electromagnetic waves, which polarization ellipses change their size over time, are also fully polarized, while their geometrical characteristics remain constant. It is obvious that the power of such waves P_0 does not remain constant over time.

In addition, the waves which have the 2δ difference of these phases constant over time, although the phases of their components change according to an arbitrary law, are also fully polarized.

Take a note of one rather delicate moment, which is not given significant attention in the literature and often plays a very important role with the development of practical radio polarimetry.

Equations (1.4) and (1.8) define the type of the polarization ellipse. Inside the polarization ellipse, the vector $\vec{E}$ performs regular rotation with a periodically changing angular velocity, where the $\vec{E}$ vector makes the complete revolution during the $T = 2\pi/\omega_0$ carrier cycle.

For two electromagnetic waves of the same frequency and the same polarization type, the instantaneous positions of their electric vectors do not generally coincide

(Fig. 1.1). Herewith, the angle between them is continuously changing due to the angular velocity inequality of each of these vectors. This means that the coincidence of the polarization ellipses of electromagnetic waves does not mean their equivalence in polarization characteristics. In the radio band, where there is a hardware possibility of controlling both types of polarization of the emitted radio wave and changing the receiving antenna polarization characteristics, failure to take into account this fact will lead to decrease and even to a complete loss of those gains that can be given by the use of various methods of radio polarimetry.

The foregoing means that the electromagnetic waves polarization state is generally characterized by not three, but four numbers. To the above-mentioned three parameters E_{xm}, E_{ym}, δ or P_0, γ, δ, a fourth parameter shall be added—the initial phase of one of the components, for example $\varphi_0 \equiv \varphi_x$.

The quoted hereinabove definition of radio wave polarization is related to the change of the vector $\vec{E}$ at a fixed point in space. Another representation of polarization is possible, when the polarization analysis fixes the moment of time $t = t_0$ and determines the spatial change of the vector along the coordinate axes.

Figure 1.2 conditionally shows the distribution of the instantaneous values of the vector $\vec{E}(t_0, z)$ at a time $t = t_0$. The wave propagation can be represented as the movement of such an image along the OZ-axis at phase velocity $v = c/\sqrt{\mu\varepsilon}$, where c is the light velocity.

The vector $\vec{E}$ rotates along an elliptic curve around the direction of propagation and simultaneously moves along that direction to form an elliptical spiral.

Figure 1.3 shows an image of the polarization ellipse in the XOY-coordinate system. The known from elementary mathematics formulas allow to determine by writing (1.4) the greater semiaxis of the ellipse (a), the ellipticity ratio equal to the ratio of the small semiaxis (b) to the large one ($r = b/a$), the angle of inclination of the large semiaxis to the axis OX–β, while $0 \le \beta \le \pi$.

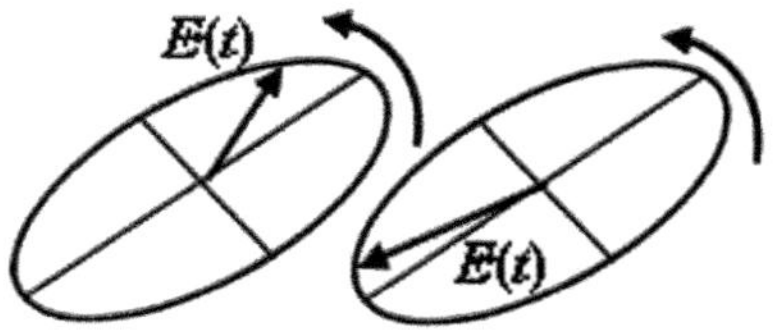

Fig. 1.1 Instantaneous position of the electric vectors of two electromagnetic waves of the same polarization

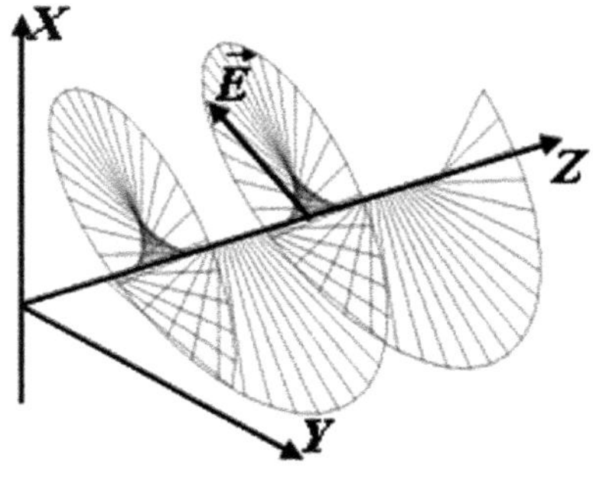

Fig. 1.2 To an illustration of an elliptically polarized radio wave

$$\left.\begin{aligned} a^2 &= 0.5P_0\left(1+\sqrt{1-(\sin 2\gamma \sin 2\delta)^2}\right)\\ b^2 &= 0.5P_0\left(1-\sqrt{1-(\sin 2\gamma \sin 2\delta)^2}\right)\\ r &= \frac{1-\sqrt{1-(\sin 2\gamma \sin 2\delta)^2}}{\sin 2\gamma \sin 2\delta}\\ \tan 2\beta &= \tan 2\gamma \cos 2\delta \end{aligned}\right\}. \tag{1.10}$$

In practice, the ellipticity angle α is often introduced with equation $\alpha = \operatorname{arctg} r$ instead of the ellipticity ratio, while it is believed that $-0.25\pi \le \alpha \le 0.25\pi$. The negative value of the ellipticity ratio, which takes place when $-0.25\pi \le \alpha < 0$, is a certain generalization of this ratio concept in comparison with its definition as $r = b/a$. As will be discussed below, this approach will take into account the direction of the electric vector rotation.

The introduction of the ellipticity angle gives the possibility to represent the first formulas in the relations (1.10) as:

$$\left.\begin{aligned} a &= \sqrt{0.5P_0}\cos\alpha\\ b &= \sqrt{0.5P_0}\sin\alpha\\ r &= \tan\alpha \end{aligned}\right\}. \tag{1.11}$$

In light of the foregoing the second of the (1.11) formulas gives possibility to talk about negative values of the parameter b, i.e., the small axis of the polarization ellipse.

Finally, let us pay attention to two more useful relations. The sum of the squares of the large and small semi-axes of the polarization ellipse is equal to the signal strength $\left(a^2+b^2\right) = P_0$, and the area of this ellipse is equal to $0.25\pi\ P_0|\sin 2\alpha|$.

Figure 1.3 quite clearly shows that the type of polarization ellipse is uniquely determined by a single number, for example r, α. If we have to specify position of the ellipse in space, besides its type, then we need to have two numbers, for example, β and r; β and α; β and γ; and γ and δ. In case it is necessary to consider the wave power, i.e., size of the ellipse, it is necessary to add the third number, e.g., P_0, a or b. Finally, when it is important to know the instantaneous position of vector $\vec{E}$, it is required to introduce another one parameter, for example, the initial phase φ_0.

Fig. 1.3 Polarizing ellipse

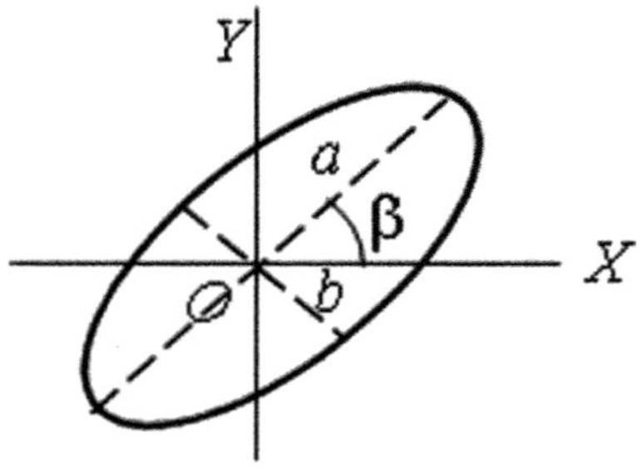

If the vector $\vec{E}$ bypasses the polarization ellipse counterclockwise when observed in the direction of wave propagation, the ellipticity ratio r is considered positive ($0 \leq \alpha \leq 0.25\pi$), and the wave is called right-polarized. During the clockwise rotation of the field vector, the wave will be left-polarized, and the ratio r will be negative ($-0.25\pi \leq \alpha < 0$).

Special cases of elliptically polarized waves are the waves of linear polarization (LP), for which $r = 0$ ($\alpha = \delta = 0$). For this case, the polarization ellipse degenerates into a line segment. Horizontally polarized (HP) waves, in which $\beta = \gamma = 0$ or π, vertically polarized (VP) waves, in which $\beta = \pi/2$, $\gamma = \pi/2$ or $\gamma = 3\pi/2$, waves polarized at an angle of 45° (SP), in which $\beta = \pi/4$, $\gamma = \pi/4$ or $\gamma = 5\pi/4$, are distinguished among the waves of linear polarization. Waves of rotary polarization (RP), sometimes called circularly polarized, play an important role. For this case, the polarization ellipse is a circle ($|r| = 1$, $\delta = \pm\pi/2$, $\gamma = \pi/4$ or $\gamma = 5\pi/4$, the β angle is not defined). RP is right-polarized when $\delta = 3\pi/2$ and left-polarized when $\delta = \pi/2$.

In the literature which covers the issues related to the radio wave polarization, depending on certain considerations, a variety of ways for describing the polarization state of electromagnetic waves are used. At the same time, which is extremely important, the measuring equipment is also adapted to measure a wide variety of wave parameters and their combinations. That is why the following is a certain systematization of such descriptions and the relationship between their quantitative characteristics is shown.

Let's start the discussion with two *parametric characteristics* of radio wave polarization. Among them, the most widespread gets the description of the polarization state of radio waves using phasor and Poincare sphere.

The phasor is the ratio of complex amplitudes defined by a Formula (1.5):

$$p = \frac{E_{2m}}{E_{1m}} \mathrm{e}^{2i\delta} = \mathrm{e}^{2i\delta} \tan\gamma. \quad (1.12)$$

The geometric parameters of the polarization ellipse are determined by the given phasor as follows:

$$\left.\begin{aligned} r &= \frac{\left(1+|p|^2\right) - \sqrt{\left(1+|p|^2\right)^2 + (p-p^*)^2}}{i(p^*-p)} \\ \tan 2\beta &= \frac{p+p}{1-|p|^2} \end{aligned}\right\}. \quad (1.13)$$

(see details in Sect. 4.1).

Now let's move on to one of the most common devices—the Poincare sphere. So, any polarization ellipse can be put in one-for-one matching pair of numbers, such as α and β or γ and 2δ. This means that the set of all possible ellipses can be one-for-one depicted to two-dimensional sets: plane and sphere. The latter is more convenient, since there are no infinitely remote points on the sphere.

If the longitude of some point on the unit sphere Fig. 1.4 assume equal to twice the orientation angle of the polarization ellipse 2β, and the latitude to twice the ellipticity angle 2α, then each possible state of the fully polarized wave corresponds to one point on the sphere and vice versa. With the help of this sphere, known in the literature as the Poincare sphere, it is easy to take into account the direction of rotation of the polarization ellipse. The right rotation is put in accordance with the points of the northern hemisphere and the left with the southern.

The parallels on the sphere correspond to ellipses with a constant ellipticity angle $\alpha = \text{const}$, and the meridians correspond to ellipses with one and the same orientation angle $\beta = \text{const}$. For the longitudes, reference (angles of orientation β) usually takes the meridian FAG relevant to all the ellipses of polarization of a horizontally oriented longer axis. The opposite meridian FDG shows all ellipses, the longer axis of which is oriented vertically. The zero parallel (sphere equator), for which $\alpha = 0$, is a representation of all linear polarizations, whereas the north pole of the sphere (point F, $2\alpha = 90°$, $r = 1$) corresponds to the right and south pole (point G, $2\alpha = -90°$, $r = -1$) to the left circular polarization.

Sphere in Fig. 1.4 oriented so that the zero meridian ($\beta = 0$) passes through point A corresponding to horizontal polarization (HP). In this case, the point D ($\beta = \pi/2$) will correspond to the vertical polarization (VP), and points C and E to the linear polarization with the angle of inclination of the electric vector in relation to the original HP equal to 45° (45P) and 135° (135P). Polarization ellipse with parameters in Fig. 1.4 corresponds to point P with coordinates 2α, 2β. Diametrally opposite point P' corresponds to an ellipse with the same ellipticity ratio, but rotated in space at an angle of 90°. Herewith, the direction of rotation of the vector $\vec{E}$ changes to the opposite. Ellipses corresponding to points P and P' are called orthogonal, and the corresponding waves are orthogonally polarized.

As it is seen, the corners α and β have a clear graphical representation on the Poincare sphere.

It is of natural interest to obtain a geometric interpretation on this sphere, of another pair of angles 2γ and 2δ. First of all, note that, as it follows from Fig. 1.4, the cosine of the angle POA is equal to the $\cos 2\alpha \cos 2\beta$ product. From the same figure, it is seen that the tangent of the angle PKL is equal to the $\text{tg}2\alpha\ \text{cosec}\ 2\beta$ product.

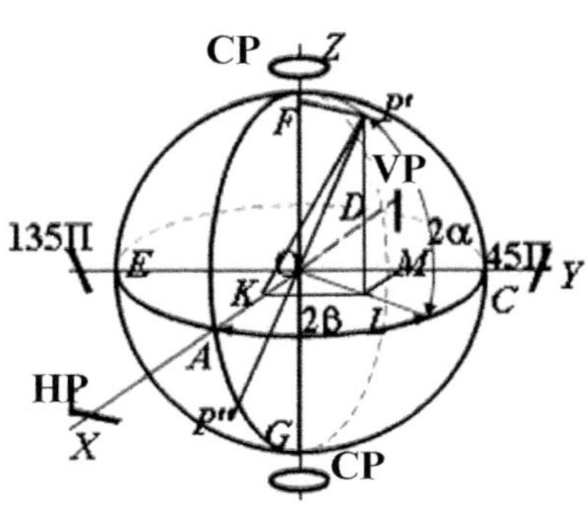

Fig. 1.4 Poincaré sphere

On the other hand, direct calculations by Formula (1.9) lead to the following equations:

$$\left.\begin{aligned}\sin 2\alpha &= \sin 2\gamma \sin 2\delta \\ \operatorname{tg} 2\beta &= \operatorname{tg} 2\gamma \cos 2\delta\end{aligned}\right\}. \quad (1.14)$$

The correlations (1.14) make it possible to obtain the next relations after simple, though rather cumbersome, calculations:

$$\left.\begin{aligned}\cos 2\gamma &= \cos 2\alpha \cos 2\beta \\ \operatorname{tg} 2\delta &= \operatorname{tg} 2\alpha \operatorname{cosec} 2\beta\end{aligned}\right\}. \quad (1.15)$$

The resulting equations show that $\angle POA = 2\gamma$ and $\angle PKL = 2\delta$, which determines the geometric meaning of the angles 2γ and 2δ, an illustration of which is shown in Fig. 1.5.

For more clarity, Fig. 1.6 shows two Poincare spheres, on which lines of the same angles α and β are drawn in the first case, and lines of the same angles γ and δ in the second case. As it is seen, when the sphere is rotated by 90°, mutual replacement of the corresponding angles takes place.

The Poincare sphere "does not distinguish" waves of the same polarization, but of different intensity, and is not sensitive to the initial phase φ_0. Sometimes, the sphere attributed a certain radius proportional to the intensity of the wave P_0, thereby taking into account the three parameters of an elliptically polarized wave.

In the practice of polarization measurements, it is often more convenient to determine the polarization using the quadratic parameters of the electric field. This follows from the fact that the measurement of the electromagnetic field intensity is carried out using power meters.

As the frequency of the studied oscillations increases, phase and correlation measurements become more and more complicated, and therefore polarization analysis

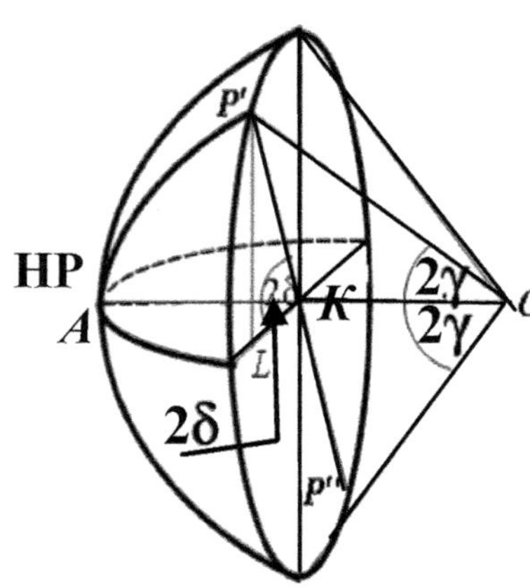

Fig. 1.5 To an illustration of angles 2γ and 2δ on the Poincaré sphere

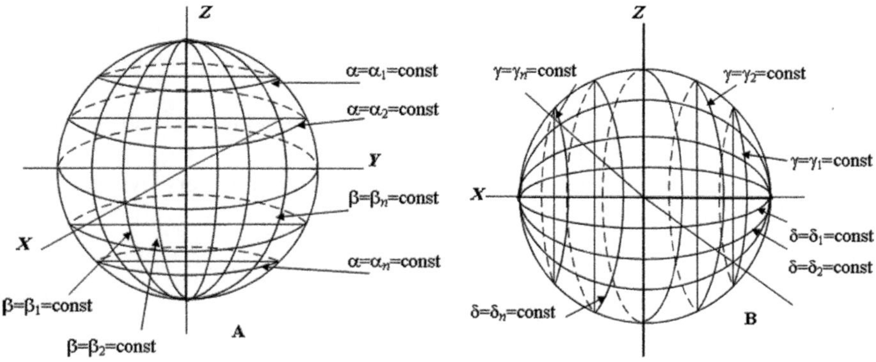

Fig. 1.6 **a** Lines of identical angles α and β; **b** lines of identical angles γ and δ

methods based on the measurement of quadratic values relative to the quantitative field play an increasingly important role.

Stokes parameters (see Sect. 3.2 for details) are widely used, three of which are Cartesian coordinates of a point on the Poincare sphere of the radius P_0 corresponding to the polarization of the analyzed wave:

$$\left.\begin{aligned} S_1 &= P_0 \cos 2\gamma \\ S_2 &= P_0 \sin 2\gamma \cos 2\delta \\ S_3 &= P_0 \sin 2\gamma \sin 2\delta \end{aligned}\right\}. \tag{1.16}$$

The fourth Stokes parameter S_0 for FPV is determined by the following equation:

$$S_0 = \sqrt{S_1^2 + S_2^2 + S_3^2} = P_0, \tag{1.17}$$

i.e., just equal to the radio wave intensity (in the above-mentioned figure, the fourth Stokes parameter is equal to the radius of the Poincare sphere).

The Formulae (1.5), (1.6), and (1.13) allow to relate the entered parameters to the complex amplitudes:

$$\left.\begin{aligned} S_0 &= |E_x|^2 + \left|E_y\right|^2 \\ S_1 &= |E_x|^2 - \left|E_y\right|^2 \\ S_2 &= E_x E_y^* + E_x^* E_y \\ S_3 &= \mathrm{i}\left(E_x E_y^* - E_x^* E_y\right) \end{aligned}\right\}. \tag{1.18}$$

As it is seen, the Stokes parameters are expressed through the quadratic characteristics of the electric field.

Another widely used apparatus for describing the radio wave polarization state using three parameters is the coherence matrix K (see Sect. 3.1 for details), defined as follows:

$$K = \begin{pmatrix} |E_1|^2 & E_1 E_2^* \\ E_1^* E_2 & |E_2|^2 \end{pmatrix} = \begin{pmatrix} E_{1m}^2 & E_{1m} E_{2m} \mathrm{e}^{2\mathrm{i}\delta} \\ E_{1m} E_{2m} \mathrm{e}^{-2\mathrm{i}\delta} & E_{2m}^2 \end{pmatrix}, \tag{1.19}$$

In the record above, there are indices 1 and 2 instead of indices "x" and "y", which, as will be shown below, is more common.

The coherence matrix can be expressed in Stokes parameters by taking into account the Formula (1.17):

$$K = \frac{1}{2} \begin{pmatrix} S_0 + S_1 & S_2 - \mathrm{i}S_3 \\ S_2 + \mathrm{i}S_3 & S_0 - S_1 \end{pmatrix}. \tag{1.20}$$

As already mentioned, the most complete description of the radio wave polarization state can be carried out with the help of a four-parameter apparatus. Four such representations are known in the literature as the Jones vector, the tetra-vector, the complex vector, and the previously considered complex amplitudes (1.7).

The Jones vector is a record of the polarization state in the form of a matrix-column:

$$\vec{E} = \begin{pmatrix} E_{1m} \\ E_{2m} \mathrm{e}^{2\mathrm{i}\delta} \end{pmatrix} \mathrm{e}^{\mathrm{i}\varphi_0} = \sqrt{P_0} \begin{pmatrix} \cos\gamma \\ \mathrm{e}^{2\mathrm{i}\delta} \sin\gamma \end{pmatrix} \mathrm{e}^{\mathrm{i}\varphi_0}. \tag{1.21}$$

Each of the complex field amplitudes $E_1 = E_{1m}\mathrm{e}^{\mathrm{i}\varphi_0}$ and $E_2 = E_{2m}\mathrm{e}^{2\mathrm{i}\delta}\mathrm{e}^{\mathrm{i}\varphi_0}$, again, can be represented as the sum of two summands, called quadrature components:

$$\left. \begin{aligned} E_1 &= x_1 + \mathrm{i}x_2 \\ E_2 &= x_3 + \mathrm{i}x_4 \end{aligned} \right\}. \tag{1.22}$$

In this case, the polarization of the field $\vec{E}(t)$ is determined by four real numbers, which can be written as a tetra-component vector-column:

$$\vec{E} = \begin{pmatrix} x_1 & x_2 & x_3 & x_4 \end{pmatrix}^{\mathrm{T}} \tag{1.23}$$

given Eq. (1.21):

$$\vec{E} = \sqrt{J_0}(\cos\gamma\cos\varphi_0 \quad \cos\gamma\sin\varphi_0 \quad \sin\gamma\cos(\varphi_0 + 2\delta) \quad \sin\gamma\sin(\varphi_0 + 2\delta))^{\mathrm{T}}, \tag{1.24}$$

where T sign means matrix transpose operation.

If case of move to the time representation in Formula (1.22), then $E_j (j = 1, 2)$ can be written in the form $E_j(t) = \left|E_j(t)\right| \exp\left(\omega_0 t + \varphi_j\right)$ where $\left|E_j(t)\right| = \left(x_m^2 + x_{m+1}^2\right)^{0.5}$, $\varphi_j = \arctan(x_{m+1}/x_m)$, $m = 1{,}3$, which is elementary followed by $x_{1,3} = \left|E_{1,2}\right| \cos\varphi_{1,3}$, $x_{2,4} = \left|E_{1,2}\right| \cos\varphi_{2,4}$. The value $\left|E_j\right|$ is called the envelope, and φ_j orthogonal component phase.

The representation of the field in the form of Eq. (1.23) as opposed to the complex representation (1.21) will be called real. Since the components of the real representation x_1 and x_2, and x_3 and x_4 are shifted in phase by $\pi/2$ relative to each other, components x_1 and x_2 will be called phasic, and components x_3 and x_4 non-phasic components of the tetra-dimensional vector (four vectors) $\vec{E}$ (for details, see Sect. 2.1).

For the mathematical recording of fully polarized radio waves, the concept of a complex vector $\vec{E}$ is used sometimes, the real part of which is a vector describing the desired polarization ellipse over time:

$$\vec{E} = \left(\vec{E}_1 + \mathrm{i}\vec{E}_2\right) \mathrm{e}^{\mathrm{i}\omega_0 t}, \tag{1.25}$$

where $\vec{E}_1 = \vec{E}_1\left(E_{1x}, E_{1y}\right), \quad \vec{E}_2 = \vec{E}_2\left(E_{2x}, E_{2y}\right)$.

The real part of the expression (1.25) is easy to form as (1.5):

$$\left.\begin{aligned} E_x &= \sqrt{E_{1x}^2 + E_{2x}^2}\exp(-\mathrm{iarc\ tanE}_{2x}/E_{1x}) \\ E_y &= \sqrt{E_{1y}^2 + E_{2y}^2}\exp\left(-\mathrm{iarc\ tanE}_{2y}/E_{1y}\right) \end{aligned}\right\}, \tag{1.26}$$

where E_x, E_y with relevant indexes are the Cartesian components of the electric vectors.

The polarization ellipse parameters can be determined using the relations (1.9) while comparing the expressions (1.5) and (1.26).

To change the direction of rotation of the polarization ellipse, the change of the angle $\alpha = \mathrm{arctg}\, r$ sign is needed, which requires, as it follows from the Formula (1.26), a change in the 2δ sign, i.e., the maintaining of complex conjugation operation in the Formulae (1.25) and (1.26). Therefore, the ellipse corresponding to the complex vector $\vec{E}^*$ (* means the complex conjugation) will differ from the ellipse corresponding to the vector $\vec{E}$ only in the direction of the bypass. Complex vectors are also convenient because all vector and algebraic transformations remain valid for them.

Let us consider the basic properties of a complex vector $\vec{E}$.

The square of its modulus, numerically equal to $|E_x|^2 + \left|E_y\right|^2$, is proportional to the electromagnetic wave power flux density at the observation point. The module of the vector $\vec{E}$ is equal to its length, i.e., $\vec{E}\vec{E}^*$. The scalar product of two complex vectors $\vec{E}_1$ and $\vec{E}_2$, defined $\vec{E}_1\vec{E}_2^*$, is the projection of one vector onto another. If this product is zero, then the vectors $\vec{E}_1$ and $\vec{E}_2$ are called orthogonal. If a vector $\vec{E}_2$ can be represented as $\vec{E}_2 = k\vec{E}_1$ (where k is a constant complex factor), then it is said that $\vec{E}_1$ and $\vec{E}_2$ are parallel. In this case, the polarization ellipses do not differ in shape or orientation. The difference is only in their size (in $|k|$ times) and shift of the initial phase of the vector forming an ellipse by an angle arg k.

Orthogonal vectors play an important role. Let's construct a complex vector $\vec{E}_{\text{ort}} = \left(\vec{A} + \mathrm{i}\vec{B}\right)\mathrm{e}^{\mathrm{i}\omega_0 t}$ (where $\vec{A}$ and $\vec{B}$ are some real vectors), which is orthogonal to vector $\vec{E} = \left(\vec{E}_1 + \mathrm{i}\vec{E}_2\right)\mathrm{e}^{i\omega_0 t}$. Then, $\vec{E}_{\text{ort}}\vec{E}_1^* = \vec{A}\vec{E}_1 + \vec{B}\vec{E}_2 - \mathrm{i}\left(\vec{A}\vec{E}_2 - \vec{B}\vec{E}_1\right) = 0$.

The above-mentioned equation takes place only when the real and imaginary parts are equal to zero, which correspond to the conditions $\vec{A}\perp\vec{E}_1$, $\vec{B}\perp\vec{E}_2$, $|\vec{A}| = \left|\vec{E}_1\right|$, $|\vec{B}| = \left|\vec{E}_2\right|$.

Therefore, the desired complex vector $\vec{E}_{\text{ort}}$ is determined by vectors $\vec{A}$ and $\vec{B}$, which are perpendicular to $\vec{E}_1$ and $\vec{E}_2$. This suggests that the corresponding polarization ellipse is rotated for 90° without changing the shape, and the sign "minus" in brackets indicates the opposite direction of rotation. Such polarizations are called orthogonal.

Let's consider the conditions of different types of polarization formation.

The condition of the linearly polarized field is $2\delta = 0$ equation, which corresponds to the condition $E_{2x}/E_{1x} = E_{2y}/E_{1y} = n$ where n is an arbitrary real number; i.e., taking into account the above-mentioned, we have $\vec{E}_2 = n\vec{E}_1$.

Thus, the condition for obtaining a linear polarization is the parallelism of vectors $\vec{E}_1$ and $\vec{E}_2$ (for real n). In this case, $\vec{E} = \vec{E}_1(1 + \mathrm{i}n)\mathrm{e}^{\mathrm{i}\omega_0 t}$. The angle of inclination of the linear polarized field vector relevant to the X-axis is determined by the equation:

$$\beta = \text{arc tan}\frac{\sqrt{E_{1y}^2 + E_{2y}^2}}{\sqrt{E_{1x}^2 + E_{2x}^2}} = \text{arc tan}\frac{E_{1y}}{E_{1x}}. \tag{1.27}$$

For linear polarization at an angle of 45° (45P), we have $E_{1y} = E_{1x}$; $E_{2y} = E_{2x}$, i.e., $n = 1$.

Let us define the conditions for obtaining a circular polarization of the vector $\vec{E}$.

The condition of providing the required phase difference between E_x and E_y equal to $0.5\pi + \pi l$ leads to $E_{2x}/E_{1x} = -E_{1y}/E_{2y} = \text{const}$ equation.

On the other hand, the requirement of equality of E_x and E_y amplitudes also leads to the condition: $E_{1x}^2 + E_{1y}^2 = n^2 E_{2x}^2 + n^{-2} E_{2y}^2$, which can be fulfilled only if $n = 1$. It follows that $E_{1x} = E_{2x} = E_{1y} = -E_{2y}$, and therefore the vectors $\vec{E}_1$ and $\vec{E}_2$ should be presented in the form of:

$$\vec{E}_1 = E_{1x}(1 + \mathrm{i})\vec{e}_x \quad ; \quad \vec{E}_2 = E_{1y}(1 - \mathrm{i})\vec{e}_y, \tag{1.28}$$

where $\vec{e}_x$, $\vec{e}_y$ are the unit vectors of Cartesian coordinates.

Thus, the condition for obtaining a circular polarization of the full $\vec{E}$ field vector is the complex conjugation of its components $\vec{E}_1$ and $\vec{E}_2$ and the corresponding position of these vectors on the complex plane, for which $\varphi_1 = \pi/4$ and $\varphi_2 = \pi/4 \pm \pi/2(2l + 1)$. Vector representation of polarization allows to determine the geometrical parameters of the wave polarization ellipse. At the same time, the large and small semi-axes of the ellipse are determined by the Formula (1.9):

$$\left.\begin{aligned} a^2 &= 0.5\left(\left|\vec{E}\right|^2 + \left|\vec{E}^2\right|\right) \\ b^2 &= 0.5\left(\left|\vec{E}\right|^2 - \left|\vec{E}^2\right|\right) \end{aligned}\right\}, \tag{1.29}$$

A special conclusion that in the case of circular polarization ($a = b$) the complex $\vec{E}$ wave vector satisfies the condition $\vec{E}^2 = 0$ follows from the expression (1.29).

The field vector direction of rotation is determined by the sign of the mixed product of vectors $\xi = \vec{n}\left(\vec{E} \times \vec{E}^*\right)$.

When $\xi > 0$, the $\vec{E}$ field vector rotates clockwise relative to the observer looking in the normal direction $\vec{n}$. Thus, the sign of the ellipticity ratio r coincides with the ξ sign. Therefore, the value ξ can be called a characteristic of the field vector direction of rotation.

As follows from the above, the representation of the wave polarization with the help of complex vectors allows to determine the invariant with respect to the choice of the coordinate system characteristics of the polarization ellipse: the ratio of the axes and the direction of rotation. To determine the orientation of the large axis of the ellipse in the selected coordinate system, it is necessary to use the decomposition of the complex vector along the axes:

$$\beta = \frac{1}{2}\arctan\left|\frac{\vec{E}_1\vec{E}_2^* - \vec{E}_1^*\vec{E}_2}{\vec{E}_1\vec{E}_2^* + \vec{E}_1^*\vec{E}_2}\right|, \tag{1.30}$$

where $\vec{E}_1$ and $\vec{E}_2$ are the $\vec{E}$ vector projections onto X- and Y-axes, respectively.

If the normal vector $\vec{n}$ does not coincide with any of the coordinate axes, then each of the real vectors $\vec{E}_1$ and $\vec{E}_2$ in the expression (1.26) can be represented by three projections:

$$\left.\begin{aligned} \vec{E}_1 &= E_{1x}\vec{e}_x + E_{1y}\vec{e}_y + E_{1z}\vec{e}_z \\ \vec{E}_2 &= E_{2x}\vec{e}_x + E_{2y}\vec{e}_y + E_{2z}\vec{e}_z \end{aligned}\right\}. \tag{1.31}$$

From the expression (1.31) and the Formula (1.25), we obtain:

$$\vec{E}_t = (E_{1x} + \mathrm{i}E_{2x})\vec{e}_x + \left(E_{1y} + \mathrm{i}E_{2y}\right)\vec{e}_y + (E_{1z} + \mathrm{i}E_{2z})\vec{e}_z \tag{1.32}$$

The ratio (1.32) shows that in an arbitrarily oriented system of rectangular coordinates an elliptically polarized wave can be considered as a superposition of three elliptically polarized waves propagating along the coordinate axes.

Three complex numbers, standing as multipliers of the unit vectors in the Formula (1.31), can be considered as a set of three spatial components of the complex vector in the Cartesian coordinate system.

If the normal vector $\vec{n}$ coincides with one of the coordinate axes, for example, with the unit vector $\vec{k}$, then the decomposition of the polarization ellipse is greatly simplified.

It is obvious that the complexity of analytical transformations in the study of the electromagnetic waves polarization depends on the choice of the coordinate system.

1.2 Interconnection and Comparison of Different Forms of Signals Polarization State in Radio Polarimetric Navigation Systems

In the previous paragraph, various ways of representing and describing the polarization structure of radio waves were given. Let's summarize the results.

So, in a fully polarized radio wave, the end of the $\vec{E}$ vector at each point in space writes an ellipse whose shape and orientation depend on time. If not to distinguish waves by energy and the initial phase, then two numbers are enough to describe the polarization, for example, the angle of the major axis of the ellipse β and the ellipticity coefficient r (1.9), phasor (1.11). In this case, the Poincare sphere also gives "all information" about the wave state. If it is necessary to take into account the radio wave energy, describing the radio wave polarization requires three numbers. The Poincare sphere "does not distinguish" waves of the same polarization, but of different intensities. The three-numerical apparatus is the coherence matrix K (1.18), Stokes parameters (1.15) and (1.16). Finally, the necessity to take into account the initial phase leads to the necessity of describing the radio wave polarization using four numbers. The corresponding apparatus is the Jones vector $\underline{\vec{E}}$ (matrix-column) (1.20), tetra-vector $\vec{E}$ (1.22), complex envelopes E_1, E_2 (1.7), and complex vector $\underline{\vec{E}}$ (1.24).

Since the practice of radio polarimetry uses almost all representations of the radio wave polarization state, Table 1.1 summarizes the relationships making connections between the parameters defining named views.

In a number of columns of Table 1.1 against a particular studied parameter, the phrase "not determined" is written. This means that the considered system of parameters, which includes the studied parameter, does not include the phase φ_0, and therefore the studied parameter cannot be determined. However, if the system of parameters is determined by the phase φ_0, the studied parameter can be determined using the formulas given in the same graph.

One or another way of representation can be used depending on the necessity of solving a specific problem.

Table 1.1 The relationships making connections between the parameters defining named views

	$a,\ \alpha,\ \beta$	$a,\ r,\ \beta$
E_{1m}	$a\sqrt{(1+\cos 2\alpha\cos 2\beta)/(1+\cos 2\alpha)}$	$a\sqrt{\cos^2\beta + r^2\sin^2\beta}$
E_{2m}	$a\sqrt{(1-\cos 2\alpha\cos 2\beta/(1+\cos 2\alpha))}$	$a\sqrt{\sin^2\beta + r^2\cos^2\beta}$
$\tan 2\delta$	$\tan 2\alpha\csc 2\beta$	$2r/\lfloor(1-r^2)\sin 2\beta\rfloor$
$\tan\varphi_0$	**Not determined** ($\tan\varphi_0$)	**Not determined** ($\tan\varphi_0$)
P_0	$a^2\sec^2\alpha$	$a^2(1+r^2)$
$\tan^2\gamma$	$(1-\cos 2\alpha\cos 2\beta)/(1+\cos 2\alpha\cos 2\beta)$	$(r^2+\tan^2\beta)/(1+r^2\tan^2\beta)$
$\cos 2\gamma$	$\cos 2\alpha\cos 2\beta$	$[(1-r^2)\cos 2\beta]/(1+r^2)$
$\tan 2\delta$	$\tan 2\alpha\csc 2\beta$	$2r/[(1-r^2)\sin 2\beta]$
X_1	**Not determined** $a\sqrt{\frac{1+\cos 2\alpha\cos 2\beta}{1+\cos 2\alpha}}\cos\varphi_0$	**Not determined** $a\sqrt{\cos^2\beta + r^2\sin^2\beta}\cos\varphi_0$
X_2	**Not determined** $a\sqrt{\frac{1+\cos 2\alpha\cos 2\beta}{1+\cos 2\alpha}}\sin\varphi_0$	**Not determined** $a\sqrt{\cos^2\beta + r^2\sin^2\beta}\sin\varphi_0$
X_3	**Not determined** $a^2\frac{\cos 2\alpha\sin 2\beta\cos\varphi_0 - \sin 2\alpha\sin\varphi_0}{\sqrt{2(1+\cos 2\beta\cos 2\alpha)}\cos\alpha}$	**Not determined** $a\frac{(1-r^2)\sin 2\beta\cos\varphi_0 - 2r\sin\varphi_0}{2\sqrt{\cos^2\beta + r^2\sin^2\beta}}$
X_4	**Not determined** $a^2\frac{\cos 2\alpha\sin 2\beta\sin\varphi_0 + \sin 2\alpha\cos\varphi_0}{\sqrt{2(1+\cos 2\beta\cos 2\alpha)}\cos\alpha}$	**Not determined** $a\frac{2r\cos\varphi_0 + (1-r^2)\sin 2\beta\sin\varphi_0}{2\sqrt{\cos^2\beta + r^2\sin^2\beta}}$
K_{11}	$a^2(1+\cos 2\alpha\cos 2\beta)/(1+\cos 2\alpha)$	$a^2(\cos^2\beta + r^2\sin^2\beta)$
$K_{12} = K_{21}^*$	$a^2\frac{\sin 2\beta\cos 2\alpha + i\sin 2\alpha}{1+\cos 2\alpha}$	$0.5a^2\lfloor(1-r^2)\sin 2\beta + 2ir\rfloor$
K_{22}	$a^2(1-\cos 2\alpha\cos 2\beta)/(1+\cos 2\alpha)$	$a^2(\sin^2\beta + r^2\cos^2\beta)$
P_0	$a^2\sec^2\alpha$	$a^2(1+r^2)$
$\|p\|$	$\sqrt{\frac{1-\cos 2\alpha\cos 2\beta}{1+\cos 2\alpha\cos 2\beta}}$	$\sqrt{\frac{r^2+\tan^2\beta}{1+r^2\tan^2\beta}}$
$tg2\delta$	$\tan 2\alpha\csc 2\beta$	$2r/\lfloor(1-r^2)\sin 2\beta\rfloor$
S_0	$a^2\sec^2\alpha$	$a^2(1+r^2)$
S_1	$a^2(1-\tan^2\alpha)\cos 2\beta$	$a^2(1-r^2)\cos 2\beta$
S_2	$a^2(1-\tan^2\alpha)\sin 2\beta$	$a^2(1-r^2)\sin 2\beta$

(continued)

Table 1.1 (continued)

	$a,\ \alpha,\ \beta$	$a,\ r,\ \beta$
S_3	$2a^2 \tan\alpha$	$2a^2 r$
a^2	a^2	a^2
$\tan 2\beta$	$\tan 2\beta$	$\tan 2\beta$
$r \tan\alpha$	$\tan\alpha$	r
$\tan\alpha$	$\tan\alpha$	$\tan\alpha$
$\sin 2\alpha$	$\sin 2\alpha$	$\sin 2\alpha$

1.3 Polarization Bases Used to Analyze the Signals

1.3.1 Expansion of Polarized Radio Navigation Systems Signals in Orthogonal Polarization Bases

As it is known, the decomposition of the vector in space in the chosen coordinate system can be carried out in countless ways. This position can be generalized to the case of complex vectors.

Let us introduce the concept of complex unit vectors, which are a generalization of the concept of real unit vectors. The system of complex vectors $\vec{e}_1, \vec{e}_2, \vec{e}_3$ is called orthonormal if

$$\vec{e}_m \vec{e}_n^* = \delta_{mn} = \begin{cases} 0, \text{ if } m \neq n \\ 1, \text{ if } m = n. \end{cases} \quad (1.33)$$

The scalar product (1.31) shows that the orthonormal system consists of unit vectors, each of which is orthogonal to the rest. If the number of vectors of an orthonormal system coincides with the number of space dimensions, the system (1.30) is said to form an orthonormal basis of space.

In three-dimensional space, any complex vector $\vec{E}$ of arbitrary orientation can be represented by a decomposition of three unit vectors $\vec{e}_1, \vec{e}_2, \vec{e}_3$. When considering plane waves, it makes sense to carry out such a decomposition in two unit vectors $\vec{e}_1, \vec{e}_2$:

$$\vec{E} = E_1 \vec{e}_1 + E_2 \vec{e}_2. \quad (1.34)$$

Scalarly multiplying the scalar vector $\vec{E}$ sequentially by $\vec{e}_1^*$ and $\vec{e}_2^*$ taking into account (1.31), we obtain:

$$\vec{E} = \left(\vec{E}\vec{e}_1^*\right)\vec{e}_1 + \left(\vec{E}\vec{e}_2^*\right)\vec{e}_2, \quad (1.35)$$

where scalar products $\left(\vec{E}\vec{e}_1^{\,*}\right) = E_1$ and $\left(\vec{E}\vec{e}_2^{\,*}\right) = E_2$, representing the projections of the vector $\vec{E}$ in the direction of the unit vectors $\vec{e}_1$ and $\vec{e}_2$, respectively, are called complex coordinates of the vector $\vec{E}$ in the polarization basis $\left[\vec{e}_1, \vec{e}_2\right]$. Moduli and arguments of the complex coordinates are the amplitudes and phases of the elliptically polarized components of the wave described by a complex vector $\vec{E}$. In this case, each component is polarized parallel to the corresponding unit vector of the basis.

It shall be noted that orthogonality of vectors implies that

$$|\vec{E}|^2 = |E_1|^2 + |E_2|^2, \tag{1.36}$$

i.e., the power of an elliptically polarized wave is the sum of the powers of two orthogonal elliptically polarized components.

Equation (1.35) can be interpreted as a decomposition of a fully polarized wave into two orthogonal fully polarized components. In other words, Eq. (1.35) means that any elliptically polarized electromagnetic wave can be unequivocally represented as the sum of two other orthogonally together polarized electromagnetic waves (e.g., as the sum of two circular polarization waves, right and left rotation or two waves which polarization ellipses have the same shape, with their major axes mutually perpendicular, and the directions of rotation of the electric vector are opposite).

The problem of transformation of the polarization basis (PB) plays an important role in polarization analysis. Let the matrix-columns of the unit vectors of the old basis be $\vec{e}_C = (\vec{e}_{1C} \quad \vec{e}_{2C})^{\mathrm{T}}$ and the new one $\vec{e}_H = (\vec{e}_{1H} \quad \vec{e}_{2H})^{\mathrm{T}}$. In this case, it is obvious that with the change of PB between vectors $\vec{e}_C$ and $\vec{e}_H$ there should be a linear connection:

$$\vec{e}_H = Q\vec{e}_C, \tag{1.37}$$

where $Q = \left(q_{ij}\right)$, $(i, j = 1, 2)$ is some square matrix with complex elements q_{ij}.

The new basis must also be orthonormalized, i.e., obey the condition (1.33), which in matrix form gives the following equation: $\vec{e}_H\vec{e}_H^{\,+} = I$, where $I = \begin{pmatrix} 1 & 0 \\ 0 & 1 \end{pmatrix}$ is a single matrix, and the sign "+" means the simultaneous operation of transposition and complex conjugation (i.e., for an arbitrary matrix A, an identical equation $(A^*)^{\mathrm{T}} \equiv A^+$ takes place).

The obtained equation taking into account the ratio (1.35) takes the form:

$$\vec{e}_H\vec{e}_H^{\,+} = Q\vec{e}_C\vec{e}_C^{\,+}Q^+ = QIQ^+ = QQ^+ = I. \tag{1.38}$$

It follows from the Formula (1.37) that the transition matrix Q must have the property of unitarity:

$$QQ^+ = I. \tag{1.39}$$

From the unitarity of the matrix Q follows some more properties that will be used in the subsequent analysis.

First of all, let us note that in accordance with Eq. (1.39) the matrix inverse to the matrix Q is equal to the complex-conjugate transposed matrix Q, i.e., $Q^{-1} = Q^{+}$.

The matrix Q determinant modulus, as it follows from Eq. (1.39), is obviously equal to 1. The following system of four equations for determination of the required matrix Q elements q_{ij} follows from the same equation:

$$\left.\begin{array}{l} |q_{11}|^2 + |q_{12}|^2 = 1; \quad |q_{21}|^2 + |q_{22}|^2 = 1 \\ q_{11}q_{21}^* + q_{12}q_{22}^* = 0; \quad q_{21}q_{11}^* + q_{22}q_{12}^* = 0 \end{array}\right\}. \tag{1.40}$$

The first two equations in the system (1.39) give the possibility to represent each of the matrix Q elements q_{ij} as a product of an exponent with a purely imaginary exponent and one of the trigonometric functions from some arbitrary real numbers. The foregoing makes it possible to get the following views:

$$\left.\begin{array}{l} |q_{11}|^2 = \cos^2\chi \quad ; \; q_{11} = \mathrm{e}^{\mathrm{i}\xi}\cos\chi \\ |q_{12}|^2 = \sin^2\chi \quad ; \; q_{12} = -\mathrm{e}^{\mathrm{i}\psi}\sin\chi \\ |q_{21}|^2 = \sin^2\chi_1 \quad ; \; q_{21} = -\mathrm{e}^{-\mathrm{i}\psi_1}\sin\chi_1 \\ |q_{22}|^2 = \cos^2\chi_1 \quad ; \; q_{22} = \mathrm{e}^{-\mathrm{i}\xi_1}\cos\chi_1 \end{array}\right\}. \tag{1.41}$$

The parameters $\chi, \chi_1, \xi, \xi_1, \psi, \psi_1$ introduced in the Formula (1.41) are some real numbers, between which, due to the presence of condition (1.41), there is a certain functional relation, which can be revealed by the substitution of relation (1.40) in the last two equations of the system (1.41), which gives:

$$\cos\chi\sin\chi_1\exp[\mathrm{i}(\xi - \xi_1)] = \sin\chi\cos\chi_1\exp[\mathrm{i}(\pi + \psi - \psi_1)], \tag{1.42}$$

where from the equality of moduli follows that $\chi - \chi_1 = n\pi$, and from the equality of phases $\xi - \xi_1 = \psi - \psi_1 + (2m + 1)\pi$; $n, m = 0, \pm 1, \pm 2, \ldots$.

Without any loss of generality, it can be assumed that $m = n = 0$, i.e., $\chi = \chi_1$; $\xi - \xi_1 = \psi - \psi_1 + \pi$.

As it will be seen in the following, instead of the parameters $\chi, \chi_1, \xi, \xi_1, \psi, \psi_1$ will be more convenient to use some linear combination. To this end, we move on to the new parameters $\varphi_0, \gamma, \delta, \eta$ using the following linear transformations: $\xi = \eta + \varphi$; $\xi_1 = \eta - \varphi$; $\psi = \varphi + \eta + 2\delta - \pi$; $\psi_1 = -\varphi + \eta + 2\delta$; $\chi = \gamma$.

Taking into account Eq. (1.42), we get:

$$Q = \mathrm{e}^{\mathrm{i}\varphi}\begin{pmatrix} \mathrm{e}^{\mathrm{i}\eta}\cos\gamma & \mathrm{e}^{\mathrm{i}(\eta+2\delta)}\sin\gamma \\ -\mathrm{e}^{-\mathrm{i}(\eta+2\delta)}\sin\gamma & \mathrm{e}^{-\mathrm{i}\eta}\cos\gamma \end{pmatrix}. \tag{1.43}$$

Thus, the unitary matrix is determined by four parameters $\eta, \delta, \gamma, \varphi$. Consider their geometric meaning.

Let the Cartesian basis, i.e., the basis formed by the pair of horizontal–vertical polarization $\vec{e}_C = \left(\vec{e}_{1C} \equiv \vec{i} \quad \vec{e}_{2C} \equiv \vec{j}\right)$, act as the old polarization basis. Unit vectors of the new basis in this case in accordance with Eqs. (1.38) and (1.43) take the form:

$$\left.\begin{aligned} \vec{e}_{1H} &= \left(e^{i\eta}\cos\gamma\vec{e}_{1C} + e^{i(\eta+2\delta)}\sin\gamma\vec{e}_{2C}\right)e^{i\varphi} \\ \vec{e}_{2H} &= \left(-e^{-i(\eta+2\delta)}\sin\gamma\vec{e}_{1C} + e^{-i\eta}\cos\gamma\vec{e}_{2C}\right)e^{i\varphi} \end{aligned}\right\}. \tag{1.44}$$

The components of the first unit vector along the *X*- and *Y*-axes are:

$$\left.\begin{aligned} (\vec{e}_{1H})_x &= e^{i\varphi}e^{i\eta}\cos\gamma \\ (\vec{e}_{1H})_y &= e^{i\varphi}e^{i\eta}e^{2i\delta}\sin\gamma \end{aligned}\right\} \tag{1.45}$$

and it is the component of the polarization ellipse $\vec{e}_{1H}$.

Similarly, you can write an expression for the components of an orthogonal polarization ellipse $\vec{e}_{2H}$:

$$\left.\begin{aligned} (\vec{e}_{2H})_x &= -e^{i\varphi}e^{-i\eta}e^{-2i\delta}\sin\gamma \\ (\vec{e}_{2H})_y &= e^{i\varphi}e^{-i\eta}\cos\gamma \end{aligned}\right\}. \tag{1.46}$$

Let us compare the expression (1.45) with the view (1.7). As it can be seen, in both relations, the angle γ in the expressions for the PB conversion coincides with the angle γ characterizing the ratio of the amplitudes of the orthogonal components of the polarized electromagnetic wave. The sum of the phases $(\eta+\varphi)$ in the Formula (1.45) coincides with the initial phase φ_0 in the Formula (1.7). If we take away from the orthogonal ellipse of polarization determined by Eq. (1.46), it is sufficient to denote $(\eta+\varphi)$ by φ_0, so that the expressions (1.7) and (1.46) will become identical (within the accuracy of up to the factor $\sqrt{P_0}$ playing no role in the ongoing arguments). Therefore, the parameter η is not determinative and can be set equal to an arbitrary number, in particular, for example, equal to zero.

If both polarization ellipses defined by Eqs. (1.45) and (1.46) are considered together, the parameter η manifests itself as the difference of the initial phases in these ellipses. For the first ellipse, this phase, as already mentioned, is equal to $(\varphi+\eta)$, and for the second it will be equal to $(\varphi-\eta+\pi)$. Thus, the parameter η determines the difference of initial phases for orthogonal polarization ellipses.

In connection with the above in the foregoing, where we are talking about only one of the two orthogonal polarization ellipses, the parameter η will be related to the initial phase φ_0.

In this case, the unitary matrix Q will have the following common view:

$$Q = e^{i\varphi_0}\begin{pmatrix} \cos\gamma & e^{2i\delta}\sin\gamma \\ -e^{-2i\delta}\sin\gamma & \cos\gamma \end{pmatrix}. \tag{1.47}$$

The geometric meaning of the transformation of the basis (1.37) using the matrix (1.47) is that it is used to transfer the Poincare sphere from points *A* and *D* (HP-VP) in

Fig. 1.4 to points P and P' having a polarization ratio equal to $\tan\gamma$, for the 2δ phase difference of their orthogonal components. For clarity, Fig. 1.5 presents a spherical segment, which shows the angles 2γ and 2δ.

The geometric locus of the points representing the unit vectors of the basis with an angle $\gamma = \text{const}$ is a small circle, the plane of which is perpendicular to the axis connecting the center of the Poincare sphere with the point A (Fig. 1.6). Its position is determined by the angle POA which is equal to 2γ. The geometric locus of the points for which $2\delta = \text{const}$ will be a large circle, the plane of which forms an angle 2δ with the plane of the equator AMN (this is $\angle PKL$ angle in Fig. 1.4). In this case, we have $2\delta = 0$ for the plane AMN.

The given arguments are more common. So, if you choose some other bases, not Cartesian, as the old basis, then in this case the relations (1.45) and (1.47) will also take place. As for the geometric interpretation, the angle 2δ will play the role of the angle formed by two radii of the Poincare sphere drawn to the points representing the first unit vectors of the new and old bases. In other words, in Fig. 1.4 point A will be the "image" of the unit vector of the old basis, and point P will be the new one, and in this case the angle PKL will be equal to 2δ.

It shall be noted that two parameters of the angles γ and δ transformation matrix are sufficient to display the basis transformation on the Poincare sphere.

The remaining two parameters η and φ ensure the transition to the unit vectors of the polarization basis, which initial phases corresponding to the angle φ_0 in the record (1.7) are set.

The introduction of the matrix of transformation of the polarization basis Q gives the possibility to degree the entire set of possible polarization bases, since each of them receives a specific four-digit number—a set of four parameters γ, δ, η, and φ_0, defining the matrix Q. However, this approach requires the introduction of a zero polarization basis, in relation to which the above parameters are determined. Any polarization basis can be zero. On the Poincare sphere, for which the correspondence of the parameters defining the matrix Q to the parameters defining the polarization ellipse is given, the zero polarization basis in this case will be represented by points for which $\gamma = \delta = \beta = 0$. Such points on Figs. 1.4 and 1.5 are the points A and B that serve in these figures as "images" of horizontal and vertical polarization.

The matrix Q in the form (1.42) can be represented as a product of two matrices, one of which (1.46) defines only the geometric parameters of the basis (γ and δ), and the other only the phase state of the basis:

$$Q = F(\varphi_0, \eta) Q_0(\gamma, \delta), \tag{1.48}$$

where

$F(\varphi_0, \eta) = e^{i\varphi_0} \begin{pmatrix} e^{i\eta} & 0 \\ 0 & e^{-i\eta} \end{pmatrix}$ is the diagonal matrix of the phase state of the basis;

Q_0 indicates a matrix Q in the form (1.46) without factor $e^{i\varphi_0}$.

As it can be seen, the elements of the matrices $F(\varphi_0, \eta)$ and $Q_0(\gamma, \delta)$ are independent from each other.

Sometimes, the transition from one basis to another is carried out using the transition to some intermediate basis firstly. We will call it zero and denote it as a symbol (0, 0, 0, 0) (such basis can be a basis, for example, formed by horizontal and vertical polarization). Any other basis, which polarization parameters are obtained by the transformation from the zero basis using the correlation (1.36), will be denoted by a symbol $(\gamma, \delta, \varphi_0, \eta)$. The symbolic record of such transformation has the form:

$$(\gamma, \delta, \varphi_0, \eta) = Q(\gamma, \delta, \varphi_0, \eta)(0, 0, 0, 0). \tag{1.49}$$

The relation between the parameters of two bases (1 and 2) can be represented as follows:

$$\left.\begin{aligned} (\gamma_1, \delta_1, \varphi_{01}, \eta_1) &= Q(\gamma_1, \delta_1, \varphi_{01}, \eta_1)(0, 0, 0, 0) \\ (\gamma_2, \delta_2, \varphi_{02}, \eta_2) &= Q(\gamma_2, \delta_2, \varphi_{02}, \eta_2)(0, 0, 0, 0) \end{aligned}\right\}. \tag{1.50}$$

It follows from (1.49) that $(0,0,0,0) = Q^{-1}(\gamma_2, \delta_2, \varphi_{02}, \eta_2)(\gamma_2, \delta_2, \varphi_2, \eta_2)$, where Q^{-1} is the matrix inverse to the matrix Q. Using the property (1.39) of the unitary matrix and substituting the value of the zero basis parameters from Eq. (1.50), we have:

$$(\gamma_1, \delta_1, \varphi_{01}, \eta_1) = Q(\gamma_1, \delta_1, \varphi_{01}, \eta_1)Q^{+}(\gamma_2, \delta_2, \varphi_{02}, \eta_2)(\gamma_2, \delta_2, \varphi_{02}, \eta_2). \tag{1.51}$$

If the initial basis is not zero, but some arbitrary basis $(\gamma_1, \delta_1, \varphi_{01}, \eta_1)$, then the basis $(\gamma_2, \delta_2, \varphi_{02}, \eta_2)$ will have parameters related to it:

$$\left.\begin{aligned} \gamma &= \arccos\sqrt{\cos^2(\gamma_1 - \gamma_2) + \sin 2\gamma_1 \sin 2\gamma_2 \cos^2(\delta_2 - \delta_1)} \\ \eta &= \eta_2 - \eta_1 + \operatorname{arctg}\frac{\sin\gamma_1 \sin\gamma_2 \sin 2(\delta_2 - \delta_1)}{\cos\gamma_1 \cos\gamma_2 + \sin\gamma_1 \sin\gamma_2 \cos 2(\delta_2 - \delta_1)} \end{aligned}\right\}. \tag{1.52}$$

Let us consider the examples of construction of some matrices of transformation of the polarization basis.

1. The phase change of the unit vectors of the polarization basis. In this case, it is necessary to assume $\gamma = \delta = 0$. For this case, the transition matrix has the form $Q_1 = F(\varphi_0, \eta)$.
   The transformation of the polarization basis with the help of such matrix changes only the phases of the unit vectors without changing their appearance, and since the arguments φ_0 and η are arbitrary, there can be arbitrarily many transition matrices of this type.
2. Transition from the initial linear basis to another linear polarization basis (without changing the phase of the unit vectors), rotated by an angle β_0. In this case $\delta = \varphi_0 = \eta = 0$, only the angle γ changes. At the same time, $\gamma = \beta_0$; thus:

$$Q_2 = \begin{pmatrix} \cos\beta_0 & \sin\beta_0 \\ -\sin\beta_0 & \cos\beta_0 \end{pmatrix}. \tag{1.53}$$

As it is seen, the matrix Q_2 is the matrix of rotation of the coordinate system.

3. The transition from a linearly polarized pair, for example, horizontally and vertically polarized (HP–VP), to another linearly polarized pair, rotated by an angle β_0 (with the change of the phase of the unit vectors).
   For this case, it is necessary to assume that $\delta = 0$; $\gamma = \beta_0$, φ_0 and η are arbitrary. The common view of the transition matrix here will be $Q_3 = F(\varphi_0, \eta)Q_2$.
4. The transition from a pair of HP–VP to circular (right circle–left circle).
   For this case, it is necessary that $\gamma = \delta = \pi/4$, and φ_0 and η are arbitrary. The common view of these matrices will be:

$$Q_4 = \frac{\mathrm{e}^{\mathrm{i}\varphi_0}}{\sqrt{2}} \begin{pmatrix} \mathrm{e}^{\mathrm{i}\eta} & \mathrm{i}\mathrm{e}^{\mathrm{i}\eta} \\ \mathrm{i}\mathrm{e}^{-\mathrm{i}\eta} & \mathrm{e}^{-\mathrm{i}\eta} \end{pmatrix}. \tag{1.54}$$

   In view of the arbitrariness of the choice of the argument of the representations of the matrices Q can be arbitrarily many. Note of this set only the most common representations in the literature, leading to the simplification of its form of record.
   When $\varphi_0 = -\eta = -\pi/4$, we have $Q_4 = \frac{1}{\sqrt{2}}\begin{pmatrix} 1 & i \\ 1 & -i \end{pmatrix}$.
   When $\varphi_0 = \eta = 0$, we have $Q_4 = \frac{1}{\sqrt{2}}\begin{pmatrix} 1 & i \\ i & 1 \end{pmatrix}$.
5. The transition from a pair of HP–VP to circular (left circle–right circle).
   In this case, it is necessary to assume that $\gamma = -\delta = \pi/4$. The common view of the matrix corresponding to this transformation will be:

$$Q_5 = \frac{\mathrm{e}^{\mathrm{i}\varphi_0}}{\sqrt{2}} \begin{pmatrix} \mathrm{e}^{\mathrm{i}\eta} & -\mathrm{i}\mathrm{e}^{\mathrm{i}\eta} \\ -\mathrm{i}\mathrm{e}^{-\mathrm{i}\eta} & \mathrm{e}^{-\mathrm{i}\eta} \end{pmatrix}. \tag{1.55}$$

   Then, $\varphi_0 = \eta = 0$, and matrix Q takes the form given: $Q_5 = \frac{1}{\sqrt{2}}\begin{pmatrix} 1 & -i \\ -i & 1 \end{pmatrix}$.
6. The transition from a linearly polarized pair to the elliptical basis, the major unit vectors of which coincide with the unit vectors of the original basis.

For this case, it is necessary to assume that $\delta = \pi/2$, $\gamma = \gamma_0$ and φ_0 and η are arbitrary. The common form of the unitary transformation matrix here will be:

$$Q_6 = \mathrm{e}^{\mathrm{i}\varphi} \begin{pmatrix} \mathrm{e}^{\mathrm{i}\eta} \cos\gamma_0 & -\mathrm{e}^{\mathrm{i}\eta} \sin\gamma_0 \\ \mathrm{e}^{-\mathrm{i}\eta} \sin\gamma_0 & \mathrm{e}^{-\mathrm{i}\eta} \cos\gamma_0 \end{pmatrix}. \tag{1.56}$$

Finally, let us consider how the components of the vector $\vec{E}$ change when the polarization basis changes.

Suppose that in the old polarization basis ($\vec{e}_C$) these components are presented as a matrix-column $E_C^{\mathrm{T}} = (E_{1C} \;\; E_{2C})^{\mathrm{T}}$, and in the new as a matrix-column $E_H^{\mathrm{T}} = (E_{1H} \;\; E_{2H})^{\mathrm{T}}$. In this case, the following chain of equations is obvious:

$$\vec{E} = E_C^T \vec{e}_C = E_H^T \vec{e}_H = E_H^T Q \vec{e}_C. \tag{1.57}$$

In Eq. (1.57), the correlation (1.38) is taken into account.

Using the matrix property (1.39) and the matrix product transposition rules, from the Formula (1.57) we obtain the desired transformation of the coordinate components:

$$E_H = Q^* E_C. \tag{1.58}$$

1.3.2 Expansion of Polarized Radio Navigation Systems Signals in Affine Polarization Bases

In previous sections, the decomposition of an elliptically polarized wave in orthogonal polarization bases was considered. It is also possible to decompose the vector into non-orthogonal bases; thus, there are all kinds of pairs of unit vectors, for which the correlation (1.33) is not satisfied. For example, decompositions of a wave of any polarization on a few components having different polarizations are possible; i.e., one unit vector is ellipse, and another is a circle, circle and linear unit vector, and linear and elliptic. Figure 1.7 clearly explains the transition to non-orthogonal decomposition, and it shows the decomposition of the vector in the orthogonal (a) and non-orthogonal (b) or, as often called, in the affine basis, in this case, the linear affine polarization basis. Note that in the affine decomposition the components of the vector $\vec{E}$ can be greater than its length OE.

Analyzing the transformation of the polarization basis with the help of the correlation (1.43) and comparing it with (1.44), it can be seen that the first unit vector of the

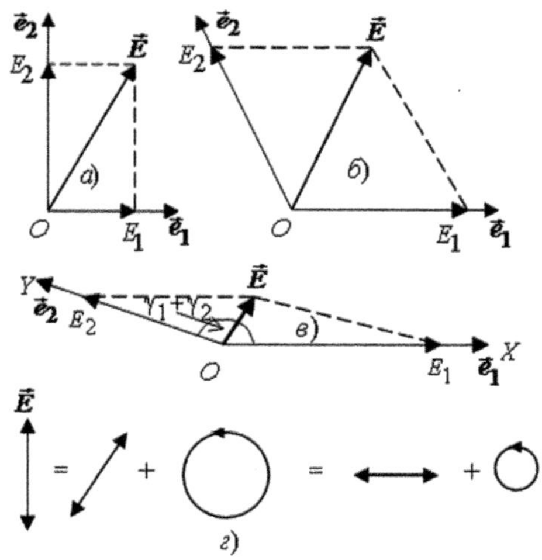

Fig. 1.7 Expansion of vector $\vec{E}$ in **a** linear orthogonal and **b**, **c** linear and **d** non-linear non-orthogonal (affine) polarization bases

transformation is defined by the first line of the transition matrix Q, and its second line determines the second unit vector of the transformation. For the affine basis, the position and the size of one unit vector are not related to the second position; therefore, the polarization parameters of the first row of the transformation matrix in common shall be different from the parameters of the second row. In this case, six parameters $(\gamma_1, \gamma_2, \delta_1, \delta_2, \varphi_{01}, \varphi_{02})$ are required to construct an affine basis from the initial one.

As we have already noted, the first four are used to determine the position of the base ellipses corresponding to the first and second lines, and the last four to take into account the instantaneous position of the vector in these various ellipses. The transition to the affine basis shall be carried out using different signs of the imaginary part of the transition matrix components, located in different lines, although for the affine basis this is certainly not mandatory. Therefore, the transformation matrix by analogy with (1.42) can be written as:

$$Q_{12} = \begin{pmatrix} e^{i\varphi_{01}} \cos\gamma_1 & e^{i2\delta_1} e^{i\varphi_{01}} \sin\gamma_1 \\ -e^{-i2\delta_2} e^{i\varphi_{02}} \sin\gamma_2 & e^{i\varphi_{02}} \cos\gamma_2 \end{pmatrix}. \tag{1.59}$$

If we assume that $\gamma_1 = \gamma_2 = \gamma$, $\delta_1 = \delta_2 = \delta$, $\varphi_{01} = \varphi_0 + \eta$, $\varphi_{02} = \varphi_0 - \eta$, then the Formula (1.59) goes into the Formula (1.44).

If the unit vectors of the Cartesian basis are identified as $\vec{e}_0 = \left(\vec{e}_x \quad \vec{e}_y\right)^{\mathrm{T}}$, and the unit vectors of affine as $\vec{e}_{12} = (\vec{e}_1 \quad \vec{e}_2)^{\mathrm{T}}$, then

$$\vec{e}_{12} = Q_{12}\vec{e}_0. \tag{1.60}$$

Let us introduce a matrix Q_{12}^{-1} that is inverse to the matrix Q_{12} such that $Q_{12}Q_{12}^{-1} = I$. Then,

$$Q_{12}^{-1} = C\begin{pmatrix} e^{i\varphi_{02}} \cos\gamma_2 & -e^{i(\varphi_{01}+2\delta_1)} \sin\gamma_1 \\ e^{i(\varphi_{02}-2\delta_2)} \sin\gamma_2 & e^{i\varphi_{01}} \cos\gamma_1 \end{pmatrix}, \tag{1.61}$$

where $C = e^{-i(\varphi_{01}+\varphi_{02})}\left[\cos\gamma_1 \cos\gamma_2 + e^{i2(\delta_1-\delta_2)} \sin\gamma_1 \sin\gamma_2\right]^{-1}$.

From Eq. (1.60), we can obtain the expression for $\vec{e}_0$:

$$\vec{e}_0 = Q_{12}^{-1}\vec{e}_{12}, \tag{1.62}$$

The modulus of the determinant of the matrix Q_{12} of transition will be equal to:

$$\det Q_{12} = \sqrt{\cos^2(\gamma_1 + \gamma_2) + \sin 2\gamma_1 \sin 2\gamma_2 \cos^2[(\delta_1 - \delta_2)]}, \tag{1.63}$$

it vanishes only when the following conditions are met:

$$\left.\begin{aligned} \gamma_1 + \gamma_2 &= 0.5(2n+1)\pi \\ \delta_2 - \delta_1 &= 0.5(2n+1)\pi \end{aligned}\right\}, \tag{1.64}$$

which corresponds to the case where the first and second polarizations are the same.

Expressions (1.60)–(1.62) allow to find the relations for the calculation of affine components E_1, E_2 of a fully polarized wave $\vec{E} = (E_1 \quad E_2)^{\mathrm{T}}$.

Let these components be equal to E_{10}, E_{20} in the initial basis, i.e., $\vec{E}_0 = (E_{10} \quad E_{20})^{\mathrm{T}}$.

From the condition of determining of the same vector in different bases, the equation below is obvious:

$$\vec{E} = \vec{E}_0^{\mathrm{T}}\vec{e}_0 = \vec{E}_1^{\mathrm{T}}\vec{e}_{12} = \vec{E}_{12}^{\mathrm{T}}Q_{12}\vec{e}_0, \tag{1.65}$$

whence it follows

$$\vec{E}_{12} = \left(Q_{12}^{\mathrm{T}}\right)^{-1}\vec{E}_0. \tag{1.66}$$

Equation (1.65) gives the possibility to find the absolute value of the components of a fully polarized wave in the basis (1-2). If we use the Formula (1.64), then for the wave specified in the Cartesian basis as a matrix-column of type (1.7):

$$E_0 = \sqrt{P_0}\mathrm{e}^{\mathrm{i}\varphi_0}\left(\cos\gamma_0 \quad \mathrm{e}^{2\mathrm{i}\delta_0}\sin\gamma_0\right)^{\mathrm{T}} \tag{1.67}$$

projections for the first and second unit vectors will be, respectively:

$$|E_1| = \sqrt{P_0\frac{\cos^2(\gamma_0+\gamma_2)+\sin 2\gamma_0\sin 2\gamma_2\cos^2(\delta_0-\delta_2)}{\cos^2(\gamma_1+\gamma_2)+\sin 2\gamma_1\sin 2\gamma_2\cos^2(\delta_1-\delta_2)}}, \tag{1.68}$$

and for the second:

$$|E_2| = \sqrt{P_0\frac{\sin^2(\gamma_0-\gamma_1)+\sin 2\gamma_0\sin 2\gamma_1\sin^2(\delta_0-\delta_1)}{\cos^2(\gamma_1+\gamma_2)+\sin 2\gamma_1\sin 2\gamma_2\cos^2(\delta_1-\delta_2)}}. \tag{1.69}$$

If the first unit vector of the affine basis coincides with the wave ($\gamma_0 = \gamma_1, \delta_0 = \delta_1$) polarization vector, then $|E_1| = \sqrt{P_0}$, and most importantly, the projection to the second affine unit vector regardless of the choice of its parameters is always (!) equal to zero.

In affine decomposition, the polarization basis can be chosen such that the components of the wave can become arbitrarily large. To illustrate, let us consider one particular case. Assume that $\gamma_0 = 0, \delta_1 = \delta_0, \delta_2 = \delta_0 + 0.5\pi$.

In this case, the expressions for the projection take a very simple form:

$$\left.\begin{aligned} |E_1| &= |\cos\gamma_2 \sec(\gamma_1+\gamma_2)| \\ |E_2| &= |\sin\gamma_1 \sec(\gamma_1+\gamma_2)| \end{aligned}\right\}. \tag{1.70}$$

Correlation (1.69) shows that the closer the sum $(\gamma_1 + \gamma_2)$ to $\pi/2$ projection $|E_1|$ and $|E_2|$ unlimitedly increase. For illustration, in Fig. 1.7 this situation is presented on an example of the decomposition of a linearly polarized wave in a linear affine polarization basis. As it is seen, as the angle *YOX* approaches 180°, the projections of the vector $\vec{E}$ on the axes *OX* and *OY* increase unlimitedly.

From a physical point of view, the unlimited growth of the projection of the wave in the affine bases does not mean unlimited growth of power of the wave. It is to determine the power in the real measurement conditions that the transition from the affine basis of the general form to the orthogonal basis is necessary, in which the sum of the power of the components is equal to the total power of the wave.

As another illustration of the obtained correlations, let us present the decomposition of a vertically polarized wave into a linear wave with an angle $\gamma = \gamma_0$ and a circular polarization. Let us limit ourselves to bases where $\varphi_{01} = \varphi_{02} = 0$. In accordance with the task at hand, it is necessary to require that $2\delta_1 = \pi/2$, $2\gamma_1 = \pi/2$, $2\delta_2 = \pi$, $\gamma_2 = \gamma_0$.

Under the above conditions, the matrix (1.60) will take the form:

$$Q_{12}^{-1} = e^{i\gamma_0}\begin{pmatrix} \sqrt{2}\cos\gamma_0 & -i \\ -\sqrt{2}\sin\gamma_0 & 1 \end{pmatrix}. \tag{1.71}$$

Using Eq. (1.66) and the representation of the column matrix for the VP wave $\begin{pmatrix} 0 & 1 \end{pmatrix}^{\mathrm{T}}$, let us find the required expansion coefficients:

$$\left.\begin{aligned} E_{K\Pi} &= -\sqrt{2}\, e^{i\gamma_0} \sin\gamma_0 , \\ E_{Л\gamma_0} &= e^{i\gamma_0} \end{aligned}\right\}. \tag{1.72}$$

Note that in this example, the horizontal polarization can act as a linear polarization, which is a component of the vertical polarization decomposition. In this case, it is necessary to assume that the angle $\gamma_0 = \pi/2$. Symbolic illustration of the example above is presented in Fig. 1.7.

The transition from one (1-2) to another (3-4) affine basis can be carried out directly, bypassing the intermediate transition to the zero (0-0) basis. Of course, it shall be remembered that each of the bases is given in some zero basis with respect to which its parameters $(\gamma_1, \gamma_2, \delta_1, \delta_2, \varphi_{01}, \varphi_{02})$ are defined.

We obtain the transition formulae from some affine (1-2) to another affine (3-4) basis. By analogy with the correlation (1.60), we have:

$$\vec{e}_{34} = Q_{34}\vec{e}_0, \tag{1.73}$$

taking into account Eq. (1.59), it gives the desired formula of the transition:

$$\vec{e}_{34} = Q_{34} Q_{12}^{-1} \vec{e}_{12}. \tag{1.74}$$

Having done the calculations similar to (1.65) and (1.66), we elementarily obtain the relation between the coordinates of the $\vec{E}$ vector components in the bases (1-2) and (3-4):

$$E_{34} = \left(Q_{34}^{\mathrm{T}}\right)^{-1} Q_{12}^{\mathrm{T}} E_{12}. \tag{1.75}$$

The unit vectors of an arbitrary elliptic affine polarization basis can be represented as follows:

$$\begin{aligned}\vec{e}_j\left(\alpha_j, \beta_j\right) = &\left(\cos\alpha_j \cos\beta_j - i \sin\alpha_j \sin\beta_j\right)\vec{e}_x \\ &+ \left(\cos\alpha_j \sin\beta_j + i \sin\alpha_j \cos\beta_j\right)\vec{e}_y, \quad (j = 1, 2).\end{aligned} \tag{1.76}$$

Let us present special cases corresponding to different values of the ellipticity angle α_j and slope angle of the major axis of the polarization ellipse β_j:

- Horizontal polarization: $\vec{e}(\alpha = 0, \beta = 0) = \vec{e}_x$; vertical polarization: $\vec{e}(\alpha = 0, \beta = \pi/2) = \vec{e}_y$.
- Linear polarization with a slope angle equal to $\pi/4$: $\vec{e}(\alpha = 0, \beta = \pi/4) = \frac{\sqrt{2}}{2}\left(\vec{e}_x + \vec{e}_y\right)$; linear polarization with a slope angle equal to $-\pi/4$: $\vec{e}(\alpha = 0, \beta = -\pi/4) = \frac{\sqrt{2}}{2}\left(\vec{e}_x - \vec{e}_y\right)$.
- Right circular polarization: $\vec{e}(\alpha = \pi/4, \beta = 0) = \frac{\sqrt{2}}{2}\left(\vec{e}_x + i\vec{e}_y\right)$; left circular polarization: $\vec{e}(\alpha = -\pi/4, \beta = 0) = \frac{\sqrt{2}}{2}\left(\vec{e}_x - i\vec{e}_y\right)$.

An arbitrary elliptic (affine) polarization basis consists of two unit vectors $\vec{e}_1(\alpha_1, \beta_1)$ and $\vec{e}_2(\alpha_2, \beta_2)$ that can be represented in matrix form:

$$\left(\vec{e}_1(\alpha_1, \beta_1) \quad \vec{e}_2(\alpha_2, \beta_2)\right)^{\mathrm{T}} = Q(\alpha_1, \alpha_2, \beta_1, \beta_2)\left(\vec{e}_1(0, 0) \quad \vec{e}_2(0, \pi/2)\right)^{\mathrm{T}}. \tag{1.77}$$

Given into account (1.76), the matrix $Q(\alpha_1, \alpha_2, \beta_1, \beta_2)$ accepts the following record:

$$Q(\alpha_1, \alpha_2, \beta_1, \beta_2) = \begin{pmatrix} \cos\alpha_1 \cos\beta_1 - \mathrm{i}\sin\alpha_1 \sin\beta_1 & \cos\alpha_1 \sin\beta_1 + \mathrm{i}\sin\alpha_1 \cos\beta_1 \\ \cos\alpha_2 \cos\beta_2 - \mathrm{i}\sin\alpha_2 \sin\beta_2 & \cos\alpha_2 \sin\beta_2 + \mathrm{i}\sin\alpha_2 \cos\beta_2 \end{pmatrix}. \tag{1.78}$$

The matrix $Q(\alpha_1, \alpha_2, \beta_1, \beta_2)$ thus makes a change in the angle of ellipticity and rotation of the axial reference of the elliptic basis.

1.4 Basic Characteristics of Probability Repres

In practice, they usually deal with partially polarized electromagnetic waves. In this case, the parameters describing the polarization structure of the wave are random functions of time and space. Therefore, statistical methods of analysis shall be used to describe the polarization state of a partially polarized wave. Consequently, let us give the necessary data from the theory of random processes for further use of this data.

Let us consider some random experiment or experiment, which is repeated under constant conditions a large number of times. We believe that the result of each individual trial or experiment is given by a certain number x. If the result of an individual trial is given by a set of numbers $x_1, x_2, \ldots, x_K$, then a vector $\vec{\xi} = \{x_1, x_2, \ldots, x_K\}$ can be introduced in K-dimensional space and $\vec{\xi}$ can be called as K-dimensional random variable or K-dimensional random vector.

For a complete description of a random variable, the possible values taken by the random variable and the probability of these values must be specified. The correlation that establishes in one form or another the relation between the possible values of a random variable and their probabilities is called the distribution law. In the foregoing, we will consider mainly continuous random variables that take countless values, and the probability of entering into any infinitely small area is infinitely small. The distribution law of a continuous random variable ξ is defined by the probability density function $W(\xi)$; sometimes, this function is simply called the probability distribution of a random variable ξ, and often acronym PDF is used for it. The probability density function $W(\xi)$ of a continuous random variable ξ is defined as the limit of the correlation of the probability of getting the values of a random variable in a small interval $[x, x + \Delta x]$ to the length of this interval Δx when $\Delta x \to 0$:

$$W(x) = \frac{\lim\limits_{\Delta x \to 0} \{P(x) \leq \xi \leq x + \Delta x\}}{\Delta x} \tag{1.79}$$

Note that the probability density $W(\xi)$ is dimensional and its dimension is inverse to the dimension of a random variable.

A distribution function of a random variable ξ is also introduced, which is a function $F(x)$ that is equal to the probability that the random variable ξ will take a value less than x:

$$F(x) = P(\xi < x), \quad -\infty < x < \infty$$

There is a known relation also between the distribution function and the distribution density $W(\xi)$ recorded in the form $W(x) = \frac{\mathrm{d}F(x)}{\mathrm{d}x}$.

A complete description of any scalar or vector random variables is given by the probability distribution laws (probability densities), distribution functions, or other functions associated by certain transformations with above-mentioned. Thus, a multidimensional random variable, or a random vector $\vec{\xi}$ with components or projection $\xi_1, \xi_2, \ldots, \xi_n$, is fully described by the distribution function of this random vector $\vec{\xi}$:

$$F_n(x) \equiv F_n(x_1, x_2, \ldots, x_n) = P\{\xi_1 < x_1; \xi_2 < x_2; \ldots; \xi_n < x_n\}$$
$$= \int_{-\infty}^{x_n} \ldots \int_{-\infty}^{x_1} W_n(u_1, u_2, \ldots, u_n) du_1 du_2 \ldots du_n. \quad (1.80)$$

The probability density is introduced similarly
$W_n(\vec{x}) = W_n(x_1, x_2, \ldots, x_n)$, of random vector $\vec{\xi} = \{\xi_1, \xi_2, \ldots, \xi_n\}$:

$$W_n(\vec{x}) = W_n(x_1, x_2, \ldots, x_n)$$
$$= \lim_{\substack{\Delta x_1 \to 0 \\ \Delta x_2 \to 0 \\ \vdots \\ \Delta x_n \to 0}} \frac{P\{x_1 \le \xi_1 < x_1 + \Delta x_1, \ldots, x_n \le \xi_n < x_n + \Delta x_n\}}{\Delta x_1 \Delta x_2 \cdots \Delta x_n}. \quad (1.81)$$

From the n-dimensional probability density, it is always possible to obtain any probability density of a smaller dimension by integrating the initial over not used arguments, i.e.,

$$W_m(x_1, x_2, \ldots, x_n) = \int_{-\infty}^{\infty} \ldots \int_{-\infty}^{\infty} W(x_1, \ldots, x_m, x_{m+1}, \ldots, x_n)\, dx_{m+1} \cdots dx_n, \quad (1.82)$$

where $m < n$.

As noted, if the n-dimensional probability density of a random variable is known, then there is no need in other characteristics of the random variable, since in this case there is a complete description of its probability. However, in the real environment, the n-dimensional probability density is either unknown or so complex that its practical use is impossible. Therefore, in the implementation of practical measurements it is advisable to use simpler probabilistic characteristics of a random variable, which include:

- Expectation of a random variable x:

$$< x >= M\{x\} \equiv m_x \equiv \bar{x} = \int_{-\infty}^{\infty} xW(x)dx, \quad (1.83)$$

which characterizes the average value of a random variable x. In the correlation (1.82), all symbols to the left of the equal sign indicate statistical averaging operations. In the foregoing, these symbols will be used.

– Variance D_x of a random variable x:

$$D_x = \langle\{x^2 - (\langle x\rangle)^2\}\rangle = \int\limits_{-\infty}^{\infty} (x - m_x)^2 W(x)\mathrm{d}x, \tag{1.84}$$

which characterizes the measure of the spread of values of a random variable relative to its mathematical expectation. Since the variance has the dimension of a square of a random variable, then the concept of the mean square deviation (MSD) d_x is introduced, having the dimension of a random variable and associated with the variance by correlation $d_x = \sqrt{D_x}$.

– The correlation moment (correlation function):

$$\begin{aligned}\mu_{12} &= \langle(x_1 - m_1)(x_2 - m_2)\rangle \\ &= \int\limits_{-\infty}^{\infty}\int\limits_{-\infty}^{\infty} (x_1 - m_1)(x_2 - m_2)W(x_1, x_2)\,\mathrm{d}x_1\mathrm{d}x_2,\end{aligned} \tag{1.85}$$

characterizing the correlation between random variables x_1 and x_2.

– The covariance moment (covariance function):

$$m_{12} = \langle x_1x_2\rangle = \int\limits_{-\infty}^{\infty}\int\limits_{-\infty}^{\infty} x_1x_2W(x_1, x_2)\,\mathrm{d}x_1\mathrm{d}x_2, \tag{1.86}$$

characterizing the correlation between random variables x_1 and x_2, if their expectations are zero.

The normalized correlation moment (normalized correlation function) or correlation factor (radius):

$$r_{12} = \mu_{12}/(d_1d_2). \tag{1.87}$$

The mentioned characteristics are determined much easier in practice than the multidimensional distribution density, and knowledge of these characteristics is often sufficient for the analysis of random variables.

If the random variable depends on time and during one observation changes randomly over time, then such a random variable is called a random process. In the foregoing, both a scalar random process $x(t)$, the range of values of which is a set in the space of real numbers, and a vector random process $\vec{x}(t)$, the range of values of which is a set in the corresponding coordinate space, are considered.

A complete and detailed description of the random process is given by multidimensional probability densities (distribution functions) $W_n(x_1, \ldots, x_n; t_1, \ldots, t_n)$, which allow to judge the relation between the probable values of the process $\vec{x}(t)$ at n arbitrary moments of time.

For random processes, as well as for random variables, the partial probabilistic characteristics of random processes are introduced, for example:

– Expectation of random process:

$$m_x = \langle x(t) \rangle = \int\limits_{-\infty}^{\infty} x W(x; t) \, \mathrm{d}x; \tag{1.88}$$

– Variance of the random process:

$$D_x = \left\langle x^2(t) \right\rangle - \langle x(t) \rangle^2 = \int\limits_{-\infty}^{\infty} (x - m_x)^2 W(x; t) \, \mathrm{d}x \tag{1.89}$$

– Covariance function of the random process:

$$K_x(t_1, t_2) = \langle x(t_1) x(t_2) \rangle = \int\limits_{-\infty}^{\infty} \int\limits_{-\infty}^{\infty} x_1 x_2 W_2(x_1, x_2; t_1, t_2) \, \mathrm{d}x_1 \mathrm{d}x_2. \tag{1.90}$$

An important characteristic of the random process is stationarity. The random $x(t)$ process is called stationary if the finite-dimensional probability distribution functions of any degree are invariant with respect to the time shift; i.e., for any n and t_0, the equation below is true:

$$F_n(x_1, \ldots, x_n; t_1 - t_0, \ldots, t_n - t_0) = F_n(x_1, \ldots, x_n; t_1, \ldots, t_n). \tag{1.91}$$

Consequently, the processes $x(t)$ and $x(t - t_0)$ have the same probabilistic characteristics at any t_0. Otherwise, random processes are called non-stationary. Besides, the process $x(t)$ is called periodically stationary with a period T, if the given condition of the stationarity is observed only when $t_0 = mT$, $m = 1, 2, \ldots$; i.e., random processes have the same probability distribution (probability density).

A stationary random process $x(t)$ is ergodic if all its probabilistic characteristics can be obtained by one implementation of the process with probability one, i.e.,

$$\left.\begin{aligned} m_x &= \lim_{T\to\infty} \tfrac{1}{T} \int\limits_0^T x(t)\,\mathrm{d}t \\ D_x &= \lim_{T\to\infty} \tfrac{1}{T} \int\limits_0^T [x(t) - m_x]^2\,\mathrm{d}t \\ K_x(\tau) &= \lim_{T\to\infty} \tfrac{1}{T} \int\limits_0^T x(t+T)x(t)\,\mathrm{d}t \end{aligned}\right\}. \tag{1.92}$$

When considering various devices and systems, it is often necessary to operate with probability density transformations (distribution functions); i.e., it is necessary to determine the probability density $W_1(y)$ of a random variable y by a known probability density $W_1(x)$ of a random variable x and a known functional relation $y = g(x)$ of random variables y and x.

The rule of transformation of probability densities of random variables is determined by the formula

$$W_1(y) = W_1(x)\left|\frac{\mathrm{d}y}{\mathrm{d}x}\right| = W_1[h(y)]\left|\frac{\mathrm{d}h(y)}{\mathrm{d}y}\right|, \tag{1.93}$$

where $x = h(y)$ is the function inverse to the $y = g(x)$ function.

For n-dimensional densities of probabilities $W_x(x_1, x_2, \ldots, x_n)$ of random variables $x_1 \ldots x_n$ and $W_n(y_1, y_2, \ldots, y_n)$ of random values $y_1 = g_1(x_1, x_2, \ldots, x_n), \ldots, y_n = g_n(x_1, x_2, \ldots, x_n)$, the transformation law of distribution density of probabilities has the form:

$$\begin{aligned} W_n(y_1, y_2, \ldots, y_n) &= W_\xi(x_1, x_2, \ldots, x_n)|J_n(x_1, x_2, \ldots, x_n)|^{-1} \\ &= W_\xi[h(y_1, \ldots, y_n), \ldots, h_n(y_1, \ldots, y_n)]|J_n(y_1, y_2, \ldots, y_n)|^{-1}, \end{aligned} \tag{1.94}$$

where $x_1 = h_1(y_1, y_2, \ldots y_n)$; $x_2 = h_2(y_1, y_2, \ldots, y_n), \ldots, x_n = h_n\left(y_1, y_2, \ldots, y_{n_n}\right)$ are unequivocal inverse functions;

J_n is the Jacobian of the variable transformation, numerically equal to the modulus of the following determinant:

$$J_n(y_1, y_2, \ldots, y_n) = \frac{\partial(x_1, x_2, \ldots, x_n)}{\partial(y_1, y_2, \ldots, y_n)} = \begin{vmatrix} \partial h_1/\partial y_1 & \cdots & \partial h_1/\partial y_n \\ \vdots & & \vdots \\ \partial h_n/\partial y_1 & \cdots & \partial h_n/\partial y_n \end{vmatrix} \tag{1.95}$$

Having a common correlation (1.94) for the transformation of n-dimensional probability densities $W_x(x_1, x_2, \ldots, x_n)$, let us present the formulae by which the probability densities of the sum, difference, product, and quotient of two random variables η and χ are found:

$$\left.\begin{aligned} W(\eta) &= \int_{-\infty}^{\infty} W(x, \eta - x)\mathrm{d}x;\ \eta = x + y; \\ W(\eta) &= \int_{-\infty}^{\infty} W(x, \eta + x)\mathrm{d}x;\ \eta = y - x \\ W(\eta) &= \int_{-\infty}^{\infty} W(x, \eta/x)|x|^{-1}\mathrm{d}x;\ \eta = xy; \\ W(\eta) &= \int_{-\infty}^{\infty} W(x, \eta x_1)|x|\mathrm{d}x;\ \eta = y/x. \end{aligned}\right\}. \tag{1.96}$$

In the mathematical statistics, the so-called independent random variables are widely used, which are understood as such random variables x and y, for which their two-dimensional probability distribution density $W(x, y)$ is equal to the product of one-dimensional density distributions of probabilities for each of the random variables, i.e.,:

$$W(x, y) = W(x)W(y). \tag{1.97}$$

In this case, the Formula (1.96) can be represented as follows:

$$\left.\begin{aligned} W(\eta) &= \int_{-\infty}^{\infty} W(x)W(\eta - x)\mathrm{d}x;\ \eta = x + y; \\ W(\eta) &= \int_{-\infty}^{\infty} W(x)W_4(\eta + x)\mathrm{d}x;\ \eta = y - x; \\ W(\eta) &= \int_{-\infty}^{\infty} W(x)W(\eta/x)|x|^{-1}\mathrm{d}x;\ \eta = xy; \\ W(\eta) &= \int_{-\infty}^{\infty} W(x)W(\eta x)|x|\mathrm{d}x;\ \eta = y/x. \end{aligned}\right\}. \tag{1.98}$$

It is easy to see from the given formulae that searching the transformed probability density is connected with the calculation of the convolution integral. Such calculations are sometimes difficult, and therefore to find transformations of probability densities the method of characteristic functions is often used. The characteristic function of a random variable x is the average value of a random variable $\mathrm{e}^{\mathrm{i}\vartheta x}$, which from the Formula (1.83) is defined as

$$\Theta_1(\vartheta) = \langle \mathrm{e}^{\mathrm{i}\vartheta x} \rangle = \int_{-\infty}^{\infty} W(x)\mathrm{e}^{\mathrm{i}\vartheta x}\mathrm{d}x, \tag{1.99}$$

where $W(x)$ is probability distribution density of the random variable x.

Since the characteristic function $\Theta_1(\vartheta)$ and the distribution function $W(x)$ are a pair of Fourier transforms, it can be written as:

$$W(x) = (2\pi)^{-1} \int_{-\infty}^{\infty} \Theta_1(\vartheta) \mathrm{e}^{-\mathrm{i}\vartheta x} \mathrm{d}\vartheta. \quad (1.100)$$

The characteristic function of independent random variables is equal to the product of the characteristic functions of individual values, which, taking into account the relation of the characteristic function $\Theta_1(\vartheta)$ and the probability density $W(x)$, greatly simplifies the calculation of the probability density of the sum of independent random variables. That is, the characteristic function of the sum of independent random variables is a simple multiplication of the characteristic functions of the summands and then is calculated from the result of the Fourier transform. This fact will be widely used further.

Among many existing types of random variables, the exceptional place is occupied by the Gaussian random variable x, which is described by the normal probability density.

For the multidimensional case, the probability density of Gaussian (normal) random variables $\vec{x}(x_1, x_2, \ldots, x_n)$ has the form:

$$W_n(\vec{x}) = \frac{1}{(2\pi)^{0.5n}\sqrt{|\det R|}} \exp\left\{-\frac{1}{2}(\vec{x} - \vec{m})^{\mathrm{T}} R^{-1} (\vec{x} - \vec{m})\right\}, \quad (1.101)$$

where R^{-1} is the matrix, inverse to R; $R = \langle(\vec{x} - \vec{m})(\vec{x} - \vec{m})^{\mathrm{T}}\rangle$ is the correlation matrix; $\vec{x} = (x_1 x_2 \cdots x_n)^{\mathrm{T}}$; $\vec{m} = (m_1, m_2, \ldots, m_n)$.

In particular cases of two-dimensional and one-dimensional probability densities, we obtain relations which will be used further:

$$W_2(x_1, x_2) = \frac{1}{2\pi d_1 d_2 (1 - r^2)} \times \exp\left\{-\frac{1}{2(1 - r^2)}\left[\left(\frac{x_1 - m_1}{d_1}\right)^2 + 2r\left(\frac{x_1 - m_1}{d_1} \cdot \frac{x_2 - m_2}{d_2}\right) + \left(\frac{x_2 - m_2}{d_2}\right)^2\right]\right\} \quad (1.102a)$$

$$W_1(x) = \frac{1}{\sqrt{2\pi} d} \exp\left[-\frac{1}{2}\left(\frac{x - m}{d}\right)^2\right] \quad (1.102b)$$

where r is the correlation ratio.

The distribution function of a Gaussian random variable has the form:

$$F(x) = \int_{-\infty}^{\infty} W(y)\mathrm{d}y = \frac{1}{\sqrt{2\pi}} \int_{-\infty}^{\frac{x-m}{\sigma}} \mathrm{e}^{-0.5t^2} \mathrm{d}t = \Phi\left(\frac{x - m}{\sigma}\right), \quad (1.103)$$

where $\Phi(x) = \frac{1}{\sqrt{2\pi}} \int_{-\infty}^{x} e^{-0.5t^2} dt$ is the probability integral.

It can be shown that by integrating the multidimensional normal probability density $(n-1)$ times in infinite limits, we obtain each of the sets of n-dependent random variables obeying the normal law.

Given into account that partially polarized waves can be represented in different polarization bases, the issue of the corresponding transformations of multidimensional probability densities is important.

In the initial polarization basis, a random vector $\vec{x}_C = (x_{1C}, x_{2C}, \ldots, x_{nC})^{\mathrm{T}}$ is considered, which corresponding multidimensional probability distribution density has the form $W(\vec{x}_C)$. Let this vector have the form $\vec{x}_H = (x_{1H}, x_{2H}, \ldots, x_{nH})^{\mathrm{T}}$ in the new polarization basis, then it is necessary to determine the multidimensional probability distribution density $\vec{x}_H$. In accordance with the Formula (1.94) is the desired probability density function will be written in the following way:

$$W(\vec{x}_H) = W(\vec{x}_C)|J_n(x_C)|^{-1} = W\big(Q^{+}\vec{x}_H\big)|J_n(x_H)|, \tag{1.104}$$

where, in the accordance with (1.58), $\vec{x}_H = Q^{*}\vec{x}_C$.

The Jacobian of transformation $|J_n(\vec{x}_H)|$ can be written as:

$$|J_n(\vec{x}_n)| = \left| \frac{\partial\big[Q^{+}(\vec{x}_H)\big]}{\partial \vec{x}_H} \right| = \left|Q^{*}\right| = 1. \tag{1.105}$$

Therefore, the probability densities of the vectors $\vec{x}_H$ and $\vec{x}_C$ will coincide; i.e., they will be invariant in relation to the change of the polarization basis.

Thus, if the transformation of variables when changing the polarization basis is carried out using a unitary matrix, then the multidimensional probability density of the polarization vector is invariant with respect to the polarization basis, i.e., $W(\vec{x}_C) = W(\vec{x}_H)$.

The consequence of this statement is the following important case for practice. If any multidimensional distribution density of the vector $\vec{X}$ is described by the Gaussian law, then in any other polarization basis the corresponding probability distribution density will also have a Gaussian character. Equally, this statement applies to the one-dimensional distribution laws of the components of this vector.

At the end of this section, let us present Table 1.2 of the basic probability densities of random variables, which will be widely used in the future.

1.5 Quaternion Representation of the Signal Polarization State in Navigation Systems

Quaternions as a generalization of complex numbers were proposed in 1843. It was found that a numerical system that fully preserves all the properties of addition and multiplication of real and complex numbers cannot be constructed from the points of the space of three or more dimensions. However, in the four-dimensional space there

Table 1.2 Basic laws of distribution and their numerical characteristics

The law of distribution	Range of the value of a random variable	Probability density	The mathematical expectation (m) and variance (D)	Distribution function
1. Uniform distribution	$a < x < b$	$\frac{1}{b-a}$	$m = \frac{1}{2}(a+b)$ $D = \frac{1}{12}(b-a)^2$	$\begin{cases} 0, & x < a \\ \frac{x-a}{b-a}, & a < x < b \\ 1, & x > b \end{cases}$
2. Gaussian distribution	$-\infty < x < \infty$	$\frac{1}{d\sqrt{2\pi}} \exp\left(-\frac{(x-m)^2}{2d^2}\right)$	$m = m;$ $D = d^2$	$\Phi\left(\frac{x-m}{d}\right)$
3. Lognormal distribution	$0 < x < \infty$	$\frac{\lg e}{xd\sqrt{2\pi}} \exp\left(-\frac{(\lg x-m)^2}{2d^2}\right)$	$m = M(\lg x)$ $D = D(\lg x)$	$\begin{cases} 0, & x < 0 \\ \Phi\left(\frac{\lg x-m}{d}\right), & x > 0 \end{cases}$
4. Rayle distribution	$0 < x < \infty$	$\frac{x}{k^2} \exp\left(-\frac{x^2}{2k^2}\right)$	$m = k\sqrt{\frac{\pi}{2}}$ $D = k^2\left(2 - \frac{\pi}{2}\right)$	$\begin{cases} 0, & x < 0 \\ 1 - \exp\left(-\frac{x}{2k^2}\right); & x > 0 \end{cases}$
5. Nakagami distribution	$0 < x < \infty$	$\frac{2}{\Gamma(m)} \left(\frac{m}{k^2}\right)^m x^{2m-1}$ $\times \exp\left(-\frac{m}{k^2}x^2\right); m \geq \frac{1}{2}$	$m = \frac{\Gamma\left(m+\frac{1}{2}\right)}{\Gamma(m)} \left(\frac{k^2}{m}\right)^{\frac{1}{2}}$ $D = \frac{\Gamma(m+1)}{\Gamma(m)} \left(\frac{k^2}{m}\right)$	$\begin{cases} 0, & x < 0 \\ \frac{\Gamma\left(m, \frac{mx^2}{k^2}\right)}{\Gamma(m)}, & x > 0 \end{cases}$
7. Exponential distribution	$0 < x < \infty$	$\lambda e^{-\lambda x}$	$m = 1/\lambda$ $D = 1/\lambda^2$	$\begin{cases} 0, & x < 0 \\ 1 - e^{-\lambda x}, & x > 0 \end{cases}$

is a non-commutative under multiplication algebra, called the quaternion algebra. Note that in the space of three, five, or more dimensions there is not even such a system of numbers.

A quaternion is a linear combination of four basis unit elements $1, a_1, a_2, a_3$ of (hypercomplex unit vectors):

$$X = x_0 \cdot 1 + x_1 \cdot a_1 + x_2 \cdot a_2 + x_3 \cdot a_3, \tag{1.106}$$

where x_0, x_1, x_2, x_3 are real numbers (coordinates of the point of the four-dimensional space corresponding to the quaternion X). Actions on quaternions are performed according to the usual rules satisfying axioms of non-commutative under multiplication algebra over the field of real numbers, taking into account the following law of multiplication of hypercomplex unit vectors:

$$\left.\begin{array}{llll} 1 \cdot 1 = 1 & 1 \cdot a_1 = a_1 & 1 \cdot a_2 = a_2 & 1 \cdot a_3 = a_3 \\ a_1 \cdot 1 = a_1 & a_1 \cdot a_1 = -1 & a_1 \cdot a_2 = a_3 & a_1 a_3 = -a_2 \\ a_2 \cdot 1 = a_2 & a_2 \cdot a_1 = -a_3 & a_2 \cdot a_2 = -1 & a_2 \cdot a_3 = a_1 \\ a_3 \cdot 1 = a_3 & a_3 \cdot a_1 = a_2 & a_3 \cdot a_2 = -a_1 & a_3 \cdot a_3 = -1 \end{array}\right\}. \tag{1.107}$$

From the above formulae, it is seen that unit vector 1 plays the role of a usual unit element and in the quaternion record can be omitted.

Using the introduced rules, let us determine the multiplication of quaternions. Assume:

$$\left.\begin{array}{l} X = x_0 + x_1 \cdot a_1 + x_2 \cdot a_2 + x_3 \cdot a_3 \\ Y = y_0 + y_1 \cdot a_1 + y_2 \cdot a_2 + y_3 \cdot a_3 \end{array}\right\}. \tag{1.108}$$

Then in accordance with the property of distributiveness, we have:

$$\begin{aligned} XY = {} & x_0 \cdot y_0 + x_0(y_1 \cdot a_1) + x_0(y_2 \cdot a_2) + x_0(y_3 \cdot a_3) \\ & + (x_1 \cdot a_1)y_0 + (x_1 \cdot a_1)(y_2 \cdot a_2) + (x_1 \cdot a_1)(y_3 \cdot a_3) \\ & + (x_2 \cdot a_2)y_0 + (x_2 \cdot a_2)(y_1 \cdot a_1) + (x_2 \cdot a_2)(y_2 \cdot a_2) \\ & + (x_2 \cdot a_2)(y_3 \cdot a_3) + (x_3 \cdot a_3)y_0 + (x_3 \cdot a_3)(y_1 \cdot a_1) \\ & + (x_3 \cdot a_3)(y_2 \cdot a_2) + (x_3 \cdot a_3)(y_3 \cdot a_3). \end{aligned} \tag{1.109}$$

By transforming the resulting expression according to the axioms of the algebra over the field of real numbers, we find:

$$\begin{aligned} XY = {} & (x_0 \cdot y_0 - x_1 \cdot y_1 - x_2 \cdot y_2 - x_3 \cdot y_3) \\ & + (x_0 \cdot y_1 + x_1 \cdot y_0 + x_2 \cdot y_3 - x_3 \cdot y_2)a_1 \\ & + (x_0 \cdot y_2 + x_2 \cdot y_0 + x_3 \cdot y_1 - x_1 \cdot y_3)a_2 \\ & + (x_0 \cdot y_3 + x_3 \cdot y_0 + x_1 \cdot y_2 - x_2 \cdot y_1)a_3. \end{aligned} \tag{1.110}$$

By identifying the units a_1, a_2, a_3 to the unit vectors of the three-dimensional space, in the quaternion X it is possible to distinguish the scalar x_0 and vector $\vec{x} = x_1\vec{a}_1 + x_2\vec{a}_2 + x_3\vec{a}_3$ parts, and the product of the quaternions can be represented as:

$$XY = x_0 \cdot y_0 - \vec{x}\vec{y} + x_0\vec{y} + y_0\vec{x} + \vec{x} \times \vec{y}, \tag{1.111}$$

where

$$\begin{aligned} \vec{x}\vec{y} &= x_1 \cdot y_1 + x_2 \cdot y_2 + x_3 \cdot y_3 \\ \vec{x} \times \vec{y} &= \det \begin{pmatrix} a_1 & a_2 & a_3 \\ x_1 & x_2 & x_3 \\ y_1 & y_2 & y_3 \end{pmatrix} \end{aligned} \tag{1.112}$$

are scalar and vector products of three-dimensional vectors, respectively.

From the obtained expression, it is seen that the non-commutativity of the quaternion product is due to the non-commutativity of the vector product. Therefore, the repositioning of the quaternion product cofactors is valid only if the quaternion is multiplied by a scalar or in the case of the proportionality of the vector parts of the cofactors: $\vec{x} \times \vec{y} = \vec{x} \times \alpha\vec{x} = \alpha(\vec{x} \times \vec{x}) = 0$. The scalar part of the quaternion product obviously does not depend on the order of the cofactors.

By analogy with complex numbers in the quaternion algebra, the concept of the conjugate $\bar{X} = x_0 - \vec{x}$ and the norm of quaternion is introduced

$$\|X\| = X\bar{X} = x_0^2 + \vec{x}\vec{x} = \sum_{j=0}^{3} x_j^2, \tag{1.113}$$

which is a scalar value. When $\|X\| = 1$, we speak of a normalized quaternion. The norm of quaternion is zero if and only if $x_0 = x_1 = x_2 = x_3 = 0$.

Half the degree of the norm

$$\|X\| = |X| = \sqrt{\sum_{j=0}^{3} x_j^2} \tag{1.114}$$

it is called a modulus or tensor of quaternion, and the value

$$x = \frac{X}{|X|} = \frac{x}{|X|} + \frac{\vec{x}}{|X|} \tag{1.115}$$

is a versor of quaternion.

Therein, any quaternion can be represented as $X = |X|x$.

Using the equation $\vec{x} = \vec{\eta}|\vec{x}|$ where $\vec{\eta}$ is the unit vector of the vector part of quaternions in three-dimensional space, and the identity $x_0^2 + |\vec{x}|^2 = |X|^2$, we introduce designations

$$\cos\gamma = \frac{x_0}{|X|} \quad ; \quad \sin\gamma = \frac{|\vec{x}|}{|x|}, \tag{1.116}$$

where $\gamma = \operatorname{Arg} X$ is the argument of quaternion. As a result, we come to the trigonometric formula

$$\bar{X} = |X|(\cos\gamma + \vec{\eta}\sin\gamma), \tag{1.117}$$

for which Moivre and Euler formulae are true:

$$\left.\begin{array}{l} X^n = |X|^n(\cos n\gamma + \vec{\eta}\sin n\gamma), \\ X = |X|\mathrm{e}^{\vec{\eta}\gamma}. \end{array}\right\} \tag{1.118}$$

The direct check shows that the conjugate to the sum of quaternions is equal to the sum of the conjugates, i.e.,

$$\overline{X+Y} = \bar{X} + \bar{Y}, \tag{1.119}$$

and conjugate to the product of quaternions is equal to the product of conjugates taken in the reverse order:

$$\overline{XY} = \bar{Y}\bar{X}. \tag{1.120}$$

Due to the non-commutativity of multiplication in the quaternion algebra, there are two ratios for each pair of elements: left and right, defined from the equations

$$X_2 X_L^{(1,2)} = X_1; \quad X_P^{(1,2)} X_2 = X_1, \tag{1.121}$$

respectively. Multiplying both parts of the first equation from left to $\bar{X}_2$, and then on $\|X\|^{-1}$, we obtain:

$$X_L^{(1,2)} = \|X\|^{-1}\bar{X}_2 X_1. \tag{1.122}$$

On substituting this expression into the original equation, it turns into the identity $X_1 \equiv X_1$; i.e., the solution is found correctly. Analogously,

$$X_P^{(12)} = \|X_2\|^{-1} X_1 \bar{X}_2. \tag{1.123}$$

Thus, the system of quaternions is an algebra with division.

Let us consider now two pure vector quaternions

$$X(0,\vec{x}) = \sum_{j=1}^{3} x_j a_j \quad \text{and} \quad Y(0,\vec{y}) = \sum_{j=1}^{3} y_j a_j, \tag{1.124}$$

the product of which

$$XY = -(x_1 \cdot y_1 + x_2 \cdot y_2 + x_3 \cdot y_3) + (x_2 \cdot y_3 - x_3 \cdot y_2)a_1 \\ + (x_3 \cdot y_1 - x_1 \cdot y_3)a_2 + (x_1 \cdot y_2 - x_2 \cdot y_1)a_3 \quad (1.125)$$

contains vector and scalar parts:

$$\left.\begin{aligned} \mathrm{sqal}(XY) &= -\sum_{i=1}^{3} x_j y_j = -\vec{x}\bar{y}, \\ \mathrm{vect}(XY) &= (x_2 \cdot y_3 - y_2 \cdot x_3)a_1 + (x_3 \cdot y_1 - x_1 \cdot y_3)a_2 \\ &\quad + (x_1 \cdot y_2 - x_2 \cdot y_1)a_3 = \bar{x} \times \vec{y}. \end{aligned}\right\} \quad (1.126)$$

It follows from the obtained equations that the scalar part of the product of vector quaternions is equal to their scalar product taken with a minus sign, and the vector part to the vector product of quaternions.

In the quaternion algebra, the rotation group of a three-dimensional space contains elements which form is determined by the following theorem:

If X and Y are non-scalar quaternions, then the value $X' = YXY^{-1}$ is a quaternion, the norm and the scalar part of which are equal to the norm and the scalar part of the quaternion X, and the vector part X' is formed by a conal rotation at the angle $\varphi = 2\,\mathrm{Arg}Y$ of the vector part X around the unit vector of the vector part Y.

This theorem will be used in the foregoing when considering the issues related to the description of the polarization plane rotation.

Chapter 2
Analysis of the Signals' Polarization of Radiopolarimetric Navigation Systems Using Coordinate Components

2.1 Quadrature Components

2.1.1 Basic Properties and Transformations

In Sect. 1.2, a four-dimensional vector $\vec{E} = (x_1 x_2 x_3 x_4)^{\mathrm{T}}$ defined by Eqs. (1.21)–(1.22) is introduced as one of the possible representations of radio wave polarization. The use of vector $\vec{E}$ is convenient, firstly, because the consideration is conducted in the real domain and, secondly, because it is easy to build correct mathematical models of real physical processes for it. However, the use of the vector $\vec{E}$ is suitable to carry out when the researcher has a good computing technique.

Let us start the consideration of the main properties of the tetra-vector $\vec{E}$ with the clarification of the laws of its change within the polarization basis changing. (To further distinguish the matrix-column representation, the tetra-vector $\vec{E}$ will be denoted $\vec{X}$.)

When changing the polarization basis (1.34), the vector $\vec{X}$ is obviously.

If the four-dimensional vector $\vec{X} = (x_1 + ix_2 \quad x_3 + ix_4)^{\mathrm{T}}$ is represented as a two-element matrix-column, then for any complex vector $\vec{E}$ the following chain of equations is valid: $\vec{E} = X_c^{\mathrm{T}} \vec{e}_c = X_H^{\mathrm{T}} \vec{e}_H = X_H^{\mathrm{T}} Q \vec{e}_C$, where Eq. (1.37) is subjected to some linear transformation. Let us denote a real square matrix of size 4×4, performing this transformation, through R. Thus, we have:

$$\vec{X}_H = R \vec{X}_C, \tag{2.1}$$

where $\vec{X}_H$ and $\vec{X}_C$ are vector $\vec{X}$ representations in new and old bases, respectively.

Relying on the representation for the unitary 2×2 matrix Q introduced by the Formula (1.47), we obtain an explicit representation for the matrix R.

If the four-dimensional vector $\vec{X} = \left(x_1 + ix_2 \; x_3 + ix_4\right)^{\mathrm{T}}$ is represented as a two-element matrix-column, then for any complex vector the following chain of equalities holds: $\vec{E} = X_c^{\mathrm{T}} \vec{e}_c = X_H^{\mathrm{T}} \vec{e}_H = X_H^{\mathrm{T}} Q \vec{e}_C$, where equality (1.37) is taken into account.

Kozlov A. I. et al., *Introduction to the Theory of Radiopolarimetric Navigation Systems*, Springer Aerospace Technology, https://doi.org/10.1007/978-981-13-8395-3_2

Therefore, taking into account the property of the unitarity of the matrix Q (1.39), we will have:

$$\vec{X}_H = Q^* \vec{X}_C = Q^* \begin{pmatrix} x_1 + ix_2 \\ x_3 + ix_4 \end{pmatrix}. \tag{2.2}$$

Substituting the value of the matrix of transformation Q in the form (1.43) and multiplying the matrices, we obtain an expression for the matrix R in the following form:

$$R = \mathrm{e}^{\mathrm{i}\varphi_0} \begin{pmatrix} \cos\gamma\ \cos\eta & \cos\gamma\ \sin\eta & \sin\gamma\ \cos(\eta+2\delta) & \sin\gamma\ \sin(\eta+2\delta) \\ -\cos\gamma\ \sin\eta & \cos\gamma\ \cos\eta & -\sin\gamma\ \sin(\eta+2\delta) & \sin\gamma\ \cos(\eta+2\delta) \\ -\sin\gamma\ \cos(\eta+2\delta) & \sin\gamma\ \sin(\eta+2\delta) & \cos\gamma\ \cos\eta & -\cos\gamma\ \sin\eta \\ -\sin\gamma\ \sin(\eta+2\delta) & -\sin\gamma\ \cos(\eta+2\delta) & \cos\gamma\ \sin\eta & \cos\gamma\ \cos\eta \end{pmatrix}. \tag{2.3}$$

The transformation (2.1) is restricted to the field of real numbers the matrix R, at the same time, in the common case when $\varphi_0 \neq 0$ is a complex number. In this regard, taking into account the above, it is necessary to put $\varphi_0 = 0$ in the expression (2.3).

As will be seen from the following, a number of decompositions can be carried out for a somewhat modified vector $\vec{X}_m$, which sequence of its coordinates is changed (the first two places are the real components, and the third and fourth place are the imaginary components of the complex vector $\vec{X}$), i.e., $\vec{X}_m = (x_1 x_3 x_2 x_4)^{\mathrm{T}}$.

In this case, instead of the matrix R defined by Eq. (2.3), a somewhat modernized R_m matrix will be used, which can be obtained from the matrix R by appropriate replacement of the second row by the third row, the second column by the third column and conversely:

$$R_m = \begin{pmatrix} \cos\gamma\ \cos\eta & \sin\gamma\ \cos(\eta+2\delta) & \cos\gamma\ \sin\eta & \sin\gamma\ \sin(\eta+2\delta) \\ -\sin\gamma\ \cos(\eta+2\delta) & \cos\gamma\ \cos\eta & \sin\gamma\ \sin(\eta+2\delta) & -\cos\gamma\ \sin\eta \\ -\cos\gamma\ \sin\eta & -\sin\gamma\ \sin(\eta+2\delta) & \cos\gamma\ \cos\eta & \sin\gamma\ \cos(\eta+2\delta) \\ -\sin\gamma\ \sin(\eta+2\delta) & \cos\gamma\ \sin\eta & -\sin\gamma\ \cos(\eta+2\delta) & \cos\gamma\ \cos\eta \end{pmatrix}. \tag{2.4}$$

The matrix R_m can be represented as a product of two unitary block matrices, the first of which F_b determines the phase state of the basis, and the second Q_b the geometric parameters of the transformation:

$$R_m = F_b Q_b, \tag{2.5}$$

and in this case, the block matrix F_b has the form $F_b = \frac{1}{\sqrt{2}} \begin{pmatrix} F & F^* \\ iF & -iF^* \end{pmatrix}$, the block matrix Q_b is represented in the form $Q_b = \frac{1}{\sqrt{2}} \begin{pmatrix} Q_0 & -iQ_0 \\ Q_0^* & iQ_0^* \end{pmatrix}$, and the matrices F and Q Q_0 are determined by the expression (1.48).

Let us show that the matrix R_m has the orthogonality property, a special case of the unitarity property for real matrices, which are matrices (2.3) and (2.4), for which we calculate $R_m R_m^+ = F_b Q_b (F_b Q_b)^+ = F_b Q_b Q_b^+ F_b^+ = F_b I F_b^{\mathrm{T}} = F_b F_b^{\mathrm{T}} = I$.

Since the matrix R_m is purely real, then $R_m \equiv R_m^*$, and thus, $R_m^+ \equiv R_m^{\mathrm{T}}$. Therefore,

$$R_m R_m^{\mathrm{T}} = I. \tag{2.6}$$

Matrices having this property are called orthogonal, and their elements r_{km} (k, m = 1, 2, 3, 4) satisfy the relations:

$$\sum_{k=1}^{4} r_{km} r_{kn} = \delta_{mn}. \tag{2.7}$$

Since the matrix R_m is obtained from the matrix R by above-mentioned moving the rows and columns, the orthogonality of the matrix R follows from the orthogonality of the matrix R_m.

The representation of the real transformation matrix by the product of two complex matrices invariant to each other's parameters is very useful for determining the parameters of the functional units of real technical polarization converters.

The purpose of the introduction of the four vectors was to represent the polarization by real numbers. Therefore, the product of complex matrices may be inconvenient. Division into independent factors is possible also in real form.

Representing the matrix of cofactors of expression (2.5) as the sum of the imaginary and the real part, we obtain:

$$R = \begin{pmatrix} \mathrm{Re}F & 0 \\ 0 & \mathrm{Im}F \end{pmatrix} \begin{pmatrix} 1 & 1 \\ -1 & -1 \end{pmatrix} \begin{pmatrix} \mathrm{Re}Q_0 & 0 \\ 0 & \mathrm{Im}Q_0 \end{pmatrix} - \begin{pmatrix} \mathrm{Im}F & 0 \\ 0 & \mathrm{Re}F \end{pmatrix} \begin{pmatrix} 1 & -1 \\ 1 & -1 \end{pmatrix} \begin{pmatrix} \mathrm{Im}Q_0 & 0 \\ 0 & \mathrm{Re}Q_0 \end{pmatrix}. \tag{2.8}$$

Let us define some common forms of the basis transformation, for which we use the Formulae (2.1) and (2.3), taking into account the fact that in the Formula (2.3) $\varphi_0 = 0$ shall be put.

Assume there is a horizontally polarized (HP) radio wave (the initial vector $\vec{X}$). As the initial (old) polarization basis, we choose a pair of horizontal–vertical (HP-VP) polarization.

In this case, the vector $\vec{X}$ will have the following representation $\vec{X} = \begin{pmatrix} 1 & 0 & 0 & 0 \end{pmatrix}^{\mathrm{T}}$.

Consider the representation of HP in a circular polarization (CP) basis, which for the case of the right circle–left circle will be determined by the parameters ($\gamma = \pi/4$, $2\delta = \pi/2$)

$$\vec{X}_{\mathrm{CP}} = \frac{1}{\sqrt{2}} (\cos\eta \quad -\sin\eta \quad \sin\eta \quad -\cos\eta)^{\mathrm{T}}, \tag{2.9}$$

for the case of the left circle–right circle ($\gamma = \pi/4$, $2\delta = -\pi/2$):

$$\vec{X}_{\mathrm{CP}} = \frac{1}{\sqrt{2}}(\cos\eta \quad -\sin\eta \quad -\sin\eta \quad \cos\eta)^{\mathrm{T}}. \tag{2.10}$$

If we put the value of the phase angle $\eta = 0$, the HP wave in the named circular polarization bases will have the form: $\frac{1}{\sqrt{2}}\begin{pmatrix}1 & 0 & 0 & -1\end{pmatrix}$ and $\frac{1}{\sqrt{2}}\begin{pmatrix}1 & 0 & 0 & 1\end{pmatrix}$ accordingly.

When you move to the other linearly polarized pair rotated by angle β_0 against HP, you must put $\gamma = \beta_0$, $\delta = 0$, which gives:

$$\vec{X}_{\mathrm{LP}} = (\cos\beta_0\cos\eta - \cos\beta_0\sin\eta - \sin\beta_0\cos\eta - \sin\beta_0\sin\eta)^{\mathrm{T}} \tag{2.11}$$

In a particular case when $\eta = 0$, $\vec{X}_{\pi\Pi} = (\cos\beta_0 \quad 0 \quad -\sin\beta_0 \quad 0)^{\mathrm{T}}$.

In the transition from a linearly polarized pair to an elliptic basis, the main unit vectors of which coincide with the efforts of the original basis, it is necessary to put $\delta = 0$, $\gamma = \gamma_0$:

$$\vec{X}_{\mathrm{EP}} = (\cos\gamma_0\cos\eta \quad -\cos\gamma_0\sin\eta \quad -\sin\gamma_0\cos\eta \quad -\sin\gamma_0\sin\eta)^{\mathrm{T}}. \tag{2.12}$$

In a particular case when $\eta = 0$, $\vec{X}_{\mathrm{EP}} = (\cos\gamma_0 \quad 0 \quad -\sin\gamma_0 \quad 0)^{\mathrm{T}}$.

The foregoing real representation of polarization allows us to move to a variety of geometric representations of polarization, greatly facilitating in some cases the analysis of polarization of electromagnetic waves.

2.1.2 Probabilistic Descriptions

So far, the consideration has been conducted for the case when the characteristics of the wave (the components of the tetra-vector $\vec{X}$) have remained unchanged over time. In practice, this situation is extremely rare. It is more realistic to have a change in the vector $\vec{X}$ in time (usually randomly). This leads to the appearance of partially polarized waves (PPW), to describe the polarization state of which it is necessary to apply statistical methods of analysis, allowing the most complete research of the radio wave structure.

This means that for a complete description of the PPW, it is necessary to know at the point where the polarization of the radio wave is determined, for a fixed moment of time, the joint four-dimensional density of the probability distribution of the components of the random vector $\vec{X}$, which we denote $W(\vec{X})$, or $W(x_1, x_2, x_3, x_4)$.

Before analyzing the peculiarities of the laws of the PPW, let us pay attention to the applicability of the results obtained in p. 1.5 (1.104), (1.105), in accordance with which there is an invariance of the distribution law $W(\vec{X})$ to changing of the polarization basis, since the change of the vector $\vec{X}$ when the basis changes is orthogonal transformation (2.1), using the matrix R (2.3).

In radio polarimetry, multidimensional Gauss distribution for the components of the vector $\vec{X}$ has got the most widespread use as a four-dimensional law. The basis of this approach is the ability to represent each of the components of the vector $\vec{X}$ in the form of a stationary narrowband random process. In this case, each of the components $\vec{X}$ is the sum of a large number of partial summands $x_i = \sum_j^N x_{ij}$ $(i = \overline{1, 4})$, which in turn makes it possible to use the central limit theorem, according to which the distribution asymptotically converges to the Gauss law when $N \to \infty$. In this regard, an exhaustive characteristic of the process is its covariance matrix D_2^K and the matrix of first-order moments D_1.

The matrix D_2^K can be obtained by the following equation:

$$D_2^K = < \vec{X}\vec{X}^{\mathrm{T}} > . \tag{2.13}$$

If you use the correlation (2.1), it is easy to obtain the law of transformation of the matrix D_2^K when changing the polarization basis:

$$D_{2H}^K = < \vec{X}_H \vec{X}_H^{\mathrm{T}} > = < R\vec{X}_C \left(R\vec{X}_C\right)^{\mathrm{T}} > = R D_{2C}^K R^{\mathrm{T}}. \tag{2.14}$$

Thus, the covariance matrix under the change of PB is subjected to orthogonal transformation with the help of the matrix R. Let us show that the same transformation is also applied to the correlation matrix D_2, which is defined as follows:

$$D_2 = D_2^K - \overline{\vec{X}\vec{X}^{\mathrm{T}}}. \tag{2.15}$$

In the transition to a new basis have:

$$D_{2H} = R D_{2C} R^{\mathrm{T}} - \overline{R\vec{X}_C \vec{X}_C^{\mathrm{T}} R^{\mathrm{T}}} = R D_{2C} R^{\mathrm{T}}. \tag{2.16}$$

The matrix-column $\overline{\vec{X}} = D_1$ introduced in Eq. (2.15) is the matrix of moments of the first-order mentioned above. If Eq. (2.15) is written in the form $D_2^K = D_2 + \overline{\vec{X}\vec{X}^{\mathrm{T}}} >$, it can be given the following meaning. The covariance matrix D_2^K is the sum of two real matrices, one of which $M = \overline{\vec{X}\vec{X}^{\mathrm{T}}} >$ is describing the polarization properties of the monochromatic wave and the other D_2 is describing the partially polarized quasi-monochromatic wave.

From the common properties of matrices D_2 and D_{2H}, the following shall be noted. The sum of the diagonal elements of the matrix D_2, i.e., its trace, is a value proportional to the P_0 power of the radio wave. Naturally, when the polarization basis changes, it shall remain constant. Let us show it analytically.

$$P_0 = S_P D_2 = \sum_{j=1}^{4} \overline{x_j^2} = \overline{X_H^{\mathrm{T}} \vec{X}_H} = \overline{\left(R_{04}\vec{X}_C\right)^{\mathrm{T}} R_{04}\vec{X}_C} = \overline{\vec{X}_C^{\mathrm{T}} R_{04}^{\mathrm{T}} R_{04} \vec{X}_C} = \overline{\vec{X}_C^{\mathrm{T}} \vec{X}_C}. \tag{2.17}$$

The same invariance takes place also for the product $\overline{\vec{X}^{\mathrm{T}}\vec{X}}$:

$$\overline{\vec{X}_H^{\mathrm{T}}\vec{X}_H} = \overline{\left(R_{04}\vec{X}_C\right)^{\mathrm{T}}R_{04}\vec{X}_C} = \overline{\vec{X}_C^{\mathrm{T}}R_{04}^{\mathrm{T}}R_{04}\vec{X}_C} = \overline{\vec{X}_C^{\mathrm{T}}\vec{X}_C}. \tag{2.18}$$

Subtracting Eq. (2.17) from Eq. (2.18), we obtain that the sum of the variances d_j^2 of the random components of the vector $\vec{X}$ when the polarization basis changes also remains unchanged, i.e.,

$$\sum_{j=1}^{4}\left(\overline{x_j^2} - \overline{x_i}^2\right) = \sum_{j=1}^{4} d_j^2 = \text{const} \tag{2.19}$$

The Formula (2.14) implies another property of matrices D_2 and D_2^K—the invariance of their determinants:

$$\left.\begin{array}{l} \det D_{2H} = \det\left(RD_{2C}R^{\mathrm{T}}\right)^{\mathrm{T}} = \det R\ \det D_{2C}^{\mathrm{T}}\ \det R^{\mathrm{T}} = \det D_{2C}, \\ \det D_{2H}^K = \det\left(RD_{2C}^K R^{\mathrm{T}}\right)^{\mathrm{T}} = \det R\ \det D_{2C}^{KT}\ \det R^{\mathrm{T}} = \det D_{2C}^K \end{array}\right\}. \tag{2.20}$$

For polarized waves, when $< \vec{X} >= \vec{X}$ it follows from Eq. (2.13) that $\det D_2^K = \det D_2 = 0$.

Equation $\det D_2^K = \det D_2 = 0$ can serve as a definition of a polarized wave.

The obtained relations are universal, because they are not related to a specific type of distribution law $W(\vec{X})$. Let us now consider the case when $W(\vec{X})$ is the Gaussian distribution law.

Without taking into account the physical nature of the radiation source (scattering, reflection) of the wave, let us present a plane homogeneous quasi-monochromatic wave in the form of a superposition of two orthogonally polarized Gaussian components in accordance with (1.22), (1.23), and (1.35):

$$\vec{E}(t) = E_1(t)\vec{e}_1 + E_2(t)\vec{e}_2, \tag{2.21}$$

where $\vec{e}_1, \vec{e}_2$ are individual unit vectors, which determine an arbitrary polarization basis of decomposition.

By decomposing $\vec{E}_j(t)$ $(j = 1, 2)$ into quadrature components $x_{1,3} = \left|E_{1,2}(t)\right| \cos\varphi_{1,3}(t)$ and $x_{2,4} = \left|E_{1,2}(t)\right| \cos\varphi_{2,4}(t)$, and introducing notations $x_1 = Z_1$, $x_2 = Z_2$, $x_3 = Z_3$, $x_4 = Z_4$, we obtain the following expression for the joint probability distribution function (PDF) of Gaussian quadrature components of orthogonally polarized components of PPW:

$$W(Z_1, Z_2, Z_3, Z_4) = \frac{1}{\prod_{n=1}^4 d_n\sqrt{(2\pi)^4 D}} \exp\left[-\frac{1}{2D}\sum_{l,n=1}^{4} D_{ln}\frac{(Z_l - m_l)(Z_n - m_n)}{d_l d_n}\right], \tag{2.22}$$

where $m_j =< Z_j >$; $d_j^2 = \left(< Z_j^2 > - m_j^2\right)$ variance; $j = 1, 2, 3, 4$;

D—the determinant of the fourth order, having the following form:

$$D = \begin{vmatrix} 1 & \rho_{12} & \rho_{13} & \rho_{14} \\ \rho_{21} & 1 & \rho_{23} & \rho_{24} \\ \rho_{31} & \rho_{32} & 1 & \rho_{34} \\ \rho_{41} & \rho_{42} & \rho_{43} & 1 \end{vmatrix}, \tag{2.23}$$

and D_{ln} value is an algebraic complement of the correlation matrix element (2.10); $\rho_{ln} = (< Z_l Z_n > -m_l m_n)/(d_l d_n)$—correlation coefficient ($l \neq n, l, n = \overline{1,4}$).

The parameters describing the joint probability density of the quadrature components Z_j orthogonally polarized components of PPW in accordance with the phenomenological approach are determined by the type of joint probability densities of amplitudes and phases of elementary waves and the number of the last [1]. In turn, mentioned probability densities depend on the physical nature of the radiation source or object of scattering (reflection) of radio waves. Herewith in common case, $m_l \neq m_n, d_l^2 \neq d_n^2, \rho_{ln} \neq 0$. In other words, the matrix (2.13) can be written taking into account (2.23) in the following expanded form:

$$D_2^K = \begin{pmatrix} m_1^2 & m_1 m_2 & m_1 m_3 & m_1 m_4 \\ m_1 m_2 & m_2^2 & m_2 m_3 & m_2 m_4 \\ m_1 m_3 & m_2 m_3 & m_3^2 & m_3 m_4 \\ m_1 m_4 & m_2 m_4 & m_3 m_4 & m_4^2 \end{pmatrix} + \begin{pmatrix} d_{11}^2 & 0.5\sqrt{1-\alpha_1^2}\rho_{12} & 0.5 d_{11} d_{22}\sqrt{(1+\alpha_1)(1+\alpha_2)}\rho_{13} & 0.5 d_{11} d_{22}\sqrt{(1+\alpha_1)(1-\alpha_2)}\rho_{14} \\ 0.5 d_{11}^2\sqrt{1-\alpha_1^2}\rho_{12} & d_{11}^2 & 0.5 d_{11} d_{22}\sqrt{(1-\alpha_1)(1+\alpha_2)}\rho_{23} & 0.5 d_{11} d_{22}\sqrt{(1-\alpha_1)(1-\alpha_2)}\rho_{24} \\ 0.5 d_{11} d_{22}\sqrt{(1+\alpha_1)(1+\alpha_2)}\rho_{13} & 0.5 d_{11} d_{22}\sqrt{(1-\alpha_1)(1+\alpha_2)}\rho_{23} & d_{22}^2 & 0.5 d_{22}^2\sqrt{1-\alpha_2^2}\rho_{34} \\ 0.5 d_{11} d_{22}\sqrt{(1+\alpha_1)(1-\alpha_2)}\rho_{14} & 0.5 d_{11} d_{22}\sqrt{(1-\alpha_1)(1-\alpha_2)}\rho_{24} & 0.5 d_{22}^2\sqrt{1-\alpha_2^2}\rho_{34} & d_{22}^2 \end{pmatrix}, \tag{2.24}$$

where

$$d_{11}^2 = d_1^2 + d_2^2;\ d_{22}^2 = d_3^2 + d_4^2;$$
$$\alpha_1 = \frac{d_1^2 - d_2^2}{d_{11}^2};\ \alpha_2 = \frac{d_3^2 - d_4^2}{d_{22}^2}. \tag{2.25}$$

The expression (2.24) is convenient to use in the analysis of statistical properties of quadrature components of orthogonally polarized components of PPW in an arbitrary fixed basis using polarimeters with coherent receivers.

The parameters m_j ($j = \overline{1,4}$) characterize the quadrature components of orthogonally polarized components of the corresponding monochromatic wave, and $(Z_j - m_j)$ characterizes the quadrature components of orthogonally polarized components of the corresponding quasi-monochromatic wave. The parameters d_j^2, D, D_{ln} characterize the polarization structure of the fluctuating component of the wave at the coinciding moments of time, and the Formula (2.22) itself is a mathematical description of the generalized probabilistic model of PPW. It serves as an initial point for obtaining the densities of the probabilities of envelopes, phases, and all groups of polarization parameters of PPW in any polarization basis. In general, the PPW wave is periodically non-stationary. Its non-stationarity is due to the non-stationarity of orthogonally polarized components, which moments are functions of time and, therefore, do not satisfy the necessary and sufficient conditions of stationarity.

Indeed, if we imagine orthogonally polarized wave components (2.21) through the quadrature components, it is easy to show that for the common case $\left(d_{2j-1}^2 \neq d_{2j}^2, \rho_{2j-1,2j} \neq 0,\ j = 1, 2\right)$ their moments periodically change in time. For example,

$$\overline{E_j^2(t)} = 0.5 d_{jj}^2 \left[1 + \alpha_j \cos 2(\omega_0 t + \varphi_j) + \sqrt{1 - \alpha_j^2} \rho_{2j-1,2j} \sin(\omega_0 t + \varphi_j)\right]\left(1 + E_{0j}^2 d_{jj}^{-2}\right), \tag{2.26}$$

where $E_{0j}^2 = m_{2j-1}^2 + m_{2j}^2, j = 1, 2$.

Further in formulas instead $\rho_{2j-1,\,2j}$, we will write simply ρ_0.

Consequently, the process at hand belongs to periodically non-stationary processes, the properties of which are described in detail in [1, 2]. Periodically, non-stationary, as follows from the Formula (2.24), will also be elements of the matrix D_2 (in case of $\alpha_j \neq 0, \rho_0 = 0$).

It follows from the expression (2.22) that the joint probability density determined by it can be interpreted as a generalized probabilistic model of the total oscillation, which is a set of fully polarized (useful) signal and partially polarized normal noise, which has no deterministic component. Indeed, in this case, the monochromatic polarized component will correspond to the signal, and the quasi-monochromatic partially polarized component will correspond to the noise.

Note that the non-stationarity of orthogonally polarized components can be due to the nature of the radiation source of the wave and the conditions of its propagation and reception.

The analysis of the expression (2.26) shows that temporary independence of $\overline{E_j^2}$ from the time is performed only when $\alpha_j = 0$ and $\rho_0 = 0$. Consequently, the parameters α_j and ρ_0 characterize orthogonally polarized wave components in terms of their stationarity.

In practice, we often have to deal with stationary narrowband random processes in which, as is known, their quadrature components x_1 and x_2, x_3 and x_4 are connected by the Hilbert transform. The correlation coefficients of these processes at coincident moments of time are zero, i.e., $\rho_{12} = \rho_{34} = 0$. In addition, the following equations become true $\rho_{13} = \rho_{24} = \rho$; $\rho_{14} = \rho_{23} = g$; $d_1 = d_2$; $d_3 = d_4$.

Taking into account the above, we write the law $W(\vec{X})$ in explicit form, where the assumption $\overline{\vec{X}} = 0$ is made:

$$W(\vec{X}) = \frac{1}{(2\pi)^2 d_1^2 d_2^2 [1 - R^2]} \times \exp\left\{-\frac{0.5}{(1 - R^2)}\left[\frac{x_1^2 + x_1^2}{d_1^2} + \frac{x_3^2 + x_4^2}{d_2^2} - \frac{2\rho(x_1 x_3 + x_2 x_4)}{d_1 d_2} - \frac{2g(x_1 x_4 - x_2 x_3)}{d_1 d_2}\right]\right\}, \tag{2.27}$$

where $R^2 = \rho_{13}^2 + \rho_{14}^2 = \rho^2 + g^2$ is the generalized correlation coefficient.

The Formula (2.27) is the main for calculating the polarization properties of a narrowband stationary radio wave defined by a tetra-vector $\vec{X}$. d_2 is introduced in the Formula (2.27) instead of d_3 for convenience of recording.

Having integrated the law (2.27) into three variables, we obtain one-dimensional laws, which will also be Gaussian.

As follows from the definition of a non-polarized wave, the power per E_X and E_Y components must be equal. In addition, there should be no correlation between the complex amplitudes.

This means that for such waves $d_1 = d_2 = d$, $R = 0$. Thus, for the NPW the respective distribution law $W(\vec{X})$ will take the form:

$$W(\vec{X}) = \frac{1}{(2\pi)^2 d^4} \exp\left\{-\frac{1}{2d^2}\left(x_1^2 + x_2^2 + x_3^2 + x_4^2\right)\right\}. \tag{2.28}$$

In conclusion, we note, as already mentioned, within the framework of the considered model of PPW one-dimensional PDF $W(x_j)$ will also be a Gaussian process. As for the PDF of the modulus of the value x_j, as shown by the direct calculations by the Formula (1.92) in the case of $\overline{x_j} \neq 0$, the Rice law acts as $W\left(\left|x_j\right|\right)$, and when $\overline{x_j} = 0$, the Rayleigh law acts (see Table 1.2).

2.2 Complex Components

2.2.1 *Basic Properties and Transformations*

In Sect. 1.2, as one of the possible representations of the polarization of radio waves, the Jones vector is introduced by the Formula (1.20), which coordinates represent the complex projection of an arbitrary vector $\vec{E}$ at an arbitrary polarization basis $\{\vec{e}\}$, i.e., $\vec{E} = E^{\mathrm{T}}\vec{e}$. This implies the following obvious equations:

$$\vec{E} = E_H^{\mathrm{T}}\vec{e}_H = E_C^{\mathrm{T}}\vec{e}_C = E_C^{\mathrm{T}} Q^{-1}\vec{e}_H = E_C^{\mathrm{T}} Q^{+}\vec{e}_H, \tag{2.29}$$

the consequence of which is:

$$E_H = Q^* E_C, \tag{2.30}$$

where the property of unitarity of the matrix Q was used in the Formula (2.30) derivation.

As it can be seen, the transformation of the Jones vector differs from the similar expression relating to the polarization basis (1.36).

The Formula (2.30) allows the decomposition of a given polarized wave in an arbitrary polarization basis.

As examples, we present different types of polarization of the wave in the linear and circular bases.

Let a linearly polarized wave in the initial linear basis, the first unit vector of which coincides in the direction with the field vector (HP), have a column-vector $E_L^{\mathrm{T}} = \sqrt{P_0}(1 \quad 0)^{\mathrm{T}}$ representation. When you turn a linear basis to the corner, γ_0 components of the field are subjected to a linear rotation transformation defined by the matrix (1.42) when $\delta = 0$. In the new basis, the components of the field E_1 and E_2 are determined by the equation:

$$E_L = \sqrt{P_0}\mathrm{e}^{-\mathrm{i}\varphi_0}\left(\mathrm{e}^{-\mathrm{i}\eta}\cos\gamma_0 \quad -\mathrm{e}^{\mathrm{i}\eta}\cos\gamma_0\right)^{\mathrm{T}}.$$

To represent HP in a circular basis (left circle–right circle) from the same formulae, we obtain ($\gamma_0 = \delta = \pi/4$): $E_{\mathrm{CP}} = \sqrt{0.5P_0}\mathrm{e}^{-\mathrm{i}\varphi_0}\left(\mathrm{e}^{-\eta} \quad -\mathrm{i}\mathrm{e}^{\mathrm{i}\eta}\right)^{\mathrm{T}}$. When the basis is transformed by the matrix Q, which have $\gamma = \delta = 0$, we have $E_H = \sqrt{P_0}\mathrm{e}^{-\mathrm{i}\varphi_0}\left(E_{1C}\mathrm{e}^{-\mathrm{i}\eta} \quad E_{2C}\mathrm{e}^{\mathrm{i}\eta}\right)^{\mathrm{T}}$, from where it is seen that the coordinates of the new vector differ from the corresponding old ones only by the phase. The phases $(\varphi_0 + \eta)$ and $(\varphi_0 - \eta)$ of the coordinates E_{1H} and E_{2H} can be chosen arbitrary, in particular zero. In other words, there is always such a polarization basis with a given type of polarization ellipses forming new unit vectors, in which the values of the orthogonal components of any fully polarized wave will be real numbers.

2.2.2 *Probabilistic Description—The Gaussian Case*

In the framework of the considered models of PPW and in accordance with the correlations (2.22) and (2.27), the coordinates of the components of the tetra-vector are described by the normal law. In practice, envelope processes, i.e., E_1 and E_2, and their phases φ_1 and φ_2 are often found as a result of measurements. In this regard, it is of interest to define a joint four-dimensional PDF $W(E_1, E_2, \varphi_1, \varphi_2)$ and a PDF of smaller dimensionality.

Using the rule of functional transformations of the laws of random variables described in p. 1.5, and considering the correlations $E_j^2 = x_j^2 + x_{j+1}^2$, $\tan\varphi_j = x_{j+1}/x_j$ $(j = 1, 2)$, it is not difficult to go from Formula (2.22) to expression for $W(E_1, E_2, \varphi_1, \varphi_2)$. In the common case for a periodically non-stationary PPW, this problem is solved, where sufficiently cluttered appearance for $W(E_1, E_2, \varphi_1, \varphi_2)$, $W(E_1, E_2, 2\delta)$, $W(E_1, E_2)$ and $W(2\delta)$ when $2\delta = \varphi_2 - \varphi_1$ are obtained. In this monograph, we consider only the case of stationary PPW, which has an important place in many applications of polarimetry.

Assuming in formulas (2.22), (2.24), and (2.25) $d_{2j-1}^2 = d_{2j}^2$, i.e., $\alpha_j = 0$; $\rho_{12} = \rho_{34} = 0$; $\rho_{13} = \rho_{24} = \rho$; $\rho_{14} = -\rho_{23} = g$, and conducting appropriate functional transforms and then integrating by variable φ within limits of $[0 \div 2\pi]$, we get the three-dimensional density distribution of envelopes and phase difference of the stationary PPW in an arbitrary basis:

$$
\begin{aligned}
&W_3(E_1, E_2, 2\delta, a_1, b_1, 2\delta_0, R, \beta) \\
&= \tfrac{E_1 E_2}{(2\pi)^2 d_{11}^2 d_{22}^2 (1-R^2)} \exp\Big\{ -\tfrac{a_1^2[1-2bR\cos(2\delta_0-\beta)+b^2]}{2(1-R^2)} - \tfrac{1}{2(1-R^2)} \Big[\tfrac{E_1^2}{d_{11}^2} + \tfrac{E_2^2}{d_{22}^2} \\
&- \tfrac{2RE_1E_2}{d_{11}d_{22}} \cos(2\delta_0 - \beta) \Big] \Big\} I_0 \Big\{ \tfrac{1}{(1-R^2)} \big[a_1^2 \big(1 - 2bR\cos(2\delta_0 - \beta) + b^2 R^2 \big) \\
&\times \Big(\tfrac{E_1^2}{d_{11}^2} + \chi^2 \tfrac{E_2^2}{d_{22}^2} + 2\chi \tfrac{E_1 E_2}{d_{11} d_{22}} \Big) \big]^{0.5} \Big\},
\end{aligned}
\tag{2.31}
$$

where

$$
\chi^2 = \frac{b^2 - 2bR\cos(2\delta_0 - \beta) + R^2}{1 - 2bR\cos(2\delta_0 - \beta) + b^2 R^2};
$$

$$
\mu = \arctan\left\{ \frac{R\sin(2\delta_0 - \beta)\big[1 - 2bR\cos(2\delta_0 - \beta) - b^2\big]}{b\big[1 + R^2\cos(2\delta_0 - \beta)\big] - R\big(1 + b^2\big)\cos(2\delta_0 - \beta)} \right\}.
$$

In the expression (2.31), the following notations are entered:

$a_j^2 = E_{0j}^2 / d_{jj}^2$; $j = 1, 2$—the ratio of the power of a component of a monochromatic wave (a deterministic component) to the power of the corresponding component of a partially polarized wave (a fluctuating component);

$b = a_2/a_1 = (E_{02}/E_{01})(d_{11}/d_{22}) = r_0(d_{11}/d_{22})$—some parameter of the combined wave, where $r_0 = E_{02}/E_{01}$ considers the polarization properties of the deterministic component of the wave, and d_{11}/d_{22} are the polarization properties of the fluctuating component of the wave;

$2\delta_0 = (\varphi_{02} - \varphi_{01})$—the phase difference between the deterministic components of the components, $\varphi_{0j} = \arctan\big(m_{j+1}/m_j\big)$;

$R^2 = \rho^2 + g^2$, $\tan\beta = g/\rho$—the modulus and the argument of the complex coefficient of mutual correlation characterizing the statistical relation between the quadrature components of the fluctuating part of orthogonally polarized components.

As follows from the expression (2.31), the considered probability density is completely determined by the parameters $a_j, 2\delta_0, d_{jj}^2$, R and β. Parameters a_1 and a_2 determine the properties of the combined wave $2\delta_0$, deterministic component, and d_{11}^2, d_{22}^2, R and β characterize the fluctuating properties of orthogonally polarized wave $\big(d_{11}^2, d_{22}^2\big)$ components and their statistical relation (R, β). For this probability model of the combined partially polarized wave (2.31), the real matrix (2.24) with regard to the equation $\alpha_j = \rho_0 = 0$ takes the following form:

$$
D_2^K = M + \begin{pmatrix}
d_{11}^2 & 0 & \rho d_{11} d_{22} & g d_{11} d_{22} \\
0 & d_{11}^2 & -g d_{11} d_{22} & \rho d_{11} d_{22} \\
\rho d_{11} d_{22} & -g d_{11} d_{22} & d_{22}^2 & 0 \\
g d_{11} d_{22} & \rho d_{11} d_{22} & 0 & d_{22}^2
\end{pmatrix},
\tag{2.32}
$$

where M is a real matrix describing the polarization properties of a monochromatic wave and determined by the first summand in the expression (2.24): $d_{jj}^2 = d_{2j-1}^2 + d_{2j}^2$, $j = 1, 2$.

If there are no deterministic components in orthogonally polarized components ($a_1 = a_2 = 0$), then for the four-dimensional distribution law, we have:

$$W\left(E_1, E_2, \varphi_1, \varphi_2, d_{11}^2, d_{22}^2, R, \beta\right) = \left[(2\pi)^2 d_{11}^2 d_{22}^2 \left(1 - R^2\right)\right]^{-1} E_1 E_2 \times \exp\left\{ -\frac{1}{2\left(1 - R^2\right)} \left[\frac{E_1^2}{d_{11}^2} + \frac{E_2^2}{d_{22}^2} - \frac{2E_1 E_2}{d_{11} d_{22}} R \cos(\varphi_1 - \varphi_2 - \beta) \right] \right\}, \quad (2.33)$$

and instead of the expression (2.31), we get the form of law as (2.33), where 2δ is instead of $(\varphi_1 - \varphi_2)$, i.e., $W(E_1, E_2, \varphi_1, \varphi_2) = W_3(E_1, E_2, 2\delta = \varphi_1 - \varphi_2)$. The matrix M vanishes upon that.

To determine the two-dimensional joint probability density of the envelopes of orthogonally polarized components $W(E_1, E_2)$, you need to integrate the Formula (2.31) by the variable 2δ within the limits of $[0 \div 2\pi]$.

In the absence of deterministic components ($a_j = 0$, $\varphi_{0j} = 0$), the expression (2.34) takes a simpler form:

$$W_2(E_1, E_2, d_{11}, d_{22}, R) = \frac{E_1 E_2}{\left(1 - R^2\right) d_{11}^2 d_{22}^2} \exp\left[-\frac{1}{2\left(1 - R^2\right)} \left(\frac{E_1^2}{d_{11}^2} + \frac{E_2^2}{d_{22}^2} \right) \right] I_0 \left[\frac{R E_1 E_2}{\left(1 - R^2\right) d_{11} d_{22}} \right]. \quad (2.34)$$

The resulting expressions (2.31) and (2.34) are of particular interest as generalizing cases; in addition, they are initial for obtaining the density distribution of the probabilities of polarization parameters of the PPW, which will be discussed below.

In practical radio polarimetry, as a rule, envelope and phases of orthogonally polarized components are used. Therefore, in practice, it is essential to know both their one-dimensional probability density function and other statistical characteristics. Let us consider this issue in more detail.

As it was mentioned above, any quasi-monochromatic PPW using the phenomenological approach can be represented as a sum of n elementary waves with amplitudes l_{ji} and phases ψ_{ji}, where $j = 1, 2$ is the number of orthogonally polarized component and $i = 1, 2, 3 \ldots$ is the order number of elementary waves.

We believe that the amplitudes l_{ji} and phases ψ_{ji} are statistically independent, and their probability densities $W_i(l_j)$ and $W_i(\psi_j)$ are the same for different indices i; that is, the equations $W_i(l_j) = W(l_j)$ and $W_i(\psi_j) = W(\psi_j)$ are observed. In this case, the expressions for the mean values of the quadrature components m_{2j-1}, m_{2j}, their variance d_{2j-1}^2, d_{2j}^2, and the coefficient of mutual correlation ρ_0 can be represented as follows:

$$\left.\begin{aligned}
&m_{2j-1} = n\overline{l_j} \int_{-\infty}^{\infty} W(\psi_j)\cos\psi_j \mathrm{d}\psi_j;\ d_{2j-1}^2 = n\overline{l_j^2}\left[\int_{-\infty}^{\infty} W(\psi_j)\cos^2\psi_j \mathrm{d}\psi_j - m_j^2/n\right]\\
&m_{2j} = n\overline{l_j} \int_{-\infty}^{\infty} W(\psi_j)\sin\psi_j \mathrm{d}\psi_j;\ d_{2j}^2 = n\overline{l_j^2}\left[\int_{-\infty}^{\infty} W(\psi_j)\sin^2\psi_j \mathrm{d}\psi_j - m_j^2/n\right]\\
&\rho_0 = \left[0.5n\overline{l_j^2}\int_{-\infty}^{\infty} W(\psi_j)\sin 2\psi_j \mathrm{d}\psi_j + n(n-1)\overline{l_j}^2\int_{-\infty}^{\infty} W(\psi_j)\cos\psi_j \mathrm{d}\psi_j\right.\\
&\quad\left.\times\int_{-\infty}^{\infty} W(\psi_j)\sin\psi_j \mathrm{d}\psi_j - m_j m_{j+1}\right]/d_{2j-1}d_{2j}.
\end{aligned}\right\}. \tag{2.35}$$

From correlations, (2.35) follows that mean values and variances of the quadrature components of the orthogonally polarized components of the waves are the same as in the common case, not equal to each other, and the coefficients of mutual correlation between the non-homogeneous quadrature components ρ_0 are not equal to zero. Therefore, the joint density of the envelope and phase probabilities for each orthogonally polarized wave component with respect to Eq. (2.22) can be written as:

$$\begin{aligned}
W_2(A_j,\varphi_j,\alpha_j,\rho_0,a_j,\varphi_{0j}) &= \frac{A_j}{2\pi\sqrt{1-\rho_0^2}}\exp\left\{-\frac{a_j^2[1-B(\alpha_j,\rho_0)\cos(2\varphi_{0j}-\beta(\alpha_j,\rho_0))]}{2(1-\rho_0^2)\sqrt{1-\alpha^2}}\right\}\\
&\times\exp\left\{-\frac{A^2[1-B(\alpha_j,\rho_0)\cos(2\varphi_j-\beta(\alpha_j,\rho_0))]}{2(1-\rho_0^2)\sqrt{1-\alpha^2}}\right.\\
&-\frac{a_jA_j}{\left(1-\rho_{j,j+1}^2\right)}[\rho_0\sin(\varphi_j+\varphi_{0j})\\
&-C(\alpha_j,\varphi_{0j})\cos(\varphi_j-\gamma(\alpha_j,\varphi_{0j}))]\Big\},
\end{aligned} \tag{2.36}$$

where $\gamma(\alpha_j,\varphi_{0j}) = \arctan\lfloor(1+\alpha_j)(1-\alpha_j)^{-1}\tan\varphi_{0j}\rfloor$;

$A_j = E_j/\sqrt{d_{2j}d_{2j-1}}$; $a_j = E_{0j}/\sqrt{d_{2j}d_{2j-1}}$—normalized envelopes; entering in the Formula (2.36), values B, C, and β are defined as follows:

$$\left.\begin{aligned}
&B^2(\alpha_j,\rho_0) = \alpha_j^2 + \rho_0^2\left(1-\alpha_j^2\right),\\
&\beta(\alpha_j,\rho_0) = \operatorname{arctg}\left[\rho_0\sqrt{1-\alpha_j^2}/\alpha_j\right],\\
&C(\alpha_j,\varphi_{0j}) = \left[(1-\alpha_j)(1+\alpha_j)^{-1}\cos^2\varphi_{0j} + (1+\alpha_j)(1-\alpha_j)^{-1}\sin^2\varphi_{0j}\right],\\
&j = 1,2.
\end{aligned}\right\}. \tag{2.37}$$

The expression (2.36) generalizes the previously known special cases and is initial for determination of one-dimensional densities of probabilities of envelopes $W(A_j)$ and phases $W(\varphi_j)$ of orthogonally polarized wave components which can be found, having carried out the corresponding integration by φ within the limits of $[0 \div 2\pi]$ or by A within the limits of $[0 \div A]$. Direct calculations give:

$$W\left(A_j, \alpha_j, \rho_0, a_j, \varphi_{0j}\right) = \frac{A_j}{\sqrt{1-\rho_0^2}} \exp\left\{-\frac{A_j^2 + a_j^2[1 - B\cos(2\varphi_0 - \beta)]}{2\left(1-\rho_0^2\right)\sqrt{1-\alpha_j^2}}\right\}$$

$$\times \sum_{n=0}^{\infty} \varepsilon_n I_n \left[\frac{A_j^2 B}{2\left(1-\rho_0^2\right)\sqrt{1-\alpha_j^2}}\right]$$

$$I_{2n}\left[\frac{aA}{1-\rho_0^2}\sqrt{\rho_0^2 + B^2 - \frac{2\rho_0 \sin 2\varphi_0}{\sqrt{1-\alpha_j^2}}}\right] \cos 2nv, \quad (2.38)$$

where $\varepsilon_0 = 0$, $\varepsilon_n = 2$ for $n > 0$.

It follows from the Formula (2.38) that the density of the probabilities of the envelopes of orthogonally polarized components in the considered common case depends on four parameters: α_j, ρ_0, a_j and φ_{0j}, when changing of which the shape of the probability density curve and its numerical characteristics change.

An explicit form of expression for $W\left(\varphi_j, \alpha_j, \rho_0, a_j, \varphi_{0j}\right)$ in view of its extreme awkwardness is not given.

It shall be noted that the phase distribution depends on the same parameters as the envelope distribution. Moreover, the unsteadiness of orthogonally polarized components leads to the fact that the envelopes and phases are dependent between themselves. Note that the expressions (2.36) and (2.38) are true for arbitrary laws of distribution of amplitudes and phases of elementary waves. The relation of $W(A, \alpha, \rho, a, \varphi_0)$ and $W(\varphi, \alpha, \rho, a, \varphi_0)$ with both densities $W(l)$ and $W(\psi)$ is expressed in accordance with the expression (2.36) through four parameters $\alpha, \rho, a, \varphi_0$.

Table 2.1 presents the dependence of the parameters $m_j, m_{j+1}, d_{j+1}^2, \rho_{j,j+1}, a_j, \varphi_0$ on the numerical characteristics of the two laws of distribution of elementary phases:

– proportional

$$W(\psi, l_1, l_2) = 1/(l_2 - l_1) \text{ for } l_1 \le \psi \le l_2, \quad (2.39)$$

– normal

$$W\left(\psi, m_\psi, d_\psi^2\right) = \frac{1}{\sqrt{2\pi} d_\psi} \exp\left[-\frac{\left(\psi - m_\psi\right)^2}{2d_\psi^2}\right]. \quad (2.40)$$

Table 2.1 uses the notation $\sin i(z) = \sin z/z$.

The formulae given in Table 2.1 allow to determine the numerical values of the parameters of the generalized probabilistic model $\rho_{j,j+1}, \alpha_j, a_j, \varphi_{0j}$ with known numerical characteristics (mean values and variances) of the laws of distribution of elementary phases, mean values $n\bar{l}$, and variances $n\overline{l^2}$ of elementary amplitudes.

Table 2.1 Dependence of the parameters m_{2j-1}, m_{2j}, d^2_{2j-1}, d^2_{2j}, ρ_0, α_j, φ_{0j}, a_j on the numerical characteristics of the distribution of elementary phases

	Uniform distribution of elementary phases $W(\psi) = 1/(l_2 - l_1)$	Normal distribution of elementary phases $W(\psi) = \frac{1}{\sqrt{2\pi}d_\psi}\exp\left[-\frac{(\psi - m_\psi)^2}{2d^2_\psi}\right]$
1	2	3
m_{2j-1}	$n\bar{l}\cdot\sin c\left(\frac{l_2-l_1}{2}\right)\cos\left(\frac{l_1+l_2}{2}\right)$	$n\bar{l}\cdot\exp\left[-\frac{d^2_\psi}{2}\right]\cos m_\psi$
m_{2j}	$n\bar{l}\cdot\sin c\left(\frac{l_2-l_1}{2}\right)\sin\left(\frac{l_1+l_2}{2}\right)$	$n\bar{l}\cdot\exp\left[-\frac{d^2_\psi}{2}\right]\sin m_\psi$
d^2_{2j-1}	$\frac{n\overline{l^2}}{2}\left\{[1+\sin c(l_2-l_1)\cos(l_2+l_1)]-2\left[\sin c\left(\frac{l_2-l_1}{2}\right)\right]^2\cos^2\left(\frac{l_2+l_1}{2}\right)\right\}$	$\frac{n\overline{l^2}}{2}\left[1+\exp\left(-2d^2_\psi\right)\cos 2m_\psi-2\exp\left(-d^2_\psi\right)\cos^2 m_\psi\right]$
d^2_{2j}	$\frac{n\overline{l^2}}{2}\left\{[1-\sin c(l_2-l_1)\cos(l_2+l_1)]-2\left[\sin c\left(\frac{l_2-l_1}{2}\right)\right]^2\sin^2\left(\frac{l_2+l_1}{2}\right)\right\}$	$\frac{n\overline{l^2}}{2}\left[1-\exp\left(-2d^2_\psi\right)\cos 2m_\psi-2\exp\left(-d^2_\psi\right)\sin^2 m_\psi\right]$
ρ_0	$\frac{\sin(l_1+l_2)}{\sqrt{\overline{l^2}\left[1+\sin c(l_2-l_1)\cos(l_1+l_2)-2\sin c^2\left(\frac{l_2-l_1}{2}\right)\cos^2\left(\frac{l_1+l_2}{2}\right)\right]}} \times \frac{\left[\overline{l^2}\sin c(l_2-l_1)-2\sin c^2[0.5(l_2-l_1)]\right]}{\sqrt{\overline{l^2}\left[1-\sin c(l_2-l_1)\cos(l_1+l_2)-2\sin c^2\left(\frac{l_2-l_1}{2}\right)\sin^2\left(\frac{l_1+l_2}{2}\right)\right]}}$	$\frac{\sin 2m_\psi}{\sqrt{\overline{l^2}\left[1+\exp\left(-2d^2_\psi\right)\cos 2m_\psi-2\exp\left(-d^2_\psi\right)\cos^2 m_\psi\right]}} \times \frac{\overline{l^2}\left[\exp\left(-2d^2_\psi\right)-2\exp\left(-d^2_\psi\right)\right]}{\sqrt{\overline{l^2}\left[1-\exp\left(-2d^2_\psi\right)\cos 2m_\psi-2\exp\left(-d^2_\psi\right)\sin^2 m_\psi\right]}}$
α_j	$\frac{\cos(l_1+l_2)\left[\sin c(l_2-l_1)-\sin c^2[0.5(l_2-l_1)]\right]}{1-\sin c^2[0.5(l_2-l_1)]}$	$\frac{\overline{l^2}\cos 2m_\psi\left[\exp\left(-2d^2_\psi\right)-2\exp\left(-d^2_\psi\right)\right]}{1-\exp\left(-d^2_\psi\right)}$
φ_{0j}	$0.5(l_1+l_2)$	m_ψ

(continued)

Table 2.1 (continued)

	Uniform distribution of elementary phases $W(\psi) = 1/(l_2 - l_1)$	Normal distribution of elementary phases $W(\psi) = \frac{1}{\sqrt{2\pi}d_\psi} \exp\left[-\frac{(\psi - m_\psi)^2}{2d_\psi^2}\right]$
a_j	$\frac{\sqrt{2}n\bar{l} \cdot \sin c[0.5(l_2 - l_1)]}{\sqrt[4]{\bar{l^2}\left[1 + \sin c(l_2 - l_1)\cos(l_1 + l_2) - 2\sin c^2\left(\frac{l_2 - l_1}{2}\right)\cos^2\left(\frac{l_1 + l_2}{2}\right)\right]}} \times \frac{1}{\sqrt[4]{\bar{l^2}\left[1 - \sin c(l_2 - l_1)\cos(l_1 + l_2) - 2\sin c^2\left(\frac{l_2 - l_1}{2}\right)\sin^2\left(\frac{l_1 + l_2}{2}\right)\right]}}$	$\frac{\sqrt{2}n\bar{l} \cdot \exp\left(-0.5d_\psi^2\right)}{\sqrt[4]{\bar{l^2}\left[1 + \exp\left(-2d_\psi^2\right)\cos 2m_\psi - 2\exp\left(-d_\psi^2\right)\cos^2 m_\psi\right]}} \times \frac{1}{\sqrt[4]{\bar{l^2}\left[1 - \exp\left(-2d_\psi^2\right)\cos 2m_\psi - 2\exp\left(-d_\psi^2\right)\sin^2 m_\psi\right]}}$

Let us now consider particular probability models of the envelopes and phases of the orthogonally polarized components of the wave resulting from the expressions (2.37) and (2.38). Therein, without loss of generality, we assume that the probability densities of elementary phases $W(\psi)$ are symmetric functions with respect to zero, that is, in the Formulae (2.39) and (2.40) parameters $l_2 = l_1 = l$, and $m_\psi = 0$.

In order not to overcharge the following formulae, the index j of the letters E, φ, α, m, if it does not lead to two interpretations, is not written.

1. If the parameter $l = k\pi$, where $k = 1, 2, 3 \ldots$ (or $d_\psi^2 \to \infty$), then from Table 2.1 it follows that $\alpha = \rho_0 = a = \varphi_0 = 0$. Therein, the expressions (2.36) and (2.38) are significantly simplified, taking the well-known form:

$$\left.\begin{array}{l} W(E,\varphi) = \frac{1}{2\pi}\frac{E}{d^2}\exp\left(-\frac{E^2}{2d^2}\right), \\ W(E) = \frac{E}{d^2}\exp\left(-\frac{E^2}{2d^2}\right), \\ W(\varphi) = 1/(2\pi), \end{array}\right\}, \tag{2.41}$$

where $d^2 = d_{2j-1}^2 = d_{2j}^2$.

As it can be seen, the envelope and phase are independent, while the envelope is distributed by Rayleigh, phase by uniform laws, and orthogonally polarized wave components are normal stationary processes with an average power equal to $\overline{E^2} = d^2$. The mean value and the envelope variance are equal, respectively:

$$m_E = \sqrt{0.5\pi}d; \quad d_E^2 = (2 - 0.5\pi)d^2. \tag{2.42}$$

2. If $m_{2j-1} = m_{2j} = 0$, and parameters l and d_ψ^2 are arbitrary (except special points $l = k\pi, d_\psi^2 \to \infty$), then it is easy to show that $a = \rho_0 = \varphi_0 = 0$ and $\alpha \neq 0$. The probability density (2.36) takes the form therein:

$$\left.\begin{array}{l} W_2(E,\varphi,\alpha) = \dfrac{E}{\pi d^2(1-\alpha^2)}\exp\left[-\dfrac{E^2}{d^2(1-\alpha^2)}(1-\alpha\cos 2\varphi)\right], \\ d^2 = d_{2j}^2 + d_{2j-1}^2 \end{array}\right\}. \tag{2.43}$$

As you can see, the envelope and the phase in this case are dependent on each other. By integrating Eq. (2.43) by the eliminating parameters, we obtain:

$$\left.\begin{array}{l} W(E,\varphi) = \dfrac{2E}{(1-\alpha^2)d^2}\exp\left[-\dfrac{E^2}{d^2(1-\alpha^2)}\right]I_0\left[\dfrac{\alpha E^2}{(1-\alpha^2)d^2}\right], \\ W(\varphi,\alpha) = \dfrac{\sqrt{1-\alpha^2}}{2\pi(1-\alpha\cos 2\varphi)} \end{array}\right\}. \tag{2.44}$$

The average power in this case is a periodic temporary function:

$$\overline{E^2} = 0.5d^2[1 + \alpha\cos 2(\omega_0 t + \varphi)] \tag{2.45}$$

That is, orthogonally polarized wave components are not stationary. The mean value and the variance of the envelopes are calculated by the formulae:

$$\left.\begin{aligned} m_E(\alpha) &= d\sqrt{\frac{1+\alpha}{\pi}}E\left(\sqrt{\frac{2\alpha}{1+\alpha}}\right), \\ d_E^2(\alpha) &= d^2\left[1-\frac{1+\alpha}{\pi}E^2\left(\sqrt{\frac{2\alpha}{1+\alpha}}\right)\right] \end{aligned}\right\}, \tag{2.46}$$

where $E(\xi)$ is fully elliptic integral of the second kind.

The Formula (2.44) are the Hoyt probability densities, describing the distributions of the module and the vector argument, the quadrature constituents of which have zero mean values and unequal to each other variances. When $\alpha = 0$ (stationary process), the expressions (2.44) and (2.46) turn into the expressions (2.41) and (2.42), respectively.

In the extreme case, when $\alpha = 1$, the probability density $W(E, \alpha)$ is converted to a one-way normal density:

$$W(E, \alpha = 1) = \sqrt{\frac{2}{\pi}} \cdot \frac{1}{d^2} \exp\left(-\frac{E^2}{2d^2}\right) \tag{2.47}$$

Herewith $m_E = \sqrt{\frac{2}{\pi}}d$, $d_E^2 = d^2\left(1 - \frac{2}{\pi}\right)$, and probability density $W(\varphi; \alpha = 1)$ turn into the delta function.

Figures 2.1 and 2.2, respectively, show the graphs of densities of probabilities $W(E, \alpha)$ and $W(\varphi, \alpha)$ for some values of the parameter α calculated by the Formula (2.44). It is seen that $W(E, \alpha)$ is a single-mode asymmetric function, the mode of which decreases with increasing of α. The phase distribution is symmetric with respect to $\varphi = k\pi$, where $k = 0, 1, 2, 3 \ldots$ The function $W(\varphi, \alpha)$ reaches its maximum when $\varphi = k\pi$, and the minimum when $\varphi = 0.5\pi(1 + 2k)$.

3. If the conditions of p. 1 (Rayleigh model) are met, and in addition, there are n vectors with constant moduli and arguments forming a deterministic constituent of an orthogonally polarized component, then the functions $B(\alpha, \rho_0)$, $\beta(\alpha, \rho_0)$ vanish, $\gamma(\alpha, \varphi_0) = \varphi_0$; $B(\alpha, \varphi_0) = 1$, and the expressions (2.36) and (2.38)

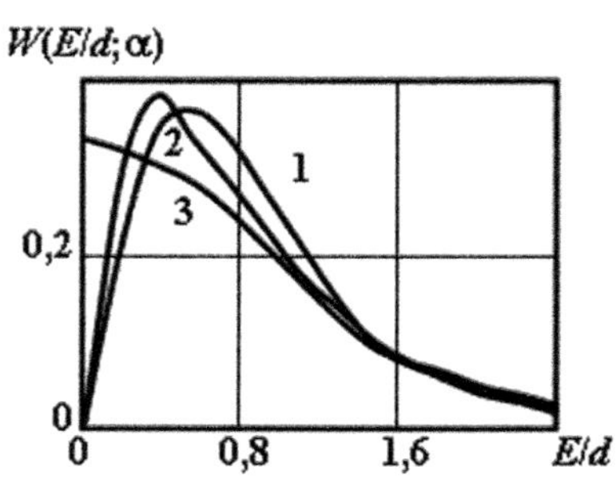

Fig. 2.1 Density of probability distribution $W(E, \alpha)$. 1—$\alpha = 0.7$; 2—$\alpha = 0.9$; 3—$\alpha = 0.9$

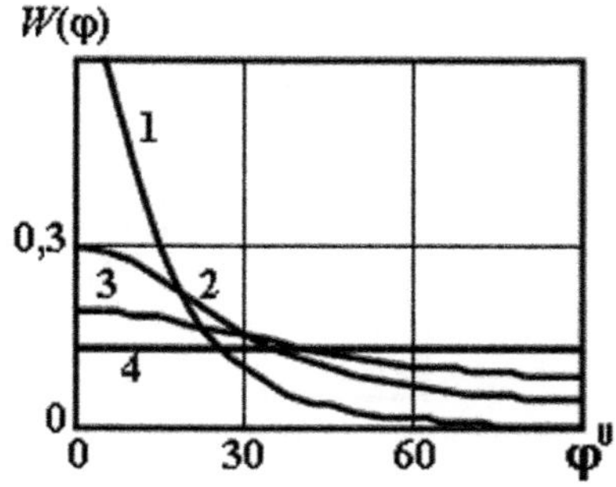

Fig. 2.2 Probability distribution density of normalized phase. 1—$\alpha = 0.9$; 2—$\alpha = 0.6$; 3—$\alpha = 0.3$; 4—$\alpha = 0$

are simplified, taking the known form of the Rice model (generalized Rayleigh vector):

$$\left.\begin{array}{l} W_2(A,\varphi,a,\varphi_0)=\frac{A}{2\pi}\exp\left[-0.5\left(A^2-2aA\cos(\varphi-\varphi_0)+a^2\right)\right],\\ W(A,a)=A\exp\left[-0.5\left(a^2+A^2\right)\right]I_0(aA),\\ W(\varphi,a,\varphi_0)=\frac{1}{2\pi}\exp\left(-0.5a^2\right)+\frac{a\cos(\varphi-\varphi_0)}{2\sqrt{2\pi}}\\ \times\Phi[a\cos(\varphi-\varphi_0)]\exp\left\{-0.5a^2[1-\cos(\varphi-\varphi_0)]\right\} \end{array}\right\}, \quad (2.48)$$

where $\varphi = \arctan(Y_0/X_0)$, $a = E_0/d$; $A = E/d$; $E_0 = \sqrt{X_0^2 + Y_0^2}$; $d^2 = d_{2j-1}^2 = d_{2j}^2$.
In this case, orthogonally polarized components are a normal stationary process, the average power of which is equal to $\overline{E^2} = d^2\left(1 + 0.5a^2\right)$, and the mean value and the variance of the envelopes are determined by the expressions:

$$\left.\begin{array}{l} m_E(a)=d\sqrt{\frac{\pi}{2}}\left[\left(1-0.5a^2\right)I_0\left(0.5a^2\right)+0.5a^2\exp\left(-0.25a^2\right)I_1\left(0.25a^2\right)\right],\\ d_E^2(a)=2d^2\left(1+0.5a^2\right)-m_E^2 \end{array}\right\}. \quad (2.49)$$

4. Under the conditions of p. 2 (Hoyt model $\rho_0 = 0$) and the presence of a deterministic component in the Formulae (2.36) and (2.38), we shall assume $\rho_0 = 0$. The resulting functional dependencies are illustrated quite clearly in Figs. 2.3 and 2.4.
5. If $\rho_0 \neq 0, \alpha = a = \varphi_0 = 0$, the expressions (2.36) and (2.38) take the following form:

$$\left.\begin{array}{l} W_2(A,\varphi,\rho_0)=\frac{A}{2\pi\sqrt{1-\rho_0^2}}\exp\left[-\frac{A^2}{2\left(1-\rho_0^2\right)}\left(1-\rho_0\sin 2\varphi\right)\right],\\ W(A,\rho_0)=\frac{A}{\sqrt{1-\rho_0^2}}\exp\left[-\frac{A^2}{2\left(1-\rho_0^2\right)}\right]I_0\exp\left[\frac{A^2\rho_0}{2\left(1-\rho_0^2\right)}\right],\\ W(\varphi,\rho_0)=\sqrt{1-\rho_0^2}[2\pi(1-\rho_0\sin 2\varphi)]^{-1} \end{array}\right\}, \quad (2.50)$$

where $A = E/d$; $d^2 = d_{2j}^2 = d_{2j-1}^2$.

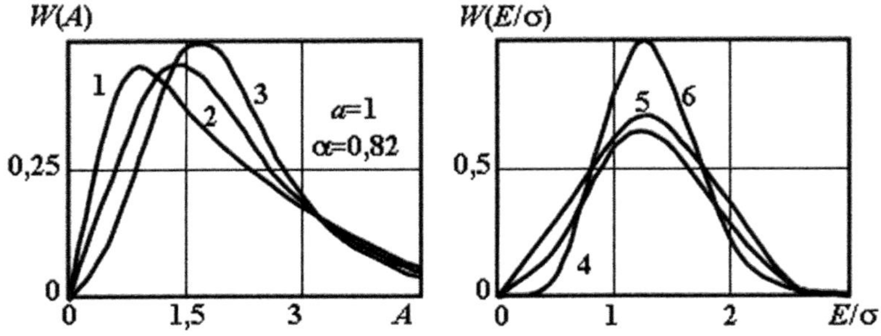

Fig. 2.3 Probability distribution density of the envelope. 1—$\varphi_0 = \pi/2$; 2—$\varphi_0 = \pi/4$; 3—$\varphi_0 = 0$; 4—$\alpha = 0$; 5—$\alpha = 0.5$; 6—$\alpha = 0.9$

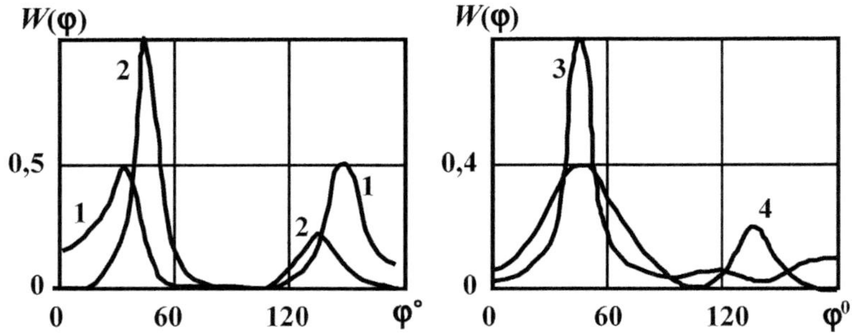

Fig. 2.4 Probability distribution density of phase 1—$\varphi_0 = \pi/2$; 2—$\varphi_0 = 0$; 3—$a = 1$; 4—$\varphi_0 = \pi/2$

These formulae describe the probability density of the modulus and the argument vector (we will call it ψ-vector), and the quadrature components which are correlated between themselves and its structure coincide with the expressions (2.43) and (2.44), if for the last to replace α with ρ_0, and $\cos 2\varphi$ with $\sin 2\varphi$ and introduce the appropriate normalization $A = E/d$.

Figures 2.5 and 2.6 show, respectively, graphs of densities of probabilities $W(A, \rho_0)$ and $W(\varphi, \rho_0)$ for some values of ρ_0. It is seen that parameter ρ_0 change lead to a significant transformation of the probability densities of the envelope and phase of ψ-vector. Note that the influence of parameters ρ_0 and α on the form of $W(A)$ and $W(\varphi)$ is equal.

The envelope and the phase of the ψ-vector are dependent on each other, and orthogonally polarized components in this case are periodically non-stationary processes with average power:

$$\overline{E^2} = d^2[1 + \rho\, \sin 2(\omega_0 t + \varphi)]. \tag{2.51}$$

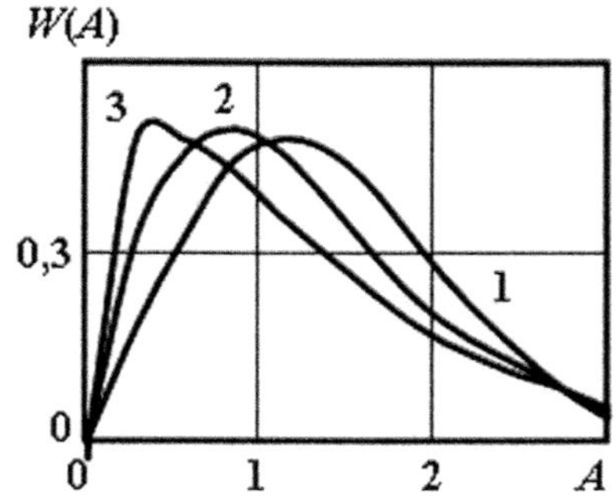

Fig. 2.5 Probability distribution density $W(A, \rho_0)$ with values of different ρ. 1—$\rho_0 = 0$; 2—$\rho_0 = 0.8$; 3—$\rho_0 = 0.9$

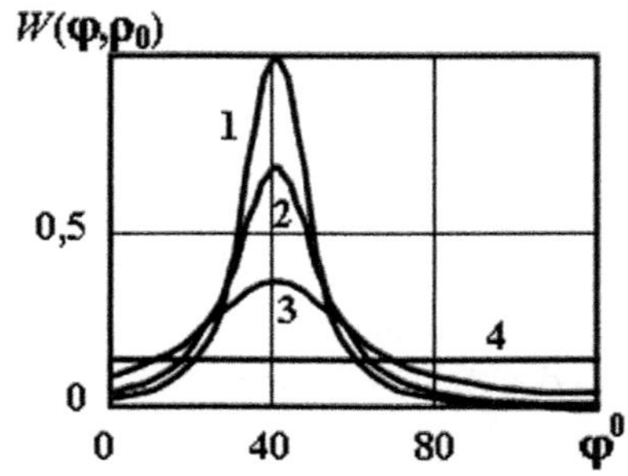

Fig. 2.6 Probability distribution $W(\varphi; \rho_0)$ density of phase. 1—$\rho_0 = 0.95$; 2—$\rho_0 = 0.9$; 3—$\rho_0 = 0.7$; 4—$\rho_0 = 0$

The initial moments of the envelope based on the Formula (2.50) can be expressed in the final form:

$$m_A^{(v)}(\rho_0) = 2^{0.5v}\Gamma(1+0.5v)\left(1-\rho_0^2\right)^{0.5(v+1)} {}_2F_1\left[0.25(v+2); 0.25(v+4); 1; \rho_0^2\right], \tag{2.52}$$

where ${}_2F_1(\xi)$ is the hypergeometric function.

6. If the mean values m_{2j-1} and m_{2j} are zero, that is, $a = \varphi_0 = 0$ and the parameters $\alpha \neq 0$ and $\rho_0 \neq 0$, then the expressions (2.36) and (2.38) are significantly simplified and take the following form:

$$\left.\begin{aligned} W(A,\varphi,\alpha,\rho_0) &= \frac{A}{2\pi\sqrt{1-\rho_0^2}}\exp\left\{-\frac{0.5A}{\left(1-\rho_0^2\right)\sqrt{1-\alpha^2}}\right. \\ &\times\left.\left[1-\sqrt{\alpha^2+\rho_0^2\left(1-\alpha^2\right)}\cos\left(2\varphi-\arctan\frac{\rho_0\sqrt{1-\alpha^2}}{\alpha}\right)\right]\right\} \\ W(A,\alpha,\rho_0) &= \frac{A}{\sqrt{1-\rho_0^2}}\exp\left[-\frac{0.5A^2}{\left(1-\rho_0^2\right)\sqrt{1-\alpha^2}}\right]I_0\left[\frac{A^2\sqrt{\alpha^2+\rho_0^2\left(1-\alpha^2\right)}}{2\left(1-\rho_0^2\right)\sqrt{1-\alpha^2}}\right] \\ W(\varphi,\alpha,\rho_0) &= \frac{\sqrt{\left(1-\rho_0^2\right)\left(1-\alpha^2\right)}}{2\pi\left[1-\sqrt{\alpha^2+\rho_0^2\left(1-\alpha^2\right)}\cos\left(2\varphi-\operatorname{arctg}\frac{\rho_0^2\sqrt{1-\alpha^2}}{\alpha}\right)\right]} \end{aligned}\right\} \tag{2.53}$$

These formulae describe the probability density of the vector module and argument, the quadrature constituents of which are correlated between themselves and have different variances. The envelopes and phases of this vector are dependent quantities, and orthogonally polarized wave components are periodic non-stationary processes with average power:

$$\overline{E^2} = 0.5d^2\left[1 + \alpha\,\cos 2(\omega_0 t + \varphi) + \rho_0\sqrt{1-\alpha^2}\sin 2(\omega_0 t + \varphi)\right]. \tag{2.54}$$

Thus, the particular cases considered above (pp. 1–6) and their analysis allow to represent a generalized probabilistic model of envelopes and phases of orthogonally polarized components of PPW in the form of a vector sum of the deterministic component with parameters and fluctuating component described in p. 6. Figure 2.7 shows the vector diagram of the generalized probabilistic model.

This model and its particular cases (pp. 1–6) can be interpreted with the help of Fig. 2.8, where certain points, edges, surface planes, and sections correspond to different models of orthogonally polarized component of the Z wave:

- point A—to Rayleigh model ($\alpha = \rho_0 = a = 0$);
- edge AA'—to Rice model ($\alpha = \rho_0 = 0, a \neq 0$);
- edge AB—to Hoyt model ($\rho_0 = a = 0, \alpha \neq 0$);
- surface plane $AA'BB'$—to Hoyt generalized model law ($\rho_0 = 0, \alpha \neq 0; a \neq 0$);
- edge AD—to ψ-vector model ($\rho_0 \neq 0, a = \alpha = 0$);
- surface plane $AA'DD'$—to generalized ψ-vector model ($\rho_0 \neq 0, a \neq 0, \alpha = 0$);
- surface plane $ABCD$—to model, described in p. 6 ($\rho_0 \neq 0, \alpha \neq 0, a = 0$);
- section planes, parallel to surface plane $AA'B'B$—generalized model when $\rho_0 =$ const, $\alpha \neq 0, a \neq 0$;

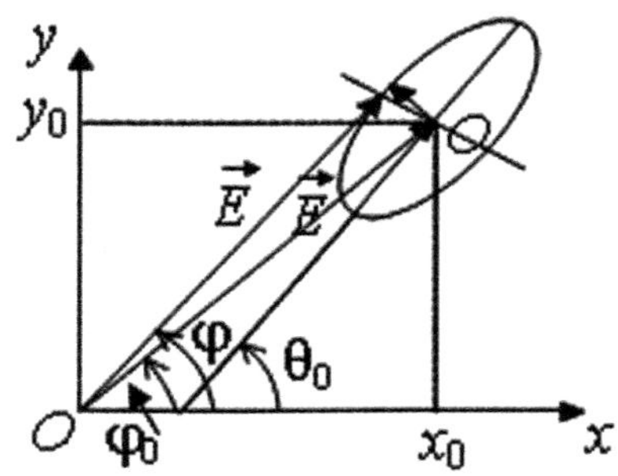

Fig. 2.7 Vector diagram of generalized distribution model

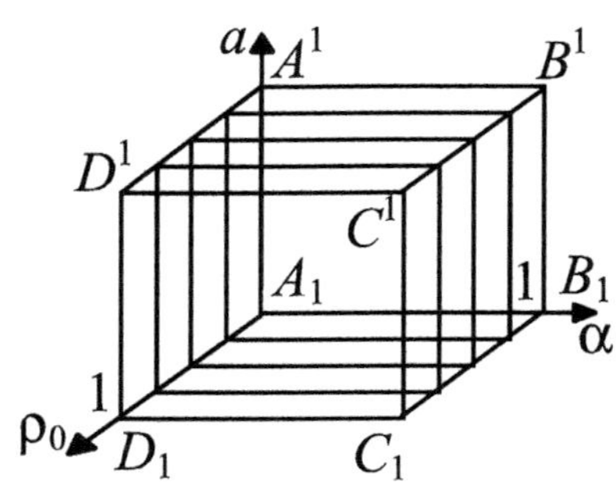

Fig. 2.8 Presentation of generalized distribution model and its specific cases

- section planes, parallel to surface plane $AA'DD'$—generalized model when $a =$ const, $\alpha \neq 0$, $\rho_0 \neq 0$;
- section planes, parallel to surface plane $ABCD$—generalized model when $a =$ const, $\alpha \neq 0$, $\rho_0 = 0$.

Without violating the generality, here we assume that the parameter $\varphi_0 = 0$ and the parameter $\alpha \geq 0$. Similar constructions can be made for $\varphi_0 \neq 0$ and $\alpha \leq 0$. It is enough to fix one of the parameters a or ρ_0 (e.g., $a =$ const) and build a rectangular parallelepiped in the coordinate system $A, \rho, \varphi_0, \alpha$.

2.2.3 Approximation of Probabilistic Models

Along with the probabilistic models of the envelopes of orthogonally polarized wave components considered above, other distributions can be used, in particular, the m-distribution (the Nakagami law):

$$W(E, m, \Omega) = \frac{2}{\Gamma(m)}\left(\frac{m}{\Omega}\right)^m E^{2m-1} \exp\left(-\frac{m}{\Omega}E^2\right), \tag{2.55}$$

where $m = \overline{E^2}\left[\overline{\left(E^2 - \overline{E^2}\right)^2}\right]^{-1} > 0.5, \quad \Omega = \overline{E^2}$.

The physical interpretation of this distribution is the same as of the generalized probabilistic model (2.38).

Based on the definition of the parameters m and Ω, it is easy to find the relation of these parameters with the parameters of the probability density (2.38):

$$\left.\begin{aligned} m &= \frac{\left(1+a^2\right)^2}{1+\left[\alpha^2+\rho_0^2(1-\alpha^2)\right]+2a^2\left[1+\sqrt{\alpha^2+\rho_0^2(1-\alpha^2)}\cos\left(2\varphi_0-\arctan\frac{\rho_0\sqrt{1-\alpha^2}}{\alpha}\right)\right]} \\ \Omega &= \left(1+a^2\right)d^2; \quad d^2 = d_{2j-1}^2 + d_{2j}^2 \end{aligned}\right\} \tag{2.56}$$

Initial νth moments of the envelope are equal therein:

$$m_E^\nu = \frac{\Gamma(m+0.5)}{\Gamma(m)} \cdot \left(\frac{m}{\Omega}\right)^{0.5\nu}. \tag{2.57}$$

Thus, rather complex analytical expressions for the probability density (2.38) and the νth moments of the envelope can be successfully approximated by simpler expressions (2.55) and (2.57), respectively. The error of such approximation for the solution of a number of applied problems of statistical polarimetry is quite satisfactory.

Using expression (2.56), let us define the relation of the parameters m, Ω with the parameters α, ρ_0, a and φ_0 (and conversely) for particular probabilistic models described above (pp. 1–6 of Sect. 2.2.2). The specified relation must be known when moving from one distribution to another.

1. The Rayleigh model corresponds to the case when the parameters $\alpha = \rho_0 = a = \varphi_0$. It follows from the correlation (2.56) that $m = 1, \Omega = 2d^2, d^2_{2j-1} = d^2_{2j} = d^2$. Thus, the Rayleigh distribution corresponds to the particular case of the m-distribution;

$$W(E, m, \Omega)|_{m=1,\Omega=2d^2} \leftrightarrow W\left(E, d^2\right)|_{d^2=0.5\Omega}. \tag{2.58}$$

2. The Hoyt model, characterized by the parameter α for $\rho_0 = a = \varphi_0 = 0$, taking into account the Formula (2.56), is easily transformed into the m-distribution, which parameters are: $m = \left(1 + d^2\right)^{-1}$; $\Omega = d^2 = d_j^2 + d^2_{2j-1}$; $0.5 < m < 1$. For reverse transition from the m-distribution to the Hoyt distribution, the correlations $\alpha = \sqrt{(1-m)/m}$; $d^2 = \Omega$ shall be used. Thus, we have a transition scheme:

$$W(E, m, \Omega)|_{0.5<m<1,\Omega=d^2} \leftrightarrow W\left(E, \alpha, d^2\right)|_{-1<\alpha<1,d^2=\Omega}. \tag{2.59}$$

3. The Rice model (2.48) after the transition to the non-normalized envelope $E = Ad$, as follows from the Formula (2.56) when $\alpha = \rho_0 = 0$, is converted to an m-distribution, with parameters $m = \left(1 + a^2\right)^2/\left(1 + 2a^2\right)$; $\Omega = 2\left(1 + a^2\right)/d^2$; $1 < m < \infty$ where $d^2 = d^2_{2j} = d^2_{2j-1}$.
   To move from the m-distribution to the Rice distribution, the correlations $E_0 = \frac{\Omega}{m}\sqrt{m^2 - m}$, $d^2 = \frac{\Omega}{m}\left(m - \sqrt{m^2 - m}\right)$ shall be used, whence it follows that the parameter a is associated with the parameter m by the formula $a^2 = \sqrt{m-1}/(\sqrt{m} - \sqrt{m-1})$. Thus, the mutual transition from one distribution to another can be represented as:

$$W(E, m, \Omega)|_{1<m<\infty} \leftrightarrow W\left(E, a, d^2\right)|_{0<a<\infty}. \tag{2.60}$$

4. If in the previous cases the transition was made from one type of distribution to another, having the same number of parameters with the first, then for the approximation of the generalized Hoyt distribution by the m-distribution problem is reduced to the transition from the distribution, depending on the three parameters (α, a_0, φ_0) to the distribution, depending only on two parameters (m, Ω). Indeed, assuming in the Formula (2.56) $\rho_0 = 0$, we obtain:

$$\left.\begin{array}{l} m = \dfrac{\left(1 + a^2\right)^2}{\left(1 + 2a^2\right) + \alpha^2 + 2\alpha a^2 \cos 2\varphi_0} \\ \Omega = \left(1 + a^2\right)d^2 \quad ; \quad 0.5 < m < \infty \end{array}\right\}. \tag{2.61}$$

These correlations do not allow to determine the inverse functions $\alpha = f_1(m, \Omega)$, $a = f_2(m, \Omega)$, $\varphi_0 = f_3(m, \Omega)$ unambiguously and, consequently, to make the transition from the probability density $W(E, m, \Omega)$ to the probability density $W(E, \alpha, a, \varphi_0)$.

Therefore, the transition from the generalized Hoyt distribution to the m-distribution is one way, that is

$$W\big(E, d^2, \alpha, a, \varphi_0\big)\Big|_{\substack{-1 < \alpha < 1 \\ 0 < a < \infty \\ 0 \le \varphi_0 < 2\pi}} \leftrightarrow W(E, m, \Omega)|_{0.5 < m < \infty}. \tag{2.62}$$

5. The distribution of the modulus of the ψ-vector (or ψ-distribution), characterized by the parameter ρ_0 for $\alpha = a = \varphi_0 = 0$, taking into account the Formula (2.56) can easily be approximated by the m-distribution, the parameters of which are $m = \left(1 + \rho_0^2\right)^{-1}$; $\Omega = 2d^2, d_{2j}^2 = d_{2j-1}^2 = d^2; 0.5 < m < 1$. From these correlations, follow an inverse relation $\rho_0 = \sqrt{(1 - m)/m}, d^2 = 0.5\Omega$.
Thus, the transition from ψ-distribution to the m-distribution and conversely can be represented in the form:

$$W(E, m, \Omega)\Big|_{\substack{0.5 < m < 1 \\ \Omega = 2d^2}} \leftrightarrow W\big(E, d^2, \rho_0\big)\big|_{0 < \rho_0 < 1} \tag{2.63}$$

6. The density of the envelope probability (2.53), characterized by the parameters α and ρ_0 $(a = 0, \varphi_0 = 0)$, goes into m-distribution, and the parameters of which in accordance with the correlation (2.56) are determined as $m = \left[1 + \alpha^2 + \rho_0^2\left(1 - \alpha^2\right)\right]^{-1}$; $\Omega = d^2 = d_{2j}^2 + d_{2j-1}^2$. It follows from these equations that the parameter m changes within limits of $0.5 < m < 1$ and an unequivocal transition is possible only in one direction:

$$W\left(E, d^2, \alpha, \rho_0\right)\Big|_{\substack{-1 < \alpha < 1 \\ 0 < \rho_0 < 1}} \leftrightarrow W(E, m, \Omega)|_{\substack{0.5 < m < 1 \\ \Omega = 2d^2}}. \tag{2.64}$$

The generalized results of the considered approximation of the probability models of the envelopes of the orthogonal polarized components of the wave by the m-distribution with regard to Fig. 2.8 are presented in Table 2.2. Here, without breaking the generality, the phase of the deterministic component φ_0 is assumed to be zero.

When approximating the correlated envelopes of orthogonally polarized wave components by m-distribution, their joint probability density can be represented by the expression:

$$W_2(E_1, E_2, m, \Omega_i, R_2) = (1 - R_2) \sum_{n=0}^{\infty} \frac{(m)_n}{n!} R_2^n W(E_1) W(E_2). \tag{2.65}$$

Table 2.2 Approximation of probabilistic models

Type of distribution	Notations in Fig. 2.8	Distribution parameters	Parameter Ω	Parameter m	Limits of the parameter change m
The Rayleigh distribution	Point A	$d^2 = d_{2j-1}^2 = d_{2j}^2$	$\Omega = 2d^2$	–	$m = 1$
The generalized Rayleigh distribution	Edge AA'	a $d^2 = d_{2j-1}^2 = d_{2j}^2$	$\Omega = 2d^2(1 + a^2)$	$m = \frac{(1+a^2)^2}{1+2a^2}$	$1 < m < \infty$
Hoyt distribution	Edge AB	α $d^2 = d_{2j-1}^2 = d_{2j}^2$	$\Omega = d^2$	$m = \frac{1}{1+\alpha^2}$	$0.5 < m < 1$
The generalized distribution of Hoyt	Edge $AA'B'B$	α, a $d^2 = d_{2j-1}^2 = d_{2j}^2$	$\Omega = d^2(1 + a^2)$	$m = \frac{(1+a^2)^2}{1+2a^2(1+\alpha)+\alpha^2}$	$0.5 < m < \infty$
ψ-distribution	Edge AD	ρ_0 $d^2 = d_{2j-1}^2 = d_{2j}^2$	$\Omega = 2d^2$	$m = \frac{1}{1+\rho_0^2}$	$0.5 < m < 1$
Generalized ψ-distribution	Edge $AA'ДД$	$\rho_0, \quad a$ $d^2 = d_{2j-1}^2 = d_{2j}^2$	$\Omega = 2d^2(1 + a^2)$	$m = \frac{(1+a^2)^2}{1+2a^2+\rho_0^2}$	$0.5 < m < \infty$
The distribution $W(E, \alpha, \rho_0)$, Formula (2.53)	*Edge* $ABCD$	$\alpha, \quad \rho_0$ $d^2 = d_{2j-1}^2 = d_{2j}^2$	$\Omega = d^2$	$m = \frac{1}{1+\rho_0(1-\alpha^2)+\alpha^2}$	$0.5 < m < 1$

(continued)

Table 2.2 (continued)

Type of distribution	Notations in Fig. 2.8	Distribution parameters	Parameter Ω	Parameter m	Limits of the parameter change m
The generalized distribution, the Formula (2.75) with $\rho_0 = \text{const}$	Section planes, parallel $AA'B'B$	$\alpha, \quad a$ $d^2 = d_{2j-1}^2 = d_{2j}^2$ $\rho_0 = \text{const}$	$\Omega = d^2(1 + a^2)$	The Formula (2.56) $\varphi_0 = 0; \quad \rho_0 = \text{const}$	$0.5 < m < \infty$
The generalized distribution, the Formula (2.54) with $\alpha = \text{const}$	Section planes, parallel $AA'DD'$	$\rho_0, \quad a$ $d^2 = d_{2j-1}^2 = d_{2j}^2$ $\alpha = \text{const}$	$\Omega = d^2(1 + a^2)$	The Formula (2.56) $\varphi_0 = 0; \quad \alpha = \text{const}$	$0.5 < m < \infty$
The generalized distribution, the Formula (2.54) with $\alpha = \text{const}$	Section planes, parallel $ABCD$	$\alpha, \quad \rho_0$ $d^2 = d_{2j-1}^2 = d_{2j}^2$ $\alpha = \text{const}$	$\Omega = d^2(1 + a^2)$	The Formula (2.56) with $\varphi_0 = 0; \quad a = \text{const}$	$0.5 < m < \infty$
The generalized distribution, the Formula (2.54)	Объем V	a, α, ρ_0 $d^2 = d_{2j-1}^2 = d_{2j}^2$	$\Omega = d^2(1 + a^2)$	Formula (2.56) with $\varphi_0 = 0$	$0.5 < m < \infty$
One-dimensional normal distribution	Points B, C, D	$\alpha = 1, \quad \rho_0 = 1$ $d^2 = d_{2j-1}^2$	$\Omega = d_{2j-1}^2$	–	$m = 0.5$

where $(m)_n = m(m+1)\cdots(m+n-1)$; $m_1 = m_2 = m$; R—the coefficient of mutual correlation between the squares of envelopes E_1 and E_2, and the probabilities $W(E_1)$ and $W(E_2)$ density follow the m-distribution with the parameters $m' = m+n$; $\Omega_i' = (1 - R_2)\Omega_i$; $i = 1, 2$.

If the envelopes E_1 and E_2 are not correlated and $m_1 \neq m_2$, then according to the Formula (2.55) can be written:

$$W(E_1, E_2, m_1, m_2, \Omega_1, \Omega_2) = \frac{4}{G(m_1)G(m_2)}\left(\frac{m_1}{\Omega_1}\right)^{m_1}\left(\frac{m_2}{\Omega_2}\right)^{m_2} E_1^{2m_2-1}E_2^{2m_2-1}\exp\left(-\frac{m_1E_1^2}{\Omega_1} - \frac{m_1E_2^2}{\Omega_1}\right). \quad (2.66)$$

Thus, the m-distribution is a convenient form of approximation, which allows describing the probabilistic models of the orthogonally polarized components of partially polarized waves in respect of different by physical nature sources of radiation, reflection, or scattering.

For some types of sources of scattering or reflection of radio waves, the characteristic probabilistic model of the envelopes of orthogonally polarized components is the lognormal distribution, which can be explained by the m-distribution only in the particular case when its parameters m and Ω slowly change in lognormal law [3]. However, in the common case, contrary to the envelopes probability densities described in Sect. 2.2.2, mentioned law is very poorly approximated by the m-distribution and therefore the statistical characteristics of the polarization parameters of partially polarized radio waves, the envelopes of which are subjected to the lognormal distribution:

$$W\left(E_j, \mu_j, d_j^2\right) = \frac{1}{\sqrt{2\pi}d_jE_j}\exp\left[-\frac{\left(\ln E_j - \mu_j\right)^2}{2d_j^2}\right], \quad (2.67)$$

where $j = 1, 2$; $\mu_j = \overline{\ln E_j}$; $d_j^2 = \overline{\left(\ln E_j - \mu_j\right)^2}$—the mean value and the envelope variance, accordingly, cannot be obtained by the probabilistic models described in 2.2.2.

Probabilities' density (2.67), as shown by theoretical and experimental studies, is used in the description of the envelopes of radio waves of different origin. For example, the propagation of very-high-frequency radio waves in the city, when there is a linear dependence of the field at a given point on the field strengths of closely spaced scatterers; when the radio waves are scattered by ionized meteor tracks and atmospheric formations, etc. In radio detection, the lognormal distribution of the envelopes of orthogonally polarized components of radio waves is typical for the underlying ground and marine surfaces.

The physical nature of the mechanism of radio wave scattering leads to lognormal distributions of the envelopes of radio waves when they propagate in inhomogeneous environments (e.g., due to the presence of ionospheric meteor tracks or atmospheric noises). As for the radar detection, the physical nature of the mechanism of reflection of radio waves, leading to a lognormal distribution of envelopes, is not fully understood. One of the possible explanations is given, where the radar signal is considered as a signal reflected from one plane selected from a certain set and having irregularities, the depth of which is normally distributed.

The joint density of the probabilities of the envelope of orthogonally polarized components of the radio wave in the case under consideration can be determined as follows. Using the joint probability density of two normal random variables $Z_1 = \ln E_1$ and $Z_2 = \ln E_2$, it is easy to find the desired probability density in an arbitrary polarization basis:

$$W_2\left(E_1, E_2, \mu_j, d_j^2, \rho_{12}\right) = \frac{1}{2\pi d_1 d_2 E_1 E_2 \sqrt{1-\rho_{12}}} \exp\left\{-\frac{1}{2(1-\rho_{12}^2)} \times \left[\frac{(\ln E_1 - \mu_1)^2}{d_1^2} + \frac{(\ln E_2 - \mu_2)^2}{d_2^2} - \frac{2\rho_{12}}{d_1 d_2}(\ln E_1 - \mu_1)(\ln E_2 - \mu_2)\right]\right\}, \tag{2.68}$$

where d_j^2, μ_j^2—the variances and mean values of the envelopes' logarithms, $\rho_{12} = \overline{(\ln E_1 - \mu_1)(\ln E_2 - \mu_2)}$—the coefficient of mutual correlation between the logarithms of the envelope components of the radio wave, $j = 1, 2$.

Using the expression for νth initial moments of the envelopes, subject to the law (2.67):

$$m_E^{(\nu)}(\mu, d) = \exp\left(0.5\nu^2 d^2 + \nu\,\mu\right), \tag{2.69}$$

it is easy to determine expressions for the mean value and variance directly using the statistical characteristics of the envelopes:

$$\left.\begin{aligned} \mu_j\left(m_{E_j}, d_{E_j}^2\right) &= \ln\left[m_{E_j}\left(1 + d_{E_j}^2 m_{E_j}^{-1}\right)^{-0.5}\right] \\ d_j^2\left(m_{E_j}, d_{E_j}^2\right) &= \ln\left[1 + d_{E_j}^2 m_{E_j}^{-2}\right]. \end{aligned}\right\} \tag{2.70}$$

Therein, the mutual correlation coefficient is:

$$\rho_{12}\left(m_{E_j}, d_{E_j}^2, \rho_{E_1 E_2}\right) = \frac{\ln\left(1 + \rho_{E_1 E_2} d_{E_1} d_{E_2} / m_{E_1 E_2}\right)}{\sqrt{\ln\left(1 + d_{E_1}^2 / m_{E_1}^2 / m_{E_1}^2\right) \ln\left(1 + d_{E_2}^2 / m_{E_2}^2\right)}}, \tag{2.71}$$

where $\rho_{E_1 E_2}$—the mutual correlation coefficient between the envelopes of orthogonally polarized wave components.

Thus, the lognormal distribution is a useful form of approximation of the envelope of orthogonally polarized components and can be used in the statistical analysis of partially polarized radio waves.

2.2.4 Probabilistic Description—Non-Gaussian Case

In p. 2.2.2, as a model of a partially polarized wave, the case was considered when each of the quadrature components x_j was the sum of a sufficiently large number of elementary signals, which allowed, based on the central limit theorem, to consider a four-dimensional distribution law $W(\vec{X})$ defined by the Formula (2.22), as Gaussian.

Consider the situation when on the input of the receiving antenna at the same time comes N such PPW independent from each other. In this case, let N be insufficiently large. In this case, the power per each of the orthogonal components will be proportional to the sum of $2N$ summands, each of which is a square of the Gaussian random variable, i.e.,

$$E_j^2 = \sum_{k=1}^{N} \left[\left(x_{2j-1}^2\right)_k + \left(x_{2j}^2\right)_k \right], \tag{2.72}$$

where $j = 1, 2$ is specified.

As it is known, the sum of squares of independent Gaussian random variables (2.72 type) with the mathematical expectation not equal to zero is subjected to non-central χ^2-distribution. This allows to record the distribution density for each of the $\sum_{k=1}^{N} \left(x_{2j-1}^2\right)_k = X_{2j-1}^2$ and $\sum_{t=1}^{N} \left(x_{2j}^2\right)_k = X_{2j}^2$ sums in the form:

$$W\left(Z_j^2\right) = \left(\frac{Z_{0j}^2}{4d_j^2}\right) \exp\left(-\frac{Z_{0j}^2}{4d_j^2}\right) \sum_{p=0}^{\infty} \frac{W_0\left(Z_j^2\right)}{p!}, \tag{2.73}$$

where $Z_{0j}^2 = \overline{Z_j^2}$, by Z_j^2 is meant X_{2j-1}^2 or X_{2j}^2,

$$d_j^2 = N\left(d_{2j-1}^2\right)_k = N\left(d_{2j}^2\right)_k = N\left[\overline{\left(Z_{2j-1}^2\right)_k} - \overline{\left(Z_{2j-1}^2\right)_k}^{2}\right] = N\left[\overline{\left(z_{2j}^2\right)_k} - \overline{\left(Z_{2j}^2\right)_k}^{2}\right].$$

The function included in the Formula (2.73) is a central χ^2-distribution with (N + $2p$) degrees of freedom:

$$W_0\left(Z_{0j}^2\right) = \frac{2^{-(p+0.5N)}\left(Z_j^2\right)^{0.5N+p+1}}{G(p+0.5N)d_j^{2p+N}} \exp\left(-\frac{Z_j^2}{2d_j^2}\right). \tag{2.74}$$

Using the equation $I_m(X) = \sum_{p=0}^{\infty} \frac{1}{p!G(m+p+1)}\left(\frac{X}{2}\right)^{m+2p}$ and the transition rule from variable Z_j^2 to Z_j, we obtain:

$$W_0(Z_j) = \frac{2^{0.25(N-6)}\left|Z_j\right|^{0.5N}}{d_j^2 Z_{0j}^{0.5N-1}} \exp\left[-\frac{1}{2}\left(\frac{Z_{0j}^2}{2d_j^2} + \frac{Z_j^2}{d_j^2}\right)\right] I_{0.5N-1}\left(\frac{Z_{0j}\left|Z_j\right|}{\sqrt{2}d_j^2}\right). \tag{2.75}$$

As it can be seen from the Formula (2.75), the distribution of the quadrature components of orthogonally polarized components of the signal X_{2j-1}, X_{2j} is non-Gaussian; therefore, the orthogonally polarized components themselves will be $E_j(t)$. Note that the properties of the function (2.75) do not change if the parameter N takes non-integer values, i.e., if $n = \alpha > 0$.

Let us introduce new parameters $\gamma_{Z_j} = Z_{0j}/d_j^2$; $\beta_j = 1/d_j^2$, $N = \alpha$. In the foregoing, we will not write the index j ($j = 1, 2$), i.e., assume $\gamma_{Z_j} \equiv \gamma_Z$; $\beta_j \equiv \beta$, etc. In this case, instead of the Formula (2.75) we will have:

$$W(Z) = 2^{0.25(\alpha-6)}|Z|^{0.5\alpha}\gamma_Z^{1-0.5\alpha}\exp\left[-\frac{1}{2}\left(\frac{\gamma_Z^2}{2\beta} + \beta Z^2\right)\right]I_{0.5\alpha-1}\left(\frac{\gamma_Z|Z|}{\sqrt{2}}\right). \tag{2.76}$$

It is not difficult to find the probability density of the envelope E and the phase φ of the orthogonally polarized component of the considered PPW.

For this purpose, following up the independence of the quadrature components, we define taking into account Eq. (2.76) and their joint distribution $W\left(X_{2j-1}, X_{2j}\right) \equiv W(x, y) = W(x)W(y)$, and then move it to the new variables $x = E\cos\varphi$, $y = E\sin\varphi$. (For convenience of writing $X_{2j-1} \equiv x$, $X_{2j} \equiv y$ is specified.) As a result of elementary transformations, we obtain:

$$\begin{aligned} W(E,\varphi) = {} & \frac{2^{0.5(\alpha-6)}\beta^{\alpha}E}{\left(\gamma_x\gamma_y\right)^{0.5\alpha-1}}\exp\left(-\frac{\gamma_x^2+\gamma_y^2}{4\beta} - \frac{\beta}{2}E^2\right)|E\cos\varphi|^{\alpha} \\ & \times |E\sin\varphi|^{\alpha} I_{0.5\alpha-1}\left(\frac{\gamma_x|E\cos\varphi|}{\sqrt{2}}\right)I_{0.5\alpha-1}\left(\frac{\gamma_y|E\sin\varphi|}{\sqrt{2}}\right). \end{aligned} \tag{2.77}$$

The expression (2.77) describes the joint density of the distribution of the envelope and phase of the orthogonally polarized component probabilities of the considered PPW, generalizes the previously known particular cases, and is initial for the determination of one-dimensional $W(E)$ and $W(\varphi)$ densities.

By integrating Eq. (2.77) on eliminating variables and assuming for simplicity $\gamma_x = \gamma_y = \gamma$, we obtain:

$$\left.\begin{aligned} W(E) &= \frac{\beta^{\alpha}E}{\gamma^{\alpha-1}}\exp\left[-\frac{1}{2}\left(\frac{\gamma^2}{\beta} + \beta E^2\right)\right]I_{\alpha-1}(\gamma E) \\ W(\varphi) &= 0.5\sum_{k=0}^{\infty}\sum_{p=0}^{\infty}H_{kp}|\sin\varphi|^{\alpha+2p-1}|\cos\varphi|^{\alpha+2k-1} \end{aligned}\right\}, \tag{2.78}$$

where $H_{kp} = \frac{(0.5\gamma)^{2k+2p}\exp\left(-0.5\gamma^2/\beta\right)}{\beta^{k+p}k!p!B(0.5\alpha+k, 0.5\alpha+p)}$, $B(a, b) = \Gamma(a)\Gamma(b)/[\Gamma(a + b)]$—is beta function.

As follows from the Formula (2.78), the envelope E and phase φ are statistically dependent, and the phase distribution is unequiprobable.

Included in the Formula (2.78), the parameter α considering the number of resulting PPW characterizes the depth of the fluctuations of the orthogonal polarized components of the wave. The parameter γ describes the ratio of the deterministic component to its variance, and the parameter β is inversely proportional to the variance.

In practice, an important special case relates to a situation where mathematical expectations $\overline{Z_{2j-1}} = \overline{Z_{2j}} = 0$. The formulas (2.77) and (2.78) will be significantly simplified and take the form:

$$\left.\begin{aligned} W(E_0, \varphi_0) &= \left(\frac{\beta_0}{2}\right)^{\alpha_0} \frac{E^{2\alpha_0 - 1}|\sin 2\varphi_0|^{\alpha_0 - 1}\exp\left(-0.5\beta_0 E_0^2\right)}{2^{\alpha_0 - 1}G^2(0.5\alpha_0)} \\ W(E_0) &= 2(0.5\beta_0)^{\alpha_0} G^{-1}(\alpha_0)\exp\left(-0.5\beta_0 E_0^2\right) \\ W(\varphi_0) &= 2^{\alpha_0}\Gamma(\alpha_0)G^{-2}(0.5\alpha_0)|\sin 2\varphi_0|^{\alpha_0 - 1} \end{aligned}\right\} \quad (2.79)$$

In the Formula (2.79), to emphasize the fact that they relate to the particular case under consideration, all letters have a zero index.

Figure 2.9 represents the families of probability density curves $W(E_0)$ expressed by the Formula (2.79) for different values of the parameters α_0 and β_0. It is seen that these parameters significantly affect the shape of the curves and the numerical characteristics of the envelope E_0.

The considered probabilistic models of orthogonally polarized components of non-Gaussian partially polarized waves are of basic character, and it is not difficult to obtain different probabilistic models as particular cases from them. The distribution parameters (2.77) and (2.78) or (2.79) need to be assigned certain values to do this. For example, the following distributions follow from Formulae (2.78) and (2.79):

- Rayleigh for $\alpha = 1, \beta = d^{-2}, \gamma = 0$;
- Rice for $\alpha = 1, \beta = d^{-2}, \gamma = E_q d^{-2}$
- unidominant + Rayleigh for $\alpha = 2, \beta = 4/\overline{E^2}, \gamma = 0$;
- Maxwell–Boltzmann, for $\alpha = 1.5, \beta = d^{-2}, \gamma = 0$;
- Nakagami (m-distribution), for $\alpha = m \geq 0.5$; $\beta = 2m/\Omega$; $\Omega = \overline{E^2}, \gamma = 0$ and others.

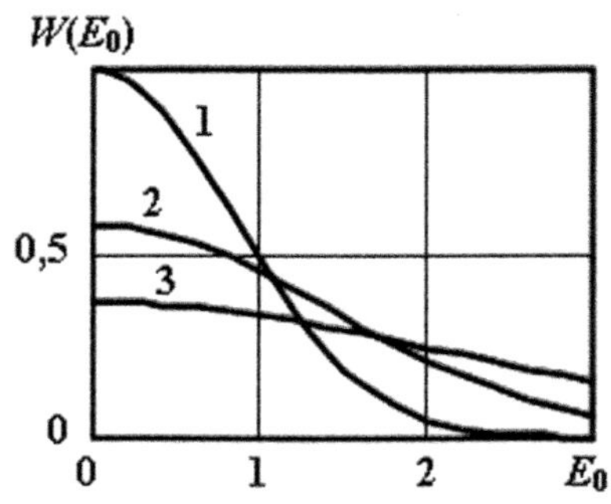

Fig. 2.9 Probability distribution density $W(E_0)$. $\alpha_0 = 0.5$; 1—$\beta_0 = 0.2$; 2—$\beta_0 = 0.5$; 3—$\beta_0 = 1.5$

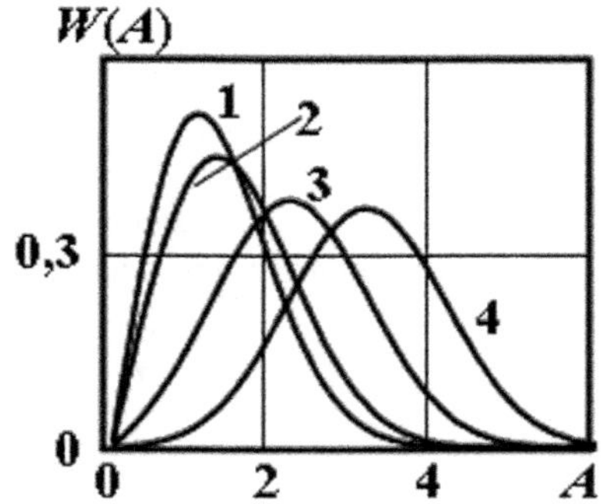

Fig. 2.10 Families of curves $W(A)$. 1—$\alpha_0 = 1, a = 1$; 2—$\alpha_0 = 1, a = 0.5$; 3—$\alpha_0 = 1, a = 2$; 4—$\alpha_0 = 1, a = 3$

Here, $E_q = \sqrt{x_0^2 + y_0^2}$ is the determined amplitude of the signal; d^2 is the variance of the signal or disturbance.

As already mentioned, any PPW can be unequivocally decomposed into the sum of polarized and non-polarized waves. The last exactly corresponds to just consider a particular case.

Let us present the first of the Formula (2.78) in such a way that the parameters relating to the polarized component appear in it explicitly. Obviously, the equations $\beta = \beta_0,\ \alpha = \alpha_0,\ \gamma = \beta_0 E_P$ where E_P is the envelope of the polarized component will be valid.

Thus, we will have the following representation for the conditional probability density distribution of the envelope of the orthogonal component E of the mixture of polarized and non-polarized waves corresponding envelope of which is distributed by law (2.79);

$$W(E|E_P) = \beta_0 E^{\alpha_0 - 1} E_P^{1-\alpha_0} \exp\left[-0.5\left(E^2 + E_P^2\right)\right] I_{\alpha_0 - 1}(\beta_0 E_P E). \tag{2.80}$$

Figure 2.10 represents the families of curves $W(E|E_P)$ expressed by the Formula (2.80) at different values $\alpha_0,\ \beta_0,$ and E_P.

The Formula (2.80) can be somewhat simplified by introducing non-dimensional values $A = \sqrt{\beta_0}E$ and $a = \sqrt{\beta_0}E_P$:

$$W(A) = a^{1-\alpha_0} A^{\alpha_0} \exp\lfloor -0.5\left(a^2 + A^2\right)\rfloor I_{\alpha_0 - 1}(aA). \tag{2.81}$$

Direct calculations allow to determine the moments of the νth order, which can be represented in the final form:

$$\begin{aligned} m_A^\nu &= 2^{0.5\nu} G\binom{\alpha_0 + 0.5\nu}{\alpha_0} \exp\left(-0.5a^2\right) {}_1F_1\left(\alpha_0 + 0.5\nu; 0.5a^2\right) \\ &= 2^{0.5\nu} G\binom{\alpha_0 + 0.5\nu}{\alpha_0} {}_1F_1\left(-0.5\alpha_0; -0.5a^2\right). \end{aligned} \tag{2.82}$$

The considered probabilistic characteristics can be generalized as particular cases to other types of probabilistic models by varying the values of the α_0 and β_0 parameters. Summary information is presented in Table 2.3, where $F(A)$ is the distribution function, and $\gamma(x, y)$ is the incomplete gamma function.

If the antenna input receives an additive mixture of two waves, orthogonal components of which are non-Gaussian signals with the corresponding densities of the probabilities of the form (2.78) and (2.79), then to determine the density of the envelope probabilities of this mixture, the usual Bayes rule can be used:

$$W(E) = \int\limits_0^\infty W(E|E_P)W(E_P)\mathrm{d}E_P. \tag{2.83}$$

Substituting the expressions (2.78) and (2.80) into the Formula (2.83) and performing the integration, we obtain:

$$\begin{aligned} W(E) &= \frac{2\beta^\alpha}{G(\alpha_0)}\left(\frac{\beta_0}{2}\right)^{\alpha_0}\frac{E^{2\alpha_0-1}}{(\beta+\beta_0)^\alpha}\exp\left[-\frac{1}{2}\left(\frac{\gamma^2}{\beta}+\beta_0E^2\right)\right] \\ &\times\sum_{k=0}^\infty\left\{\frac{\gamma^{2k}}{k![2(\beta+\beta_0)]^k}{}_1F_1\left[k+\alpha,\alpha_0,\frac{\beta_0^2E^2}{2(\beta+\beta_0)}\right]\right\}. \end{aligned} \tag{2.84}$$

As it is seen, the required probability density is determined by five parameters $(\alpha, \beta, \gamma, \alpha_0, \beta_0)$. The first three characterize the statistical properties of one wave, and the second—the other.

In the absence of deterministic components in the components of the first wave $(\gamma = 0)$, the expression (2.84) is significantly simplified and takes the form:

$$W(E) = \frac{2\beta^\alpha}{G(\alpha_0)}\left(\frac{\beta_0}{2}\right)^{\alpha_0}\frac{E^{2\alpha_0-1}}{(\beta+\beta_0)^\alpha}\exp\left(-0.5\beta_0E^2\right){}_1F_1\left[\alpha,\alpha_0;\frac{\beta_0^2E^2}{2(\beta+\beta_0)}\right]. \tag{2.85}$$

If n case $\alpha = \alpha_0$, we have a very simple function:

$$W(E) = \frac{2}{G(\alpha)}\left(\frac{\beta\beta_0}{2(\beta+\beta_0)}\right)^{\alpha_0}E^{2\alpha-1}\exp\left[-\frac{\beta\beta_0^2E^2}{2(\beta+\beta_0)}\right]. \tag{2.86}$$

It can be shown that for $\beta_0 \to \infty$ $\left(d_0^2 \to 0\right)$ expressions (2.84) and (2.85), they pass into Formulae (2.78) and (2.79) to replaced E by E_0 in the last.

An analysis of Formula (2.86) shows that the change parameters $\alpha_0, \alpha, \beta, \beta_0$ and γ lead to the transformation of the density distribution of the probabilities of the envelopes of orthogonally polarized components of the additive mixture under consideration: the shape, scale, and distribution center of the envelope of the mixture change.

Table 2.3 Probabilistic models

Statistical characteristics		$W(A)$	m_A^ν	$F(A)$
1	The generalized distribution (2.79), $\alpha_0 > 0;\ \beta_0 > 0$	Formula (2.81)	Formula (2.82)	
2	Single-dominant + Rayleigh distribution $\alpha_0 = 2;\ \beta_0 = 4/\overline{E_0^2}$	$\frac{A^2}{a}\exp\left[-\frac{a^2+A^2}{2}\right] \times I_1(aA)$	$2^{\frac{\nu}{2}}\Gamma\left(2+\frac{\nu}{2}\right) \times {}_1F_1\left(-\frac{\nu}{2}, 2, \frac{a^2}{2}\right)$	$e^{-\frac{a^2}{2}}\sum_{k=0}^{\infty}\frac{\Gamma^{-1}(2+k)}{k!}\left(\frac{a^2}{2}\right)^k \times \gamma\left(2+k, \frac{A^2}{2}\right)$
3	The Maxwell–Boltzmann Distribution $\alpha_0 = 1.5;\ \beta_0 = 1/\alpha_0^2$	$\sqrt{\frac{2}{\pi}\frac{A}{a}}\exp\left[-\frac{a^2+A^2}{2}\right] \times$ ch(aA)	$2^{\frac{\nu}{2}}\Gamma\begin{pmatrix}1.5+0.5\nu\\1.5\end{pmatrix} \times {}_1F_1\left(-\frac{\nu}{2}, \frac{3}{2}, \frac{a^2}{2}\right)$	$e^{-\frac{a^2}{2}}\sum_{k=0}^{\infty}\frac{\Gamma^{-1}\left(\frac{3}{2}+k\right)}{k!}\left(\frac{a^2}{2}\right)^k \times \gamma\left(\frac{3}{2}+k, \frac{A^2}{2}\right)$
4	The Nakagami distribution of $\alpha_0 = m_0;\ \beta_0 = 2m_0/\Omega_0$ $\Omega_0 = \overline{E_0^2};\ m_0 \geq 0.5$	$\frac{A^{m_n}}{a^{m_n-1}} \times \exp\left[-\frac{a^2+A^2}{2}\right] \times$ $I_{m_n-1}(aA)$	$2^{\frac{\nu}{2}}\Gamma\begin{pmatrix}m_n+0.5\nu\\m_n\end{pmatrix} \times {}_1F_1\left(-\frac{\nu}{2}, m_n, \frac{a^2}{2}\right)$	$e^{-\frac{a^2}{2}}\sum_{k=0}^{\infty}\frac{\Gamma^{-1}(m_n+k)}{k!}\left(\frac{a^2}{2}\right)^k \times \gamma\left(m_n+k, \frac{A^2}{2}\right)$
5	The Rayleigh distribution $\alpha_0 = 1;\ \beta_0 = 1/\alpha_0^2$	$A\exp\left[-\frac{a^2+A^2}{2}\right]I_0(aA)$	$2^{\frac{\nu}{2}}\Gamma\left(1+\frac{\nu}{2}\right) \times {}_1F_1\left(-\frac{\nu}{2}, 1, \frac{a^2}{2}\right)$	$e^{-\frac{a^2}{2}}\sum_{k=0}^{\infty}\frac{1}{k!}\left(\frac{a^2}{2}\right)^k \times \gamma\left(1+k, \frac{A^2}{2}\right)$

Let us introduce non-dimensional parameters $A = \sqrt{\beta_0}E, a = \sqrt{\beta_0}\gamma/\beta$, and then instead of the expression (2.84), we will have:

$$W(A) = \frac{2^{1-\alpha_0}A^{2\alpha_0-1}\beta^\alpha}{G(\alpha_0)(\beta+\beta_0)^\alpha}\exp\left[-\frac{1}{2\beta_0}\left(\beta a^2+\beta_0^2A^2\right)\right]\times\sum_{k=0}^{\infty}\left\{\frac{1}{k!}\left[\frac{a^2\beta^2}{2\beta_0(\beta+\beta_0)}\right]^k\right\}. \quad (2.87)$$

Direct calculations give the following expression for the νth initial moments:

$$m^\nu = \frac{G(0.5\nu+\alpha_0)}{G(\alpha_0)}\left(\frac{2}{\beta_0}\right)^{0.5\nu}\left(\frac{\beta}{\beta+\beta_0}\right)^\alpha\exp\left(-\frac{\gamma^2}{2\beta}\right)$$
$$\times\sum_{k=0}^{\infty}\left\{\frac{1}{k!}\left[\frac{a^2\beta^2}{2\beta_0(\beta+\beta_0)}\right]^k {}_2F_1\left(k+\alpha;\,0.5\nu+\alpha_0;\,\frac{\beta}{\beta+\beta_0}\right)\right\}. \quad (2.88)$$

The Formula (2.88) is significantly simplified in the absence of a deterministic component in the first wave ($\gamma = 0$ and $a = 0$):

$$m^\nu = \frac{G(0.5\nu+\alpha_0)}{G(\alpha_0)}\left(\frac{2}{\beta_0}\right)^{0.5\nu}\left(\frac{\beta}{\beta+\beta_0}\right)^\alpha {}_2F_1\left(\alpha;\,0.5\nu+\alpha_0;\,\alpha_0;\,\frac{\beta}{\beta+\beta_0}\right). \quad (2.89)$$

The representation (2.87) provides a distribution function for E in closed form:

$$F(E) = \left(\frac{\beta_0}{2}\right)^{\alpha_0}\frac{\beta^\alpha E^{2\alpha_0}}{(\beta+\beta_0)^\alpha}\exp\left[-\frac{1}{2}\left(\frac{\gamma^2}{2\beta}-\beta_0E^2\right)\right]$$
$$\times\sum_{k,n=0}^{\infty}\left\{\frac{\beta_0^k\gamma^{2k}(\beta_0E)^{2n}{}_1F_1\left(1,\,1+n+\alpha_0;\,0.5\beta_0E^2\right)}{2^{n+k}n!k!G(\alpha_0+n+1)(\beta+\beta_0)^{n+k}}\right\}. \quad (2.90)$$

The resulting expressions for the probability distribution density (2.84), νth initial moments (2.89), and the integral distribution functions (2.90) are basic and can serve as an initial in determining mentioned probabilistic characteristics of the orthogonally polarized components envelopes of the additive mixture of a wide class of combinations of both Gaussian and non-Gaussian probabilistic models of partially polarized radio waves.

To determine particular cases, it is sufficient to substitute certain values of the parameters $\alpha_0, \alpha, \beta, \beta_0$ and γ in the expressions obtained in accordance with Table 2.4. Thirty-five different combinations of probabilistic models of the envelopes of orthogonally polarized components of the considered radio waves are presented as examples in Table 2.4, for which the probabilistic characteristics of the envelope of the additive mixture can be determined. For certainty, the first radio wave, which has $\gamma \neq 0$, we will call a useful signal and provide a "c" index, and the second, which has $\gamma = 0$ a jamming signal, or just a signal, and provide an "o" index.

Let us look at some of these examples.

Table 2.4 Combinations of probabilistic models

	Model no.	1	2	3	4	5
№	$W(E_0)$ $\Rightarrow$					
	$W(E_C)$ $\Downarrow$	Generalized (2.79) $\alpha_0, \beta_0 > 0$ $\gamma_0 = 0$	Single-dominant + Rayleigh $\alpha_0 = 2, \gamma_0 = 0$ $\beta_0 = 4/\overline{E_0^2}$	Maxwell–Boltzmann $\alpha_0 = 1,5; \gamma_0 = 0$ $\beta_0 = 1/d_0^2$	Nakagami $\alpha_0 = m_0;$ $\gamma_0 = 0$ $\beta_0 = 2m_0/\Omega_0$ $\Omega_0 = \overline{E_0^2}$	Rayleigh $\alpha_0 = 1; \gamma_0 = 0$ $\beta_0 = 1/d_0^2$
1	Generalized (2.78) $\alpha_C, \beta_C > 0, \gamma_C > 0$	1-1	1-2	1-3	1-4	1-5
2	Generalized $\alpha_C, \beta_C > 0$ $\gamma_C = 0$	2-1	2-2	2-3	2-4	2-5
3	Single-dominant + Rayleigh $\alpha_C = 2, \gamma_C = 0$ $\beta_C = 4/\overline{E_C^2}$	3-1	3-2	3-3	3-4	3-5
4	Maxwell–Boltzmann $\alpha_C = 1.5$ $\gamma_C = 0$ $\beta_C = 1/d_C^2$	4-1	4-2	4-3	4-4	4-5

(continued)

Table 2.4 (continued)

	Model no.	1	2	3	4	5
№	$W(E_0)$ $\Rightarrow$					
	$W(E_C)$ $\Downarrow$	Generalized (2.79) $\alpha_0, \beta_0 > 0$ $\gamma_0 = 0$	Single-dominant + Rayleigh $\alpha_0 = 2, \gamma_0 = 0$ $\beta_0 = 4/\overline{E_0^2}$	Maxwell–Boltzmann $\alpha_0 = 1,5; \gamma_0 = 0$ $\beta_0 = 1/d_0^2$	Nakagami $\alpha_0 = m_0;$ $\gamma_0 = 0$ $\beta_0 = 2m_0/\Omega_0$ $\Omega_0 = \overline{E_0^2}$	Rayleigh $\alpha_0 = 1; \gamma_0 = 0$ $\beta_0 = 1/d_0^2$
5	Nakagami $\alpha_C = m_C; \gamma_C = 0$ $\beta_C = 2m_C/\Omega_C$ $\overline{\Omega_C} = \overline{E_C^2}$	5-1	5-2	5-3	5-4	5-5
6	Райса $\alpha_C = 1; \beta_C = \frac{1}{d_C^2}$ $\gamma_C = E_C/d_C^2$	6-1	6-2	6-3	6-4	6-5
7	Рэлея $\alpha_C = 1; \gamma_C = 0$ $\beta_C = 1/d_C^2$	7-1	7-2	7-3	7-4	7-5

Let the non-Gaussian interference envelope function $W(E_0)$ be described by the Nakagami distribution (m-distribution) that follows from the Formula (2.79) for $\alpha_0 = m_0 \geq 0.5$, $\beta_0 = 2m_0/\Omega_0$, and the signal is Gaussian with a Rayleigh envelope distribution (see Table 2.4, column 7-4). The last is obtained from the correlation (2.78) for $\alpha = \alpha_C = 1$, $\beta = \beta_C = 1/d^2$ and $\gamma_C = 0$.

Then according to the Formulae (2.86), (2.89) and (2.90), we have:

$$W(E) = \frac{2}{G(m_0)}\left(\frac{m_0}{\Omega_0}\right)^{m_0}\frac{E^{2m_0-1}}{\left(1+C_1^{-1}\right)}\exp\left(-\frac{m_0}{\Omega_0}E^2\right){}_1F_1\left[1, m_0, \frac{m_0E^2}{\Omega_0(1+C_1)}\right]$$

$$m^{\nu} = \frac{G(m_0+0.5\nu)}{G(m_0)}\left(\frac{\Omega_0}{m_0}\right)^{0.5\nu}\frac{{}_2F_1[1; 0.5\nu+m_0; 1; 1/(1+C_1)]}{\left(1+C_1^{-1}\right)}$$

$$F(E) = \left(\frac{m_0}{\Omega_0}\right)^{m_0}\frac{E^{2m_0}\exp\left(-\frac{m_0}{\Omega}E^2\right)}{\left(1+C_1^{-1}\right)}\sum_{n=0}^{\infty}\left\{\left[\frac{m_0E^2}{\Omega_0(1+C_1)}\right]^n{}_1F_1\left[1, m_0+n+1, \frac{m_0E^2}{\Omega_0}\right]\right.$$
$$\left.\times\frac{1}{G(m_0+n+1)}\right\}, \quad (2.91)$$

where $C_1 = \beta/\beta = \Omega_0/\left(2d_C^2m_C\right)$.

If $W(E_0)$ and $W(E_C)$ are described by m-distribution (see Table 2.4, column 5-4), then the probabilistic characteristics (2.86), (2.89), and (2.90) take the following form:

$$W(E) = \frac{2}{G(m_0)}\left(\frac{m_0}{\Omega_0}\right)^{m_0}\frac{E^{2m_0-1}\exp\left(-\frac{m_0}{\Omega_0}E^2\right)}{\left(1+C_2^{-1}\right)^{m_C}}{}_1F_1\left[m_C, m_0; \frac{m_0E^2}{\Omega_0(1+C_2)}\right]$$

$$m^{\nu} = \frac{G(m_0+0.5\nu)}{G(m_0)}\left(\frac{\Omega_0}{m_0}\right)^{0.5\nu}\frac{{}_2F_1\left[m_C; m_0+0.5\nu; m_0; (1+C_2)^{-1}\right]}{\left(1+C_2^{-1}\right)^{m_C}}$$

$$F(E) = \left(\frac{m_0}{\Omega_0}\right)^{m_0}\frac{E^{2m_0}\exp\left(-\frac{m_0}{\Omega}E^2\right)}{\left(1+C_2^{-1}\right)^{m_C}G(m_C)}\sum_{n=0}^{\infty}\frac{G(m_C+n)}{n!G(m_0+n+1)}\left[\frac{m_0E^2}{\Omega_0(1+C_2)}\right]^n$$
$$\times{}_1F_1\left[1, m_0+n+1, \frac{m_0E^2}{\Omega_0}\right]. \quad (2.92)$$

where $C_2 = m_C\Omega_0/(m_0\Omega_C)$.

If the interference is Gaussian with $W(E_0)$ described by the Rayleigh distribution and $W(E_C)$ of the non-Gaussian signal is determined by the generalized distribution (2.78) for $\gamma_C = 0$ (see Table 2.4, column 2-5), then from Eqs. (2.86), (2.89), and (2.90) follows:

$$W(E) = \frac{E\exp\left(-E^2/2d_0^2\right)}{d_0^2\left(1+C_3^{-1}\right)^{\alpha_C}}{}_1F_1\left[\alpha_C, 1; \frac{E^2}{2d_0^2(1+C_3)}\right]$$

$$m^{\nu} = \frac{G(1+0.5\nu)\left(2d_0^2\right)^{0.5\nu}}{\left(1+C_3^{-1}\right)^{\alpha_C}}2F_1\left[1; 1+0.5\nu; 1; (1+C_3)^{-1}\right]$$

$$F(E) = \frac{E^2 \exp(-E^2/2d_0^2)}{2d_0^2\left(1 + C_3^{-1}\right)^{\alpha_C} G(\alpha_C)^{n=0}} \sum_{n=0}^{\infty} \left\{ \frac{G(\alpha_C + n)}{n! G(n+2)} \left[\frac{E^2}{2d_0^2(1 + C_3)} \right]^n \right.$$
$$\left. \times\, {}_1F_1 \left[1, n+2, \frac{E^2}{2d_0^2(1 + C_3)} \right] \right\}. \quad (2.93)$$

where $C_3 = \beta_C d_0^2$.

If the interference and the signal are Gaussian with $W(E_0)$ and $W(E_C)$, complying to the Rayleigh distribution (see Table 2.4, column 7-5), then the Formulae (2.86), (2.89), and (2.90) imply:

$$\left.\begin{aligned} W(E) &= \frac{E}{d_C^2 + d_0^2} \exp\left[-\frac{E^2}{2\left(d_C^{2+} d_0^2\right)} \right] \\ m^{\nu} &= G(1 + 0.5\nu)\left[2\left(d_C^2 + d_{\Pi}^2\right)\right]^{0.5\nu} \\ F(E) &= 1 - \exp\left[0.5E^2/\left(d_C^2 + d_{\Pi}^2\right)\right] \end{aligned}\right\}. \quad (2.94)$$

It is possible to continue the number of examples studied for other combinations of probability models of the envelope of the useful signal and interference that occur in different geophysical and radiophysical problems.

In the writing Sect. 2.2.4 participated A. I. Mosionzhik.

2.3 Coordinate Components with Broadband Signals

So far, there was talk only about bandlimitedness and often also about monochromaticity of considered oscillations. Moreover, the temporary dependence of polarization components was the same—in their spectrum, there were harmonic oscillations of the same frequencies. However, in recent years, there is an increasing interest in broadband and ultra-broadband oscillations of the type of impulses of very short duration (of the order of nano- and picoseconds), Walsh functions, etc. Radio polarimetry of such waves, the authors suggest to consider in future books of the proposed monograph.

When considering these broadband oscillations, it is appropriate to describe the signal in time representation without the use of relevant Fourier transforms. A typical example demonstrating this appropriateness is a set of point reflectors arranged from each other on time intervals of long direct impulse duration. In this case, the analysis of the scattered signal by the reflectors from the point of view of the time structure will create prerequisites for the identification of the scatterer.

In the time representation, a polarized signal propagating along a direction $\vec{r}$ can be represented as $\vec{E}(\vec{r}, t) = E_x(r - ct)\vec{e}_x + E_y(r - ct)\vec{e}_y$, where the unit vectors $\vec{e}_x$ and $\vec{e}_y$, together with the unit vector $\vec{e}_z$ determining the direction of propagation $\vec{r}$, form the right Cartesian coordinate system.

Independence and arbitrariness in the selection of components E_x and E_y of the signal lead to the independence of the description of each component. It is possible to represent the signal, in which one component is the length of the noise fluctuations, and the other is a monochromatic oscillation, or the first of these is the impulses of short duration, and the second is the signal of a sawtooth configuration, etc. Remaining even in the framework of the linearly polarized basis, while turning it occur very inconvenient for the analysis of the function as a sum of two noise samples or a combination of noise and sine waves, etc. Thus, it is necessary to determine the constraints on the time structure of polarization components, i.e., to establish some functional relation between them:

$$E_y(t) = G[E_x(t)]. \tag{2.95}$$

A crucial role in determining this relation is played by the requirement of analyticity or hyperanalyticity in the isomorphic representation of the polarization of the signal by complex or quaternion functions.

The functional operator $G(\xi)$ shall meet a number of conditions to ensure continuity with the methods used for narrowband signals.

Condition 1 The operator shall be linear, i.e.,
$G\left[\sum_n C_n E_{xn}(t)\right] = \sum_n C_n G[E_{xn}(t)] = \sum_n C_n E_{yn}(t)$, where C_n are numbers belonging to the field of real or complex numbers, or the quaternion body, $E_{xn}(t)$ and $E_{yn}(t)$ are polarization components of the nth component of the signal.

Condition 2 Operator $G(\xi)$ correspondence between polarizing components shall be maintained when changing the polarization basis caused by the rotation of the coordinate system.

As it is known, the rotation of the coordinate system by an angle β_0 leads to the transformation (2.30) using the matrix (1.52).

The requirement of Condition 2 will then be presented as follows: $E_{yn}(t) = G[E_{xn}(t)] = G\lfloor E_{cx}(t)\cos\beta_0 + E_{cy}(t)\sin\beta_0\rfloor$. Using the principle of superposition (Condition 1), this equation is transformed to form:

$$E_y(t) = G[E_{cx}(t)]\cos\beta_0 + G\lfloor E_{cy}(t)\rfloor \sin\beta_0 = E_{cy}(t)\cos\beta_0 + G\lfloor E_{cy}(t)\rfloor \sin\beta_0.$$

Correspondence to Eq. (2.30) will be provided if we assume that the inverse isomorphic functional operator $G(\xi)$ is determined by the rule:

$$E_{cx}(t) = G^{-1}\left[E_{cy}(t)\right] = -G\left[E_{cy}(t)\right]. \tag{2.96}$$

Condition 3 The transformation $G(\xi)$ shall not change the time structure of the monochromatic oscillation $E_x(t) = E\ \cos(\omega t + \varphi)$, and it is only possible to allow the appearance of an additional phase shift Φ:

$$E_y(t) = G[E\ \cos(\omega t + \varphi)] = E\ \cos(\omega t + \varphi + \Phi). \tag{2.97}$$

This condition provides the inclusion of elliptically polarized monochromatic waves polarimetry as a particular case.

Condition 4 If we replace $E_x(t)$ with $E_x(t - t_0)$ (where t_0 is an arbitrary delay), then $E_y(t)$ shall experience the same delay: $E_y(t - t_0)$. This condition is equivalent to the commutativity of operator of differentiation $G(\xi)$:

$$\frac{\mathrm{d}}{\mathrm{d}t}G(\xi) = G\left[\frac{\mathrm{d}\xi}{\mathrm{d}t}\right]. \tag{2.98}$$

It follows from the theory of analytic functions that only Hilbert transform satisfies the conditions (1–4):

$$E_y(t) = \frac{1}{\pi}\int_{-\infty}^{\infty}\frac{E_x(\tau)}{t-\tau}\mathrm{d}\tau, \tag{2.99}$$

where the integral is understood in the sense of the Cauchy principal value. The inverse transformation is then written as:

$$E_x(t) = -\frac{1}{\pi}\int_{-\infty}^{\infty}\frac{E_y(\tau)}{t-\tau}\mathrm{d}\tau, \tag{2.100}$$

what is in full compliance with (2.96).

The Hilbert transform, for which we keep the former operator designation $G(\xi)$, leads to the fact that in Eq. (2.97), the phase shift Φ is equal to $(-\pi/2)$, so that the component $E_y(t)$ will be equal to:

$$E_y(t) = E\ \sin(\omega t + \varphi). \tag{2.101}$$

This component together with the component $E_x(t) = E\ \cos(\omega t + \varphi)$ determines, as already mentioned, the wave of the right circular polarization.

The consequence of the fact that the polarization components must be related by the Hilbert transform is the validity of two more conditions.

Condition 5 If the polarization component $E_x(t)$ can be represented in the form $E_x(t) = E_{xM}(t)E_{xB}(t)$, where the spectrum of the slow multiplier $E_{xM}(t)$ does not intersect with the spectrum of the fast multiplier $E_{xB}(t)$, then the following equation is true:

$$E_v(t) = G[E_{xM}(t)E_{xB}(t)] = E_{xM}(t)\Gamma[E_{xB}(t)] = E_{xM}(t)E_{vB}(t). \tag{2.102}$$

This condition ensures the existence of modulation for polarized waves.

Condition 6 If we make a complex function of polarization components related by Hilbert transformation:

$$E(t) = E_x(t) + jE_y(t), \tag{2.103}$$

then it will be analytical in the upper half of the complex plane. This condition is equivalent to the restriction on the spectrum of the function $E(t)$:

$$E(\omega) = F[E(t)] = 0, \text{ when } \omega < 0 \tag{2.104}$$

($F(\xi)$—operator notation for the Fourier transform).

The last condition allows to completely explain the physical meaning of the used Hilbert transform. The oscillation $E_x(t)$ can be matched by an expansion into a Fourier series:

$$E_x(t) = \sum_k \left(E_k^C \cos\omega_k t + E_k^S \sin\omega_k t\right), \tag{2.105}$$

where E_k^C and E_k^S are coefficients of expansion.

According to (2.97) and (2.101) this oscillation, you can immediately specify its Hilbertian conjugating:

$$E_y(t) = \sum_k \left(E_k^C \sin\omega_k t - E_k^S \sin\omega_k t\right). \tag{2.106}$$

Taking into account the mutual perpendicularity of the directions of oscillations (2.105) and (2.106), it can be concluded that each spectral component at the frequency ω_k corresponds to the harmonic right circularly polarized oscillations.

Thus, Condition 6 of analyticity means that in the spectrum of oscillation, whose linear polarization components are $E_x(t)$ and $E_y(t) = G[E_x(t)]$, there are only right circularly polarized components. This allows to apply the concept of clockwise polarized broadband oscillation and use complex writing (2.103) for its defining. It is also easy to verify that the products of these oscillations are also right circularly polarized oscillation. Accordingly, fluctuation:

$$E(t) = E_x(t) - jE_y(t) = E_x(t) - jG(E_x(t)) \tag{2.107}$$

shall be associated with the left circularly polarized wave.

The oscillation (2.107) defines a function analytical by the spectrum in the lower half of the complex semiplane. Its spectrum contains only the left-polarized harmonic oscillations, which is equivalent to the consideration of negative frequencies (since there is equation $\cos[(-\omega)t] + j\,\sin[(-\omega)t] = \cos\omega t - j\,\sin\omega t$).

All of the above allows formulating the basic concepts for broadband signals in the language of polarized waves in the following way.

An arbitrary linearly polarized wave $E_x(t)$ which spectral density is $E_x(\omega)$ can be unequivocally represented by the sum of two components, each containing only positive or only negative frequencies:

$$E_x(t) = \frac{1}{2\pi}\int\limits_{-\infty}^{\infty} E_x(\omega)e^{i\omega t}d\omega = \frac{1}{2\pi}\int\limits_{-\infty}^{0} E_x(\omega)^{i\omega t}d\omega + \frac{1}{2\pi}\int\limits_{0}^{\infty} E_x(\omega)e^{i\omega t}d\omega. \tag{2.108}$$

Component:

$$E_P(t) = \frac{1}{\pi}\int\limits_{0}^{\infty} E_x(\omega)e^{i\omega t}d\omega. \tag{2.109}$$

contains only positive frequencies and therefore represents a right circularly polarized component and is connected with another (left circularly polarized component) as follows:

$$E_L(t) = \frac{1}{\pi}\int\limits_{\infty}^{0} E_x(\omega)e^{i\omega t}d\omega = E_P^*(t). \tag{2.110}$$

Thus, the following relation between the signals $E_x(t)$, $E_P(t)$ and $E_L(t)$ is established:

$$E_x(t) = 0.5[E_P(t) + E_L(t)] = 0.5\lfloor E_P(t) + E_P^*(t)\rfloor, \tag{2.111}$$

that allows to set the following correlations:

$$\left.\begin{array}{l} E_x(t) = \mathrm{Re}[E_P(t)] = \mathrm{Re}[E_L(t)] \\ E_y(t) = \mathrm{Im}[E_P(t)] = -\mathrm{Im}[E_L(t)] \\ E_y(t) = G[E_x(t)];\ E_x(t) = -G[E_x(t)] \end{array}\right\}. \tag{2.112}$$

These relations are interpreted as a representation of linearly polarized oscillations by the sum of right and left circularly polarized oscillations.

It is also possible to determine the relation between the spectra of the considered oscillations:

$$E_P(\omega) = \begin{cases} 2E_x(\omega) & \text{for } \omega > 0 \\ 0 & \text{for } \omega < 0 \end{cases},$$

$$E_y(\omega) = -j\ \mathrm{sign}(\omega)E_x(\omega), \tag{2.113}$$

where sign(ω) the sign function taking values 1 if $\omega > 0$, zero if $\omega = 0$, and -1 if $\omega < 0$.

Hence follows the method of obtaining a broadband right-circle polarization signal: the source oscillation is input to a system, which rotates each spectral component in space at (-0.5π) for positive frequencies and 0.5π for negative. Such a system can be called a spatial quadrature filter.

From the spectral representations also follows the following procedure of forming broadband circularly polarized fluctuations. At the core of this are the Sokhotski–Plemelj correlations that transform to easily implemented in practice algorithm: calculate the spectrum of the signal $E_x(t)$; i.e., find $F[E_x(t)] = E_x(\omega)$; select spectral components at positive frequencies, i.e., $E_x^+(\omega) = h(\omega)E_x(\omega)$, where $h(\omega)$ is the Heaviside function equal to 1 when $\omega > 0$ and zero when $\omega < 0$; make the inverse Fourier transform of a function $E_x^+(\omega)$, i.e., $F^{-1}\lfloor 2E_x^+[\omega]\rfloor = E_x(t) + j\Lambda(t)$; select the imaginary part $\Lambda(t)$, which is the Hilbert transform of a function $E_x(t)$; identify the found function $\Lambda(t)$ with the spectral component $E_y(t)$.

The block diagram of the spatial quadrature filter, functioning according to this algorithm, can be formed without much difficulty and therefore is not given here.

Let us define now the polarization parameters of the broadband fluctuations with the help of the Hilbert transformations. The amplitude of the circularly polarized wave will be expressed by the following equation:

$$|E_P(t)| = |E_L(t)| = \sqrt{E_x^2(t) + E_y^2(t)} = \sqrt{E_x^2(t) + \{\Gamma[E_x(t)]\}^2}, \tag{2.114}$$

and its phase

$$\varphi_\Pi(t) = \arg E_P(t) = \arg E_y(t)/[E_x(t)]. \tag{2.115}$$

For an instantaneous frequency equal to the derivative phase of time, the representation below will obviously be true:

$$\omega(t) = \frac{\mathrm{d}\varphi_P(t)}{\mathrm{d}t} = \frac{E_x(t)\frac{\mathrm{d}E_y(t)}{\mathrm{d}t} - E_y(t)\frac{\mathrm{d}E_x(t)}{\mathrm{d}t}}{E_x^2(t) + E_y^2(t)}. \tag{2.116}$$

An important property of the Hilbert transform, formulated by circularly polarized waves, is the following, if at any t component $E_x(t)$ reaches a maximum or minimum, in the neighborhood of this moment of time component $E_y(t)$ passes through zero, and inversely, the extremum of $E_y(t)$ corresponds to zero at $E_x(t)$.

Let us consider a number of examples illustrating the characteristics of circularly polarized broadband oscillations defined above.

If $E_x(t) = E_0 \cos\omega_0$, then the component $E_y(t)$ is equal to $E_0 \sin\omega_0 t$. The amplitude of this oscillation will be E_0, full phase, $\omega_0 t$, and frequency, ω_0. Consequently, the case in hand is the right circularly polarized monochromatic oscillation.

Now, let the component $E_x(t)$ represent the sum of two harmonic oscillations with different amplitudes and frequencies: $E_x(t) = E_1 \cos\omega_1 t + E_2 \cos\omega_2 t$. In this case: $E_y(t) = E_1 \sin\omega_1 t + E_2 \sin\omega_2 t$ and the amplitude is changing in time:

$$E_P(t) = \sqrt{E_1^2 + E_2^2 + 2E_1 E_2 \cos(\omega_2 - \omega_1)t}. \tag{2.117}$$

The full phase of the considered clockwise polarized oscillation in this case will be equal to:

$$\varphi_P(t) = \operatorname{arctg}\frac{E_1 \sin\omega_1 t + E_2 \sin\omega_2 t}{E_1 \cos\omega_1 t + E_2 \cos\omega_2 t}, \tag{2.118}$$

and instantaneous frequency—

$$\omega(t) = \frac{\omega_1 E_1^2 + \omega_2 E_2^2 + E_1 E_2 \cos(\omega_1 + \omega_2)t \ \cos(\omega_2 - \omega_1)t}{E_1^2 + E_2^2 + 2E_1 E_2 \cos(\omega_2 - \omega_1)t}. \tag{2.119}$$

Finally, suppose $E_x(t)$ represents a perfect bandpass signal which spectrum is constant, is equal to E_ω, and limited by the frequency interval $[\omega_1, \omega_2]$. In this case, the right circularly polarized oscillation is represented as:

$$E_P(t) = \frac{E_\omega}{\pi}\int_{\omega_1}^{\omega_2} e^{j\omega t} d\omega = \frac{E_\omega}{\pi t}[(\sin\omega_2 t - \sin\omega_1 t) - j(\cos\omega_2 t - \cos\omega_1 t)]. \tag{2.120}$$

The amplitude of this oscillation is equal to:

$$E_P(t) = \frac{E_\omega}{\pi}(\omega_2 - \omega_1)\left|\frac{\sin[0.5(\omega_2 - \omega_1)t]}{0.5(\omega_2 - \omega_1)t}\right|, \tag{2.121}$$

and instantaneous frequency: $\omega(t) = 0.5(\omega_1 + \omega_2)$ is independent of time and coincides with the central frequency of the concentration range of the signal spectrum.

We will make an array of remarks in connection with the above. First, a very special role of the Hilbert transform in determining the polarized state of the wave is seen. Second, for the imaginary unit, a different symbol is used—j than in the previous and subsequent sections. This is done in order to mark the difference between spatial and temporal quadratures. Finally, third, the fact that a specially generated signal is "good" enough in the sense of analytical functions, and the direction of wave propagation is fixed contributed to such a simple representation of polarization in the field of complex numbers without the use of quaternions. While varying not only the time structure of the oscillations, but also the direction of propagation, as well as in the absence of a rigid Hilbert connection between the polarized components, it is already necessary to recourse to quaternions and biquaternions.

2.4 The Representation of the Ultra-Broadband Signals Polarization State of Navigation Systems

The transformations and formulae, considered in p. 1.6, connecting the quaternion algebra with the algebra of three-dimensional vectors, allow to describe the origin of polarized radio waves in various environments and microwave systems [4] quite simply. For example, the correlation $X' = N^{-1}XN$ between three vector quaternions $X(0, \vec{X})$, $\vec{X}'(0, X')$ and $N(0, \vec{n})$ can be represented as $NX' = -XN$ or $\vec{n}\vec{X}' + \vec{n} + \vec{X}$. Since according to the above theorem $x_0' = -\vec{n}\vec{x}' = \vec{x}\vec{n} = x_0$, we have $\vec{n} \times \vec{x} = -\vec{x} \times \vec{n}$. The last two equations represent a vector form of the law of reflection of radio waves in the approximation of geometric optics (Snell law) for $|\vec{x}| = |\vec{n}| = 1$, relating the orientation of the three unit vectors: $\vec{n}$—environments' interface, $\vec{x}$—of falling, and $\vec{x}'$—reflected waves.

Let us now consider a quaternion description of the rotation of the polarization plane. By multiplying the normalized quaternion X by a vector quaternion $Y = (0, Y)$ such that $\vec{y}\vec{x} = 0$, we obtain the following equation $XY = Y \cos\gamma + \vec{\eta}Y \sin\gamma = \vec{Y}\cos\gamma + (\vec{\eta} \times \vec{y}) \sin\gamma$, where $\gamma = \mathrm{Arg}X$. Consequently, multiplication the vector $\vec{Y}$, orthogonal to unit vector of the vector part of quaternion, to the quaternion X from the left, leads to the rotation of this vector around the axis $\vec{\eta}$ by an angle γ, which is fully consistent with the transformation of rotation according to the theorem considered in p. 1.6. Thus, like a complex number $x = |x|e^{\mathrm{i}\arg x}$ describing the rotation of a vector with $|x|$ length in a complex plane, a quaternion $X = |X|e^{\vec{\eta}\mathrm{Arg}X}$ describes the rotation of a vector with $|X|$ length in a plane that is perpendicular to an axis $\vec{\eta}$ arbitrarily oriented in space.

Let us define the orientation of the vector part of the quaternion by directional cosines u_x and u_y with respect to the three-dimensional Cartesian coordinate system. Then $\vec{\eta} = u_x a_1 + u_y a_2 + \sqrt{1 - u_x^2 - u_y^2} a_3$, and rotation in a plane perpendicular to the direction (u_x, u_y) is described by a unit quaternion: $X = \cos\gamma + \left(u_x a_1 + u_y a_2 + \sqrt{1 - u_x^2 - u_y^2} a_3\right)\sin\gamma$. It follows, in particular, that the rotation of the polarization plane can be decomposed into three elementary rotations around the coordinate axes:

$$\left.\begin{array}{l} u_x = 1;\, u_y = 0;\, u_z = 0;\, e^{\vec{\eta}\gamma} = \cos\gamma + a_1 \sin\gamma \\ u_x = 0;\, u_y = 1;\, u_z = 0;\, e^{\vec{\eta}\gamma} = \cos\gamma + a_2 \sin\gamma \\ u_x = 0;\, u_y = 0;\, u_z = 1;\, e^{\vec{\eta}\gamma} = \cos\gamma + a_3 \sin\gamma \end{array}\right\}. \tag{2.122}$$

By analogy with the complex description of polarization, an isomorphic representation of a quaternion by a matrix below is possible:

$$Q = \begin{pmatrix} x_0 + \eta x_1 & x_2 + \eta x_3 \\ -x_2 + \eta x_3 & x_0 - \eta x_1 \end{pmatrix}, \tag{2.123}$$

which is in the basis of the Hermitian Pauli matrices

$$P_1 = \begin{pmatrix} 0 & 1 \\ 1 & 0 \end{pmatrix};\ P_2 = \begin{pmatrix} 0 & -\eta \\ \eta & 0 \end{pmatrix};\ P_3 = \begin{pmatrix} 1 & 0 \\ 0 & -1 \end{pmatrix}, \tag{2.124}$$

complemented by a unit matrix $I = \begin{pmatrix} 1 & 0 \\ 0 & 1 \end{pmatrix}$ can be brought to the form $Q = x_0 I + \eta(x_3 P_1 + x_2 P_2 + x_1 P_3)$. The quaternion algebra allows not only to describe the rotation of the polarization plane, but also to determine all the polarization parameters of the radio wave propagating in a given direction. In this case, however, the construction of a hypercomplex system of the fourth rank over the field of complex numbers will be required, which is the biquaternion algebra of $X = X_R + iX_I$ kind, where $X_R = \mathrm{Re}(x_0 + \vec{x})$, $X_I = \mathrm{Im}(x_0 + \vec{x})$ are the real and imaginary components of the biquaternion with respect to $i = \sqrt{-1}$. The axioms of this algebra keep unchanged the properties of operations, defined in the quaternion with the corresponding replacement of real numbers by complex ones. The isomorphism of matrix and biquaternion representations is also unchanged, as will be seen from the examples below. There are only some features, for example, in the properties of the tensor and the rules of normalization of biquaternions, which can be left out of account in this section.

As an illustration of the application of algebraic methods, we consider the description of the polarization of a monochromatic radio wave, for which the harmonic signal model is suitable without restrictions. Such analysis is also of great self-adjoint importance because any complex space–time oscillation can be represented as a superposition of harmonic waves with a given polarization structure. To solve this problem, we use the following E. E. Chernyshov theorem, according to which the polarization diagram of the radio wave of the unit amplitude is described by the biquaternion [5]:

$$X\left(t, u_x, u_y, \gamma\right) = \cos\gamma(t) + \left(u_x a_1 + u_y a_2 + \sqrt{1 - u_x^2 - u_y^2 a_3}\right)\sin\gamma(t), \tag{2.125}$$

determining for each moment of time t in the plane, the unit vector of which has directional cosines $\left(u_x, u_y\right)$, Cartesian coordinate system and an ellipse with semi-axes ch[Im$\gamma(t)$] and sh[Im$\gamma(t)$] rotated relative to the coordinate unit vectors by an angle $\gamma'(t) = \mathrm{Re}\gamma(t)$ in the direction of rotation of the polarization plane.

In considered case, the time dependence is not essential and the complex polarization function can be represented as $\gamma = \gamma' + i\gamma''$. Therefore, using the known formulae for the real and imaginary components of the trigonometric functions of the complex argument, from the expression (2.125) we obtain:

$$X\left(u_x, u_y, \gamma\right) = X\left(u_x, u_y, \gamma'\right)\left(\mathrm{ch}\gamma'' + i\vec{\eta}\mathrm{sh}\gamma''\right). \tag{2.126}$$

According to the previously considered properties of the quaternion product, this expression can be written in the following form:

$$X\left(u_x, u_y, \gamma\right) = BA^{-1} + CD^{-1}, \tag{2.127}$$

where $A = (0, \vec{\alpha}), B = (0, \vec{\beta}), C = (0, \vec{c}), D = (0, \vec{d})$—vector quaternions; $\vec{\alpha}, \vec{\beta}, \vec{c}, \vec{d}$—three-dimensional vectors, such that

$$\left.\begin{array}{l} |\vec{\alpha}| = 1; \left|\vec{\beta}\right| = \text{ch}\gamma''; \gamma' = \arccos[\vec{\alpha}\vec{\beta}/(|\vec{\alpha}|)(\left|\vec{\beta}\right|)] \\ \vec{\alpha}\vec{\eta} = 0; \vec{\beta}\vec{\eta} = 0; \vec{\alpha} \times \vec{\beta} = \vec{\eta}\left|\vec{\alpha} \times \vec{\beta}\right| \\ \vec{c} = i\vec{\beta}\left|\vec{\beta}\right|^{-1}\sqrt{\left|\vec{\beta}\right|^{2} - 1}; \vec{d} = \vec{\alpha}\mathrm{e}^{0.5\pi\vec{\eta}} \end{array}\right\}. \tag{2.128}$$

From Eqs. (2.128) and the correlation (2.127), it is possible to make a conclusion. First, the vectors $\vec{c}$, $\vec{\beta}$ and $\vec{\eta}$ form the right axis system, and also the vectors $\vec{\alpha}$ and $\vec{\beta}$ lie in the same plane with the vectors $\vec{c}$ and $\vec{d}$, the position of which is determined by the unit vector $\vec{\eta}$. Second, the angle between the vectors in pairs $\vec{\alpha}$, $\vec{\beta}$ and $\vec{d}$, $\vec{c}$ is the same, is counted toward the vectors, $\vec{\beta}$, $\vec{c}$, and is equal to γ'. Third, the vectors $\vec{d}$ and $\vec{\alpha}$ are both orthogonal, and the initial phases of the vectors $\vec{\beta}$ and $\vec{c}$ in the spatial complex area differ by 0.5π rad. Thus, the unit vectors and $\vec{d}$ define in a plane perpendicular to $\vec{\eta}$, the Cartesian system of coordinates where $\vec{\alpha}$ the vectors $\vec{\beta}$ and $\vec{c}$ define the orientation and settings of the polarization diagrams of the radio waves of unit amplitude.

One of the most common is the representation of an elliptically polarized wave in the orthogonal linear basis of the spatial complex plane, which contains components along the axes OY and $O\theta$, which, according to the Formula (2.126), have the form:

$$\left.\begin{array}{l} X_\varphi = \cos\gamma'\text{ch}\gamma'' - i\ \sin\gamma'\text{sh}\gamma'' \\ X_\theta = \sin\gamma'\text{ch}\gamma'' + i\ \cos\gamma'\text{sh}\gamma'' \end{array}\right\}. \tag{2.129}$$

Here, it is assumed that the axis $O\theta$ corresponds to the imaginary axis of the spatial complex plane. The representation of the radio wave in an arbitrary orthogonal elliptic basis can also be obtained from the Formula (2.126), supposing $\text{Im}\gamma = \gamma_1'' + \gamma_2''$ in it. Herewith:

$$X\left(u_x, u_y, \gamma\right) = I_1\left(u_x, u_y, \gamma_1\right)\text{ch}\gamma_2 - iI_2\left(u_x, u_y, \gamma_2\right)\text{sh}\gamma_2'', \tag{2.130}$$

where

$I_1\left(u_x, u_y, \gamma_1\right) = i_{01} + i_{11}a_1 + i_{21}a_2 + i_{13}a_3, I_2\left(u_x, u_y, \gamma_2\right) = i_{02} + i_{12}a_1 + i_{22}a_2 + i_{32}a_3$ are quaternions of an orthogonal elliptic basis that can be represented by a matrix:

$$\begin{pmatrix} I_1 \\ I_2 \end{pmatrix} = \begin{pmatrix} i_{01}\ i_{11}\ i_{21}\ i_{31} \\ i_{02}\ i_{12}\ i_{22}\ i_{32} \end{pmatrix} = \begin{pmatrix} 1\ u_x\ u_y\ \sqrt{1 - u_x^2 - u_y^2} \end{pmatrix}^{\mathrm{T}}$$

$$\times\begin{pmatrix}\sqrt{1-G_1^2}\;G_1\;G_1\;G_1\\ \sqrt{1-G_2^2}\;G_2\;G_2\;G_2\end{pmatrix} \tag{2.131}$$

where $G_1 = \sin\gamma_2$; $G_2 = \sin\gamma_2''$. The orthogonality of the obtained elliptic basis according to the considered theorem follows from the equality of the ellipticity angles, the counterphase of the unit vectors I_1 and I_2, and their orthogonal orientation in space.

Given examples can be supplemented by the representation of the wave in the orthogonal circular basis by accepting $\gamma_1'' = \pi/4$. In this case, we will have the following equation:

$$X\big(u_x, u_y, \gamma\big) = I_1\big(u_x, u_y, \gamma_1'\big)\mathrm{ch}\gamma_2 - iI_2\big(u_x, u_y, \gamma_2\big)\mathrm{sh}\gamma_2'',$$

where $\gamma_1' = \gamma' + i\pi/4$; $\gamma_2' = \big(\gamma'\big)^*$.

Thus, by changing the representation of the biquaternion argument describing the polarization of the radio wave, it is possible to obtain its decomposition in an arbitrary orthogonal basis.

In the terms of the quaternion algebra, the transformation of the polarization structure of radio waves in the propagation environments, modulator or receiver, which can be characterized by the transformation operator, is adequately described:

$$W = w_0 + w_1a_1 + w_2a_2 + w_3a_3. \tag{2.132}$$

As a result of the influence of this operator, the harmonic radio wave of the unit amplitude will take the form:

$$x' = XW = x_0 + x_1'a_1 + x_2'a_2 + x_3'a_3. \tag{2.133}$$

According to the rules of multiplication considered earlier in the hypercomplex system of the fourth rank, we find the components of this biquaternion:

$$\begin{aligned}
x_0' &= \cos(\mathrm{Arg}X)\cos(\mathrm{Arg}W) - \big(u_{xx}u_{xw} + u_{yx}u_{yw} + u_{zx}u_{zw}\big)\sin(\mathrm{Arg}X)\sin(\mathrm{Arg}W);\\
x_1' &= u_{xw}\cos(\mathrm{Arg}X)\sin(\mathrm{Arg}W) + u_{xx}\cos(\mathrm{Arg}W)\sin(\mathrm{Arg}X)\\
&\quad + \big(u_{yx}u_{zw} - u_{yw}u_{zx}\big)\sin(\mathrm{Arg}X)\sin(\mathrm{Arg}W);\\
x_2' &= u_{yw}\cos(\mathrm{Arg}X)\sin(\mathrm{Arg}W) + u_{yx}\sin(\mathrm{Arg}X)\cos(\mathrm{Arg}W)\\
&\quad + (u_{zx}u_{xw} - u_{xx}u_{zw})\sin(\mathrm{Arg}X)\sin(\mathrm{Arg}W);\\
x_3' &= u_{zw}\cos(\mathrm{Arg}X)\sin(\mathrm{Arg}W) + u_{zx}\cos(\mathrm{Arg}W)\sin(\mathrm{Arg}X)\\
&\quad + \big(u_{xx}u_{yw} - u_{yx}u_{xw}\big)\sin(\mathrm{Arg}X)\sin(\mathrm{Arg}W)
\end{aligned} \tag{2.134}$$

where $u_{\alpha\beta}$—the directional cosines of the coordinate α of biquaternion β. In the isomorphic matrix representation, the obtained transformation has the form:

$$X' = XW^{\mathrm{T}} = \left(x_0'\ x_1\ x_2\ x_3\right)^{\mathrm{T}}$$
$$= \begin{pmatrix} 1 & 0 & -u_{xx}u_{xw} - u_{xx}u_{xw} - u_{xx}u_{xw} \\ u_{xw} & u_{xx} & u_{yx}u_{zw} - u_{yw}u_{zx} \\ u_{yw} & u_{yx} & u_{zx}u_{xw} - u_{xx}u_{zw} \\ u_{zw} & u_{zx} & u_{xx}u_{yw} - u_{yx}u_{xw} \end{pmatrix} \cdot \begin{pmatrix} \cos(\mathrm{Arg}X) & \cos(\mathrm{Arg}W) \\ \cos(\mathrm{Arg}W) & \sin(\mathrm{Arg}X) \\ \sin(\mathrm{Arg}X) & \sin(\mathrm{Arg}W) \end{pmatrix} \quad (2.135)$$

As a rule, polarization transformations in the initial direction of wave propagation are considered, i.e., when the equations $u_{xx} = u_{xw} = u_x$, $u_{yx} = u_{yw} = u_y$, $u_{zx} = u_{zw} = u_z$ are fulfilled. In this case, the correlation (2.135) is simplified:

$$X' = XY^{\mathrm{T}} = \left(x_0'\ x_1\ x_2\ x_3\right)^{\mathrm{T}} = \begin{pmatrix} 1 & 0 & -1 \\ u_x & u_x & 0 \\ u_y & u_y & 0 \\ u_z & u_z & 0 \end{pmatrix} \cdot \begin{pmatrix} \cos(\mathrm{Arg}X) & \cos(\mathrm{Arg}W) \\ \sin(\mathrm{Arg}X) & \cos(\mathrm{Arg}W) \\ \sin(\mathrm{Arg}X) & \sin(\mathrm{Arg}W) \end{pmatrix}. \quad (2.136)$$

or

$$\left.\begin{aligned} x_0' &= \cos(\mathrm{Arg}X + \mathrm{Arg}W) \\ x_1' &= u_x \sin(\mathrm{Arg}X + \mathrm{Arg}W) \\ x_2' &= u_y \sin(\mathrm{Arg}X + \mathrm{Arg}W) \\ x_3' &= \sqrt{1 - u_x^2 - u_y^2}\sin(\mathrm{Arg}X + \mathrm{Arg}W) \end{aligned}\right\} \quad (2.137)$$

Therefore, any polarization transformation of the radio wave in the direction of propagation is described by the sum of the arguments of the corresponding biquaternions.

Thus, the representation of the polarization of the radio wave in quaternionic form makes it possible to simplify its mathematical analysis in various polarization bases, eliminate a number of cluttered calculations, making all the transformations physically intuitive, which is especially important in the study of the radio waves' propagation, as well as in the synthesis of space–time signals and systems.

References

1. Kanareikin DB, Potekhin VA (1968) Shishkin IF marine polarimetry. Shipbuilding, Leningrad
2. Tatarinov VN, Tatarinov SV, Lighart LP. Introduction to the modern theory of polarized radar signals. Book 1. Tomsk
3. Kozlov AI, Logvin AI, Sarychev VA (2005) Polarization of radio waves. Book 1. Polarization structure of radar signals. Radio Engineering, Moscow
4. Nikolayev AI (ed) (2016) Radar systems. MSTU them. N.E. Baumana, Moscow
5. Kozlov AI, Logvin AI, Sarychev VA (2007) Polarization of radio waves. Book 2. Radar polarimetry. Radio Engineering, Moscow

Chapter 3
Analysis of the Signal Polarization State of Navigation Systems Based on Energy Characteristics

3.1 The Coherence Matrix

3.1.1 Basic Properties and Transformations

As it was mentioned in Sect. 1.2., one of the ways to describe the polarization state of radio waves is to use the coherence matrix K, which is defined with an Eq. (1.19), obtained as a result of multiplying the matrix-column $\vec{E}$ by the row matrix $\vec{E}^{+}$:

$$K = \vec{E}\vec{E}^{+} = \begin{pmatrix} |E_1|^2 & E_1 E_2^* \\ E_1^* E_2 & |E_2|^2 \end{pmatrix}. \tag{3.1}$$

It should be noted that the diagonal elements of the coherence matrix are proportional to the power flux densities of the field components. The values of the complex that conjugate off-diagonal elements of the matrix are determined both by the value of the field components and by the phase shift between them. Due to the complex conjugacy of off-diagonal elements and the corporeity of the diagonal elements, the coherence matrix is Hermitian matrix, for which equation $K^{+} = K$ is valid.

Let us define the transformation of the coherence matrix when the polarization basis changes. In order to do this, we will use the (1.58) equation to write an expression for K in the new polarization basis:

$$K_H = \vec{E}_H \vec{E}_H^{+} = Q^* \vec{E}_C \left(Q^* \vec{E}_C \right)^{+} = Q^* K_C Q^{\mathrm{T}}, \tag{3.2}$$

which determines the desired law of transformation—the similarity transformation.

As one can see, the coherence matrix of the completely polarized electromagnetic wave K has two invariants—the matrix trace $SpK = |E_1|^2 + |E_2|^2$, proportional to the total wave intensity P_0 , and the determinant det $K = 0$. The latter, by the way, can be a definition of the completely polarized wave.

Kozlov A. I. et al., *Introduction to the Theory of Radiopolarimetric Navigation Systems*, Springer Aerospace Technology, https://doi.org/10.1007/978-981-13-8395-3_3

In the case of partially polarized waves, the elements of the coherence matrix become random functions of time, which means that statistical analysis is needed to describe it, i.e., one needs to know the corresponding three-dimensional law of distribution: $W\left(|E_1|^2, |E_2|^2, \arg E_1 - \arg E_2\right)$.

In practice, averaged or, as it is often called, the average coherence matrix\overlineK (sometimes referred to in a different denotation as $\langle K \rangle$), the elements of which are the mathematical expectations of matrix elements K:

$$\overline{K} = \overline{\vec{E}\vec{E}^{+}} = \begin{pmatrix} \overline{|E_1|^2} & \overline{E_1 E_2^*} \\ \overline{E_1^* E_2} & \overline{|E_2|^2} \end{pmatrix}. \tag{3.3}$$

Applying the averaging operation to the transformation (3.2), we obtain the law of variation of $\overline{K}$ with a change of the polarization basis:

$$\overline{K}_H = Q^* \overline{K}_C Q^{\mathrm{T}}. \tag{3.4}$$

As one can see, the matrix $\overline{K}$ also has its own trace $Sp\overline{K} = \overline{|E_1|^2} + \overline{|E_2|^2} = P_0$ and its own determinant as its invariants $\det \overline{K}_H = \det \overline{K}_C = \overline{|E_2|^2} \cdot \overline{|E_2|^2} - \left|\overline{E_1 E_2^*}\right|^2$.

From the definition of the unpolarized wave given in Sect. 1.1, it follows that in any arbitrary basis conditions, $\overline{|E_1|^2} = \overline{|E_2|^2}$ and $\overline{E_1 E_2^*} = 0$ must be satisfied for such waves.

Hence, it follows that the coherence matrix of unpolarized waves is an identity matrix I, i.e.,

$$\overline{K}_{H\Pi B} = 0.5 P_0 I. \tag{3.5}$$

Substitution of (3.5) into transformations (3.4) shows that in any AB, the coherence matrix of such waves will always be an identity matrix, i.e., power is distributed evenly between orthogonal components, and no correlation between them is provided.

The determinant of the coherence matrix of completely unpolarized electromagnetic waves of identity power is obviously equal to 0.25.

Equation (3.5), along with the above-mentioned conditions, can also serve as a definition of a completely unpolarized electromagnetic wave.

It follows from (3.4) that with the change of the polarization basis, the coherence matrix undergoes a similarity transformation. As we know, with the help of such a transformation, any Hermitian matrix can be reduced to a diagonal form, i.e., there is always a polarization basis in which the off-diagonal elements of the matrix are zero, i.e., $\overline{E_1 E_2^*} = \overline{E_1^* E_2} = 0$. Thus, there is always a polarization basis in which the orthogonal components of the wave will be completely uncorrelated (incoherent).

This applies equally to fully polarized waves as well, where the reasoning concerning the matrix (3.4.) is equally applicable to the matrix (3.1).

A basis in which the coherence matrix is diagonal is called the second singular basis. We will denote the coherence matrix in this basis $\overline{K_{\text{II}}}$, then:

$$\overline{K_{\text{II}}} = \begin{pmatrix} r_1^2 & 0 \\ 0 & r_2^2 \end{pmatrix}, \tag{3.6}$$

where r_1, r_2 are real numbers, are the eigenvalues of the matrix $\overline{K}$, numerically equal to the mean values of the orthogonal components' intensities in this basis (for fully polarized waves, one of the eigenvalues, for example, $r_2 = 0$). Invariance of the matrix determinant $\overline{K}$ enables us to obtain an important inequality

$$\det \overline{K} = \det \overline{K_{\text{II}}} = \overline{|E_1|^2} \cdot \overline{|E_2|^2} - \left|\overline{E_1^* E_2}\right|^2 = r_1^2 r_2^2 \geq 0. \tag{3.7}$$

And, obviously, the equality to zero occurs only for completely polarized waves, when $\overline{|E_1|^2} = |E_1|^2$, $\overline{|E_2|^2} = |E_2|^2$, therefore, for such waves, either r_1 or r_2 are equal to zero. If for clarity we assume that $r_1 \geq r_2$, then for a completely polarized wave:

$$\overline{K_{\text{II}}} = K_{\text{II}} = P_0 \begin{pmatrix} 1 & 0 \\ 0 & 0 \end{pmatrix}. \tag{3.8}$$

The requirement of degeneracy, the coherence matrix, i.e., condition $\det \overline{K} = 0$, can also be used to define a fully polarized wave.

The eigenvalues of the coherence matrix r_1^2 and r_2^2 can easily be expressed in terms of its invariants, as $Sp\overline{K} = P_0$, and $\det \overline{K} = r_1^2 r_2^2$, then:

$$2r_{1,2}^2 = Sp\overline{K} \pm \sqrt{(Sp\overline{K})^2 - 4\det \overline{K}} = (1 + m)P_0. \tag{3.9}$$

The parameter m introduced here, called the degree of polarization, is defined as follows, where it is assumed that $r_1 \geq r_2$:

$$m = \frac{r_1^2 - r_2^2}{r_1^2 + r_2^2} = \sqrt{1 - 4\frac{\det \overline{K}}{(Sp\overline{K})^2}}. \tag{3.10}$$

As can be seen, the invariance of $\det \overline{K}$ and $Sp\overline{K}$ implies the invariance of the degree of polarization m.

For $m = 0$, the wave is unpolarized, and for $m = 1$, on the contrary, the wave is completely polarized.

If we express the determinant of the matrix $\overline{K}$ in terms of the degree of polarization m, then we get: $\det \overline{K} = 0.25\left(1 - m^2\right)P_0$.

It follows from this equality that for waves of equal intensity, the determinant of the matrix $\overline{K}$ is the larger, the lower the degree of polarization is, it is maximal for unpolarized waves and, as already mentioned, its value in this case is 0.25.

Having chosen the second singular basis as the initial one, we can write the matrix $\overline{K}$ on an arbitrary basis if we use the explicit form of the unitary matrix Q (1.43):

$$\overline{K} = \frac{P_0}{2}\begin{pmatrix} 1 + m\cos 2\gamma & -m\mathrm{e}^{-2\mathrm{i}(\eta+\delta)}\sin 2\gamma \\ -m\mathrm{e}^{2\mathrm{i}(\eta+\delta)}\sin 2\gamma & 1 - m\cos 2\gamma \end{pmatrix}. \quad (3.11)$$

It can be seen from formulae (3.11) that the two parameters η and δ enter the corresponding expressions in the form of one variable $(\eta + \delta)$, which means that later on where only the coherence matrix is used, we can say that the coherence matrix depends only on three parameters (not taking into account, the power P_0 of course). Therefore, for convenience, it is possible to assume, for example, that $\eta = 0$.

In practice, a complex correlation index is of interest, sometimes called the coherence ratio, defined as follows:

$$\mu = |\mu|\mathrm{e}^{\mathrm{i}M} = \frac{\overline{E_1 E_2^*}}{\sqrt{\overline{|E_1|^2}\cdot\overline{|E_2|^2}}}. \quad (3.12)$$

Substituting the corresponding expressions, we get:

$$\left.\begin{aligned} |\mu| &= \frac{m\sin 2\gamma}{\sqrt{1-m^2\cos^2 2\gamma}} \\ M &= \pi - 2(\eta + \delta) \end{aligned}\right\}. \quad (3.13)$$

It follows from the introduction (3.13) that the modulus of the correlation coefficient is determined only by the angle γ. Therefore, if the Poincaré sphere (Fig. 1.4) is aligned so that its northern and southern poles (points F and G) correspond to the second singular basis, then $|\mu|$ will be the same in all bases depicted by points on parallels. It also follows from (3.13) that for $2\gamma = 0.5\pi$ (with the proposed orientation $\alpha = \gamma$), which corresponds to the points of the equator, $|\mu|$ reaches its maximum value $|\mu|_{\max} = m$. Thus, the maximum modulus value of the correlation coefficient is equal to the degree of the radio wave polarization, and the minimum value, as already mentioned, is zero.

A polarization basis, in which $|\mu| = m$, is called the first singular basis. Obviously, such bases are innumerable. (Any pair of diametrically opposite points on the equator of Fig. 1.4).

We now write the coherence matrix $\overline{K}$ in the first singular basis:

$$\overline{K}_I = \frac{P_0}{2}\begin{pmatrix} 1 & -m\mathrm{e}^{-2\mathrm{i}(\eta+\delta)} \\ -m\mathrm{e}^{2\mathrm{i}(\eta+\delta)} & 1 \end{pmatrix}. \quad (3.14)$$

The third singular basis we will refer to as a basis in which all elements of the matrix $\overline{K}$ are real non-negative numbers, for which it is necessary to set $2(\eta+\delta) = \pi$:

$$\overline{K}_{\mathrm{III}} = \frac{P_0}{2}\begin{pmatrix} 1 + m\cos 2\gamma & m\sin 2\gamma \\ m\sin 2\gamma & 1 - m\cos 2\gamma \end{pmatrix}. \quad (3.15)$$

We also pay attention to the fact that the third singular bases are also innumerable. They correspond to points lying on the meridian $2(\eta + \delta) = \pi$. When $\gamma = 0$, third and second, and when $\gamma = 0.25\pi$, third and first singular bases coincide with each other.

Having formed the difference of the diagonal elements of the matrix $\overline{K}$, we obtain a very useful equation

$$\overline{|E_1|^2} - \overline{|E_2|^2} = mP_0 \cos 2\gamma = \left(r_1^2 - r_2^2\right) \cos 2\gamma = P_0 m_K, \tag{3.16}$$

which shows that if in some polarization basis, the powers of the orthogonal components of the radio wave are equal to each other, then either they will be equal in any other basis ($m = 0$) and the wave is unpolarized, or this basis is the first singular ($\gamma = 0.25\pi$). (The ratio $\left(\overline{|E_1|^2} - \overline{|E_2|^2}\right)/P_0 = m_K$ will be called the apparent degree of polarization: $m_K = m \cos 2\gamma$).

Particularly important in the analysis of a partially polarized wave is its representation in the form of a sum of a completely polarized wave and independent of it, polarized. With respect to the coherence matrix of the wave being analyzed, this means that it can be represented as the sum of two matrices, one of which describes a completely polarized wave and the other one describes an unpolarized wave. In order to show this, we will first consider the particular problem of constructing a coherence matrix for the sum of independent waves, by which we mean those which satisfy the incoherence condition, i.e., for each of the pairs of wave components with numbers i, j (for $i \neq j$), the following equality holds:

$$\overline{E_{12}^{(i)} E_{12}^{*(j)}} = 0, \tag{3.17}$$

Suppose that there are N of such waves, we denote by $E_{\Sigma 1}$ and $E_{\Sigma 2}$ the sums of each of the waves' projections on the orthogonal axes 1 and 2:

$$E_{\Sigma 1} = \sum_{j=1}^{N} E_1^{(j)}; \quad E_{\Sigma 2} = \sum_{j=1}^{N} E_2^{(j)}. \tag{3.18}$$

We now find quadratic and bilinear values:

$$\left.\begin{aligned} \overline{E_{\Sigma 1} E_{\Sigma 2}^*} &= \sum_{i=1}^{N} \overline{E_1^{(j)} E_2^{*(j)}} \\ \overline{|E_{\Sigma 1}|^2} &= \sum_{j=1}^{N} \overline{\left|E_1^{(j)}\right|^2} \\ \overline{|E_{\Sigma 2}|^2} &= \sum_{j=1}^{N} \overline{\left|E_2^{(j)}\right|^2} \end{aligned}\right\}. \tag{3.19}$$

Equation (3.19) shows that the coherence matrix of the sum of independent waves is equal to the sum of the coherence matrices for all component waves. This assertion paves the way to all possible decompositions of partially polarized waves into independent components. The representation of a partially polarized wave in the form of a sum of polarized and unpolarized waves that do not depend on each other is of paramount importance. For this, it is necessary to show that any matrix of coherence $K = \begin{pmatrix} A & B \\ B^* & C \end{pmatrix}$ can be uniquely represented as a sum of a (3.5) type matrix and a matrix of a completely polarized wave, the determinant of which is zero, i.e.,

$$K = \begin{pmatrix} A & B \\ B^* & C \end{pmatrix} = \begin{pmatrix} A_1 & 0 \\ 0 & A_1 \end{pmatrix} + \begin{pmatrix} A_2 & B \\ B^* & C_2 \end{pmatrix} = K_0 + K_1. \tag{3.20}$$

Simple algebraic transformations give:

$$\left.\begin{array}{l} 2A_1 = SpK - \sqrt{(SpK)^2 - 4\det K} = (1-m)P_0 \\ 2A_2 = (A-C) + \sqrt{(SpK)^2 - 4\det K} = (m + m_K)P_0 \\ 2C_2 = -(A-C) + \sqrt{(SpK)^2 - 4\det K} = (m - m_K)P_0 \end{array}\right\}. \tag{3.21}$$

In doing so, it is not difficult to show that it is always $A_2 \geq 0$ and $C_2 \geq 0$

The expansion (3.20) can be carried out solely by means of the elements (3.21) of the matrices (3.20).

The trace of the matrix K_1 is the intensity $P_{\Pi\Pi B}$ of the polarized component, and the trace of the matrix K of the partially polarized wave under consideration is the wave intensity P_0. The ratio of these intensities

$$\frac{P_{\Pi\Pi B}}{P_0} = \frac{SpK_1}{SpK} = \sqrt{1 - 4\frac{\det K}{(SpK)^2}} = m, \tag{3.22}$$

which according to the definition (3.10.) is the degree of polarization of the partially polarized wave.

Using the ratio (3.21), we now write the decomposition (3.20) in an explicit form:

$$K = 0.5(1-m)P_0 I + \frac{mP_0}{2}\begin{pmatrix} 2\cos^2\gamma & -e^{-2i(\eta+\delta)}\sin 2\gamma \\ -e^{2i(\eta+\delta)}\sin 2\gamma & 2\sin^2\gamma \end{pmatrix}, \tag{3.23}$$

where (γ, η, δ) define the basis, in which the measurements are made, in respect of the second singular, in which the expansion (3.21) takes on a very simple form:

$$K = 0.5(1-m)P_0 I + mP_0\begin{pmatrix} 1 & 0 \\ 0 & 0 \end{pmatrix}. \tag{3.24}$$

Let us note a few of the most important representations of a non-polarized wave. The coherence matrix (3.23) of the completely unpolarized wave ($m = 0$) can be denoted in the form of a sum of two matrices:

$$\frac{P_0}{2}I = \frac{P_0}{2}\begin{pmatrix} 1 & 0 \\ 0 & 0 \end{pmatrix} + \frac{P_0}{2}\begin{pmatrix} 0 & 0 \\ 0 & 1 \end{pmatrix}, \tag{3.25}$$

corresponding to two orthogonal linearly polarized waves of equal intensity.

Another useful representation of the completely non-polarized wave has a form of:

$$\frac{P_0}{2}I = \frac{P_0}{4}\begin{pmatrix} 1 & i \\ -i & 1 \end{pmatrix} + \frac{P_0}{4}\begin{pmatrix} 1 & -i \\ i & 1 \end{pmatrix}. \tag{3.26}$$

It means that a completely non-polarized wave can be represented by the sum of two circularly polarized waves, of different rotations with the intensity of each wave equal to half the original one.

For practical purposes, it is of interest to determine the parameters of the polarization ellipse of a completely polarized component from the known coherence matrix of a partially polarized wave. In order to do this, we need to reconstruct the completely polarized wave in the form (1.7) in the expression (3.20) according to the matrix K_1. Obviously, we have:

$$\left.\begin{aligned} (E_{1m})_{\text{ППВ}} &= \sqrt{A_2} = \sqrt{mP_0}\cos\gamma \\ (E_{2m})_{\text{ППВ}} &= \sqrt{C_2} = \sqrt{mP_0}\sin\gamma \\ \delta_0 = \arg B &= \pi - 2(\eta+\delta) \end{aligned}\right\}. \tag{3.27}$$

The shape of the ellipse is determined not by the values E_{1m} and E_{2m} themselves, but by the ellipticity r, and the position in space is the inclination angle of the large semiaxis of the polarization axis—β. Using the ratios (3.20) and (3.27) and representations of Table 1.1., we can already denote:

$$\left.\begin{aligned} r &= \frac{1-\sqrt{1-\sin^2 2\gamma \sin^2 2(\eta+\delta)}}{\sin 2\gamma \sin 2(\eta+\delta)} \\ \tan 2\beta &= \tan 2\gamma \sin 2(\eta+\delta) \end{aligned}\right\}. \tag{3.28}$$

Elements of the coherence matrix can be determined by relatively simple experiments. It is necessary to measure only the intensity of the incident wave in several polarization bases, which is technically realized by the corresponding polarization filters. It is most convenient to use the intensity measurement of the field components $P_0(\beta,\gamma)$ in the (β,δ) representation, where β is the rotation angle of the linear basis in respect of the initial one, 2δ is the delay for the electric vector component E_y relative to the E_x component (for convenience, assume that $\eta = 0$).

Elements of the coherence matrix are expressed in terms of the intensities measured in six polarization bases:

$$\left.\begin{aligned}
&\overline{E_{xm}^2} = P_0(0^\circ, 0^\circ)\\
&\overline{E_{ym}^2} = P_0(90^\circ, 0^\circ)\\
&\overline{E_{xm}E_{ym}e^{2i\delta}} = 0.5[P_0(45^\circ, 0^\circ) - P_0(135^\circ, 0^\circ)] + 0.5i[P_0(45^\circ, 90^\circ) - P_0(135^\circ, 90^\circ)]\\
&\overline{E_{xm}E_{ym}e^{-2i\delta}} = 0.5[P_0(45^\circ, 0^\circ) - P_0(135^\circ, 0^\circ)] - 0.5i[P_0(45^\circ, 90^\circ) - P_0(135^\circ, 90^\circ)]
\end{aligned}\right\}. \tag{3.29}$$

As follows from expressions (3.29), in order to determine the diagonal elements and the real part of the off-diagonal elements of the coherence matrix, it is necessary to use a polarization filter that passes alternately horizontal, vertical, and polarizations at an angle of 45° and 135°. To determine the imaginary part of off-diagonal elements, a delay element is needed that would introduce a phase difference in a quarter period between the x- and y-components. The polarization filter must first be oriented at an angle of 45° and then 135°. The latest measurements are needed to identify the circular polarization of the right and left rotations.

For a linearly polarized field of $2\delta = \pi n (n = 0, \pm 1, \ldots)$, the coherence matrix has the form:

$$\overline{K} = \frac{P_0}{2}\begin{pmatrix} 2\cos^2\gamma & \pm\sin 2\gamma \\ \pm\sin 2\gamma & 2\sin^2 2\gamma \end{pmatrix}. \tag{3.30}$$

Coherence matrix

$$\overline{K}_{Л45} = \frac{P_0}{2}\begin{pmatrix} 1 & 1 \\ 1 & 1 \end{pmatrix} \text{ and } \overline{K}_{Л135} = \frac{P_0}{2}\begin{pmatrix} 1 & -1 \\ -1 & 1 \end{pmatrix} \tag{3.31}$$

corresponding to the linearly polarized field at an angle of 45° and $135,^\circ$ respectively.

For a circlular polarized wave, $\overline{E}_{xm} = \overline{E}_{ym}, 2\delta = 0.5\pi(1+2m)$ and the coherence matrix has the form:

$$\overline{K}_{КП} = \frac{P_0}{2}\begin{pmatrix} 1 & \pm i \\ \pm i & 1 \end{pmatrix}, \tag{3.32}$$

where the upper sign corresponds to the right and the lower sign corresponds to the left polarization.

3.1.2 Probabilistic Description

The average coherence matrix, discussed in the previous Sect. 3.1.1, naturally does not exhaust all of the statistical properties of the state of radio wave polarization. As already mentioned, the most complete characteristic in this method of description is the three-dimensional distribution law of probability.

To find it, we can use the results obtained in Sect. 2.2.2, for example, by formula (2.33). Having found the corresponding Jacobian of transformation, we get

$$W\big(|E_1|^2 = U;\,|E_2|^2 = V;\,\arg E_1 - \arg E_2 = 2\delta\big)$$
$$= \frac{1}{8\pi d_1^2 d_2^2\big(1-\rho^2\big)} \exp\left\{-\frac{1}{2\big(1-\rho^2\big)}\left[\frac{U}{d_1^2} + \frac{V}{d_2^2} - \frac{2\sqrt{UV}}{d_1 d_2}\rho\cos(2\delta - \beta)\right]\right\}. \tag{3.33}$$

Based on the obtained expressions, one can find the one-dimensional distribution densities for the elements of the coherence matrix that are of interest. Performing an integration over δ, we define the two-dimensional law

$$W(U, V) = \frac{1}{4d_1^2 d_2^2\big(1-\rho^2\big)} \exp\left\{-\frac{0.5}{1-\rho^2}\left[\frac{U}{d_1^2} + \frac{V}{d_2^2}\right]\right\} I_0\left(\frac{2\rho\sqrt{UV}}{d_1 d_2}\right), \tag{3.34}$$

and after integration over V—the one-dimensional law

$$W(U) = \frac{1}{2d_1^2} \exp\left\{-\frac{U}{2d_1^2}\right\}, \tag{3.35}$$

a similar representation is of the $W(V)$ law.

Thus, within the model (2.33) of a partially polarized wave that does not have a deterministic component, the diagonal elements of the coherence matrix have an exponential distribution law, for which, as follows from Sect. 1.5, for the mean value and variance, we have $(j = 1, 2)$:

$$\overline{\left|E_j\right|^2} = 2d_j^2; \quad \overline{\left|E_j\right|^4} - \overline{\left|E^2\right|^2} = 4d_j^4.$$

As to the off-diagonal elements of the coherence matrix, the expressions for the density of the probabilities of their real and imaginary parts turn out to be too cumbersome. We, therefore, restrict ourselves to expressions with their mean values. For this, we use the relations (3.10), (3.11), and (3.16), whence for the mean value in different polarization bases we get:

$$\overline{E_1 E_2^*} = -\frac{P_0}{2}\frac{r_1^2 - r_2^2}{r_1^2 + r_2^2} e^{-2i(\eta+\delta)} \sin 2\gamma = 0.5\big(d_2^2 - d_1^2\big) e^{-2i(\eta+\delta)} \tan 2\gamma. \tag{3.36}$$

An important role in applications is played by the problem of determining the laws for the distribution of the probabilities of the values U and V, when an additive mixture of two radio waves enters the antenna input, one of which, as a rule, carries useful information, and the other is a disturbance, interfering radio wave (their orthogonal components, as was done in Sect. 2.2, they will be called a useful and a disturbance signal, or simply a signal and an interference, respectively, and to mark them in the first case with the index "*c*" and in the second case with the index "*o*").

Let us determine the probabilistic characteristics of the square of the envelopes of the additive mixture of orthogonally polarized components of the deterministic

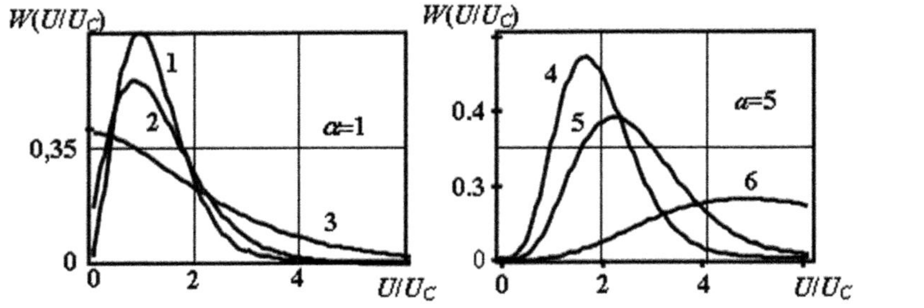

Fig. 3.1 The density of the probability distribution of the normalized square of the envelope of the mixture. 1. 4 – $q = 5$; 2. 5 – $q = 3$; 3. 6 – $q = 1$. $q = \beta_0 U_C/2$

Fig. 3.2 Density of probability distribution of the envelope square. 1—$a = 1$; 2—$a = 2$; 3—$a = 3$; 4—$a = 5$

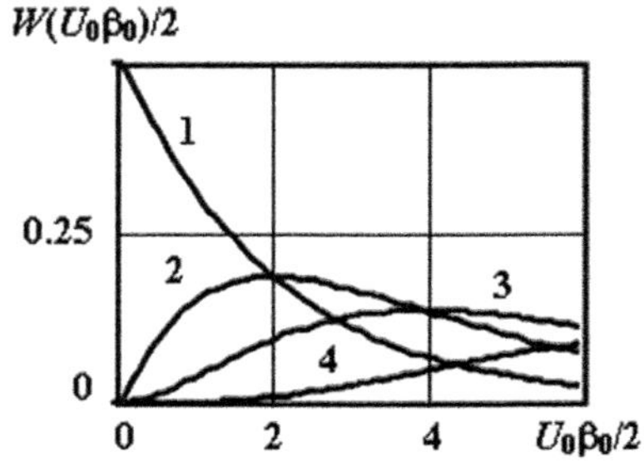

signal and the non-Gaussian interference. For this, progressing in the formula (2.81) to the new variable $U = |E|^2$ and denoting $U_C = \left|E_C^2\right|$, we obtain the expression for the probability distribution density of the envelope mixture square in the form:

$$W(U) = \frac{\beta_0}{2}\left(\sqrt{\frac{U}{U_c}}\right)^{\alpha_0 - 1} \exp\left[-\frac{\beta_0}{2}(U + U_c)\right] I_{\alpha_0 - 1}\left(\beta_0\sqrt{UU_c}\right). \tag{3.37}$$

Figure 3.1 shows families of curves $W(U)$ for different values of the interference α_0, β_0 and signal U_C.

In the absence of a signal ($U_C = 0$), formula (3.37) is transformed to the form:

$$W(U_0) = \left(\frac{\beta_0}{2}\right)^{\alpha_0} \frac{U_0^{\alpha_0 - 1}}{\Gamma(\alpha_0)} \exp\left(-\frac{\beta_0}{2}U_0\right). \tag{3.38}$$

The effect of the parameters α_0 and β_0 on the character of the probability density curves $W(U_0)$ is illustrated in Fig. 3.2.

It is not difficult to show that the normalized value $U_N = \beta_0 U$, with $a_U = \beta_0 U_C$ on the grounds of formula (3.37), will be described by the expression:

$$W(U_N) = 0.5\left(a_U^{-1} U_N\right)^{0.5(\alpha_0 - 1)} \exp[-0.5(a_U + U_N)] I_{\alpha_0 - 1}\left(\sqrt{a_U U_N}\right). \tag{3.39}$$

Using Eqs. (3.37) and (3.39) and carrying out the corresponding integration, we get an expression for the ν – th initial moments of the envelopes mixture square under consideration:

$$m_U^{(\nu)} = \left(\frac{2}{\beta_0}\right)^{\nu} \Gamma\begin{bmatrix} \alpha_0 + \nu \\ \alpha_0 \end{bmatrix} {}_1F_1(-\nu, \alpha_0; -0.5\beta_0 U_C). \tag{3.40}$$

After the calculations and taking formula (3.39) into account, the integral function takes the following form:

$$F(U_C) = \exp(-0.5\beta_0 U_C)\frac{(\beta_0 U_C)^k}{K!\Gamma(\alpha_0 + k)2^k}\gamma(\alpha_0 + k; 0.5\beta_0 U_C). \tag{3.41}$$

It can be seen that all the probabilistic characteristics considered above with $U_C = \text{const}$ depend only on two interference parameters α_0 and β_0. We will now evaluate them ($\hat{\alpha}_0$ and $\hat{\beta}_0$ accordingly). For this, we will be using the method of moments. Setting in (3.40) $U_C = 0$ and taking into account that ${}_1F_1(c, b; 0) = 1$, we obtain

$$m_{U_0}^{\nu} = \left(\frac{2}{\beta_0}\right)^{\nu} \Gamma\begin{pmatrix} \alpha_0 + \nu \\ \alpha_0 \end{pmatrix}. \tag{3.42}$$

This implies:

$$\left.\begin{array}{l} m_{U_0}^{(1)} = 2\alpha_0\beta_0^{-1}; m_{U_0}^{(2)} = 4\beta_0^{-2}\alpha_0(\alpha_0 + 1); d_{U_0}^2 = 4\beta_0^{-2}\alpha_0 \\ m_{U_0}^{(3)} = 8\beta_0^{-3}\alpha_0(\alpha_0 + 1)(\alpha_0 + 2); m_{U_0}^{(4)} = 16\beta_0^{-4}\alpha_0(\alpha_0 + 1)(\alpha_0 + 2)(\alpha_0 + 3) \end{array}\right\}. \tag{3.43}$$

Solving the system of equations together, from (3.43), we get:

$$m_{U_0}^{(1)} = 2\alpha_0\beta_0^{-1}; \quad d_{U_0}^2 = 4\alpha_0\beta_0^{-2}, \tag{3.44}$$

which makes it possible to find evaluations of the non-Gaussian interference parameters:

$$\hat{\alpha}_0 = \frac{\overline{U_0^2}}{\overline{U_0^2} - \overline{U_0}^2}; \quad \hat{\beta}_0 = \frac{2\overline{U_0}}{\overline{U_0^2} - \overline{U_0}^2}, \tag{3.45}$$

where $\overline{U_0} = \left(\sum_{j=1}^{N} U_{0j}\right)/(N-1); \quad d_{U_0}^2 = \overline{U_0^2} - \overline{U_0}^2.$

Here, the $U_{0j} = \left|E^2\right|_{oj}$ is the jth sample of the envelope interference square, N is the sample size.

From the formula, (3.45) follows the relation between the evaluations $\widehat{\alpha}_0$ and $\widehat{\beta}_0$:

$$\widehat{\beta}_0 = 2\widehat{\alpha}_0/\overline{U_0} = 2\widehat{\alpha}_0/\overline{E_0^2}. \quad (3.46)$$

On the basis of relation (3.43), it is not difficult to obtain expressions for the variation coefficients $K_{U_0}^B$, asymmetry $K_{U_0}^A$, and kurtosis:

$$K_{U_0}^B = \frac{d_{U_0}}{m_{U_0}^{(1)}} = \frac{1}{\sqrt{\alpha_0}} \quad ; \quad K_{U_0}^A = \frac{2}{\sqrt{\alpha_0}}; \quad K_{U_0}^{Э} = \frac{6}{\sqrt{\alpha_0}}. \quad (3.47)$$

These coefficients do not depend on the parameter β_0 and in a generic case vary over a wide range: with $\alpha_0 \to 0$, they tend to infinity and with $\alpha_0 \to \infty$, tend to zero. Consequently, the perceived probabilistic model of envelopes of orthogonally polarized components of non-Gaussian interference can be used to describe envelopes of a wide specter of processes. From the model (3.38), in specific cases, it is not difficult to obtain known distributions, for example, exponential, gamma distribution, χ^2-distribution, etc., by feeding some values to the α_0 and β_0 parameters.

Let us determine the probabilistic characteristics of the square of the envelopes of orthogonally polarized components of the additive mixture of the non-Gaussian signal and the non-Gaussian interference. To do this, we will be using an expression (2.84), in which we will progress to the new variable $U = |E|^2$ and will be taking equation $\gamma_C = \beta_C\sqrt{U_C}$ into account:

$$W(U) = \frac{\beta_C^{\alpha_C}}{\Gamma(\alpha_0)}\left(\frac{\beta_0}{2}\right)^{\alpha_0}\frac{U^{\alpha_0-1}}{(\beta_C+\beta_0)^{\alpha_C}}\exp\left[-\frac{1}{2}(\beta_C U_C + \beta_0 U)\right]$$
$$\times \sum_{k=0}^{\infty} A_{KU}\, {}_1F_1\left[\alpha_C + k, \alpha_0; \frac{\beta_0^2 U}{2(\beta_C+\alpha_0)}\right]; \quad (3.48)$$

where $A_{KU} = \frac{1}{k!}\left[\frac{\beta_C^2 U_C}{2(\beta_C+\beta_0}\right]^k$.

Using the formula (3.48) after the corresponding integration over variable U, we find:

$$m^{\upsilon} = \frac{\beta_C^{\alpha_C}\Gamma(\upsilon+\alpha_0)}{\Gamma(\alpha_0)}\left(\frac{2}{\beta_0}\right)^{\upsilon}\frac{\exp(-0.5\beta_C U_C)}{(\beta_0+\beta_C)^{\alpha_C}}$$
$$\times \sum_{k=0}^{\infty} A_{kU}\, {}_2F_1\left(\alpha_C + k, \upsilon + \alpha_0; \alpha_0; \frac{\beta_0}{\beta_0+\beta_C}\right)$$
$$F(U) = \left(\frac{\beta_0}{2}\right)^{\alpha_0}\frac{U^{\alpha_0}\beta_C^{\alpha_C}}{(\beta_0+\beta_C)^{\alpha_C}}\exp[-0.5(\beta_C U_C + \beta_0 U]$$
$$\times \sum_{k,n=0}^{\infty} A_{kU} A_{nU}\,\Gamma\left[\begin{matrix}\alpha_C + k + n\\ \alpha_C + k\end{matrix}\right]{}_1F_1\left(1, \alpha_0 + n + 1; \frac{\beta_0 U}{2}\right), \quad (3.49)$$

where $A_{nU} = \beta_0^{2n} U^n \left[2^n n! (\beta_0 + \beta_C)^n \Gamma(\alpha_0 + n + 1)\right]^{-1}$.

The obtained expressions (3.48) and (3.49) are basic. The probabilistic characteristics of the envelope mixture square for different combinations of probability models of a signal and interference can be obtained out of them (similar to the envelope signal mixture and interference) (for example, see Table 2.4).

Relations (3.37) through (3.49) are obtained by A. I. Mosionzhik at the authors' request.

3.2 The Stokes Parameters

3.2.1 Basic Properties and Transformations

It was noted above that the complex nature of off-diagonal elements of the coherence matrix $E_1 E_2^*$ and $E_1^* E_2$ leads to the necessity of separate measurements of their real and imaginary parts.

Free from this disadvantage is a method based on measuring not the elements of the coherence matrix, but some of their linear combinations that lead to material values. The system of four real parameters obtained with this approach, having the intensity dimension, was first introduced by Stokes.

The most convenient for practical use are the following four parameters S_0, S_1, S_2, S_3 introduced by Eqs. (1.16) and (1.17), which simply and unambiguously determine the form of the radio waves polarization. The inverse dependence, from which according to the Stokes parameters the elements of the coherence matrix are found, is determined by the formulas given in Table 1.1.

The polarization representations with the aid of the coherence matrix and Stokes parameters are called parametric, since both represent field polarization by means of a system of the quadratic, in respect of field intensity, complex, or real parameters.

In the linear Cartesian basis, the Stokes parameters are denoted conventionally as J_0, Q, U, V. These components are often considered the coordinates of a certain vector $\vec{S}$ in the corresponding four-dimensional functional space. The vector $\vec{S} = (S_0\ S_1\ S_2\ S_3)^{\mathrm{T}}$ is called the Stokes vector.

In passing from one polarization basis to another, the Stokes vector undergoes a certain linear transformation:

$$\vec{S}_H = L\vec{S}_C \tag{3.50}$$

with the aid of the quadratic 4×4 matrix L, called the Muller matrix, the explicit form of which can be obtained from the relations (1.16), (1.43), and (3.4):

$$L = \begin{pmatrix} 1 & 0 & 0 & 0 \\ 0 & \cos 2\gamma & \sin 2\gamma \cos 2\delta & -\sin 2\gamma \sin 2\delta \\ 0 & -\sin 2\gamma \cos 2\xi & \sin 2\xi \sin 2\delta + \cos 2\xi \cos 2\delta \cos 2\gamma & \sin 2\xi \cos 2\delta - \cos 2\xi \sin 2\delta \cos 2\gamma \\ 0 & \sin 2\gamma \sin 2\xi & \cos 2\xi \sin 2\delta - \sin 2\xi \cos 2\delta \cos 2\gamma & \cos 2\xi \cos 2\delta + \sin 2\xi \sin 2\delta \cos 2\gamma \end{pmatrix}. \tag{3.51}$$

where $\xi = \eta + \delta$.

It can be shown (see Table 1.1) that the Stokes parameters are related to the parameters of the polarization ellipse by the following relations:

$$\left.\begin{array}{l} Q = J_0 \cos 2\alpha \cos 2\beta \\ U = J_0 \cos 2\alpha \sin 2\beta \\ V = J_0 \sin 2\alpha \end{array}\right\}. \quad (3.52)$$

From here, the inverse relations are easy to get: $\sin 2\alpha = V/J_0$, $\tan 2\beta = U/Q$.

Equations (3.52) can also be considered as the formulas for the transition from the Stokes parameters to the geographical coordinates of the representing point on the Poincaré sphere, since the angles 2α and 2β determine, respectively, the latitude and longitude of the point on the sphere, and are the spherical angular coordinates of the point representing the polarization of the field (Fig. 1.4), and therefore the parameters S_1, S_2, S_3 are the Cartesian coordinates of point P on the sphere of radius J_0 (Fig. 1.4).

It is easy to obtain formulas relating the Stokes parameters to the polarization ratio: $\cos 2\gamma = Q/J_0$, $\tan 2\delta = U/V$, determine the amplitudes of the orthogonally polarized components of the wave: $E_{1m}^2 = 0.5(S_0 + S_1)$, $E_{2m}^2 = 0.5(S_0 - S_1)$, and the modulus of the phasor as well: $p = \tan \gamma = [(S_0 - S_1)/(S_0 + S_1)]^{0.5}$.

The ratio of the third and fourth of the Stokes parameters determines the phase of the polarization ratio. Thus, according to the Stokes parameters known in some polarization basis, the amplitudes of the wave components in this basis, their phase difference, the parameters of the polarization ellipse, and the unambiguous representation of the polarization state on the Poincaré sphere are easily found (see Table 1.1).

Let us clarify the physical meaning of the Stokes parameters. It is obvious from expression (1.17) that the parameter S_0 is the amplitude square or the intensity of the elliptically polarized wave. The parameter Q is equal to the difference in the intensities of the orthogonal components in the Cartesian basis Setting the $\gamma = 0.25\pi$, $\delta = \eta = 0$ in the formula (3.51), we obtain: $U_H = E_{1m}^2 - E_{2m}^2$, from which the physical meaning of the third Stokes parameter follows as the difference in the powers of the orthogonal components of the electric field intensity vector in a linear basis, rotated in respect of the initial one by an angle equal to 0.25π.

We now transform the Stokes vector to incorporate the physical meaning of the fourth parameter. In case of decomposition of an elliptically polarized wave in a circular basis $[\vec{e}_R, \vec{e}_L](\gamma = \delta = 0.25\pi, \eta = 0)$, we get $V_H = E_R^2 - E_L^2$.

Thus, the fourth Stokes parameter V is the intensity difference of the orthogonal components of the vector with circular polarization of the left and right rotation.

Like the elements of the coherence matrix, the Stokes parameters of any monochromatic plane wave can be determined using simple experiments. We denote $J_0(\alpha, \beta)$ with the intensity of the electromagnetic oscillations and write down a system of parameters that makes it possible to determine the polarization of the wave in practice:

$$\left.\begin{aligned} J_0(S_0) &= J_0(0^\circ, 0^\circ) + J_0(90^\circ, 0^\circ) \\ Q(S_1) &= J_0(0^\circ, 0^\circ) - J_0(90^\circ, 0^\circ) \\ U(S_2) &= J_0(45^\circ, 0^\circ) - J_0(135^\circ, 0^\circ) \\ V(S_3) &= J_0(45^\circ, 90^\circ) - J_0(135^\circ, 90^\circ) \end{aligned}\right\}. \quad (3.53)$$

All parameters have the intensity dimension and depend not only on the polarization, but also on the total intensity of the analyzed emission. This circumstance can in some cases be inconvenient. Then, it is advisable to apply a somewhat different system of parameters, obtained from the initial normalization by the total intensity J_0: $q = Q/J_0$, $u = U/J_0$, $v = V/J_0$.

The dimensionless parameters q, u, and v do not depend on the power of the received emission and are determined exclusively by the polarization of the electromagnetic field. Bearing in mind the method of their introduction, they are called the "normalized" Stokes parameters, the invariant of which (1.17) can now be represented in the form: $q^2 + u^2 + v^2 = 1$.

Let us consider the Stokes parameters for a partially polarized wave. The Stokes parameters for a partially polarized wave when it is expanded in an arbitrary basis are the following four values having the same dimension:

$$\left.\begin{aligned} S_0 &= \overline{|E_1|^2} + \overline{|E_2|^2}; \quad S_2 = 2\mathrm{Re}\overline{E_1 E_2^*} \\ S_1 &= \overline{|E_1|^2} - \overline{|E_2|^2}; \quad S_3 = 2\mathrm{Im}\overline{E_1 E_2^*}. \end{aligned}\right\}. \quad (3.54)$$

In a linear basis, these values are still denoted as J_0, Q, U, V. The expressions (3.54) relate the Stokes parameters to the elements of the average coherence matrix. The inverse relationship has the form:

$$\left.\begin{aligned} \overline{|E_1|^2} &= 0.5(J_0 + Q); \; \overline{E_1 E_2^*} = 0.5(U + iV) \\ \overline{|E_2|^2} &= 0.5(J_0 - Q); \; \overline{E_1 E_2^*} = 0.5(U - iV) \end{aligned}\right\}. \quad (3.55)$$

It follows from the expressions (3.54), that the first Stokes parameter is the total intensity of the wave, the second is the difference in the intensities of the orthogonally polarized components of the wave, and the other two parameters U and V determine the correlation between these components.

For the averaged Stokes parameters, the formulas (3.53) hold true.

We get another important practical representation of the Stokes parameters, for which we will use the relations (3.50) and (3.51), where the (γ, δ, η) should be understood as the Cartesian basis symbol $\lfloor \vec{e}_x, \vec{e}_x \rfloor$ in respect of the second singular polarization basis, and substitute them into the system of Eq. (3.54):

$$\left.\begin{aligned} Q &= mJ_0 \cos 2\gamma \\ U &= mJ_0 \sin 2\gamma \cos 2(\eta + \delta) \\ V &= mJ_0 \sin 2\gamma \sin 2(\eta + \delta) \end{aligned}\right\}. \quad (3.56)$$

From the relations (3.56) follows the obvious inequality

$$Q^2 + U^2 + V^2 = m^2 J_0^2 \leq J_0^2, \tag{3.57}$$

characteristic of a partially polarized wave (in contrast to Eq. (1.17) for a completely polarized radio wave). Substituting the expressions (3.56) into (3.13) allows us to determine the parameters of the correlation coefficient of the orthogonal components of the wave in terms of the Stokes parameters:

$$|\mu| = \sqrt{(U^2 + V^2)/(J_0^2 - Q^2)}; \quad M = \pi - \arctan(V/U). \tag{3.58}$$

Thus, measuring the intensities of a radio wave makes it possible to establish a correlation relationship between the orthogonal components of the field.

The Stokes parameters also allow us to determine the parameters γ, $(\eta + \delta)$, and the degree of polarization m:

$$\left.\begin{array}{l} \cos 2\gamma = Q/\sqrt{Q^2 + U^2 + V^2} \\ \tan 2(\eta + \delta) = V/U \\ m = \sqrt{Q^2 + U^2 + V^2}/J_0 \end{array}\right\}. \tag{3.59}$$

For the Stokes parameters, there is an additivity property that follows from the additivity property of the coherence matrix.

It follows from the (3.54) equations that:

$$J_0 = \sum_{j=1}^{N} J_{0j}; \quad Q = \sum_{j=1}^{N} Q_j; \quad U = \sum_{j=1}^{N} U_j; \quad V = \sum_{j=1}^{N} V_j, \tag{3.60}$$

where the Stokes parameters of the resulting field are denoted without indices; *N is* the number of incoherent summable fields.

Let us consider the consequences of the additivity of the Stokes parameters for independent fluxes of electromagnetic radiation.

It follows from (3.60) that any arbitrarily polarized flow characterized by the parameters (J_0, Q, U, V) can be regarded as the sum of two independent radiation fluxes, the first of which is completely polarized and characterized by the Stokes parameters $(\sqrt{Q^2 + U^2 + V^2}, Q, U, V)$, and the second is unpolarized $(m = 0)$ and has the Stokes parameters $(J_0 - \sqrt{Q^2 + U^2 + V^2}, 0, 0, 0)$.

This consequence is called the Stokes theorem and establishes the possibility of the arbitrary decomposition of polarized radiation into polarized and unpolarized components of the electromagnetic field.

The unpolarized component $(J_0 - \sqrt{Q^2 + U^2 + V^2}, 0, 0, 0)$ can be decomposed into the sum of two independent polarized fluxes, the Stokes parameters of which satisfy the following conditions:

$$Q_1 + Q_2 = 0; \quad U_1 + U_2 = 0; \quad V_1 + V_2 = 0. \tag{3.61}$$

It follows from the formula (3.61) that a partially polarized wave can be represented by the sum of three polarized waves. In addition, conditions (3.61) make it possible to obtain large information on the polarization of the field terms into which unpolarized radiation can be decomposed. Taking into account the expression (3.52), they can be rewritten as:

$$\left.\begin{array}{l} E_{1m}^2 \cos 2\alpha_1 \cos 2\beta_1 = -E_{2m}^2 \cos 2\alpha_2 \cos 2\beta_2 \\ E_{1m}^2 \cos 2\alpha_1 \sin 2\beta_1 = -E_{2m}^2 \cos 2\alpha_2 \sin 2\beta_2 \\ E_{1m}^2 \sin 2\alpha_1 = -E_{2m}^2 \sin 2\alpha_2 \end{array}\right\}. \tag{3.62}$$

Squaring the left and right sides of (3.62) equation and adding the results, we get $E_{1m}^4 = E_{2m}^4$. Consequently, $E_{1m}^2 = E_{2m}^2$.

Since the ellipticity angle of the polarization ellipse is restricted $-\pi/2 \leq \alpha \leq \pi/2$, it follows from the third Eq. (3.62) that $\alpha_1 = -\alpha_2$.

With this in mind, the first two Eqs. (3.62) allow us to obtain the relationships between the orientation angles:

$$\cos 2\beta_1 = -\cos 2\beta_2; \quad \sin 2\beta_1 = -\sin 2\beta_2. \tag{3.63}$$

The latter is only possible with $|\beta_1 - \beta_2| = 0.5\pi$.

From the analysis of Eq. (3.62), it follows that the flux of unpolarized electromagnetic radiation can be considered as the sum of two independent polarized fluxes of the same intensity, but oppositely polarized (i.e., orthogonal, oppositely rotating and having the same polarization ellipse shape). At that, the form of polarization of these fluxes can be completely arbitrary, since the equations $\alpha_1 = -\alpha_2$; $|\beta_1 - \beta_2| = 0.5\pi$ represent only orthogonality conditions. If the unpolarized stream is decomposed into two components of the same intensity, equal to $(J_0 - \sqrt{Q^2 + U^2 + V^2})/2$, the polarization of one of them is chosen to be the same as the polarization of the fully polarized flux determined earlier by the parameters $(\sqrt{Q^2 + U^2 + V^2}, Q, U, V)$, then two equally polarized fluxes form a flux with an intensity of $0.5(J_0 + \sqrt{Q^2 + U^2 + V^2}, Q, U, V)$.

Thus, an arbitrarily polarized radiation characterized by the Stokes parameter system (J_0, Q, U, V) can be considered as the result of the summation of two independent polarized fluxes, the intensities of which are $0.5\,(J_0 + \sqrt{Q^2 + U^2 + V^2})$ and $0.5\,(J_0 - \sqrt{Q^2 + U^2 + V^2})$; the polarization of the first coincides with the polarization of the polarized component in partially polarized radiation, and the polarization of the second is orthogonal.

From the aforesaid, it is clear that the intensity of the fully polarized component of the partially polarized wave is $(Q^2 + U^2 + V^2)$, but not J_0, as it was for the polarized wave. With this in mind, the parameters of the polarization ellipse can be defined as

$$\sin 2\alpha = V/\sqrt{Q^2 + U^2 + V^2}; \quad \tan 2\beta = U/Q. \tag{3.64}$$

Conditions

$$\left.\begin{array}{l} q^2 + u^2 + v^2 = 1 \\ q = u = v = 0 \\ 0 < q^2 + u^2 + v^2 < 1 \end{array}\right\} \tag{3.65}$$

determine the relationship between the Stokes parameters of a polarized, unpolarized, and partially polarized wave, respectively. The value $\left(1 + \sqrt{q^2 + u^2 + v^2}\right)$ shows the proportion of the intensity due to the unpolarized component of the wave.

Dividing the numerator and denominator of expressions (3.59) and (3.64) by the total intensity of the electromagnetic wave, we get formulas expressing the dependence of the polarization ellipse parameters and the polarized ratio on the normalized Stokes parameters:

$$\left.\begin{array}{ll} m \sin 2\alpha = v; & q \tan 2\beta = u \\ m \cos 2\gamma = q; & u \tan 2(\eta + \beta) = v \end{array}\right\}. \tag{3.66}$$

The inverse dependence—the dependence of the normalized Stokes parameters on the polarization ellipse parameters can be obtained from (3.56):

$$q = m \cos 2\alpha \cos 2\beta; \quad v = m \sin 2\alpha; \quad u = m \cos 2\alpha \sin 2\beta, \tag{3.67}$$

where the degree of polarization according to the Eq. (3.46) equals:

$$m = \sqrt{q^2 + u^2 + v^2}. \tag{3.68}$$

With partially linearly polarized radiation, when $\alpha = 0$, we have: $q = m \cos 2\beta, u = m \sin 2\beta, v = 0$. For partially circular polarization $\alpha = \pm 0.25\pi$ depending on the direction of the vector's rotation $\vec{E}$, we have $q = u = 0, v = m$.

The transformation of the Stokes parameters with a change in the basis, as in the case of a completely polarized wave, can be carried out with the aid of expression (3.50).

It should be noted that since a partially polarized wave can formally be represented either as a sum of incoherent fully polarized $\vec{E}_{\Pi\Pi B}(t)$ and completely unpolarized $\vec{E}_{H\Pi B}(t)$ waves or a sum of incoherent orthogonally polarized waves $\vec{E}_1(t)$ and $\vec{E}_2(t)$ (in this case, the polarizations $\vec{E}_1(t)$ and $\vec{E}_{\Pi\Pi B}(t)$ coincide), then depending on the particular problem, the partially polarized wave on the Poincaré sphere can be associated either with one point corresponding to polarization $\vec{E}_{\Pi\Pi B}(t)$ or two diametrically opposite points. In the process, it is necessary to take into account the "weight" of the indicated points (the intensity of the corresponding waves in a partially polarized wave). These points define, as it were, the "proper basis" of the wave, which is the polarization basis, in which the coherence matrix is diagonal. Obviously, a completely unpolarized wave can be opposed by any pair of diametrically opposite points with equal "weights".

The Stokes vector allows certain generalizations that find some practical application. Let us consider them in more detail.

First of all, we will find the length of the Stokes vector. According to the formulas (1.15), (1.16), and (3.57), we have:

$$\left|\vec{S}\right| = J_0\sqrt{1+m^2}. \tag{3.69}$$

As can be seen, for waves of constant intensity J_0, the length of the Stokes vector lies in the range from J_0 for the NPW to $\sqrt{2}J_0$ for the PPW. In a number of cases (as seen later on), variation $\left|\vec{S}\right|$ creates a few inconveniences. It is useful to introduce the generalized Stokes vector $\vec{F}$, where $\left|\vec{F}\right|$ is unchanged when the degree of polarization m varies. For this, obviously, instead of the first Stokes parameter J_0, we need to introduce a parameter $J_0\sqrt{1-m^2}$. In this case, $\left|\vec{F}\right| = J_0$. Therefore:

$$\vec{F} = \left(J_0\sqrt{1-m^2}, S_1, S_2, S_3\right). \tag{3.70}$$

The $\vec{F}$ vector is called a modified Stokes vector.

So far, the analysis has been conducted for radio waves, where $E_z = 0$. The coherence matrix and the Stokes vector can be generalized for the case when $E_z \neq 0$. Naturally, in this case, the coherence matrix K_3 will be a 3×3 matrix and will still be determined by the first of the Eq. (3.3). It is obvious that equality (3.4) and invariants (a trace and a determinant) hold true for it, although the matrix Q, remaining of a unitary size of 3×3 will take a different form than (1.47).

To generalize the concepts of Stokes parameters for the case when $E_z \neq 0$, we will proceed as follows. From the theory of matrices, it is known that any 2×2 Hermitian matrix, and such is the coherence matrix K, can be unambiguously decomposed in a system of a unit and three Pauli matrices, i.e.,

$$K = \left(0.5\sqrt{2}a_0 I + \sum_{j=1}^{3} a_j P_j^{(2)}\right), \tag{3.71}$$

where P_j^2 are Pauli matrices,

$$P_1^{(2)} = \frac{1}{\sqrt{2}}\begin{pmatrix} 1 & 0 \\ 0 & -1 \end{pmatrix}, P_2^{(2)} = \frac{1}{\sqrt{2}}\begin{pmatrix} 0 & 1 \\ 1 & 0 \end{pmatrix}, P_3^{(2)} = \frac{1}{\sqrt{2}}\begin{pmatrix} 0 & i \\ -i & 0 \end{pmatrix}.$$

An elementary comparison of formulas (3.3), (3.54), and (3.71), where $j = \overline{0,4}$ is now adopted, provides

$$S_j = \sqrt{2}\,a_j. \tag{3.72}$$

We generalize the concepts of Stokes parameters for the case under consideration on the grounds of a similar decomposition of the matrix K_3. Since the size of the matrix is 3×3, the number of terms in an equation similar to (3.71) must be 9.

$$K = \frac{1}{\sqrt{3}} b_0 I + \sum_{j=1}^{8} b_j P_j^{(3)}. \tag{3.73}$$

The following matrices will be analogues of the Pauli matrices for the three-dimensional case:

$$S_1^{(3)} = \frac{1}{\sqrt{6}} \begin{pmatrix} -1 & 0 & 0 \\ 0 & -1 & 0 \\ 0 & 0 & 2 \end{pmatrix}, \tag{3.74a}$$

and also when $j = \overline{2.8}$

$$S_j^{(3)} = \frac{1}{\sqrt{2}} \begin{pmatrix} \delta_{j2} & \delta_{j3} & \delta_{j5} + i(\delta_{j4} + \delta_{j6}) \\ \delta_{j4} & -\delta_{j5} & \delta_{j7} + i\delta_{j8} \\ \delta_{j5} - i(\delta_{j4} + \delta_{j6}) & \delta_{j7} - i\delta_{j8} & 0 \end{pmatrix}, \tag{3.74b}$$

where:

By analogy with the previously introduced Stokes parameters as generalized Stokes parameters, the following parameters are chosen.

First of all, this is the average intensity of a radio wave.

$S_0^{(3)} = J_0 = \overline{|E_1|^2} + \overline{|E_2|^2} + \overline{|E_3|^2}$, and also

$$\left.\begin{array}{ll} S_1^{(3)} = \sqrt{0.5}\left(2\overline{|E_3|^2} - \overline{|E_1|^2} - \overline{|E_2|^2}\right) & S_2^{(3)} = \sqrt{1.5}\left(\overline{|E_1|^2} - \overline{|E_2|^2}\right) \\ S_3^{(3)} = \sqrt{0.5}\left(\overline{E_1 E_2^*} + \overline{E_1^* E_2}\right) & S_4^{(3)} = -i\sqrt{1.5}\left(\overline{E_1 E_2^*} - \overline{E_1^* E_2}\right) \\ S_5^{(3)} = \sqrt{1.5}\left(\overline{E_1 E_3^*} + \overline{E_1^* E_3}\right) & S_6^{(3)} = -i\sqrt{1.5}\left(\overline{E_1 E_3^*} - \overline{E_1^* E_3}\right) \\ S_7^{(3)} = \sqrt{1.5}\left(\overline{E_2 E_3^*} + \overline{E_2^* E_3}\right) & S_8^{(3)} = -i\sqrt{1.5}\left(\overline{E_2 E_3^*} - \overline{E_2^* E_3}\right) \end{array}\right\} \tag{3.75}$$

The obtained relations allow us to write the coherence matrix K_3 in terms of S_j:

$$K_3 = \begin{pmatrix} \frac{1}{3}\left(S_0 - \frac{S_1}{\sqrt{2}} + \frac{\sqrt{3}S_2}{2}\right) & \frac{1}{\sqrt{6}}(S_3 + iS_4) & \frac{1}{\sqrt{6}}(S_5 + iS_6) \\ \frac{1}{\sqrt{6}}(S_3 - iS_4) & \frac{1}{3}\left(S_0 - \frac{S_1}{\sqrt{2}} - \frac{\sqrt{3}S_2}{2}\right) & \frac{1}{\sqrt{6}}(S_7 + iS_8) \\ \frac{1}{\sqrt{6}}(S_5 - iS_6) & \frac{1}{\sqrt{6}}(S_7 - iS_8) & \frac{1}{3}\left(S_0 + \sqrt{2}S_1\right) \end{pmatrix}. \tag{3.76}$$

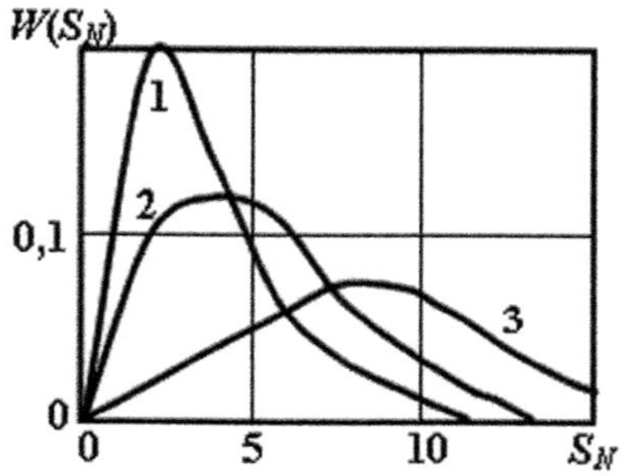

Fig. 3.3 The family of curves for different values a_1 and a_2 : 1—$a_1 = a_2 = 0$; 2—$a_1 = a_2 = 1$; 3—$a_1 = a_2 = 2$

3.2.2 Probabilistic Definition—The Gaussian Case

For a complete analysis of the polarization state of a partially polarized radio wave, just as in considering the coherence matrix, it is necessary to take into account the fact that the Stokes vector is a random function of time. Therefore, for its probabilistic description, in general, knowledge of the multidimensional probability distribution density $W(\vec{S})$ is required. However, in practice, one-dimensional statistics of the parameters S_0 and S_1 are of greatest interest. We will obtain it in the framework of mathematical models of the PPW, considered in Sect. 3.2.

We assume that the envelopes of the orthogonally polarized components of the PPW are described by the generalized Rayleigh distribution (2.48). Then, the joint probability density of the envelope of such a wave is determined by the expression (2.33).

The density of the probability distribution of the first Stokes parameter $J_0 = E_{1m}^2 + E_{2m}^2$, i.e., the power of the PPW can be found by integrating the formula (2.33) by the method described in Sect. 1.5. The resulting integrals are taken in finite form and are represented as rather complicated cumbersome expressions, containing triple series from beta, gamma, and hypergeometric functions. The resulting formula is greatly simplified for the case when the orthogonally polarized components of the PPW are statistically independent. In this case, it takes the form:

$$W\left(\frac{J_0}{d_{11}d_{22}} = S_N\right) = \frac{\sqrt{S_{0N}}}{2\sqrt{a_1^2 + a_2^2}} \exp\left(-\frac{a_1^2 + a_2^2 + S_{0N}}{2}\right) I_1\left[\sqrt{S_{0N}\left(a_1^2 + a_2^2\right)}\right]. \tag{3.77}$$

The parameters introduced into the formula (3.77) a_1, a_2, d_{11}, d_{22} are defined in (2.31) and (2.36).

Figure 3.3 shows a family of curves $W(S_N)$ calculated from (3.77) for certain values of the parameters a_1 and a_2 .

Let us consider a specific case when there is no deterministic component ($a_1 = a_2 = 0$). In this case, the joint density of the probability distribution $W(E_1, E_2)$ is

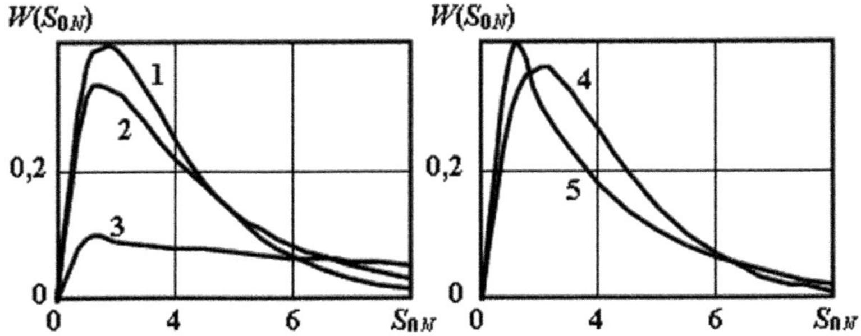

Fig. 3.4 The density of the probability distribution of the Stokes parameter S_{0N} with $a_1 = a_2 = 0$: 1. $R = 0.5$, $h = 1$; 2. $R = 0.5$, $h = 0.5$; 3. $R = 0.5$, $h = 0.1$; 4. $R = 0.5$, $h = 1$; 5. $R = 0.99$, $h = 1$

described by the expression (2.34). Proceeding similarly to the generic case, after the calculations, we get:

$$W(S_{0N}) = \frac{\exp\left\{-\frac{S_{0N}}{4\left(1-R^2\right)}\left[h+h^{-1}-\sqrt{4R^2+\left(h-h^{-1}\right)^2}\right]\right\}-\exp\left\{-\frac{S_{0N}}{4\left(1-R^2\right)}\left[h+h^{-1}+\sqrt{4R^2+\left(h-h^{-1}\right)^2}\right]\right\}}{\sqrt{4R^2+\left(h-h^{-1}\right)^2}}, \tag{3.78}$$

where $h = d_1^2/d_2^2$.

With $h = 1$, the expression (3.78) is simplified and takes the form of:

$$W(S_{0N}) = \frac{1}{4R}\left\{\exp\left[-\frac{S_{0N}}{2(1+R)}\right] - \exp\left[-\frac{S_{0N}}{2(1-R)}\right]\right\}. \tag{3.79}$$

If the parameters $R = 0$ and $h = 1$, then opening the ambiguity in (3.79) we get:

$$W(S_{0N}) = 0.25 S_{0N} \exp(-0.5 S_{0N}). \tag{3.80}$$

Figure 3.4 shows families of curves $W(S_{0N})$ for different values of h and R.

It is also possible to get the distribution function in a closed form as well:

$$\begin{aligned} 2[1 - F(S_{0N})]M = {} & \left(M + h + h^{-1}\right)\exp\left[-\frac{S_{0N}}{4(1-R^2)}\left(h + h^{-1} - M\right)\right] \\ & - \left(h + h^{-1} - M\right)\exp\left[-\frac{S_{0N}}{4(1-R^2)}\left(h + h^{-1} + M\right)\right] \end{aligned} \tag{3.81}$$

where $M = \sqrt{4R^2 + (h - h^{-1})^2}$.

To determine the ν – thinitial moments after direct calculations, we obtain the following formula:

$$m_{S_{0N}}^{(\nu)} = \frac{\nu!}{2M}\left[\left(h + h^{-1} + M\right)^{\nu+1} - \left(h + h^{-1} - M\right)^{\nu+1}\right]. \tag{3.82}$$

This implies the expressions for the mean value and variance:

$$m_{S_{0N}} = 2\left(h + h^{-1}\right);\quad d_{S_{0N}}^2 = 4\left(h^2 + h^{-2} + 2R^2\right). \tag{3.83}$$

Before proceeding to find the PPW of the other three Stokes parameters, we pay attention to the fact that each of them is the difference of the orthogonal components in the AB, formed by the pairs of HP-VP, 45°П-135°П, and RCP-LCP. In view of the invariance of the law $W_H(\vec{x})$ with respect to the change of AB in each of the bases mentioned in Sect. 1.5, the corresponding difference $E_{1m}^2 - E_{2m}^2 = \left(x_1^2 + x_2^2\right) - \left(x_3^2 + x_4^2\right)$ will obviously be distributed according to the same law. This means that to solve the required problem, it is sufficient to find one of the distribution laws $W(S_j)$, where $j = 1, 2, 3$.

It is known that in the framework of the PPW model considered in Sect. 3.2, a rigorous expression was found for $W(S_j)$, which, because of its cumbersome nature, is not given here.

When the envelopes E_{1m} and E_{2m} are distributed according to Rayleigh's law (2.34), the distribution law of a random variable $S_{jN} = S_j/d_{11}d_{22}$ will be expressed as follows:

$$W(S_{jN}) = \frac{0.5}{\sqrt{\left(h + h^{-1}\right)^2 - 4R^2}}\left\{\exp\left[-\frac{S_{jN}}{\sqrt{\left(h + h^{-1}\right)^2 - 4R^2} \mp \left(h^{-1} - h\right)}\right]\right\} \tag{3.84}$$

In this case, the minus sign refers to the case $S_{jN} < 0$, the plus—$S_{jN} > 0$. In a specific case, when $d_{11} = d_{22} = d(h = 1)$ instead of (3.84), we will have:

$$W(S_{jN}) = \frac{1}{2\sqrt{1 - R^2}}\exp\left(-\frac{|S_{jN}|}{2\sqrt{1 - R^2}}\right), \tag{3.85}$$

i.e., exponential law.

Figure 3.5 shows families of curves $W(S_{jN})$ for certain values of parameters h and R. It can be seen that in the generic case, the density $W(S_{jN})$ is one-mode and asymmetrical in respect to the ordinate axis.

For the integral distribution function of the parameter S_{jN} and the $vtextth$ initial moments in the case when $W(S_{jN})$ it is described by the expression (3.71), we get:

$$\left.\begin{array}{l} 1 - F(S_{jN}) = \frac{a}{\sqrt{a^2 - 4R^2}}\exp\left[-\frac{S_{jN}}{\sqrt{a^2 - 4R^2} - a}\right] \\ m_{S_j}^N = \frac{\nu! h}{2\sqrt{a^2 - 4h^2R^2}}\left\{\left(a + \sqrt{a^2 - 4R^2}\right)^{\nu+1} - \left(a - \sqrt{a^2 - 4R^2}\right)^{\nu+1}\right\} \\ a = h - h^{-1} \end{array}\right\}. \tag{3.86}$$

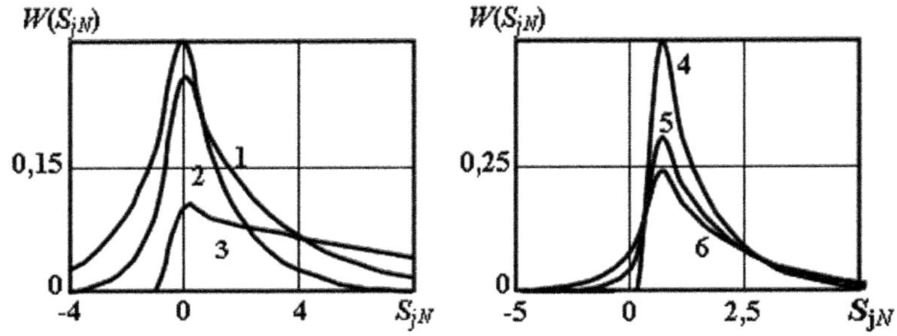

Fig. 3.5 The density of the probability distribution of the Stokes parameter S_{jN} at $a_1 = a_2 = 0$: 1. $h = 1$; 2. $h = 0.6$; 3. $h = 0.1$; 4. $R = 0.9$; 5. $R = 0.5$; 6. $R = 0$

From expression (3.86), it is easy to find the average value and variance of the parameter S_j:

$$\left.\begin{array}{l} m_{S_j} = 2\left(h^{-1} - h\right) \\ d_{S_j}^2 = 4\left(h^2 + h^{-2} - 2R^2\right) \end{array}\right\}. \tag{3.87}$$

We note again that the expressions obtained above for the statistical characteristics of the parameter S_j are invariant for the parameters S_1, S_2, S_3 in the corresponding polarization bases.

In practice, in a number of problems of polarization selection, the ratio $z = S_0/S_1$, that is, the apparent degree of polarization (3.16) is of interest. The formulas obtained make it possible to find the distribution law by the apparent degree of polarization, which is denoted instead of m_K by z for convenience. Direct calculations yield:

$$W(z) = \frac{2h^2\left(1 - R^2\right)\lfloor(1 + z) + h^2(1 - z)\rfloor}{\left\{\left[h^2(1 + z) + (1 - z)\right]^2 - 4R^2h^2\left(1 - z^2\right)\right\}^{1,5}}. \tag{3.88}$$

When $h = 1$, the formula (3.88) is substantially simplified:

$$W(z) = \frac{\left(1 - R^2\right)}{2\left[1 - R^2\left(1 - z^2\right)\right]^{1,5}}. \tag{3.89}$$

When $h = 0$, the probability distribution density $W(z)$ is a uniform law on the interval $[-1, 1]$. If the value $z_1 = |z|$ is of interest, then instead of formula (3.88), we will have:

$$W(z_1) = \frac{2(1-R^2)h^2[(1+h^2)-z_1(1-h^2)]}{\left\{[(1+h^2)-z_1(1-h^2)]^2 - 4R^2h^2(1-z_1^2)\right\}^{1,5}} + \frac{2(1-R^2)h^2[(1+h^2)+z_1(1-h^2)]}{\left\{[(1+h^2)+z_1(1-h^2)]^2 - 4R^2h^2(1-z_1^2)\right\}^{1,5}}. \quad (3.90)$$

When $h = 1$ and $R = 0$, the law $W(z_1)$ also turns into a uniform one, but on the interval [0, 1].

In Sect. 2.2.3, it was shown that in probabilistic analysis of both stationary and periodically non-stationary PPW, it is advisable to use the Nakagami probabilistic model (m-distribution) for envelopes of orthogonally polarized components. This model generalizes other types of probabilistic models of envelopes (Rayleigh, Rice, Hoyt, etc.) of radio waves and is a convenient approximation of the latter. Let us, therefore, consider the statistical characteristics of the Stokes parameters S_0 and S_3 for the case of the m-distribution of the envelopes E_{1m} and E_{2m} .

We will determine $W(S_0)$ when the envelopes E_{1m} and E_{2m} are statistically independent of each other, and the PPW is considered in an arbitrary polarization basis. Then, in accordance with the accepted conditions and (2.55), we will write:

$$W(z_1, z_2) = \left(\frac{m_1}{\Omega_1}\right)^{m_1}\left(\frac{m_2}{\Omega_2}\right)^{m_2} \frac{z_1^{m_1-1} z_2^{m_2-1}}{\Gamma(m_1)\Gamma(m_2)} \exp\left[-\left(\frac{m_1 z_1}{\Omega_1} + \frac{m_2 z_2}{\Omega_2}\right)\right], \quad (3.91)$$

where $z_1 = E_{1m}^2$; $z_2 = E_{2m}^2$.

Integrating over the variable z_1 , we get:

$$W(S_N) = \frac{m_1^{m_1} m_2^{m_2}}{\Gamma(m_1+m_2)} h_\Omega^{m_2-m_1} S_{0N}^{m_1+m_2-1} \exp(-m_2 h_\Omega S_N) \times {}_1F_1(m_1, m_1+m_2; S_N(m_2 h_\Omega - m_1/h_\Omega)). \quad (3.92)$$

where $S_N = S_0/\sqrt{\Omega_1\Omega_2}$; $h_\Omega = \sqrt{\Omega_1/\Omega_2}$.

It follows from (3.92) that the probability density $W(S_{0N})$ is determined by three parameters: m_1, m_2, h_Ω. When the equality $m_2 = m_1/h_\Omega^2$ is satisfied, the argument of the function ${}_1F_1(\xi)$ becomes zero and the expression (3.92) is simplified:

$$W(S_{0N}) = \frac{m_1^{m_1} m_2^{m_2}}{\Gamma(m_1+m_2)} h_\Omega^{m_2-m_1} S_{0N}^{m_1+m_2-1} \exp(-m_2 h_\Omega S_N). \quad (3.93)$$

Figure 3.6 shows the families of curves $W(S_{0N})$ for certain values m_1 , m_2 , h_Ω. It can be seen that the specified parameters affect the shape of the probability curve significantly.

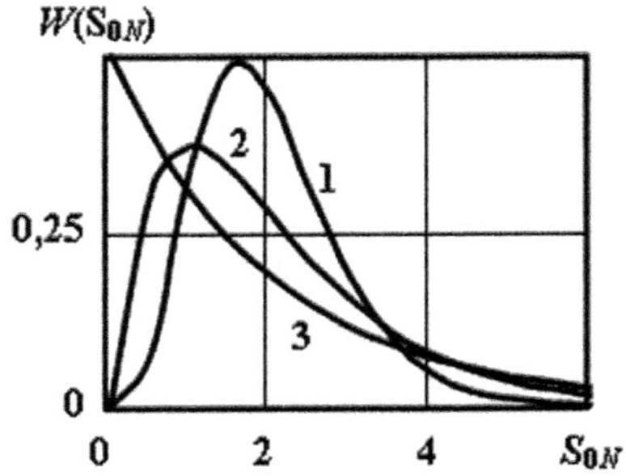

Fig. 3.6 The density of the probability distribution of the Stokes parameter S_{0N} for different values of the parameters $m_1 = m_2 = m$; $h_\Omega = 1$; 1. $m = 2.5$; 2. $m = 1$; 3. $m = 0.5$

To determine the statistical parameters of the other Stokes parameters (S_1, S_2, S_3), which are the differences in the intensities of the orthogonally polarized components of the PPW in the corresponding bases, without violating the generality of the problem, we use the circular basis for which $S_3 = E_{1m}^2 - E_{2m}^2$. Representing the parameter S_3 in terms of the envelopes squares and using formula (3.84), taking into account (3.91), after integrating over the variable z_1, we get:

$$W(S_{3N}) = \frac{m_1^{m_1} m_2^{m_2}}{\Gamma(m_2)} \left(\frac{m_1}{m_2} + m_2 h_\Omega \right)^{-\frac{m_1+m_2}{2}} h^{m_2-m_1} \exp\left[\frac{|S_{3N}|}{h_\Omega} \left(\frac{m_1}{h_\Omega} - m_2 h_\Omega \right) \right]$$
$$\times (\pm S_{3N})^{\frac{m_1+m_2-2}{2}} W_{\pm\left(\frac{m_2-m_1}{2}, \frac{1-m_1-m_2}{2}\right)} \left[|S_{3N}| \left(\frac{m_1}{h_\Omega} + m_2 h_\Omega \right) \right] \quad (3.94)$$

In formula (3.94), the plus sign corresponds to $S_{3N} > 0$ and the minus sign to $S_{3N} < 0$. The $W_{\alpha,\beta(\xi)}$ introduced into Eq. (3.94) is the Whittaker function.

Formula (3.81) allows us to obtain a number of specific cases: When $m_1 = m_2 = 1$, we have:

$$W(S_{3N}) = h_\Omega \left(1 + h_\Omega^2\right)^{-1} \exp(-|S_{3N}| h_\Omega), \quad (3.95)$$

with $m_1 = m_2 = 2$:

$$W(S_{3N}) = 4h_\Omega^2 \left(1 + h_\Omega^2\right)^{-2} |S_{3N}| \left[1 + \frac{1}{|S_{3N}| \left(h_\Omega + h_\Omega^{-1}\right)} \right] \exp\{-2h_\Omega |S_{3N}|\}. \quad (3.96)$$

Figure 3.7 shows families of curves $W(S_{3N})$ for some m_1, m_2, and h_Ω . It can be seen from figure, that when $h_\Omega = 1$ the curves $W(S_{3N})$ are symmetrical in respect of the ordinate axis. When $h_\Omega \neq 1$ they become asymmetrical. The degree of asymmetry increases, and the value of the maxima decreases with increasing (or decreasing) the parameter relative to one.

We would like to remind that the obtained expressions (3.92)–(3.96) are invariant for all three of the Stokes parameters S_1, S_2, S_3 .

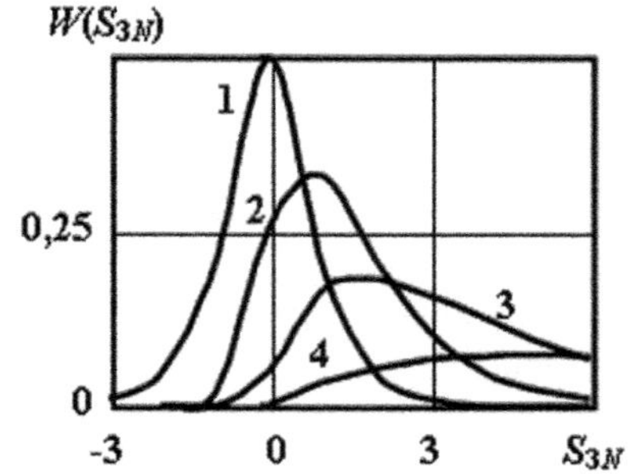

Fig. 3.7 The density of the probability distribution of the Stokes parameter S_{3N} for different values of the parameters h_Ω $m_1 = m_2 = 2$; 1. $h_\Omega = 1$; 2. $h_\Omega = 0.5$; 3. $h_\Omega = 0.25$; 4. $h_\Omega = 0.1$

3.2.3 Probabilistic Description—The Non-Gaussian Case

We will now define the probability distribution density of the first Stokes parameter S_0 in an arbitrary polarization basis, when the signal is completely polarized, and the interference is a partially polarized non-Gaussian process.

In this case, the envelopes of the orthogonally polarized components of the additive mixture of signal and interference are described by the probability distribution densities (2.81), and envelopes' squares $U_1 = E_1^2, U_2 = E_2^2$ —by the probability distribution density (3.37). Assuming that U_1 and U_2 are independent of each other and substituting the representation (3.37) into the transformation formula (1.97), we get:

$$W(S_0) = \int\limits_0^{S_0} W(U_1)W(S_0 - U_1)|J|dU_1, \tag{3.97}$$

where module of the Jacobian transformation $|J| = 1$.

To obtain the required distribution density of the first Stokes parameter, we integrate over the variable U_1 :

$$\begin{aligned} W(S_0) = {} & \left(\frac{\beta_{01}}{2}\right)^{\alpha_{01}} \left(\frac{\beta_{02}}{2}\right)^{\alpha_{02}} \exp\left[-\frac{\beta_{01}U_{C1} + \beta_{02}U_{C2}}{2}\right] S_0^{\alpha_{01}+\alpha_{02}-1} \\ & \times \exp\left(-\frac{\beta_{02}S_0}{2}\right) \times \sum_{k,n=0}^{\infty} A_k A_n S_0^{k+n} \Gamma(\alpha_{01} + \alpha_{02} + k + n)\, {}_1F_1\left[\begin{matrix} \alpha_{01} + k, \alpha_{01} + \alpha_{02} \\ +k + n; S_3\left(\frac{\beta_{02} - \beta_{01}}{2}\right) \end{matrix}\right] \end{aligned} \tag{3.98}$$

where $A_k = \frac{1}{k!}\left(\frac{\beta_{01}\sqrt{U_{01}}}{2}\right)^{2k}$, $A_n = \frac{1}{n!}\left(\frac{\beta_{02}\sqrt{U_{02}}}{2}\right)^{2n}$, $U_{Cj} = E_{Cj}^2, j = 1, 2.$

Equation (3.98) show that the probability distribution density of the first Stokes parameter depends on six parameters: $U_{0j}, \alpha_{0j}, \beta_{0j}$. The first two characterize the polarization properties of the deterministic signal, and the last four parameters are the polarization properties of the non-Gaussian partially polarized interference.

If there is no signal ($E_{Cj} = 0$), then expression (3.98) is simplified and begins to describe the probability distribution density of the first Stokes parameter S_0 of partially polarized interference:

$$W(S_{0\Pi}) = \left(\frac{\beta_{01}}{2}\right)^{\alpha_{01}} \left(\frac{\beta_{02}}{2}\right)^{\alpha_{02}} S_{0\Pi}^{\alpha_{01}+\alpha_{02}-1} \exp\left(-\frac{\beta_{02} S_0}{2}\right) \times \frac{{}_1F_1\left[\alpha_{01}, \alpha_{01}+\alpha_{02}; S_0\left(\frac{\beta_{02}-\beta_{01}}{2}\right)\right]}{\Gamma(\alpha_{01}+\alpha_{02})}. \quad (3.99)$$

If the interference is completely unpolarized, that is, $\beta_{02} = \beta_{01} = \beta_0$, then from formula (3.99), we get:

$$W(S_{0\Pi}) = \left(\frac{\beta_0}{2}\right)^{\alpha_{01}+\alpha_{02}} \frac{S_{0\Pi}^{\alpha_{01}+\alpha_{02}-1}}{\Gamma(\alpha_{01}+\alpha_{02})} \exp\left(-\frac{\beta_{02} S_{0\Pi}}{2}\right). \quad (3.100)$$

To determine the probabilistic characteristics of other Stokes parameters (S_1, S_2, S_3), which are the differences of the envelopes squares of the orthogonally polarized components of the additive mixture of the signal and the interference in the corresponding polarization bases, without violating the generality of the problem, we use the circular polarization basis. In this case, we denote $S_3 = E_1^2 - E_2^2 = U_1 - U_2$, where $U_j = E_j^2$, and index 1 corresponds to the orthogonally polarized component of the left direction of rotation of the polarization plane, and the index 2 corresponds to the right one. The formulae obtained below will obviously also be invariant for other polarization bases.

Using the rule for transforming the probability distribution densities (1.97) and taking into account the independence of U_1 and U_2, we write:

$$W(S_3) = \int_{-\infty}^{\infty} W(U_1) W(U_1 - S_3) |J| dU_1, \quad (3.101)$$

where module of the Jacobian of variables transformation $|J| = 1$.

The analysis shows that the integral (3.101) differs from zero when the conditions $U_1 \geq 0$ and $U_1 \geq S_3$ are satisfied. Therefore, expression (3.101) can be represented in the form:

$$W(S_3) = \begin{cases} \int_v^{\infty} W(U_1) W(U_1 - S_3) \mathrm{d}U_1, \text{ if } S_3 > 0 \\ \int_0^{\infty} W(U_1) W(U_1 - S_3) \mathrm{d}U_1, \text{ if } S_3 < 0 \end{cases}. \quad (3.102)$$

We will now define the probability distribution density of the Stokes parameter S_3 , when the signal is completely polarized, and the interference is a partially polarized non-Gaussian process. In this case, substituting expression (3.37) into formula (3.102) and integrating over the variable U_1, we get (the upper sign refers to the case $S_3 > 0$, the lower one to $S_3 < 0$):

$$W(S_3) = D S_3^{\frac{\alpha_{01}+\alpha_{02}+k+n-2}{2}} \exp\left[-\frac{S_3(\beta_{01}+\beta_{02})}{4}\right] W_{\pm\lambda,\pm\mu}\left[\frac{S_3(\beta_{01}+\beta_{02})}{2}\right], \quad (3.103)$$

where

$$D = \left(\frac{\beta_{01}}{2}\right)^{\alpha_{01}} \left(\frac{\beta_{02}}{2}\right)^{\alpha_{02}} \left(\frac{2}{\beta_{01}+\beta_{02}}\right)^{\frac{\alpha_{01}+\alpha_{02}}{2}} \exp\left[\frac{\beta_{02} S_3 - \beta_{01} U_{C1} - \beta_{02} U_{C2}}{k+n}\right]$$
$$\times \sum_{k,n=0}^{\infty} \frac{1}{k!n!} \left(\frac{\beta_{01}\sqrt{U_{C1}}}{2}\right)^{2k} \left(\frac{\beta_{02}\sqrt{U_{C2}}}{2}\right)^{2n} \left(\frac{2}{\beta_{01}+\beta_{02}}\right)^{k+n}$$

$W_{\pm\lambda,\pm\mu}(z)$—degenerate Whittaker hypergeometric function, $\lambda = 0.5(\alpha_{01} + \alpha_{02} + k + n)$, $\mu = 0.5(1 - \alpha_{01} - \alpha_{02} - k - n)$.

If there is no signal ($U_{Cj} = 0$), then from (3.103), it is not difficult to obtain the probability density distribution of the Stokes parameter $S_{3\Pi}$ for a partially polarized non-Gaussian interference:

$$W(S_{3\Pi}) = D_1 S_{3\Pi}^{\frac{\alpha_{01}+\alpha_{02}-2}{2}} \exp\left[\mp \frac{S_{3\Pi}}{4}(\beta_{01}+\beta_{02})\right] W_{\pm\lambda,\pm\mu}\left[S_{3\Pi}\left(\frac{\beta_{01}+\beta_{02}}{2}\right)\right], \tag{3.104}$$

where $D_1 = \left(\frac{\beta_{01}}{2}\right)^{\alpha_{01}} \left(\frac{\beta_{02}}{2}\right)^{\alpha_{02}} \left(\frac{2}{\beta_{01}+\beta_{02}}\right)^{\frac{\alpha_{01}+\alpha_{02}}{2}} \exp\left(\frac{\beta_{02}}{2} S_{\Pi}\right)$, $\lambda_1 = \frac{\alpha_{01}-\alpha_{02}}{2}$.
$\mu_1 = 0.5(1 - \alpha_{01} - \alpha_{02})$.

When the parameters of the signal $\alpha_{01} = \alpha_{02} = \alpha_0$ are equal, the Whittaker function in (3.104) becomes the MacDonald function (a modified Bessel function of the third kind) $K_\nu(z)$ [1], and expression (3.104) takes the following form:

$$W(S_{3\Pi}) = \left[\frac{\beta_{01}\beta_{02}}{2(\beta_{01}+\beta_{02})}\right]^{\alpha_0} \sqrt{\frac{\beta_{01}+\beta_{02}}{2\pi}} S_3^{\alpha_0 - 0.5} \exp\left[\frac{S_{3\Pi}}{2}\left(\beta_{02} \mp \frac{\beta_{01}+\beta_{02}}{2}\right)\right]$$
$$\times K_{\pm\nu}\left[\frac{S_{3\Pi}(\beta_{01}+\beta_{02})}{4}\right], \tag{3.105}$$

where $\nu = 0.5(1 - 2\alpha_0)$.

If the interference is Gaussian, that is, $\alpha_{0j} = 1$, $\beta_{0j} = d_{0j}^2$, then using the relation [1], we get:

$$K_{\pm 0.5}(z) = \sqrt{0.5\pi z^{-1}} \exp(-z). \tag{3.106}$$

From (3.105), it is not difficult to obtain an expression for the probability distribution density of the Stokes parameter $S_{3\Pi}$:

$$W(S_{3\Pi}) = \frac{0.5}{d_{01}^2 + d_{02}^2} \begin{cases} \exp\left(-0.5 S_{3\Pi} d_1^{-2}\right), \text{ if } & S_{3\Pi} > 0 \\ \exp\left(+0.5 S_{3\Pi} d_2^{-2}\right), \text{ if } & S_{3\Pi} < 0 \end{cases} \tag{3.107}$$

3.3 The Loss Matrix

3.3.1 Basic Properties and Transformations

Radiothermal radiation is a particular case of partially polarized radio waves, and therefore a description of its polarization properties can be performed with the aid of means developed for the generic case of PPW, i.e., in the terms of the coherence matrix. Despite the convenience of such means, it has a serious disadvantage: the elements of the matrix, firstly, contain only the characteristics of the received electromagnetic field and are not expressed through the measured radio brightness temperature; secondly, it does not explicitly contain information about any parameters generating its body. Therefore, it seems advisable to find such a representation of the coherence matrix, which would be free of the indicated shortcomings, take into account the parameters of the radiator itself and be expressed through the measured in practice radio brightness temperatures. This problem is solved by the loss matrix introduced below.

Let us consider how the radiating capacity of bodies depends on the type of polarization on which the measurement is made. In accordance with the general theory, the desired radio brightness temperature can be found by calculating the loss of an auxiliary plane wave if the measurements are taken sufficiently far away from the radiator with some polarization inside the radiator.

Let the radiator be a body of the arbitrary shape of volume V and limited by the surface S. The dielectric permittivity ε and the conductivity of the substance σ are arbitrary and vary in general from point-to-point, i.e., $\varepsilon = \varepsilon(x, y, z)$ and $\sigma = \sigma(x, y, z)$, where x, y, z are the Cartesian coordinates of the points of the body. We will first obtain some general relations for the losses of the above-mentioned auxiliary wave.

Let a plane electromagnetic wave fall on the indicated body. If the axis z is chosen to coincide with the wave propagation direction, then the orthogonal components of its electric vector $\vec{E}$ in an arbitrary polarization basis, for example, in the Cartesian basis, will be expressed by the relation (1.7) and $E_3 = 0$.

In generic case, the field inside the object will have all three components. To describe it at an arbitrary point in volume V, defined by a radius vector $\vec{r}$, we can introduce a matrix-column, and for an incident wave, we introduce a column matrix,, defined as follows:

$$\left.\begin{aligned} \vec{E}_{вн}(\vec{r}) &= (E_1 E_2 E_3)^T_{вн} \\ \vec{E}_{пад}(\vec{r}) &= (E_1 E_2 0)^T_{пад} \end{aligned}\right\}. \tag{3.108}$$

In this case, obviously, the matrix equality will be valid:

$$\vec{E}_{вн}(\vec{r}) = P(\vec{r})\vec{E}_{пад}, \tag{3.109}$$

where $P(\vec{r})$ is a square 3×3 matrix, the elements of which $P_{ij}(i, j = \overline{1,3})$ depend in a complex way on the shape and electrical properties of the radiator and surrounding bodies, the angle of incidence of a plane wave, and other characteristics, i.e., determined by solving the corresponding diffraction problem. Let us find some integral coefficients characterizing the reaction of the body to irradiation with electromagnetic wave. The required power losses of the auxiliary plane wave (1.7) inside the body can be found using the relation:

$$\varPi = \int_V \sigma(\vec{r}) \vec{E}_{вн}^{+} \vec{E}_{вн} dv. \tag{3.110}$$

Taking into account equality (3.109), expression (3.110) takes the form:

$$\varPi = \int_V \sigma(\vec{r}) \vec{E}_{пад} P^{+} P \vec{E}_{пад} dv = \vec{E}_{пад}^{+} \left(\int_V \sigma(\vec{r}) P^{+} P dv \right) \vec{E}_{пад}. \tag{3.111}$$

The matrix can be taken outside the integral sign, since its elements do not depend on the position of the point inside the body. We introduce the following notation:

$$G^{\circ} = \int_V \sigma(\vec{r}) P^{+} P dv. \tag{3.112}$$

The matrix G° is Hermitian, its elements $g_{ij}(i, j = \overline{1,3})$ represent the integral coefficients that are of interest to us. It is similar to the coherence matrix.

Thus, the loss in the power of the auxiliary plane wave, proportional to the radiative power P on the considered polarization of the body, can be calculated by the formula:

$$\varPi = \vec{E}_{пад}^{T} G^{\circ} \vec{E}_{пад}^{*}, \tag{3.113}$$

in order to use it, it is necessary to know three real g_{11}, g_{22}, g_{33} and three complex g_{12}, g_{13}, g_{23} numbers, i.e., only nine numbers.

Taking into account equality (3.111), we will write the expression for the radiating capacity in explicit form:

$$\varPi = g_{11} \left| E_{1пад} \right|^2 + g_{12} E_{1пад} E_{2пад}^{*} + g_{12}^{*} E_{1пад}^{*} E_{2пад} + g_{22} \left| E_{2пад} \right|^2. \tag{3.114}$$

In the last relation, there are no $g_{13}, g_{23}, g_{31}, g_{32}, g_{33}$ elements of the matrix G°, which allows us to write the equality (3.114) in matrix form:

$$\Pi = \vec{E} + G\vec{E}, \tag{3.115}$$

where by $\vec{E}$, a matrix-column is now meant:

$$\vec{E} = (E_1 \quad E_2)^T_{na\partial}, \tag{3.116}$$

and by G—the Hermitian matrix of order 2 × 2:

$$G = \begin{pmatrix} g_{11} & g_{12} \\ g_{12}^* & g_{22} \end{pmatrix}, \tag{3.117}$$

which can also be regarded as similar to the coherence matrix.

As follows from expression (3.117), the radiating capacity of an arbitrary radiator, when measured from such a distance at which electromagnetic waves generating thermal radio emission could be considered plane, is determined not by nine numbers, as discussed above, but only four—real g_{11}, g_{22}, and complex g_{12}. The difference between the matrix G defined by (3.117), which we will call the loss matrix, from the coherence matrix is purely formal. Therefore, the results obtained in the previous sections can easily be extended to the loss matrix.

Moving to a new basis, the matrix-column $\vec{E}_C$ is transformed in accordance with the literature. This allows the formula (3.111) to be transformed as follows:

$$\Pi = \vec{E}_C^+ G_C \vec{E}_C = \vec{E}_H^+ Q G_H Q^+ \vec{E}_H, \tag{3.118}$$

where $\vec{E}_H$ is the matrix-column of the auxilliary wave on the new basis. The last equation makes it possible to establish the law of change of the loss matrix upon transition to a new polarization basis.

Let G_H be the loss matrix in the new polarization basis, then

$$G_H = Q^+ G_C Q^{\mathrm{T}}, \tag{3.119}$$

which coincides exactly with the formula for the transformation of the coherence matrix. Relation (3.119) is a similarity transformation over a Hermitian matrix G; hence, there is always a basis where the matrix G is diagonal, and the diagonal elements are its eigenvalues. Such a basis will be called the second singular, and the loss matrix in it will be denoted by G_0 i.e.,

$$G_0 = \begin{pmatrix} \chi_1^2 & 0 \\ 0 & \chi_2^2 \end{pmatrix}. \tag{3.120}$$

Using the expression (3.119) and the representation of the unitary matrix in terms of the polarization parameters (1.47), we can write the loss matrix in an arbitrary basis determined by the parameters γ, δ, η in respect of the second singular in the form:

$$G = \frac{\chi_1^2 + \chi_2^2}{2} \begin{pmatrix} 1 + q\cos 2\gamma & -q\mathrm{e}^{-\mathrm{i}\Phi} \\ -q\mathrm{e}^{\mathrm{i}\Phi} & 1 + q\cos 2\gamma \end{pmatrix}, \tag{3.121}$$

where $\Phi = 2(\eta + \delta)$; q is the degree of polarization anisotropy of the radiator or simply the degree of anisotropy.

It is seen from expression (3.121) that the loss matrix, as well as the coherence matrix, depends only on two parameters characterizing the polarization basis γ and Φ. The invariants of the loss matrix are its determinant and trace:

$$\det G = \chi_1^2\chi_2^2; \quad SpG = \chi_1^2 + \chi_2^2. \tag{3.122}$$

If $\chi_1 = \chi_2$, then the loss matrix will be the identity matrix in any basis:

$$G_{ПИО} = \chi_1^2\begin{pmatrix} 1 & 0 \\ 0 & 1 \end{pmatrix} = \chi_1^2 \cdot I. \tag{3.123}$$

By analogy with active radar, the emitter, in which $\chi_1 = \chi_2$, we will call polarization-homogeneous, or polarization-isotropic (PIO). Continuing the analogy, we can introduce the concept of a degenerate object. For this, we must assume that $\chi_2 = 0$. The loss matrix in this case takes the form:

$$G_{BO} = \chi^2\begin{pmatrix} \cos^2\gamma & -0.5e^{-i\Phi}\sin 2\gamma \\ -0.5e^{i\Phi}\sin 2\gamma & \sin^2\gamma \end{pmatrix}. \tag{3.124}$$

Sometimes it may be useful to use the first singular polarization basis ($2\gamma = 0.5\pi + n\pi$), in which the loss matrix has the form:

$$G_I = \frac{\chi_1^2 + \chi_2^2}{2}\begin{pmatrix} 1 & -qe^{-i\Phi} \\ -qe^{i\Phi} & 1 \end{pmatrix}. \tag{3.125}$$

This means that there is always such a polarization basis in which the diagonal elements of the loss matrix will be equal to each other.

Taking into account the introduced notation for the invariants of the loss matrix, instead of the representation (3.122), we have:

$$\det G = P_0^2(1 - q^2); \quad SpG = 2P_0. \tag{3.126}$$

where $P_0 = 0.5(\chi_1^2 + \chi_2^2)$ is the total radiation power.

Proceeding from the obtained relations, we define the loss matrix. As follows from expression (3.113), the elements of the loss matrix (3.117) represent certain energy characteristics that determine the loss of an auxiliary plane wave in the volume of the body. Let us investigate the possibility of their determination from the minimum number of measurements of the radio brightness temperature. For this purpose, we will use Eq. (3.114).

The radio brightness temperature T of the radiator, which is evenly heated to the thermodynamic temperature T_0, can be expressed in terms of the power loss of the auxiliary plane wave Π by the following obvious equality:

$$T = \Pi T_0. \tag{3.127}$$

Let the first measurement be carried out on some linear polarization (conditionally horizontal). In this case, we have $E_1 = \exp(i\omega t)$, $E_2 = 0$ and the expression (3.114) yields:

$$\Pi_{\Gamma\Pi} = T_{\Gamma\Pi}/T_0 = g_{11}. \tag{3.128}$$

Similar measurements on the vertical (VP) polarization ($E_1 = 0$, $E_2 = \exp(i\omega t)$) yield:

$$\Pi_{B\Pi} = T_{B\Pi}/T_0 = g_{22}. \tag{3.129}$$

In Eqs. (3.128) and (3.129), $T_{\Gamma\Pi}$ and $T_{B\Pi}$ are the radio brightness temperatures measured, respectively, in horizontal and vertical polarizations.

The third measurement is conveniently carried out on the polarization obtained by rotating the receiving antenna about its electric axis by 45° from its original position, which will correspond to the angle γ in Eq. (1.7) equal to 45°. Then, $E_1 = 0.5\sqrt{2}\exp(i\omega t)$, $E_2 = 0.5\sqrt{2}\exp(i\omega t)$ and from equality (3.114), we have:

$$\Pi 45^\circ \Pi = T_{45^\circ M}/T_0 = 0.5\left(g_{11} + g_{22} + g_{12} + g_{12}^*\right) = 0.5\left(g_{12} + g_{12}^*\right) + T_\Sigma/T_0, \tag{3.130}$$

where T_Σ is the radio brightness temperature measured in the total channel.

The fourth measurement is expedient for conducting on a circular polarization (CP). Taking into account the expressions $E_1 = 0.5\sqrt{2}\exp(i\omega t)$, $E_2 = iE_1$ and the equality (3.114), we will have the following for $\Pi_{K\Pi}$:

$$\Pi_{K\Pi} = \frac{T_{K\Pi}}{T_0} = 0.5\left[g_{11} + g_{22} - i\left(g_{12} - g_{12}^*\right)\right] = \frac{T_\Sigma}{T_0} - i0.5\left(g_{12} - g_{12}^*\right). \tag{3.131}$$

From the expressions (3.130) and (3.131), we find the values of the modulus and phase of the element $g_{12} = |g_{12}|\exp(i\psi)$:

$$\left.\begin{aligned} |g_{12}| &= \frac{T_\Sigma}{T_0}\sqrt{\left(1 - \frac{T_{45^\circ \Pi}}{T_\Sigma}\right)^2 + \left(1 - \frac{T_{K\Pi}}{T_\Sigma}\right)^2} \\ \psi &= \arctan\left(\frac{T_\Sigma - T_{K\Pi}}{T_\Sigma - T_{45^\circ \Pi}}\right) \end{aligned}\right\}. \tag{3.132}$$

The obtained relations make it possible to represent the loss matrix, determined by the expression (3.117), through the radio brightness temperatures in the following form:

$$G = \frac{T_\Sigma}{T_0}\begin{pmatrix} \frac{T_{\Gamma\Pi}}{T_\Sigma} & \sqrt{\left(1-\frac{T_{45^\circ \Pi}}{T_\Sigma}\right)^2 + \left(1-\frac{T_{K\Pi}}{T_\Sigma}\right)^2} e^{i\psi} \\ \sqrt{\left(1-\frac{T_{45^\circ \Pi}}{T_\Sigma}\right)^2 + \left(1-\frac{T_{K\Pi}}{T_\Sigma}\right)^2} e^{-i\psi} & \frac{T_{B\Pi}}{T_\Sigma} \end{pmatrix}. \quad (3.133)$$

As can be seen from expression (3.133), in order to completely determine the radiator loss matrix, it is sufficient to measure its radio brightness temperature on four polarizations for which the corresponding points on the Poincare sphere do not lie on one large circle, in particular, they can be any three linear and a circular.

Let us now turn to the determination of the dependence of the radiating capacity of bodies on the type of polarization. The radiating capacity, as well as the radio brightness temperature, is proportional to the power loss of the auxiliary electromagnetic wave, determined by means of the relation (3.113). Since expressions (1.7) are common for the orthogonal components of an elliptically polarized wave, then E_1 and E_2 without any loss restrictions on generality, we can assume that they are given in the second singular polarization basis. This allows us to determine, using the relations (3.133) and (3.130), the losses on polarization with the parameters γ, Φ in the general form:

$$\Pi = \chi_1^2 \cos^2\gamma + \chi_2^2 \sin^2\gamma = 0.5(\chi_1^2 + \chi_2^2)(1 + q\cos 2\gamma). \quad (3.134)$$

In order to convert from the value of Π to the radiating capacity, the first must be attributed to the corresponding value for an absolutely black body. As it is known, it must absorb all of the energy incident on it, regardless of the type of polarization. The relation (3.125) in this case necessarily yields $\chi_1 = \chi_2 = 1$; i.e., $\Pi_{A\text{Ч}T} = I$ and the loss matrix for it $G_{A\text{Ч}T} = I$.

Expression (3.125) shows that the maximum radiating capacity takes place when measurements are made on the object's own polarization ($\gamma = 0$); it is equal to χ_1^2. The minimum is equal χ_2^2 (for clarity, it is accepted as $\chi_1 > \chi_2$). Thus, the physical meaning of the eigenvalues of the loss matrix becomes clear, and also the validity of their representation in the form of real numbers squares:

$$\chi_1^2 = \Pi_{\max} \geq 0; \quad \chi_2^2 = \Pi_{\min} \geq 0. \quad (3.135)$$

So, with a change in the form of polarization, the radiating capacity is continuously changing from χ_1^2 to χ_2^2 .

The total power of the thermal radio emission can be found as the sum of the values of Π corresponding to two orthogonal polarizations. To write the expression for the Π_{OPT} on the polarization orthogonal to the one under consideration, it is necessary to replace γ by $\pi/2 - \gamma$ and $2\eta + 2\delta$ by $2\eta - 2\delta$. From the relation (3.134), in this case, it follows:

$$\Pi_{\text{OPT}} = \chi_1^2 \sin^2 \gamma + \chi_2^2 \sin^2 \gamma = 0.5(\chi_1^2 + \chi_2^2)(1 - q \cos 2\gamma). \tag{3.136}$$

Then the expression for the total power loss will have the form:

$$\Pi_1 = \Pi + \Pi_{\text{OPT}} = \chi_1^2 + \chi_2^2 = SpG. \tag{3.137}$$

The relation (3.128) explains the physical meaning of the invariance of the trace of the loss matrix. However, it should be borne in mind that the value (3.137) refers to the case when the energy of an auxiliary plane wave is twice as large as that of a wave determined by the expression (1.7). Therefore, the $\Pi_{A\,\text{ЧТ}} = 2$ and the radiating capacity determined by the total radio flux will be:

$$\Pi_\Sigma = 0.5(\Pi + \Pi_{\text{OPT}}) = 0.5(\chi_1^2 + \chi_2^2) \tag{3.138}$$

and accordingly the total radio brightness is:

$$T_\Sigma = 0.5(\Pi + \Pi_{\text{OPT}})T_0 = 0.5(\chi_1^2 + \chi_2^2)_{T_0}. \tag{3.139}$$

In radiometry, the dimensionless parameter m, sometimes called the polarization coefficient, equal to the ratio of the difference between the extreme values of the radio brightness temperature T on orthogonal polarizations to their sum (as will be shown below, this value coincides with the degree of polarization of the PPW) is used often. Let us find its value, using expressions (3.134), (3.136), and (3.127):

$$m = \frac{T_1 - T_2}{T_1 + T_2} = \frac{\chi_1^2 - \chi_2^2}{\chi_1^2 + \chi_2^2} = q. \tag{3.140}$$

Suppose that two antennas provide the reception of signals on extreme polarizations. In this case, there will be no statistical connection between the orthogonal components of the electromagnetic wave $\vec{E}_1$ and $\vec{E}_2$, i.e., $< \vec{E}_1 \vec{E}_2^* >= 0$, which means that in the polarization basis under consideration, the coherence matrix K will be diagonal. The matrix of losses will also be diagonal, since the considered basis is its eigenbasis. As for the diagonal elements corresponding to each other, they will be equal to within a constant in both matrices, i.e., the radio brightness temperature in the first channel will be proportional $< |E_1|^2 >$, and in the second $< |E_2|^2 >$. Consequently, in the polarization basis under consideration, the loss and coherence matrices of the intrinsic thermal radio emission coincide with each other to within a proportionality coefficient that depends on the distance, which can be ignored later on.

The loss and coherence matrices, when changing to a new polarization basis, are changed by transforming the similarity with the help of the same unitary matrix. This indicates that in any basis both these matrices coincide, their general form is determined by the expressions (3.11) and (3.133). The calculation of the degree of polarization by formula (3.10) gives just the value of the degree of anisotropy q introduced by the relation (3.121).

To determine the polarization type of the polarized component, one can use the representation of the coherence matrix in the form of a sum of matrices corresponding to the polarized and unpolarized waves (3.20). Using such a representation, it is not difficult to prove that the polarized component is one of the extreme polarizations corresponding to the larger of the values of χ_1^2 and χ_2^2. The obtained results allow to determine the complex correlation coefficient between orthogonal components from measurements of the radio brightness temperature at some different polarizations. Let us demonstrate this. We represent the off-diagonal element of the coherence matrix in the form (3.12). In this case, μ represents the degree of coherence, M is the effective phase difference. The loss matrix (3.133) is written in a linear polarization basis. Therefore, the expression found from it for the complex correlation coefficient μe_{iM} will refer to the horizontal and vertical components. Direct calculations yield:

$$\left.\begin{aligned} \mu &= \sqrt{\left(1-\frac{T_{45^\circ \Pi}}{T_\Sigma}\right)^2+\left(1-\frac{T_{K\Pi}}{T_\Sigma}\right)^2}\left(\sqrt{\frac{T_{\Gamma\Pi}}{T_\Sigma}\cdot\frac{T_{B\Pi}}{T_\Sigma}}\right)^{-1} \\ M &= \arctan\left(\frac{T_\Sigma - T_{K\Pi}}{T_\Sigma - T_{45^\circ \Pi}}\right) \end{aligned}\right\}. \tag{3.141}$$

The relations obtained above make it possible to establish a relationship between the radio brightness temperature and the Stokes parameters. To do this, we use the representations of the coherence matrix in terms of the Stokes parameters and the loss matrix in the form (3.133). Elementary transformations allow to write down:

$$\left.\begin{aligned} J_0 &= \frac{T_{\Gamma\Pi}+T_{B\Pi}}{T_0}A \;;\quad Q=\frac{T_{\Gamma\Pi}-T_{B\Pi}}{T_0}A \\ U &= \frac{T_\Sigma - T_{45^\circ \Pi}}{T_0}A \;;\quad V=\frac{T_{K\Pi}-T_\Sigma}{T_0}A \end{aligned}\right\}, \tag{3.142}$$

where the coefficient A is the previously mentioned coefficient of proportionality between the loss and coherence matrices. Its dimension coincides with the dimension of the Stokes parameter and is expressed in W/m^2.

Expressions for the third and fourth Stokes parameters can be represented in a slightly different form if we recall that T_Σ / T_0 is the total radiating capacity, which is invariant to the choice of the polarization basis. This makes it possible to write it in the form:

$$2T_\Sigma = \left(T_{\Gamma\Pi}+T_{B\Pi}\right)=\left(T_{45^\circ \Pi}+T_{135^\circ \Pi}\right)=\left(T_{K\Pi\Pi}+T_{K\Pi\Lambda}\right), \tag{3.143}$$

where and is the radio brightness temperature measured on the right and left circular polarizations, respectively. The relationship between the third and fourth Stokes parameters with the radio brightness temperature can be represented by the following relationships:

$$\left.\begin{array}{l} 2U = T_{135^\circ П} - T_{45^\circ П} \\ 2V = T_{КПП} - T_{КПЛ} \end{array}\right\}. \tag{3.144}$$

3.3.2 *Influence of Spectral Characteristics of Radio Waves on Their Polarization Properties*

Apparatus for analyzing the polarization properties of radio signals is previously developed mainly in response to the needs of the active radar. This circumstance determined the following peculiarities: the presence of a priori information on the structure of a regular signal and its narrow band. The first made it possible to distinguish the signal from noise not only in intensity, but also in a number of other properties (the modulation law, the phase structure, the frequency content of the video pulse, etc.). The second feature allowed us to consider signals in the quasi-harmonic approximation, when the width of the frequency spectrum of the signal Δf_C is assumed to be much smaller than the carrier frequency f_0.

The enumerated above features made it possible with reference to the polarization parameters of the signal to ignore such significant factors as the width of the receiving band of the radio receiver $\Delta f_{\Pi P}$ and time of the measurement. The receiver band is matched to the signal spectrum, and the measurement time is also determined by its characteristics. Bandwidth $\Delta f_{\Pi P}$ affects only the signal-to-noise ratio, but not the signal characteristics on condition that $\Delta f_{\Pi P} \geq \Delta f_C$.

In microwave radiometry, the situation is different. By its nature, radiant heat emission is an electromagnetic field created by the chaotic thermal motion of charged microparticles of the medium, that is, it has the same nature as most radio equipment noise. The signal power is proportional to the receiver band $\Delta f_{\Pi P}$ (in the microwave range, the radiothermal radiation has a uniform frequency spectrum), which caused the broadness of the microwave radiometric receivers when the condition $\Delta f_C \ll f_0$ does not hold. This feature does not allow us to take full advantage of the methods of analyzing the polarization structure of radio signals developed for the active radar in the study of the same characteristics of radiothermal radiation. The problem of taking into account the bandwidth $\Delta f_{\Pi P}$ would arise for the active radar, if the condition $\Delta f_{\Pi P} \ll \Delta f_C$ were satisfied.

In literature, it is pointed out that in the generic case, the polarization characteristics of radio signals depend on the measurement time. This circumstance should also be taken into account when analyzing the polarization properties of radiothermal radiation, since in microwave radiometry the measurement time is determined by the required sensitivity of the receiver and is not related to the signal structure.

It follows from the aforementioned that the polarization characteristics of radiothermal radiation determined from measurements are apparently related to the receiver band $\Delta f_{\Pi P}$ and the measurement time and, consequently, the question of the polarization properties of radiothermal radiation requires additional studies.

Consider the effect of spectral characteristics on the polarization state of the received signal.

A quasi-harmonic oscillation is described by a function $E(t)$ in the form:

$$E(t) = E_m(t)\exp[i(\omega t + \varphi(t))], \tag{3.145}$$

where $E_m(t)$ and $\varphi(t)$ are usually considered slowly varying in comparison with the $\sin \omega t$ functions of time.

The representation of the signal in the form (3.145) is not limited by the requirement of a slow change in the functions $E_m(t)$ and $\varphi(t)$, but, if this condition is violated, the convenient notion of the envelope of the signal becomes ambiguous. If in the statement of the problem the concept of the signal envelope is not used separately, then the broadband signals can also be represented in the form (3.145). Let us use this circumstance to analyze the polarization characteristics of radiothermal radiation.

The properties of the broadband signal of the form (3.145) are considered using the same scheme as the light vibrations, that is, the elements of the coherence matrix are in some arbitrary orthogonal basis, and the degree of polarization and the parameters of the polarization ellipse are determined through them.

It follows from the formula (3.13) that the degree of polarization is equal to the maximum modulus value of the complex correlation coefficient of the orthogonal components E_1 , E_2, which is achieved in the first singular AB, which for clarity will be assumed as Cartesian.

As discussed in Sect. 3.1, the polarization state of radio emission is characterized by the intensity of unpolarized and polarized components, the amplitudes and phase differences of the orthogonal components for the latter. Obviously, the phase difference can be measured only if the measurement time τ_0 is shorter than the correlation time of the orthogonal components, which reaches the maximum value, as was previously mentioned, in the first special AB.

The correlation time τ_K is determined by the known relation:

$$\tau_K^2 = \left[\int\limits_{-\infty}^{\infty} (\tau - \overline{\tau})^2 K^2(\tau)\mathrm{d}\tau\right] \Bigg/ \left[\left[\int\limits_{-\infty}^{\infty} K^2(\tau)\mathrm{d}\tau\right]\right], \tag{3.146}$$

where $K(\tau)$ is the correlation function of the orthogonal components E_x, E_y,

$$\overline{\tau} = \left[\int\limits_{-\infty}^{\infty} \tau K^2(\tau)\mathrm{d}\tau\right] \Bigg/ \left[\int\limits_{-\infty}^{\infty} K^2(\tau)\mathrm{d}\tau\right], \tag{3.147}$$

$K(\tau)$ is related to the mutual spectral density by the Fourier transform, and, therefore, does not directly depend on the spectrum of the input signal.

As an illustration to what has been said, we consider the variation of the Stokes parameters U and V, set by relation (3.54), depending on the averaging time. In this case, the corresponding expressions will have the form: $S_2 = U = 2 < E_x E_y \cos 2\delta >$, $S_3 = V = 2 < E_x E_y \sin 2\delta >$. With a limited correlation time, an increase in the observation time leads to an increase in the range of variation $\varphi(t)$ and, correspondingly, to a decrease in the value $U^2 + V^2$, which is easy to demonstrate using simple trigonometric relations. Reduction of the latter means a decrease in the degree of polarization m, determined by the relation (3.22).

If $\tau_0 \geq \tau_K$, then the change in the phase difference $\varphi(\mathrm{t})$ during the measurement time becomes so large that $\langle \sin \varphi \rangle = \langle \cos \varphi \rangle = 0$, then $m = Q/J_0$. Thus, it is possible to obtain information only on the intensity of the polarized and unpolarized components.

Apparently, the polarization characteristics of radio emission turn out to depend on the relationship between the measurement time τ_0 and the correlation time of the orthogonal components, which in turn is determined by the mutual spectral density.

Let us now consider the radiothermal emission. We consider it to be a random δ—correlated electromagnetic field. Each of the orthogonal components of such a process can be represented as a set of an infinite number of harmonic oscillations with the same weight. Consider the case when the bandwidth $\Delta f_{\Pi P}$ is infinitely narrow and is equal to d*f*. The signals allocated by the receiver on both polarizations, will in this case, be completely correlated. The resulting from the addition of these oscillations in one plane intensity J will depend on the phase difference 2φ between them:

$$J = 2J_0 \cos^2 \varphi = J_0(1 + \cos 2\varphi), \tag{3.148}$$

where $2J_0 = \overline{|E_x|^2} = \overline{|E_y|}^2$, that is, the selection of components occurs in the first singular basis.

In accordance with the properties of the coherence matrix, the elements $\overline{E_x E_y^*}$ and $\overline{E_x^* E_y}$ are found as the sum of the corresponding elements for each of the harmonic components. In this case, the value $\mathrm{Re}\overline{E_x E_y^*}$ can be found from the known signal component J_0:

$$J = 2J_0 \frac{\int \cos(\varphi + \varphi_f) G(f) df}{\int G(f) df}, \tag{3.149}$$

where $2\varphi_f$ is the phase change caused by the frequency variation within the band $G(f)$, 2φ refers to the central frequency f_0 .

In the case of a rectangular band of a receiver of F width, expression (3.149) takes the form:

$$J = \frac{2J_0}{F} \int \cos^2(\varphi + \varphi_f) df. \tag{3.150}$$

We represent the phase shift 2φ in the form $2\varphi = \frac{2\pi}{\lambda}x$, where x is a certain distance corresponding to the phase shift, then, $2\varphi_f = 2\pi x(f - f_0)/c$.

Substituting this relation into (3.150) and integrating, we finally obtain:

$$J = J_0 \left[\frac{\sin 2\varphi_0}{2\varphi_0} + 1 \right]. \tag{3.151}$$

where $2\varphi_0 = \pi x F/c$.

If expression (3.151) is compared to the expression for the mutual intensity of two quasi-monochromatic light beams of equal intensity, then the value $m = \left| \frac{\sin 2\varphi_0}{2\varphi_0} \right|$ a is similar to the modulus of the degree of light oscillations coherence and also satisfies the inequality: $m \le 1$.

The concept of coherence and polarization are interrelated, but in active radar, it is easier to measure the energy characteristics of a signal. This ability has left an imprint on the apparatus with which the polarization of radio signals is investigated. Thus, the degree of polarization, which is determined by the coherence of the orthogonal components, can take only non-negative values, whereas in optics, they operate with the complex degree of coherence of light oscillations. Therefore, by investigating the polarization characteristics of the radiation, it is necessary to consider the modulus of the complex degree of coherence (3.12).

Taking what has been said into account, from expression (3.151), we get:

$$\overline{E_x E_y^*} = 0.5 J_0 m e^{-2i\delta_0}. \tag{3.152}$$

The value $\overline{E_x E_y^*}$ found in this way satisfies all of the requirements imposed on the complex analytic function, that is, its imaginary and real parts are interconnected by the Hilbert transform, and it is unambiguously determined by the function $\mathrm{Re}\overline{E_x E_y^*}$.

Conforming to what has been said, the coherence matrix under consideration is to be written in the form (3.14) with $\eta = 0$.

For $m = 0$, in accordance with the properties of the coherence matrix, we have unpolarized radiation, which corresponds to a phase advance, $2\delta_0 = \pi$. Physically, this means the presence of a sum of an infinite number of harmonic oscillations of equal amplitude, the initial phase of which is uniformly distributed over the interval 2π. Obviously, such radiation is unpolarized, which is confirmed by formula (3.14).

When $2\delta_0 = 0$, $m = 1$ the coherence matrix K degenerates, that is, the radiation is completely polarized. Physically, this corresponds to an infinitely narrow reception band, that is, to the allotment of harmonic components that are completely coherent, and hence the radiation is completely polarized.

Since the receiver bandwidth F has a finite value, the radiation is always partially polarized. The degree of polarization is determined from Eq. (3.14) with the help of relation (3.10).

Let us consider the dependence of the degree of polarization m on the bandwidth of F and the measurement time. As was mentioned before, the maximum degree of polarization is achieved when the condition $\tau_0 \ll \tau_K$ is satisfied. Taking into account the random nature of the input signal, it is necessary to average the signal for obtaining average values, at least for several periods, that is, to provide condition $\tau_0 > 1/f_0$. Therefore:

$$1/f_0 < \tau_0 \ll \tau_K. \tag{3.153}$$

The distance x introduced in the calculation of (3.150) is written in the form $x = \tau_0 \lambda f_0$.

Taking into account that $2\varphi_0 = \pi x F/c$ and using the representation x obtained above, we find:

$$2\varphi_0 = \pi \tau_0 F. \tag{3.154}$$

Thus, the relationship between the degree of polarization and the width of a rectangular reception band is determined by the relation:

$$m = \left| \frac{\sin(\pi \tau_0 F)}{\pi \tau_0 F} \right|. \tag{3.155}$$

If we do not require the attainment of the maximum degree of polarization, then the requirement $\tau_0 \ll \tau_K$ turns out to be unimportant, and, therefore, the only limitation on the applicability of relation (3.155) is the condition $\tau_0 > 1/f_0$.

The set of expressions (3.153) and (3.155) determines the conditions under which the polarization parameters associated with the phase structure of the signal can be measured. If these conditions are not met, only the energy characteristics can be estimated, in the process of measuring which averaging of the already random values of the phase of the signal occurs.

Due to the specifics of radiothermal radiation reception, the averaging time τ_0 is usually on the order of a second or more, which is much longer than the signal correlation time at the input. Consequently, the phase information for radiometric measurements is known to disintegrate. For a random field of radiothermal radiation, the mean value of the phase is zero, and the points depicting the polarization of the radiation on the Poincare sphere are located on the equator, that is, this radiation is linearly polarized. The latter in particular means that the intrinsic polarizations of the radiothermal radiation are also linear.

This refers to the random field of radiothermal radiation from an isotropic medium, when no phase changes are made by the average, in addition to the linear phase advance.

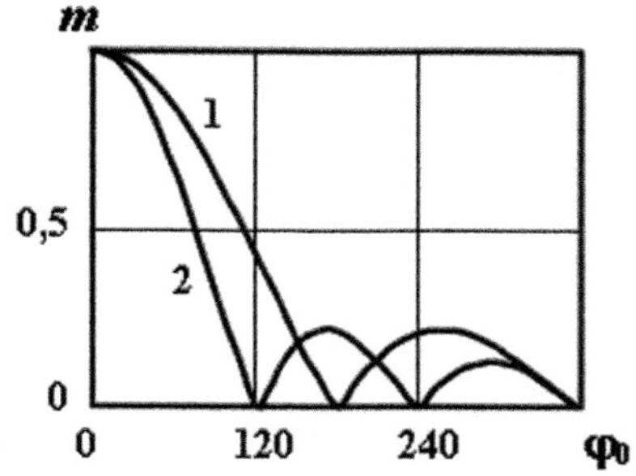

Fig. 3.8 Dependence of the degree of polarization on the parameter φ_0 for various types of frequency response of the receiving device 1—rectangular characteristic; 2—bell-shaped characteristic

If the radiative medium does not satisfy this condition and has a corresponding frequency dispersion, then the polarization of the radiothermal radiation may differ from linear. The same effect can be achieved by placing a layer with similar properties between the radiating medium and the antenna receiver.

The mutual spectral density of the orthogonal components can be found from the correlation function. For the production of two δ-correlated signals, the mutual correlation function will also be a δ function. With a finite receiving bandwidth, the correlation time of each of the orthogonal components differs from zero. Without delving into details, we note that in this case, the correlation time corresponding to the mutual spectral density is not greater than the correlation time of each of the components.

As an example, consider the case where the frequency characteristic of the receiver is described by a function $G(f) = \cos^2\left[\frac{\pi}{2}\frac{f-f_0}{\Delta f}\right]$, where $f_{\max} - f_0 = \Delta f$. This form is similar to the bell-shaped frequency characteristic.

Substituting $G(f)$ into (3.149) and integrating over the interval from $(f_0 - \Delta f)$ to $(f_0 + \Delta f)$, we finally get:

$$m = \left|\sin\left(1 + \frac{2\pi\varphi_0}{4\varphi_0^2 - \pi^2}\right)\right| \tag{3.156}$$

at that $\lim_{\varphi_0 \to 0} m(\varphi_0) = 1$, $\lim_{\varphi_0 \to \infty} m(\varphi_0) = 0$, the coherence matrix corresponds to the polarized radiation in the first case and unpolarized in the second case.

For the selected form of the frequency characteristic of the receiver, the components that are closer to the central frequency f_0 make a large contribution to the total intensity, and therefore, the degree of polarization is to be higher than in the case of a rectangular band, with the same value $2\varphi_0$. These considerations are confirmed in Fig. 3.8, which shows the dependence of the degree of polarization m on $2\varphi_0$ for both cases.

The information on phase relationships applied to the actual measurement time in microwave radiometry is also completely disintegrated here. The degree of polarization with increasing $2\varphi_0$ decreases slower than in the first case, and the term $(\pi \sin 2\delta_0)/(4\delta_0^2 - \pi^2)$ that appeared due to the non-rectitude of the receiving band decreases faster than $\sin(2\varphi_0/2\varphi_0)$, causing depolarization of the radiation due to the phase advance difference for different harmonic components.

Reference

1. Haykin S (2012) Cognitive dynamic systems: perception-action cycle, radar and radio. Cambridge University Press

Chapter 4
Analysis of the Signal Polarization of Navigation Systems in the Plane of Geometrical Parameters

4.1 Polarization Phasor

4.1.1 Basic Properties and Transformations

In Sect. 1.2, the concept of the phasor p was introduced by the relation (1.11), as the ratio of complex amplitudes E_y and E_x. A generalization of this concept is a phasor, defined in an arbitrary basis, as a relation $p = E_2/E_1$. Each polarization ellipse can be associated with a complex number or a point on the complex plane of the phasor. This representation of the electromagnetic wave polarization is sufficiently illustrative, often convenient, and allows solving a number of problems of applied significance.

The relations obtained in the previous sections make it possible to give a more general treatment of the value of p at which the phasor can be considered in an arbitrary elliptically polarized basis.

The polarization phasors p_1 and p_2 of orthogonal polarizations satisfy the relation:

$$p_1 p_2^* = 1, \tag{4.1}$$

which follows from the scalar product of two orthogonal vectors $\vec{E}_1 \vec{E}_2^* = 0$. The complex numbers p_1 and p_2 that satisfy the condition (4.1) have inverse values of the moduli, and their arguments differ by an angle π. The corresponding points p_1 and p_2 in the complex plane lie on a straight line passing through the origin, on different sides from it. Such points are called symmetric in the theory of functions of a complex variable.

The relations obtained in Sect. 1 allow to get a transformation of the phasor defined in the old polarization basis upon transition to a new polarization basis. In accordance with relations (1.11) and (1.43), the value of the phasor in the new AB will be written in the following form:

Kozlov A. I. et al., *Introduction to the Theory of Radiopolarimetric Navigation Systems*, Springer Aerospace Technology, https://doi.org/10.1007/978-981-13-8395-3_4

$$p_H = \frac{p_C - e^{2i\delta} \tan \gamma}{1 + p_C e^{-2i\delta} \tan \gamma} e^{2i\eta}. \tag{4.2}$$

Expression (4.2) is a fractional-linear function of a complex variable.

The fractional-linear function (4.2) is analytic and performs a conformal mapping. In this case, the transformation of the base unit vectors corresponds to the displacement of the origin of coordinates in the complex plane. If the basis vectors are given by the phasors p_1 and p_2, then it is obvious enough that the transformation of the corresponding basis can be carried out in a similar way. At that, if in the old polarization basis we denote the first unit vector p_{1C}^0, and the second one p_{2C}^0, then, similarly to (4.2), we obtain ($j = 1\ ,\ 2$):

$$p_{jH}^0 = \frac{p_{jC}^0 - e^{-2i\delta} \tan \gamma}{1 + p_{jC}^0 e^{2i\delta} \tan \gamma} e^{-2i\eta}, \tag{4.3}$$

In this case, the units in the new polarization basis p_{1H}^0 and p_{2H}^0, like the unit vectors p_{1C}^0 p_{2C}^0, are connected by the relation (1.36), as the unit vectors of one basis.

It can be shown that the points of the base unit vectors $p_{1C}^0 = 0$ and $p_{2C}^0 = \infty$ transition to the points $p_{1H}^0 = -e^{-2i(\eta+\delta)} \tan\ \gamma$, $p_{2H}^0 = e^{-2i(\eta+\delta)} \tan^{-1} \gamma$, which remain symmetric with respect to the origin during the transformation. Indeed, since the matrix Q is unitary, we have $p_{1H}^0 p_{2H}^{0*} = -1$.

For any fractional-linear transformation, there are two points that do not change their position on the complex plane (the so-called fixed points of the transformation). The complex numbers $p_{\Phi 1C}$ and $p_{\Phi 2C}$ corresponding to these points in the old basis satisfy condition (4.2).

The solution of this equation, taking into account the fact that the phasors in the new and old coordinate system are the same, yields:

$$p_{\kappa\Phi} = i\left[\left(\cos\ \gamma\ \sin\ \eta \pm \sqrt{1 - \cos^2 \gamma\ \cos^2 \eta}\right)e^{i(\eta - 2\delta)}\right] \csc\ \gamma. \tag{4.4}$$

For AB, which has $\eta = 0$, more convenient for graphical representation, we have $p_{\Phi\,1,2} = \pm i e^{-2i\,\delta}$.

The physical meaning of the polarization coefficients $p_{\Phi 1}$ and $p_{\Phi 2}$ lies in the fact that these coefficients determine such polarizations of the wave, where in both bases (old and new) they are expressed by the same complex number.

The phasor module can assume all possible values, including infinitely large ones. This property is inconvenient for the geometric representation of the field polarization. Therefore, often instead of the polarization relation modulus, we introduce the angle $\gamma = \arctan|p|$, which we have already came across in the representation of polarization on the Poincare sphere. Thus, the representation of the polarization state of a wave by means of a phasor on the Poincaré sphere, i.e., in fact, using a pair of numbers γ and δ is analogous to the polarization representation in Fig. 1.4.

4.1.2 The Probabilistic Description—The Gaussian Case

Let us now consider the statistical characteristics of the modulus and the argument of the polarization phasor for various probability models of partially polarized radio waves. Knowledge of such characteristics is obviously of considerable interest in practical radio polarimetry, since they are inextricably linked with the physical nature of radiation sources, scattering, and reflection of radio waves.

Let a partially polarized wave arrive at the receiving point, described by the joint probability density (2.31). Then, in accordance with (1.98), we have:

$$W_2(P,\delta) = \int_0^{\infty} W_3(E, PE, 2\delta)|E_1|dE_1, \tag{4.5}$$

where for convenience $|p| \equiv P$ is accepted.

After integration, we find the joint probability density for an arbitrary polarization basis:

$$W_2(P, 2\delta, a_1, b_1, \delta_0, R, \beta, h) = \frac{Ph^2(1-R^2)}{\pi\left[1 - 2RPh\cos(2\delta-\beta) + h^2P^2\right]} \times \exp\left\{-\frac{a_1^2}{2(1-R^2)}\left[1 - 2Rb\cos(2\delta_0-\beta) + b^2\right]\right\} \times (1+T)\exp T, \tag{4.6}$$

where

$$T = \frac{a_1^2\left[1 - 2Rb\cos(2\delta_0-\beta) + R^2b^2\right]\left[1 + 2\chi hP\cos(2\delta - 2\delta_0 - \mu) + \chi^2h^2P^2\right]}{2\left[1 - 2RhP\cos(\delta_0-\beta) + h^2P^2\right]}.$$

Comparing the expressions (2.31) and (4.6), one can note that both probability densities are determined by the same parameters. However, in contrast to equality (2.31) in expression (4.6), the variances of the quadrature components of the orthogonally polarized components are included in the form of a ratio h^2. This circumstance is a consequence of the fact that the modulus and the argument of the polarization phasor describe the wave with accuracy up to the "amplitude" of the polarization ellipse and do not carry information about the energy characteristics of this wave.

The nature of the change in the form of the joint probability density (4.6) as a function of the parameters a, b, R, h at $\beta = 0$ and $\delta_0 = 0$ is illustrated in Fig. 4.1.

In the specific case when $a_1 = a_2 = 0$ (the Rayleigh modulus), expression (4.6) is simplified:

$$W_2(P, 2\delta, R, \beta, h) = \frac{Ph^2(1-R^2)}{\pi\left[1 - 2RhP\cos(2\delta-\beta) + h^2P^2\right]^2}. \tag{4.7}$$

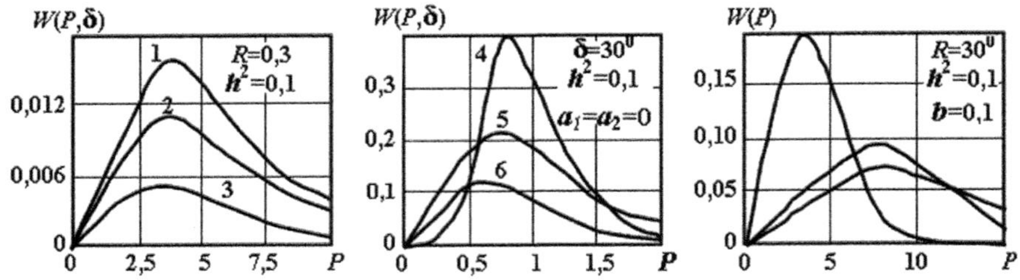

Fig. 4.1 Joint probability density of the modulus and the argument of the polarization phasor

is the sum of a monochromatic wave and a quasimonochromatic wave, the orthogonally polarized components of which are not correlated $R = 0, \beta = 0$, then it follows from (4.6):

$$W_2(P, 2\delta, a, b, 2\delta_0, h) = \frac{h^2 P}{\pi\left(1 + h^2 P^2\right)^2} \exp\left[-\frac{a_1^2\left(1 + b^2\right)}{2}\right]$$
$$\exp\left\{\frac{a_1^2\left[1 + 2bhP\,\cos 2(\delta - \delta_0) + b^2 h^2 P^2\right]}{2\left(1 + h^2 P^2\right)}\right\}$$
$$\times\left\{1 + \frac{a_1^2\left[1 + 2bhP\,\cos 2(\delta - \delta_0) + b^2 h^2 P^2\right]}{2\left(1 + h^2 P^2\right)}\right\}. \tag{4.8}$$

When the parameter b ($a_2 = 0$) is zero, expression (4.6) is simplified:

$$W_2(P, 2\delta, a_1, R, \beta, h) = \frac{h^2 P\left(1 - R^2\right)}{\pi\left[1 - 2RhP\,\cos(2\delta - \beta) + h^2 P^2\right]^2}$$
$$\times \exp\left\{-\frac{a_1^2}{2\left(1 - R^2\right)} + \frac{a_1^2\left[1 + 2Rh^2\cos(2\delta - \beta) + R^2 h^2 P^2\right]}{2\left[1 - 2RhP\,\cos(2\delta - \beta) + h^2 P^2\right]}\right\}$$
$$\left\{1 + \frac{a_1^2\left[1 + 2RhP\,\cos(2\delta - \beta) + R^2 h^2 P^2\right]}{2\left[1 - 2RhP\,\cos(2\delta - \beta) + h^2 P^2\right]}\right\}. \tag{4.9}$$

We will note that the special cases described by expressions (4.8) and (4.9) are of independent interest in the study of the statistical properties of partially polarized radio waves, respectively, in the second and first singular polarization bases.

To determine the one-dimensional probability density of the modulus of the polarization wave phasor, in accordance with Sect. 1.5 it is necessary to perform the corresponding integration in (2.31).

When the parameters R and β are equal to zero, the determination of the probability density $W(P, a_1, b, h)$ is substantially simplified. In this case, it is convenient to use

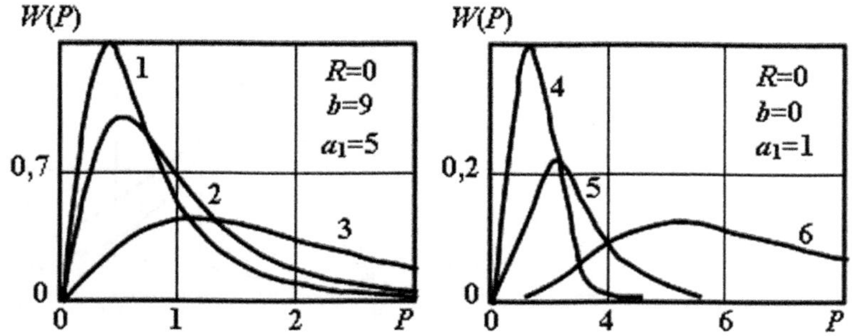

Fig. 4.2 Probability density of the polarization phasor modulus

formula (4.8), integrating it over all possible values of the argument of the polarization phasor δ. As a result of integration, we get:

$$
\begin{aligned}
&W(P, a_1, b, h) \\
&= \frac{2Ph^2}{(1+h^2P^2)} \exp\left[-\frac{a_1^2(b^2+h^2P^2)}{2(1+h^2P^2)}\right] \\
&\left\{\left[1+\frac{a_1^2(1+b^2h^2P^2)}{2(1+h^2P^2)}\right] I_0\left(\frac{a_1^2bhP}{1+h^2P^2}\right)+\frac{a_1^2bhP}{1+h^2P^2} I_1\left(\frac{a_1^2bhP}{1+h^2P^2}\right)\right\}
\end{aligned} \tag{4.10}
$$

Graphs of the probability density of the polarization phase modulus (4.10) are shown in Fig. 4.2.

It can be seen from the figure that when the parameter h^2 is changed, the form of the probability density changes significantly. In the absence of deterministic components of the orthogonally polarized components of the wave (the Rayleigh model), the expression for the probability density $W(P, R, h)$ is determined by integrating the equality (4.7) over the variable 2δ in the range $[0 \ldots \pi]$:

$$
W(p; R, h) = \frac{2Ph^2(1-R^2)(1+h^2P^2)}{\left[(1+h^2P^2)^2+4R^2h^2P^2\right]^{1,5}}. \tag{4.11}
$$

The character of the probability density (4.11) as a function of the values of the parameters R and h is illustrated in Fig. 4.3.

Approximate calculations using formula (4.11) yield the following values for the mean and dispersion of the phasor:

$$
\left.
\begin{aligned}
&< P > \equiv m_P = 1,57h^{-1}(1-0,25R^2-0,047R^4-0,019R^6) \\
&d_P^2 = (2,66+0,35R-2R^2)(0,053+1,07h-0,05h^2)^{-1}
\end{aligned}
\right\}. \tag{4.12}
$$

The corresponding relationships are shown in Fig. 4.4.

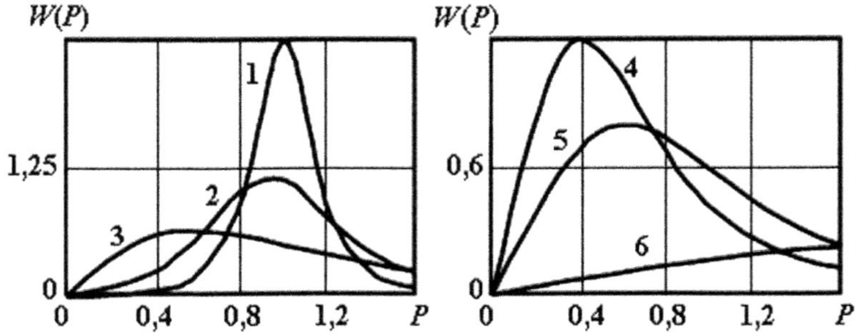

Fig. 4.3 Probability density of the phasor modulus

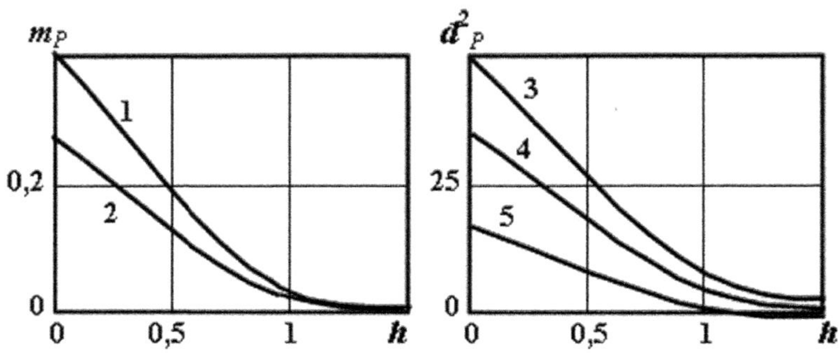

Fig. 4.4 Dependence of the mean value and dispersion of phasor modulus from h

The analysis of the obtained expressions and the curves shown in Figs. 4.1, 4.2, 4.3 and 4.4 shows the following.

The joint probability density of the modulus and the argument of the polarization phasor $W(P, 2\delta)$ and the one-dimensional probability density $W(P)$ are asymmetric functions, the degree of asymmetry of which decreases with an increasing modulus of the cross-correlation coefficient and the generalized polarization parameter $b = P_0 d_1/d_2$ of the total partially polarized wave.

For fixed values of the parameters a_1, b, R, and h, the maximum value of the function $W(P, 2\delta)$ corresponds to the zero value of the argument of the phasor 2δ. With the increase in 2δ, the maximum value $W(P, 2\delta)$ decreases. In this case, the degree of decrease essentially depends on the value of the parameter R.

The parameters a_1, b, R, and h determine the position of the modes of the considered probability densities on the abscissa axis. With increasing h, the average value m_P and the variance d_P^2 of the phasor modulus decrease, the modal values of the functions $W(P, 2\delta)$ and $W(P)$ increase, and their modes shift to the left along the abscissa axis. As the parameter a_1 (if $b_1 < 1$) increases, the modal values of these probability densities increase and shift to the left along the abscissa axis, while m_P

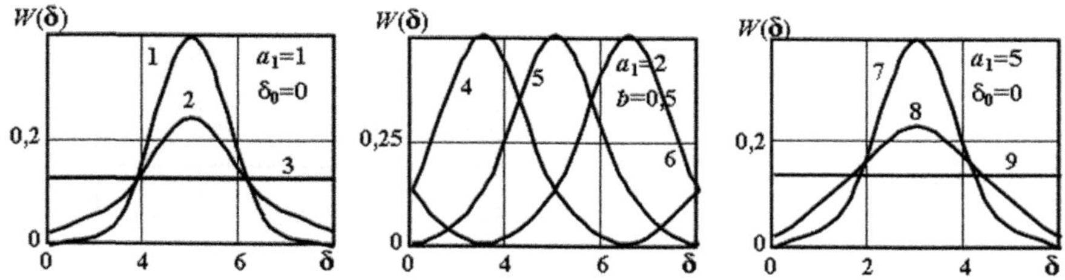

Fig. 4.5 Probability distribution density of the polarization phasor argument

and d_P^2 decrease. If parameter $b_1 > 1$, then an increase of a_1 leads to a decrease in the modal values $W(P, 2\delta)$ and $W(P)$, shifting them to the right along the abscissa axis.

Since the parameter b is related to the modulus of the polarization phasor of the deterministic component of the total wave by the relation $b = P_0 h$, where $P_0 = E_{01}/E_{02}$, it is obvious that for fixed values of h the decrease (increase) of the parameter b is equivalent to decreasing (increasing) the modulus P_0. Consequently, the position of the functions $W(P, 2\delta)$ and $W(P)$, their modes on the abscissa axis, as well as the value m_P, is to a large extent determined by the value of the modulus of the polarization phasor P_0.

Let us now consider the statistical characteristics of the argument of the polarization phasor. To determine the probability density of the polarization phasor argument $W(2\delta)$ and to determine its dependence on the parameters $a_1, b, h, 2\delta_0$, we consider the case when the orthogonally polarized components of the wave are not correlated ($R = \beta = 0$). Equating the parameters R and β to zero in formula (2.31) and performing the appropriate integration, one can obtain an expression for $W(2\delta)$ in a closed form, which being cumbersome is not given here.

The formula for determining the dependence of $W(2\delta)$ on the parameter R is more illustrative:

$$W(2\delta, R) = \frac{1 - R^2}{2\pi}\left[\frac{1}{1 - R^2\cos^2 2\delta} + \frac{\pi - \arccos(R\cos 2\delta)}{\left(1 - R^2\cos^2 2\delta\right)^{1,5}} R\cos 2\delta\right]. \tag{4.13}$$

Figure 4.5 shows a family of curves of PDD-$W(2\delta)$ calculated using exact formulas. As can be seen from Fig. 4.5, the probability density is determined by three parameters $(a, b, 2\delta_0)$ and, unlike the probability density $W(P)$, does not depend on the parameter h^2.

The probability density $W(2\delta)$ is a unimodal function, where the mode of the function coincides with the value $2\delta = \pi + 2\delta_0$ (with a change in 2δ within the limits $0 \ldots 2\pi$). When $2\delta_0 = 0$, the function $W(2\delta)$ is symmetric with respect to $2\delta = \pi$ and with increasing parameters a_1 and b is grouped around the mean value.

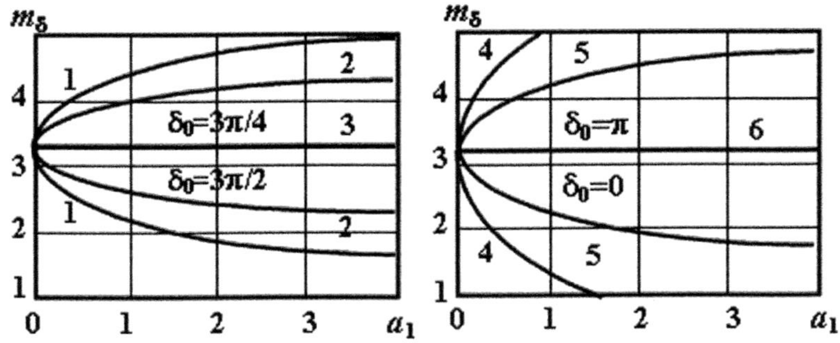

Fig. 4.6 Mean values and dispersions of the polarization phasor argument

At the limits of the interval $[0, 2\pi]$, it is less than $1/2\pi$ and with increasing a_1 and b tends to zero. At $a_1 = 0$ or $b = 0$, probability density $W(2\delta)$ becomes uniform.

When $a_1, b > 2 \ldots 3$, the exact formula for $W(2\delta)$ can well be approximated by the Gaussian law with an average value equal to $2\delta_0$ and the variance determined by the equality $d_{2\delta}^2 = (1 + b^2)/(a_1^2 b)$. With $a_1 \leq 1$ and $b \leq 1$, a good approximating function is:

$$W(2\delta; a, b, 2\delta_0) = \frac{1}{2\pi}\left\{1 + \frac{\pi\, a_1^2 b}{4} \exp\left[-\frac{a_1^2}{8}(1 + b^2)\right] \cos 2(\delta - \delta_0) + \frac{a_1^4 b^2}{8} \exp\left[-\frac{a_1^2}{6}(1 + b^2)\right] \cos[4(\delta - \delta_0)]\right\}. \tag{4.14}$$

Finally, for arbitrary values of a_1 and b it is convenient in some cases to use an approximation of the form [1]:

$$W(2\delta; a_1, b, 2\delta_0) \approx \frac{1}{2\pi} \exp\left[\frac{a_1^2 b^2}{1 + b^2} \cos 2(\delta - \delta_0)\right] / I_0\left(\frac{a_1^2 + b^2}{1 + b^2}\right). \tag{4.15}$$

The probability density (4.15) allows a Fourier expansion with a coefficient $C_n = I_n\left(\frac{a_1^2 b^2}{1+b^2}\right) I_0^{-1}\left(\frac{a_1^2 b^2}{1+b^2}\right)$.

It follows from (4.15) that for $a_1 = 0$ or $b = 0$ the probability density $W(\delta) = \frac{1}{2\pi}$, and for $a_1 \gg 1$, $b \gg 1$ a probability density, it is normalized and has a variance equal to $a_1^2 b^2/(1 + b^2)$.

Figure 4.6 shows the dependences of $m_\delta = f_1(a_1, a_2, 2\delta_0)$ and $d_{2\delta}^2 = f_2(a_1, a_2, 2\delta_0)$.

It can be seen from the figures that with increasing parameters a_1 and a_2 (for $\delta_0 \neq \delta$), we have $m_\delta \to 2\delta = \pi + 2\delta_0$, with the dispersion $d_{2\delta}^2$ decreasing.

In some cases, when solving problems related to the use of polarization effects, it is necessary to know the statistical characteristics $z_C = \cos 2\delta$ and $z_S = \sin 2\delta$.

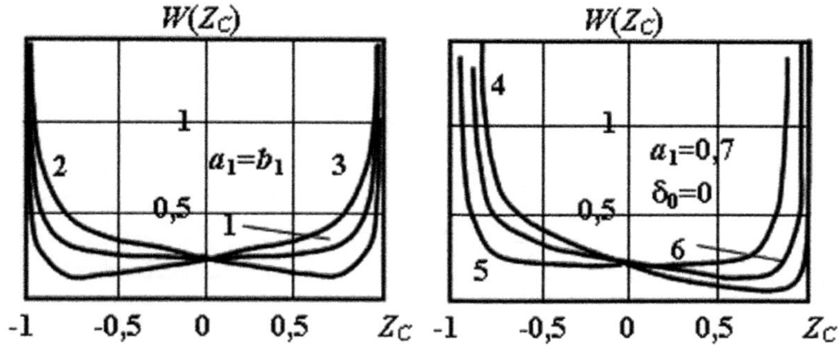

Fig. 4.7 Probability density of cosine of the argument of polarization phasor

To determine the probability density $W(z_C)$, we use the methodology outlined in Sect. 1.5.

After the transformations, we get:

$$W(z_C; a, b, 2\delta_0) = \frac{1}{\pi\sqrt{1 - z_C^2}}\left\{1 + 2\sum_{n=1}^{\infty} C_n \cos(n \arccos z_C)\cos 2n\delta_0\right\}. \tag{4.16}$$

It is not difficult to determine the expressions for the average value and variance of a random variable $z_C = \cos 2\delta$:

$$\begin{aligned} m_{z_C}(a_1, b_1, 2\delta_0) &= -\frac{\pi}{8}a_1^2 b \exp\left[-\frac{a_1^2(1+b^2)}{4}\right] \\ &\times \left[I_0\left(0.25a_1^2\right) + I_1\left(0.25a_1^2\right)\right]\left[I_0\left(0.25a_1^2b^2\right) + I_1\left(0.25a_1^2b^2\right)\right]\cos 2\delta_0 \\ d_{zC}^2(a_1, b, 2\delta_0) &= 0.5\left\{1 + 4\left[\exp\left(-0.5a_1^2\right) + 0.5a_1^2 - 1\right]\right. \\ &\times \left[\exp\left(-0.5a_1^2b^2\right) + 0.5a_1^2b^2 - 1\right]a_1^{-4}b^{-2}\cos 4\delta_0 \\ &- \frac{\pi^2}{64}a_1^4b^2\exp\left[-0.5a_1^2\left(1+b^2\right)\right]\left[I_0\left(0.25a_1^2\right) + I_1\left(0.25a_1^2\right)\right] \\ &\times \left[I_0\left(0.25a_1^2b^2\right) + I_1\left(0.25a_1^2b^2\right)\right]\cos^2 2\delta_0 \end{aligned} \tag{4.17}$$

Figure 4.7 shows the families of the cosine probability density curves of the polarization phasor argument for certain values of the parameters a_1 , b, and 2δ.

To determine the statistical characteristics of the sine of the polarization phasor argument $z_S = \sin 2\delta$, one can proceed similarly to the previous one. However, in this case it is necessary to take into account that in the interval $-\pi \leq 2\delta \leq \pi$ the function $2\delta = \arcsin z_S$ is two-valued: $2\delta_1 = 0, 5\pi - \arccos z_C$ and $2\delta_2 = 0, 5\pi + \arccos z_S$:

In this case, the above formulas (4.16) and (4.17) after replacing $\cos 2\delta_0$ in them by $\cos(0, 5\pi - 2\delta_0) = \sin 2\delta_0$ completely describe the statistical characteristics of the sine of the argument of the polarization phasor.

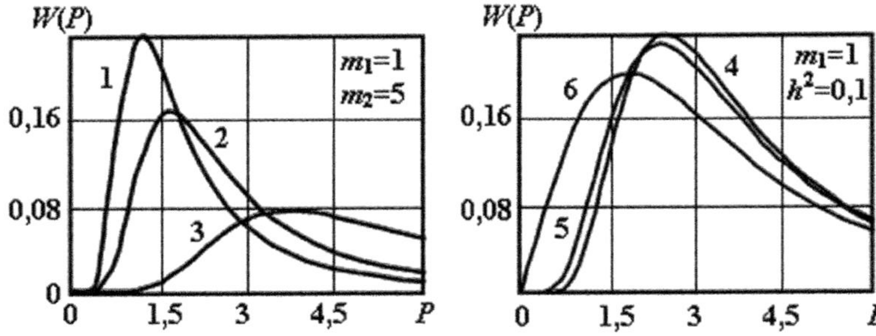

Fig. 4.8 Probability density of the polarization phasor modulus

When analyzing the statistical properties of stationary, partially polarized radio waves, in many applications it is advisable, as was shown above, to use the m-distribution, which generalizes other types of probabilistic models of envelopes of orthogonally polarized radio wave components and is a convenient approximation (2.2.3) of the latter. Let us consider the statistical characteristics of the polarization parameters of partially polarized waves in this approximation.

We assume that the envelopes of the orthogonally polarized components of the wave obey the m-distribution (2.55), whose parameters are m_j and Ω_j, where $j = 1, 2$. Suppose that $m_1 \neq m_2$ and that there is no mutual correlation between orthogonally polarized components. Then, by integrating over the variable E_1 in the expression $W(E_1, E_2)$, we find the probability density of the modulus of the polarization phase of a partially polarized wave in the form:

$$W\left(P; m_1, m_2, h_\Omega^2\right) = \frac{2r^{2m_2-1} m_1^{m_1} m_2^{m_2} h_\Omega^{2m_2}}{\left(m_1 + m_2 h_\Omega^2 P^2\right)^{m_1+m_2} B(m_1, m_2)}, \tag{4.18}$$

where $B(m_1, m_2) = \Gamma(m_1)\Gamma(m_2)\Gamma^{-1}(m_1 + m_2)$ is the beta function $h^2 = \Omega_1 \Omega_2$.

As can be seen, the probability density (4.18) is determined by three statistical parameters: m_1, m_2, and h_Ω^2. These parameters depend on the state of polarization of the wave and are functionally related to the parameters α_j, ρ_0, a_j, and φ_{0j} considered above and relation (2.56).

Figure 4.8 shows families of probability density curves (4.18) for several values of the parameters m_j and h_Ω^2.

The integral distribution function of the modulus of the polarization phasor $F(P)$, taking into account Eq. (4.18) after integration, takes the following form:

$$F\left(P; m_1, m_2, h_\Omega^2\right) = \left(\frac{m_2}{m_1} h_\Omega^2\right)^{m_2} \frac{P^2}{m_2 B(m_1, m_2)}\,{}_2F_1\left(m_1 + m_2, m_2, 1 + m_2, -\frac{m_2}{m_1} P h_\Omega^2\right). \tag{4.19}$$

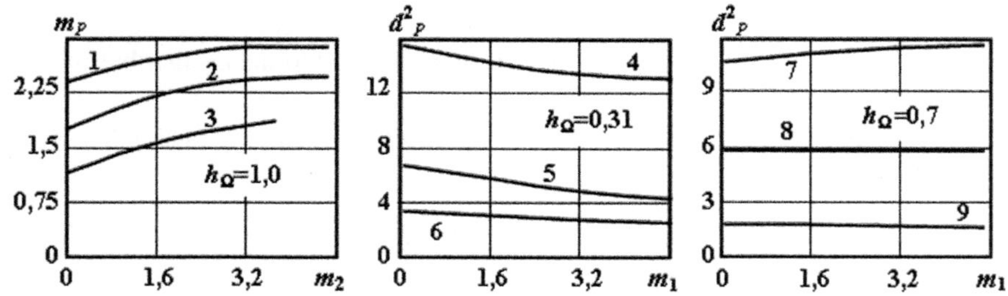

Fig. 4.9 Mean value and dispersion of the polarization phasor modulus

Direct calculations using formula (4.18) for ν-th moments give the following expression:

$$m_P^\nu\left(m_1, m_2 h_\Omega^2\right) = \frac{\Gamma(m_1 - 0.5\nu)\Gamma(m_2 + 0.5\nu)}{\Gamma(m_1)\Gamma(m_2)}\left(\frac{m_1}{m_2}h_\Omega^{-2}\right)^{0.5\nu}. \tag{4.20}$$

Hence, it is not difficult to obtain the average value and dispersion of the modulus of the polarization phasor:

$$\left.\begin{aligned} & m_P\left(m_1, m_2, h_\Omega^2\right) = \frac{\Gamma(m_1-0,5)\Gamma(m_2+0.5)}{\Gamma(m_1)\Gamma(m_2)}\left(\frac{m_1}{m_2}h_\Omega^{-2}\right)^{0.5} \\ & d_P^2\left(m_1, m_2, h_\Omega^2\right) = \left(\frac{m_1}{m_2}h_\Omega^{-2}\right)\left\{\frac{\Gamma(m_1-1)\Gamma(m_2+1)}{\Gamma(m_1)\Gamma(m_2)} - \left[\frac{\Gamma(m_1-0.5)\Gamma(m_2+0.5)}{\Gamma(m_1)\Gamma(m_2)}\right]^2\right\} \\ & m_1, m_2 > 0.5 \end{aligned}\right\}. \tag{4.21}$$

Figure 4.9 shows the dependencies (4.21) for different values of the parameters m_1, m_2, and h_Ω^2. Analysis of the obtained dependencies and graphs shows the following.

The probability density $W(P)$ is a one-modal asymmetric function with respect to its modal value. When the parameters m_j and h_Ω^2 are changed, the appearance of the curves $W(P)$ changes significantly. These parameters determine the position of the mode of the function $W(P)$ and the value of its maximum. With $h_\Omega^2 = \text{const}$ an increase in the parameters m_j, the maximum of the function $W(P)$ shifts to the right along the abscissa axis, an increase in the average value m_P and dispersion d_P^2, and a decrease in the asymmetry of the function. When the parameter h_Ω^2 is increased (the values m_j are fixed), the opposite picture is observed: The average value and dispersion of the polarization phasor modulus decrease, and the maximum of the function $W(P)$ shifts to the left along the abscissa axis.

Thus, when approximating the envelopes of uncorrelated orthogonally polarized components m-distribution, the information about the source of its radiation (scattering and reflection) is to be sought in three statistical parameters m_1, m_2, and h_Ω^2.

Let us consider the case when the orthogonally polarized components of the wave are correlated, and their envelopes are subordinated to the m-distribution, the parameters of which are $m_1 = m_2 = m$.

To determine $W(P)$ in the case under consideration, it is convenient to use the F distribution:

$$W(F) = \frac{(1-\rho_2)^m F^{m-1}(1+F)}{B(m,m)\left[(1+F)^2 - 4\rho_2 F\right]^{m+0.5}}, \tag{4.22}$$

where $F = E_1^2\Omega_2/\left(E_2^2\Omega_1\right)$, $\Omega_j =< E_j^2 >$, and ρ_2 is the cross-correlation coefficient between the squares of the envelopes E_1^2 and E_2^2.

It is not difficult to see that the variable F is related to the modulus of the polarization phasor by the relation $F = (h_\Omega P)^2$. Consequently, applying the known rule for the transformation of random variables, we obtain:

$$W\left(P; m, h_\Omega^2, \rho_2\right) = \frac{2(1-\rho_2)^m h_\Omega^{2m}\left(1+h_\Omega^2 P^2\right)P^{2m-1}}{B(m,m)\left[\left(1+h_\Omega^2 P^2\right)^2 - 4\rho_2 h_\Omega^2 P^2\right]^{m+0.5}} \tag{4.23}$$

It follows from the representation (4.23) that the probability density $W(P)$ in the case under consideration is determined by three parameters: m_1, h_Ω^2, and ρ_2. When $\rho_2 = 0$, formula (4.23) is simplified and coincides with (4.18), if in the latter we assume $m_1 = m_2 = m$.

Finding the integral distribution function and the ν-th initial moments leads to serious mathematical difficulties.

Thus, the statistical characteristics of the modulus of a polarization phasor in an arbitrary basis are completely determined in the generic case by four statistical parameters: m_1, m_2, h_Ω^2, and ρ_2.

Let us determine the probability density $W(P)$ in an arbitrary polarization basis, the envelopes of which E_1 and E_2 are distributed according to the lognormal law (2.67). To do this, we use the joint probability density (2.66). As a result of the calculations, we obtain:

$$W\left(P; \mu_j, d_j, \rho_{12}\right) = \left[P\left[2\pi\left(d_1^2 + d_2^2 - 2\rho_{12}d_1d_2\right)\right]^{0.5}\right]^{-1} \exp\left\{-\frac{[\ln P - (\mu_2 - \mu_1)]^2}{2\left(d_1^2 + d_2^2 - 2\rho_{12}d_1d_2\right)}\right\}. \tag{4.24}$$

It follows from (4.24) that the probability density of the modulus of the polarization phase is determined by five statistical parameters: μ_j, d_j, ρ_{12}, where $j = 1, 2$. In addition, a comparison of the relations (4.24) and (2.67) shows that these expressions are structurally the same. In other words, the probability density (4.24) for the distribution of envelopes E_1 and E_2 for the logarithmic-normal law with parameters

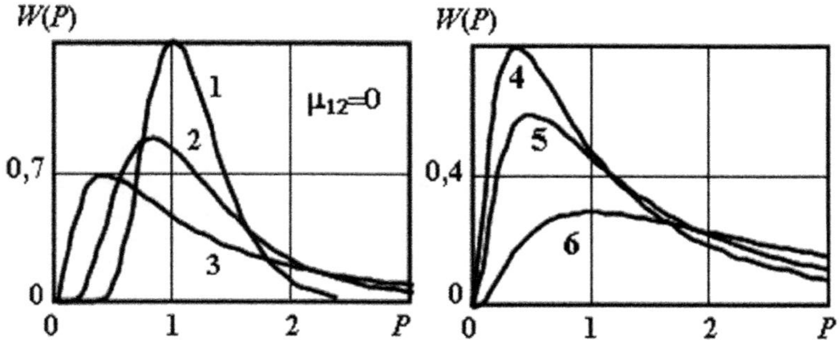

Fig. 4.10 Probability density of the polarization phasor modulus

(μ_1, d_1^2) and (μ_2, d_2^2) is also lognormal with mean $\mu_{12} = \mu_2 - \mu_1$ and variance $d_{12}^2 = d_1^2 + d_2^2 - 2\rho_{12} d_1 d_2$.

Figure 4.10 shows families of probability density curves (4.24) for several parameter μ_{12}, d_{12}^2 values.

As can be seen, $W(P)$ is a one-modal function with right-hand asymmetry. At $\mu_{12} = \text{const}$, the degree of asymmetry increases with increasing d_{12}^2. In this case, the modal value of the function $W(P)$ decreases, and the mode shifts to the left along the abscissa axis. The change of μ_{12} at $d_{12}^2 = \text{const}$ also affects the shape of the curves $W(P)$.

Calculation of the ν-th initial moments of the modulus of the polarization phasor is not difficult. Using formula (4.24) and integrating, we get:

$$m_P^\nu(\mu_j, d_j^2, \rho_{12}) = \exp\left[0.5\nu^2\left(d_1^2 + d_2^2 - 2\rho_{12} d_1 d_2\right) - \nu(\mu_2 - \mu_1)\right]. \tag{4.25}$$

From (4.25), we can obtain expressions for the average value and dispersion of the modulus of the polarization phasor:

$$\begin{aligned} m_P(\mu_j, d_j^2, \rho_{12}) &= \exp\left[0.5\left(d_1^2 + d_2^2 - 2\rho_{12} d_1 d_2\right) - (\mu_2 - \mu_1)\right] \\ d_P^2(\mu_j, d_j^2, \rho_{12}) &= \exp\left[d_1^2 + d_2^2 - 2\rho_{12} d_1 d_2 + 2(\mu_2 - \mu_1)\right] \\ &\quad \times \left[\exp\left(d_1^2 + d_2^2 - 2\rho_{12} d_1 d_2\right) - 1\right]. \end{aligned} \tag{4.26}$$

The integral distribution function of the modulus of the polarization phasor using formula (4.24) is reduced to the form:

$$F(P; \mu_j, d_j^2, \rho_{12}) = \frac{1}{2}\left\{1 + \Phi\left[\frac{\ln P - (\mu_2 - \mu_1)}{2\sqrt{2}\left(d_1^2 + d_2^2 - 2\rho_{12} d_1 d_2\right)}\right]\right\}, \tag{4.27}$$

where $\Phi(z)$ is the probability integral.

Figure 4.11 shows the family of probability density curves calculated from formula (4.24), and Fig. 4.12 shows the dependence of the mean value and dispersion of the

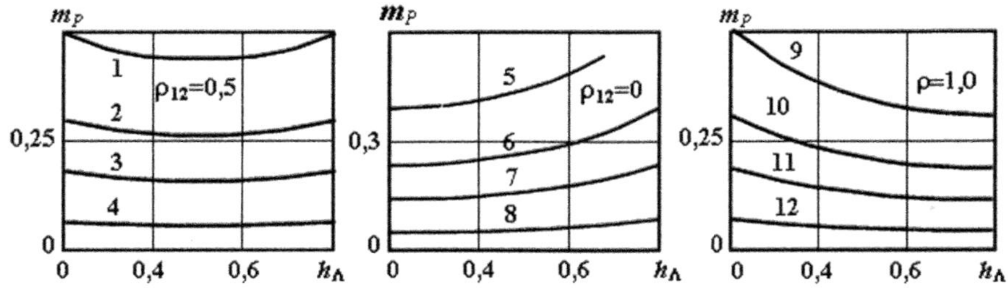

Fig. 4.11 Mean value of the polarization phasor modulus

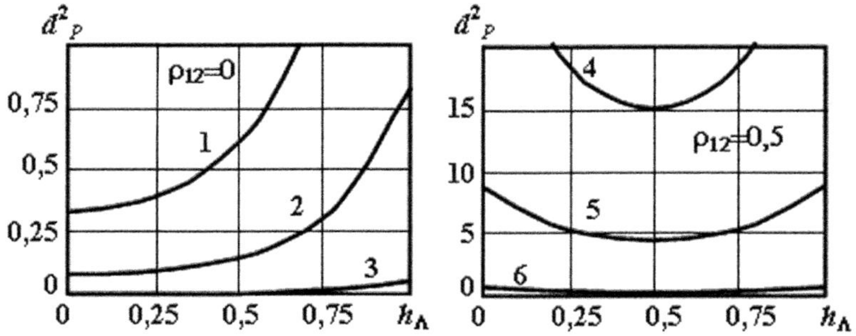

Fig. 4.12 Dispersion of the polarization phasor modulus

polarization phasor modulus on the parameter $h_\Lambda = d_1/d_2$ for certain values of $b_\Lambda = \mu_1/\mu_2$ and ρ_{12}.

4.1.3 Probabilistic Definition—The Non-Gaussian Case

Let us now consider the probabilistic characteristics of the polarization phasor for the case of non-Gaussian fields. As before, we will analyze the additive mixture of two partially polarized waves. The first of these is called a useful radio wave, and its orthogonal components are a useful signal or simply a signal. It is characterized by the difference from zero of the deterministic component (the introduced in s.2.3 coefficient $\gamma_C \neq 0$). All values that determine this wave will be assigned the index "c". The second wave is called a clutter one, and its orthogonal components are an interference signal or simply an interference. We will ascribe the index "o" to them.

We will start by considering the situation when the signal is deterministic (completely polarized), and the envelopes of the orthogonally polarized components are described by the probability distribution density (2.79). In this case, the density distributions of the envelope probabilities E_1 and E_2 the orthogonally polarized components of the additive mixture of signal and interference in accordance with (2.81) can be represented as:

$$W(E_j) = \frac{\beta_{0j} E^{\alpha_{0j}}}{E_{Cj}^{\alpha_{0j}-1}} \exp\left[-\frac{\beta_{0j}}{2}\left(E_{0j}^2 + E_j^2\right)\right] I_{\alpha_{0j}-1}\left(\beta_{0j} E_{Cj} E_j\right), \tag{4.28}$$

where α_{0j}, β_{0j} are the parameters of the j-th ($j = 1$, 2) orthogonally polarized component of the interference, defined in Eq. (2.76), and E_{Cj} are the envelopes of the orthogonally polarized component of the completely polarized signal. Taking into account that the modulus of the phasor $P = E_2/E_1$, we use the relation (1.97):

$$W(P) = \int_0^\infty W(E_1, PE_1)|E_1|dE_1. \tag{4.29}$$

Substituting (4.28) into (4.29) and setting the independence of E_1 and E_2, after integrating over the variable E_1, we obtain the probability distribution density of the modulus of the phasor in the presence of partially polarized interference in the following form:

$$W(P) = \frac{2h^{\alpha_{02}} P^{2\alpha_{02}-1} \exp\left[-0.5\left(a_{C1}^2 + a_{C2}^2\right)\right]}{\left(1 + h^2P^2\right)^{\alpha_{01}\alpha_{02}} \Gamma(\alpha_{02})} \sum_{k=0}^{\infty} \frac{\Gamma(\alpha_{01} + \alpha_{02} + k)}{k!\Gamma(\alpha_{01} + k)}$$
$$\times \left[\frac{a_{C1}^2}{2\left(1 + h^2P^2\right)}\right]^k {}_1F_1\left[\alpha_{01} + \alpha_{02} + k, \alpha_{01}; 0.5a_{C2}^2 P^2\left(1 + h^2P^2\right)^{-1}\right], \tag{4.30}$$

If one of the parameters a_{Cj} is equal to zero, for example $a_{C1} = 0$, expression (4.30) is simplified and takes the form:

$$W(P) = \frac{2h^{2\alpha_{02}} P^{2\alpha_{02}-1} \exp\left[-0.5a_{C2}^2\right] \Gamma(\alpha_{01} + \alpha_{02})}{\left(1 + h^2P^2\right)^{\alpha_{01}+\alpha_{02}} \Gamma(\alpha_{01})\Gamma(\alpha_{02})}$$
$$\times {}_1F_1\left[\alpha_{01} + \alpha_{02}, \alpha_{02}; \frac{a_{C2}^2 P^2}{1 + h^2P^2}\right]. \tag{4.31}$$

If $a_{C1} = a_{C2} = 0$ (there is no signal), then from (4.30) it follows the probability distribution density of the modulus of the non-Gaussian interference phasor:

$$W(P_0) = \frac{2h^{2\alpha_{02}} P_0^{2\alpha_{02}-1} \Gamma(\alpha_{01} + \alpha_{02})}{\left(1 + h^2P^2\right)^{\alpha_{01}+\alpha_{02}} \Gamma(\alpha_{01})\Gamma(\alpha_{02})}, \tag{4.32}$$

where $P_0 = E_{02}/E_{01}$, E_{0j} are the envelopes of the orthogonally polarized components of the interference.

Thus, the probability distribution density in the generic case is determined by five parameters: $h^2, \alpha_{01}, \alpha_{02}, a_{C1}, a_{C2}$. The change in their values leads to the transformation $W(P)$ and change in the numerical characteristics of the random modulus of the phasor. To find the latter, it is necessary to know the ν-th initial moments.

Using Eq. (4.30) and expanding the degenerate hypergeometric function ${}_1F_1(z)$ into a series, after integration we obtain:

$$ {}_Pm_{C+0}^{(\nu)} = \frac{\exp\left[-0.5\left(a_{C1}^2 + a_{C2}^2\right)\right]}{\left(h^2\right)^{\nu/2+1}} \sum_{k,n=0}^{\infty} A_k A_n \Gamma\left(\alpha_{02} + \frac{\nu}{2} + n\right)\Gamma\left(\alpha_{01} - \frac{\nu}{2} + k\right) \tag{4.33} $$

where $A_k = \left(a_{C1}^2\right)^k / \left[2^k k! \Gamma(\alpha_{01} + k)\right]$; $A_n = \left(a_{C2}^2\right)^n / [2^n n! \Gamma(\alpha_{02} + n)]$;

$$ \alpha_{01} + k > \frac{\nu}{2} > 0. $$

If in the expression (4.33) we will set $a_{C1} = 0$, then

$$ {}_Pm_{C+0}^{(\nu)} = \frac{\exp\left(-0.5a_{C2}^2\right)}{\left(h^2\right)^{\nu/2+1}\Gamma(\alpha_{01})} \sum_{k,n=0}^{\infty} A_n \Gamma\left(\alpha_{02} + \frac{\nu}{2} + n\right)\Gamma\left(\alpha_{01} - \frac{\nu}{2}\right). \tag{4.34} $$

If $a_{C1} = a_{C2} = 0$, then from the expression (4.33) it follows:

$$ {}_Pm_0^{(\nu)} = \Gamma\left(\alpha_{02} + \frac{\nu}{2}\right)\Gamma\left(\alpha_{01} - \frac{\nu}{2}\right) / \left[h^\nu \Gamma(\alpha_{01})\Gamma(\alpha_{02})\right], \quad \alpha_{01} > 0.5\nu > 0. \tag{4.35} $$

To calculate the integral distribution function of the phasor module, it is necessary to use the representation (4.30), to expand the function ${}_1F_1(z)$ into a series, and to perform the corresponding integration. As a result, we have:

$$ \begin{aligned} F(P) = \left(h^2 P^2\right)^{\alpha_{02}} \exp\left[-0.5\left(a_{C1}^2 + a_{C2}^2\right)\right] \sum_{k,n=0}^{\infty} A_k A_n \frac{\Gamma(\alpha_{01} + \alpha_{02} + k + n)}{(\alpha_{02} + n)} \\ \times {}_2F_1\left(\alpha_{01} + \alpha_{02} + k + n, \alpha_{02} + n, 1 + \alpha_{02} + n, -h^2 P^2\right), \end{aligned} \tag{4.36} $$

where A_k, A_n are defined in (4.33).

With $a_{C1} = a_{C2} = 0$, the expression (4.36) is simplified and takes the form of:

$$ F(P_0) = \frac{\left(h^2 P^2\right)^{\alpha_{02}} \Gamma(\alpha_{01} + \alpha_{02})}{\Gamma(\alpha_{01})\Gamma(\alpha_{02})\alpha_{02}} {}_2F_1\left(\alpha_{01} + \alpha_{02}, \alpha_{02}, 1 + \alpha_{02}, -h^2 P_0^2\right) \tag{4.37} $$

The obtained probabilistic characteristics (4.30), (4.33), and (4.36) of the modulus of the completely polarized signal phasor in the presence of non-Gaussian interference generalize other particular probability models of the interference.

For example, if the interference is Gaussian and $\alpha_{01} = \alpha_{02} = 1$, $\beta_{0j} = d_{0j}^{-2}$, then from the above equalities it follows:

$$W(P) = \frac{2h^2 P}{\left(1 + h^2 P^2\right)^2} \exp\left[-\frac{a_{C1}^2 + a_{C2}^2}{2}\right] \sum_{k=0}^{\infty} \frac{\Gamma(k+2)}{(k!)^2} \left[\frac{a_{C1}^2}{2\left(1 + h^2 P^2\right)}\right]^k$$

$${}_1F_1\left[2 + k, 1; \frac{a_{C1}^2 P^2}{2\left(1 + h^2 P^2\right)}\right], \quad (4.38)$$

where $h^2 = d_{01}^2/d_{02}^2$, $a_{Cj} = E_{Cj}/d_{0j}$; d_{0j}^2 is the dispersion of the j-th orthogonally polarized component of the interference,

$${}_P m_{C+0}^{(\nu)} = \frac{\exp\left[-0.5\left(a_{C1}^2 + a_{C2}^2\right)\right]}{\left(h^2\right)^{\nu/2+1}} \sum_{k,n=0}^{\infty} A_k A_n \Gamma\left(1 + \frac{\nu}{2} + n\right) \Gamma\left(1 - \frac{\nu}{2} + k\right), \quad (4.39)$$

where $A_k = \left(a_{C1}^2\right)^k / \left[2^k (k!)^2\right]$; $A_n = \left(a_{C2}^2\right)^n / \left[2^n (n!)^2\right]$.

At the same time,

$$F(P) = h^2 P^2 \exp\left[-\frac{a_{C1}^2 + a_{C2}^2}{2}\right]$$

$$\sum_{k,n=0}^{\infty} A_k A_n \frac{\Gamma(2 + k + n)}{(1 + n)} {}_2F_1\left(2 + k + n, 1 + n, 2 + n, -h^2 P^2\right). \quad (4.40)$$

Let us consider a more general case where the signal and interference are partially polarized non-Gaussian processes. In this case, the probability distribution density of the polarization phasor can be determined from the formula (4.29), taking into account the expression (2.84).

We substitute equality (2.84) into formula (4.29) and then carry out the integration E_1. As a result of the calculations, we find:

$$W(P) = \frac{2h^{\alpha_{02}} P^{2\alpha_{02}-1} \exp\left[-\frac{1}{2}\left(\frac{\gamma_{C1}^2}{\beta_{C1}} + \frac{\gamma_{C2}^2}{\beta_{C2}}\right)\right]}{\left(1 + C_1^{-1}\right)^{\alpha_{C1}} \left(1 + C_2^{-1}\right)^{\alpha_{C2}} \left(1 + h^2 P^2\right)^{\alpha_{01}+\alpha_{02}}}$$

$$\times \sum_{k,n,P=0}^{\infty} A_k A_n \frac{\Gamma(P) B^{-1}(\alpha_{01}, \alpha_{02} + P)}{P! B(\alpha_{C2} + n, P)} \left[\frac{h^2 P^2}{(1 + C_2)\left(1 + h^2 P^2\right)}\right]^P$$

$$\times {}_2F_1\left[\alpha_{C1} + k, \alpha_{01} + \alpha_{02} + P; \alpha_{01}; \left[(1 + C_1)\left(1 + h^2 P^2\right)\right]^{-1}\right], \quad (4.41)$$

where $C_j = \beta_{Cj}/\beta_{0j}$; $h^2 = \beta_{02}/\beta_{01}$;

$$A_k = \frac{1}{k!}\left[\frac{\gamma_{C1}^2}{2\beta_{C1}\left(1 + C_1^{-1}\right)}\right]^k; \; A_n = \frac{1}{n!}\left[\frac{\gamma_{C2}^2}{2\beta_{C2}\left(1 + C_2^{-1}\right)}\right].$$

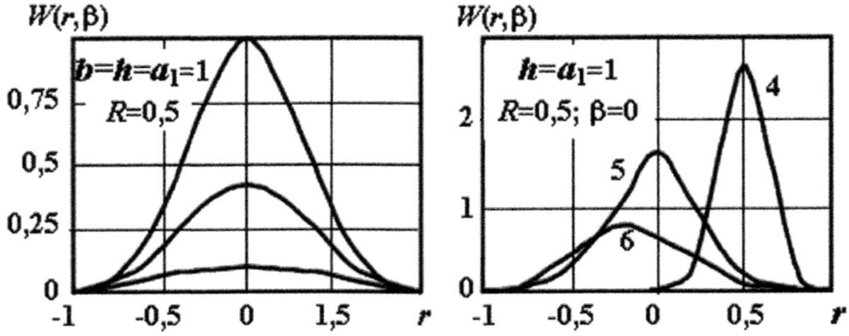

Fig. 4.13 Joint probability distribution density of the ellipticity coefficient and the orientation angle

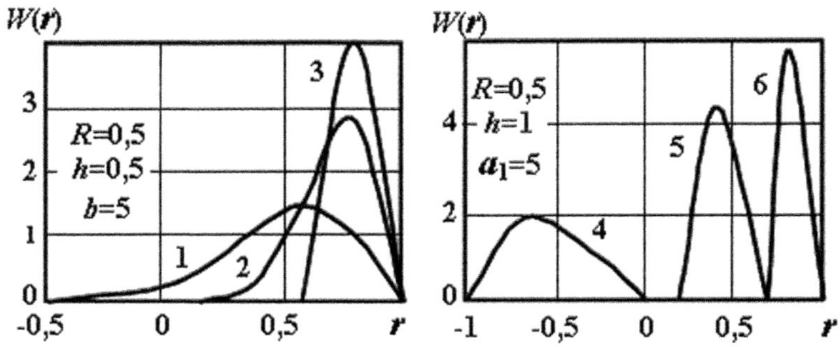

Fig. 4.14 Probability distribution density of the ellipticity coefficient

Figure 4.13 shows families of curves of functions $W_{C+0}(P)$ calculated by formula (4.41) for certain values of the parameters $\alpha_{0j}; \alpha_{Cj}, C_j, h^2, b^2, a^2$. Here, the following notation is introduced: $a_1 = \gamma_{C1}/\beta_{C1}$, $a_2 = \gamma_{C2}/\beta_{C2}$, $b^2 = a_2^2/a_1^2$.

Figures 4.14 and 4.15 are, respectively, the dependencies of the average value m_P, variance d_P^2, coefficients of variation K_B, asymmetry K_A, and kurtosis of the polarization phasor on some parameters of the probability distribution density $W_{C+0}(P)$.

The nature of the function change $F(P)$ depending on the parameters $C_j, h^2, \alpha_{0j}, b^2$ is illustrated in Fig. 4.16.

The obtained expression for the probability distribution density (4.41) is obviously the base one and can serve as the initial one for determining the indicated probabilistic characteristics of the polarization phase modulus of a wide class of combinations of probabilistic envelope models of orthogonally polarized components of useful signals and interference. To obtain particular cases, it suffices to substitute into the formula (4.41) in accordance with Table 2.4 certain values of the parameters $\alpha_{0j}; \alpha_{Cj}, \beta_{0j}, \beta_{Cj}, \gamma_{Cj} (j = 1, 2)$.

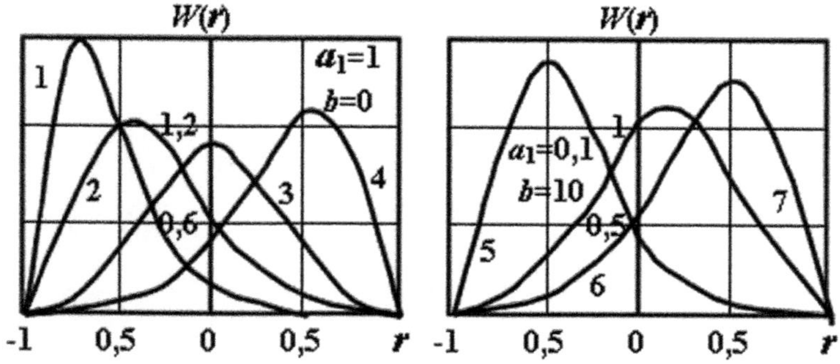

Fig. 4.15 Probability distribution density of the ellipticity coefficient

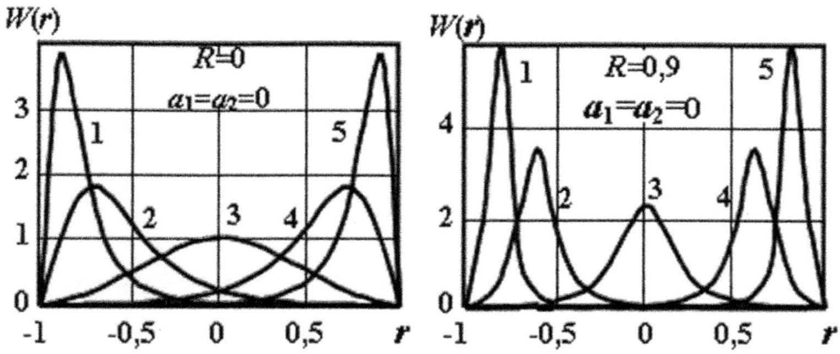

Fig. 4.16 Probability distribution density and the ellipticity coefficient

Reference

1. Yagol'nikov SV (ed) (2013) Methods for studying the radar characteristics of objects. Radiotekhnika, Moscow

Chapter 5
Graphic Representations of the Signal Polarization State in Navigation Systems

5.1 Features of Graphical Representations

In Sect. 1.1, the Poincaré sphere and the Poincaré plane were introduced as a means for representing the polarization state of the wave. From the algebraic point of view, the Poincaré sphere represents the mapping of the group of rotations of Jones vectors in the space of their stereographic projections. The radio wave polarization is represented on this sphere by a certain point P (Fig. 1.5), the position of which is uniquely determined by the angles $2\alpha, 2\beta$ or $2\gamma, 2\delta$, or Cartesian coordinates S_1, S_2, S_3 (Stokes parameters).

The orthogonal polarization basis on the Poincaré sphere is represented by a pair of diametrically opposite points and the affine one by a pair of any points. A change in the polarization basis means a transition from one pair of points to another. Note that the representation of polarization by the ellipse parameters does not reflect the initial position of the vector $\vec{E}$ inside the ellipse and is therefore incomplete. In addition, the representation using the Poincaré sphere does not take into account the wave power. A natural generalization of the Poincaré sphere would be the introduction of a "ball" of Poincaré, the radius of which would be proportional to the power of the wave.

Thus, the CPW is mapped onto a fixed point on the Poincaré sphere. This means that the coordinates of this point (Stokes parameters) do not change during the analysis period. As for the UPW, the condition $\bar{x} = \bar{y} = \bar{z} = 0$ must be satisfied for the Cartesian coordinates of the corresponding point. This leads to one of the interpretations of the UPW on the Poincaré sphere in the form of a point moving with an undetermined and unknown velocity, which during the analysis of the polarization state of the radio wave uniformly covers the entire sphere.

For PPW, the coverage of a sphere that is uneven during the analysis is characteristic, so $\bar{x} = x_0, \bar{y} = y_0, \bar{z} = z_0$ where $x_0^2 + y_0^2 + z_0^2 = 1$. Very often the trajectory of the polarization state of a point on the Poincaré sphere is localized for the PPW in a certain area occupied by the trajectory. With this interpretation, the theorem on the decomposition of the PPW into the CPW and the UPW becomes particularly

Kozlov A. I. et al., *Introduction to the Theory of Radiopolarimetric Navigation Systems*, Springer Aerospace Technology, https://doi.org/10.1007/978-981-13-8395-3_5

illustrative. The afore said means that on the Poincaré sphere, the PPW can be mapped either by one point corresponding to the UPW component (incomplete representation) or by two diametrically opposite points (the second eigenbasis of the wave), each of them should be assigned a certain "weight" (this concept is described in more detail in Sect. 2.2.1). In turn, an incomplete representation of the polarization state of the PPW can also be supplemented by an additional characteristic—the degree of polarization m that determines the contribution of the CPW to the power of the PPW.

The development of methods for representing the UPW and the PPW on the Poincaré sphere is a four-dimensional polarization sphere, where the UPW and the PPW are noted by certain points.

The methods of analysis and representation of electromagnetic fields considered above are based on the assumption of a "simple" polarization field structure—a plane homogeneous field with a known orientation of the wavefront plane. However, in real conditions of radio engineering systems operation, situations may arise where it is necessary to take into account the "complex," inhomogeneous nature of the field of received radio waves. In this case, the two-dimensional polarization basis and the apparatus of the Poincaré sphere turn out to be insufficient for an adequate representation of the polarization parameters of such fields. To represent and analyze the "complex" electromagnetic fields in space, it is proposed in this chapter to use the apparatus of a sphere S_2—a unit sphere, on which three vertices of an autopolar triangle are associated with the basis of a three-dimensional unitary space. This apparatus is less universal than the apparatus of the Poincaré sphere (in its unitary space); nevertheless, it makes it possible to simplify and clarify the formulation and solution of a number of problems.

5.2 Representation of Polarization in Cylindrical and Stereographic Projections

The graphical representation above of the polarization on the Poincaré sphere makes it possible to clearly illustrate both the polarization state of the wave and the transformation of the polarization basis.

At the same time, the sphere is inconvenient for graphical constructions, since it is a three-dimensional surface. However today—in the age of computers—the latter sentence is increasingly losing its significance. Nowadays, computer technologies make such constructions easy, and, what is more, accompanied by spectacular rotations reflecting the polarization state of the sphere. Now polarization meters are increasingly combined with a computer that displays the current polarization of the analyzed wave. Such technologies became of particular importance in organizing virtual controls of radiated signals characteristics and procedures for processing signals received by a radio polarimeter. Even more sophisticated technologies are designed to map the scattering properties of observed objects on the Poincaré sphere, both concentrated and surface or volume distributed.

However, very often, the use of computer technology included, many practical calculations are organized and conducted with the use of the projections of the points on the indicated sphere.

One of the most commonly used is the cylindrical projection of the Poincaré sphere (also called orthographic or parallel), when the points of one of the hemispheres are projected onto the equatorial plane with the help of straight lines perpendicular to this plane. The resulting polarization diagram is shown in Fig. 5.1.

The lines of equal values of β (the meridians of the sphere) are projected onto the radii of the diagram, so on this diagram the doubled orientation angle of the ellipse 2β in any linear basis is given without distortion. Lines of equal values of α (the parallel of the sphere) are also projected without distortion in the form of a system of concentric circles $\alpha = \text{const}$ the radius of which determines the latitude on the sphere and is equal to $\cos 2\alpha$.

The same diagram with lines of constant values of the parameters γ and 2δ plotted on it is shown in Fig. 5.2. Point A denotes horizontal polarization $(E_y = 0)$, point D—vertical $(E_x = 0)$. Lines of equal ratios of component amplitudes are projected onto diagrams in the form of straight lines perpendicular to the diameter of the AD. The numbers in these lines mean the ratio E_y/E_x in decibels. The lines of constant values of the phase difference 2δ in this diagram represent a family of ellipses whose major axes coincide with the diameter of the AD.

Figure 5.2 can be treated as a sketch of a pad for the transition from a system of parameters γ, 2δ to a system of parameters α, β or vice versa. The concentric circles shown in Fig. 5.1 are obtained by rotating the ruler of such a pad, calibrated directly in the ratio values of the axes of the ellipses r, or $\alpha = \arctan r$.

Cylindrical projection has the following drawbacks:

Firstly, both hemispheres of the Poincaré sphere that map the right and left rotation of the polarization ellipses are projected onto the same circle, so that to each point of the circle there correspond polarizations of the right and left rotations simultaneously.

This flaw can easily be eliminated if the diagram is presented separately for left and right polarization on two circles. One can introduce a special designation of

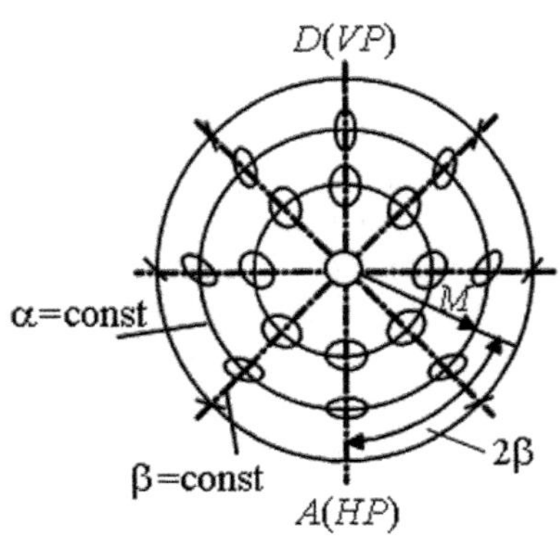

Fig. 5.1 Polarization diagram

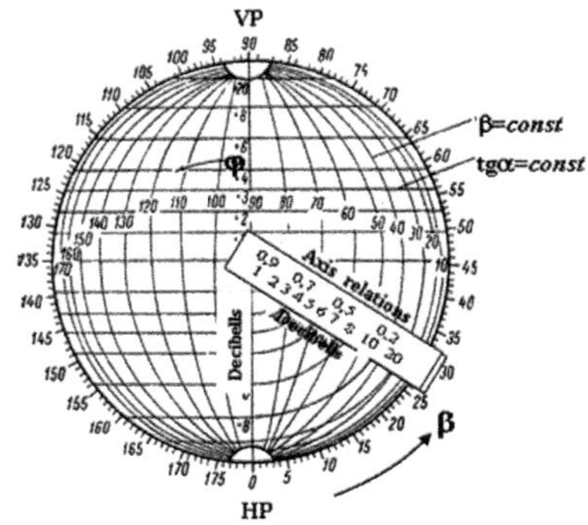

Fig. 5.2 Polarization diagram

points on the diagram, or apply two grids of constant parameter values (e.g., a solid line for right polarization, dotted for the left).

Secondly, the cylindrical projection does not provide the required accuracy of conversion when transitioning from one system of parameters to another in the case of small values of the ellipticity coefficient.

A unique correspondence between the points of the Poincaré sphere and the polarization diagram, as well as the minimum error value near the equator of the sphere, is provided by the stereographic polarization diagram, which is constructed as follows.

An arbitrary point P on the surface of the Poincaré sphere (Fig. 5.3) is associated with the point M in the plane of the equator XY, formed by the intersection of the straight line passing through the lower pole and the point P. The upper hemisphere in this case is mapped onto the plane limited by the equator, and the lower hemisphere onto the rest of the plane, with the point of the lower pole corresponding to an infinitely distant point on the plane. Nevertheless, it is practically inconvenient to use the grid lines of the diagram $2\delta = \text{const}$ and $\gamma = \text{const}$ outside the unit circle (the equator of the sphere). Therefore, for calculating the left polarization of the field, the stereographic projection of the sphere from the upper pole F is usually used. The resulting diagram is similar in form to the projection diagram from the lower pole and differs from it only by the sign of the angle α, since all the points of the unit circle represent the polarization ellipses from the point F with the left direction of rotation of the field vector.

Lines of constant values of α, of stereographic projection, as well as of cylindrical projection, are projected onto concentric circles, the scale of which essentially increases from the center to the periphery (Fig. 5.4). The lines of constant values of β are projected onto rays that coincide with the radii of the unit circle with a uniform scale in the angle. The advantage of stereographic projection is that for small values of α, a greater reading accuracy of the diagram is provided. The coordinates of the point on the sphere uniquely determine the point on the plane of the diagram. If the coordinates of the projected point P are denoted as (x, y, z) and the coordinates of the point M are denoted as $\left(x', y', z = 0\right)$, then their mutual relationship is determined by the expression:

$$x' = \frac{x}{1-z}; \quad y' = \frac{y}{1-z}. \tag{5.1}$$

When specifying projection points, the coordinates of their mappings on a sphere of unit radius will be:

$$x = \frac{2x'}{1 + x'^2 + y'^2}; \quad y = \frac{2y'}{1 + x'^2 + y'^2}; \quad z = \frac{1 - \left(x'^2 + y'^2\right)}{1 + x'^2 + y'^2}. \tag{5.2}$$

An expression for the wave phasor can be obtained by going from two real coordinates x' and y' to a complex number $\underline{P} = x' + iy'$. As was shown above, the set of values of the complex variable P in a given basis is in unique correspondence to the set of polarization states of the wave. Thus, the stereographic projection allows us to associate each elliptical polarization of the wave with a point on the unit sphere.

It can be shown that the basis transformation carried out by the unitary matrix Q leads to a transformation of the old coordinates $P\left(\hat{b}_C\right) = [x_C, y_C, z_C]^{\mathrm{T}}$ into

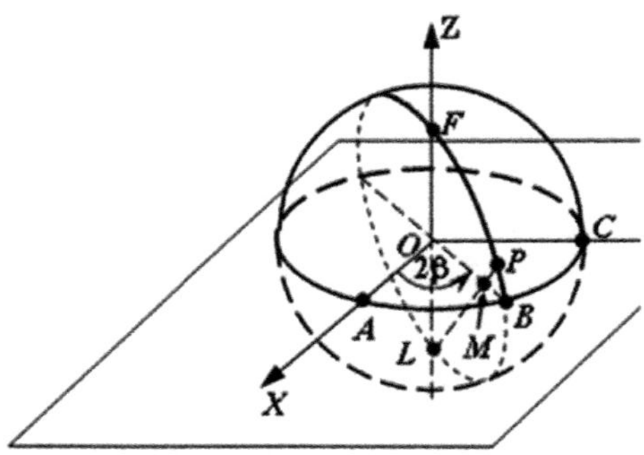

Fig. 5.3 Stereographic projection

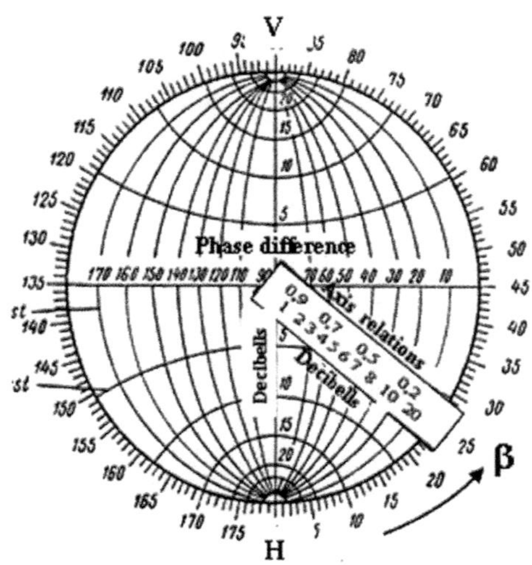

Fig. 5.4 Stereographic diagram

new coordinates $P\left(\hat{b}_H\right) = [x_H, y_H, z_H]^{\mathrm{T}}$, that corresponds to the rotation of the sphere in space. Without dwelling on the proof of this statement, we note only that such a transformation can be realized with the help of a matrix product $P\left(\hat{b}_H\right) = L\left(\hat{b}_H, \hat{b}_C\right) \cdot P\left(\hat{b}_C\right)$, where the transformation matrix $L\left(\hat{b}_H, \hat{b}_C\right)$ has the form:

$$L\left(\hat{b}_H, \hat{b}_C\right) = 2\cos^2\gamma \begin{bmatrix} \sin 2\delta \sin(\eta+\delta) & \cos 2\delta \sin 2(\eta+\delta) & -\cos 2(\eta+\delta) \\ -\cos 2(\eta+\delta) \sin 2\delta & \cos 2\delta \sin 2(\eta+\delta) & \sin 2(\eta+\delta) \\ \cos 2\delta & -\sin 2\delta & 0 \end{bmatrix}. \quad (5.3)$$

Thus, to every transformation of the basis there corresponds a certain transformation of the complex plane $\underline{P}$ (or the transformation of the real plane by means of the relation (5.3)). In turn, by virtue of the correspondence established by the stereographic projection, the linear transformations represent the rotation of the three-dimensional Euclidean space around the origin. It is known that such a rotation is determined by three independent real parameters.

The matrix $\underline{L}\left(\hat{b}_H, \hat{b}_C\right)$ is defined by two complex numbers $\mathrm{e}^{i\eta}\cos\gamma$ and $\mathrm{e}^{i(\eta+2\delta)}\cos\gamma$, combining three variables, that satisfy one relation (5.1), by means of which one of the variables can be excluded. In connection with this, $\gamma\eta$ and $\gamma(\eta+2\delta)$ (the parameters of Kaylee-Klein) completely determine the rotation of space. The more evident rotation parameters are the three Euler angles (φ_1, φ_2 and θ in Fig. 5.5) related to the Kaylee-Klein parameters by the relations:

$$\cos\gamma = \cos(\theta/2); \quad \varphi_1 = \eta + \delta; \quad \varphi_2 = \delta. \quad (5.4)$$

Formula (5.4) shows that every unitary matrix of the second order with determinant equal to 1 corresponds to a certain definite rotation of the unit sphere around the origin. Thus, when transforming a basis, it is first necessary to determine the transition matrix

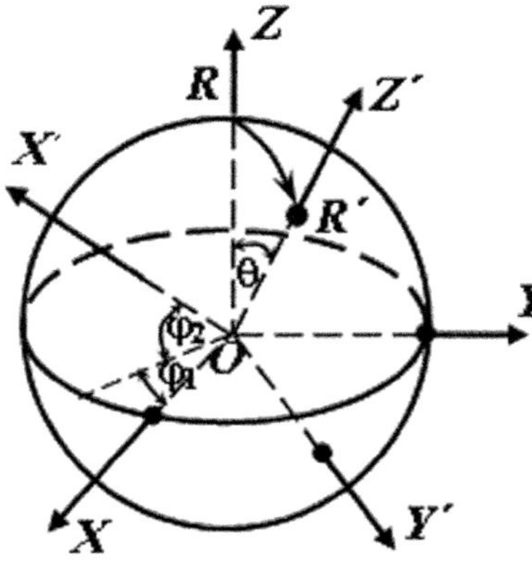

Fig. 5.5 Rotation of the Poincaré sphere

$\underline{Q}\left(\hat{b}_H, \hat{b}_C\right)$ by means of expression (1.43); in fact η, γ, δ, then using the relation (5.4), calculate the rotation angles of the sphere and perform the rotation of the sphere itself.

It should also be noted that usually stereographic projections, given in the literature and reference data, are performed in the complex plane of the polarization coefficient $\underline{P} = \underline{E}_y/\underline{E}_x$ determined in a basis of horizontally and vertically polarized components (Fig. 5.3). In addition, representations of the sphere projection onto the plane of a complex variable are used, which is the polarization relation in the circular basis $P\left(\underline{\hat{b}}_K\right) = \underline{P}(\alpha = \pi/4) = \underline{P}_{RL}$. For a circular basis, the doubled orientation angle of the ellipse 2β is given in a stereographic projection with no distortion. At the same time, the angle β and the phase of the polarization ratio are also transformed without change. The distance of an arbitrary point M of the stereographic projection from the center of the unit circle according to Eq. (5.1) is equal to tan $\gamma_{RL} = \left|\underline{P}_{RL}\right|$, which is clearly illustrated in Fig. 5.6. Thus, the length of the segment OM is equal to the modulus of the phasor $\underline{P}_{RL}$, and the angle between the direction of this segment and the projection of the zero meridian OA is equal to the argument of the same number (Fig. 5.6).

The application of a grid of constant values γ_{RL} and φ_{RL} to the stereographic projection formed by a family of concentric circles with the equator and the family of radial lines substantially expands the possibilities of graphical analysis of polarization relations.

So, summarizing the above said, one can come to the conclusion that the polarization of the electromagnetic field with stereographic (or orthographic) projection is represented by several groups of parameters. The main of them should be referred to α and β, $P_{xy}\left(\gamma_{xy}, 2\delta\right)$, $P_{RL}(\gamma_{RL}, \varphi_{RL})$, these pairs of representations are interrelated, and it is possible to recalculate the parameters of one representation of the field in the parameters of the other.

So the direct and inverse transition from the parameters of the polarization ellipse to the parameters of the phasor in a linear basis can be realized with the help of the

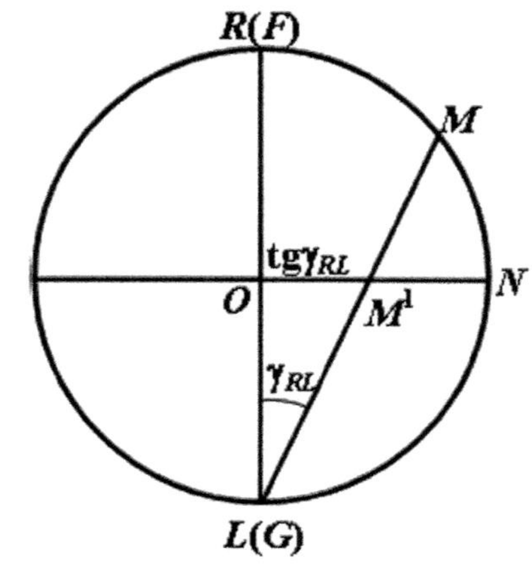

Fig. 5.6 Projection of the sphere on complex variable plane

expressions given in Table 1.1. The transformation of the coordinates of a stereographic projection from a linear basis to any other can be performed using the matrix (5.3).

Expressions for the representation of a field in a linear basis are related to the analogous expressions for circular polarization by the obvious relations:

$$\underline{P}_{xy} = \frac{1 - \underline{P}_{RL}}{1 + \underline{P}_{RL}}; \quad \underline{P}_{RL} = \frac{1 - \underline{P}_{xy}}{1 + \underline{P}_{xy}}, \tag{5.5}$$

where $\underline{P}_{xy} = iE_y/E_x$; $\underline{P}_{RL} = \underline{E}_L/\underline{E}_R$.

Thus, using the projections of the Poincaré sphere, one can analyze any kind of polarization of the electromagnetic field.

In practice, diagrams similar to the diagrams of the long lines' impedance are used somewhat less frequently than stereographic ones. This is due to the existence of a formal analogy between the expressions connecting the values of polarization ratios in linear $(\underline{P}_{xy})$ and circular $(\underline{P}_{RL})$ bases on the one hand and certain relations of the long lines' theory on the other. In this case, the polarization coefficient of the linear basis $\underline{P}_{xy}$ is equivalent to the total resistance, and for a circular basis, the same ratio is similar to the reflection coefficient.

Turning once again to the expression (5.5) for $\underline{P}_{xy}$, we will note that its relation with $\underline{P}_{RL}$ has the same form as the reflection coefficient with the normalized total impedance of the long line. As a consequence of the transformations symmetry (5.5), both $\underline{P}_{xy}$ and $\underline{P}_{RL}$ can be used to construct the diagram. For best use of existing techniques, it is advisable to consider $\underline{P}_{RL}$ as an analog of the current reflection coefficient. If we determine the polarization in terms of the complex ratio of linear orthogonal components $\underline{P}_{xy}$, then $\underline{P}_{RL}$ will be equal to the reflection coefficient obtained from the Smith–Wolpert diagram by introducing $\underline{P}_{xy}$ as the total impedance. Figure 5.7 shows the application of the Smith–Wolpert diagram as a diagram $\underline{P}_{RL}$ on which the shape and orientation of the ellipse is represented. Figure 5.8 illustrates the application of the diagram $\underline{P}_{xy}$. We will note several properties of this analogy.

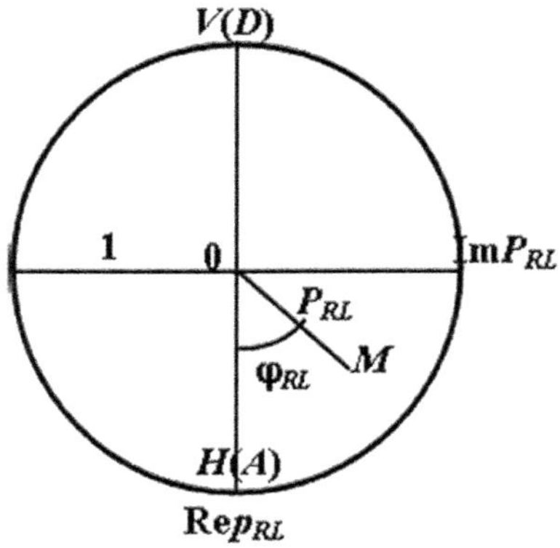

Fig. 5.7 On applications of the Smith-Voltpert diagram

Fig. 5.8 Analog of Carter diagram

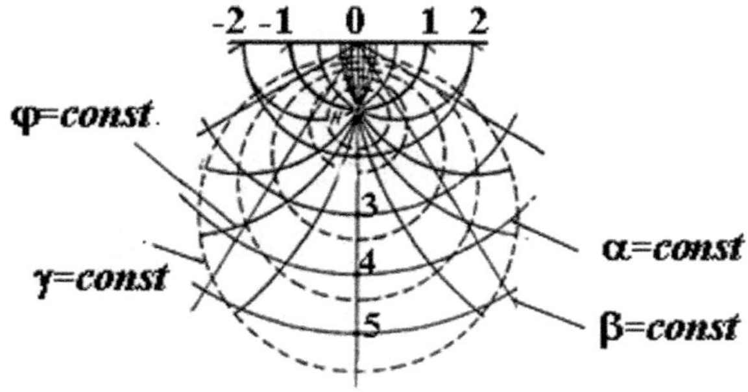

Points in the right half of the diagram represent an elliptical polarization of the left rotation, and in the left half—of the right rotation.

Points on the vertical axis represent linear polarization of different orientations.

The points (1, 0) and (−1, 0) correspond to the circular polarization.

The curves of the constant ellipticity coefficient value are the circles of constant coefficient of the standing wave in the impedances' diagram.

It follows from the diagram for $\underline{P}_{RL}$ in Fig. 5.7 that all points inside the unit circle correspond to the left elliptical polarization and outside it—to the right. The center of the diagram represents the left circular polarization, and the point at infinity is the right one. The curves of the constant ellipticity coefficient coincide with the circles of the constant SWR in the Smith–Wolpert diagram. The coefficient of ellipticity equals SWR when the reflection coefficient is replaced by a value $\underline{P}_{RL}$ in the diagram. The angle laid on the unit circle equals 2β. The geometric location of points at which the ellipticity coefficient is constant is circles with a center at a point $P_{RL} = \pm 1$ (points of short circuit and idling).

To represent all polarizations within a unit circle in diagrams, either use two diagrams (for left and right polarization) or plot points corresponding to the left polarization and the right one separately on one diagram.

An analogy with the long lines theory can be useful for determining the parameters of the polarization ellipse as a function of the polarization ratio P_{xy} which can be obtained from the formulas given in Table 1.1:

$$r = \sqrt{\frac{P^2 - \tan^2\beta}{1 - P^2\tan^2\beta}}. \tag{5.6}$$

In the $\left[\vec{r}, \beta\right]$ coordinate system, equality (5.6) yields two families of curves, in the first of which the parameter is the value of the polarization relation modulus $\left|\underline{P}_{xy}\right|$, and in the second, the phase difference of the orthogonal linearly polarized components 2δ. Both families are shown in Fig. 5.9. A nomogram covering all possible values of the variables allows one according to the measured values $\left|\underline{P}_{xy}\right|$ and 2δ to find the values of r and β or to solve the reverse problem—to determine

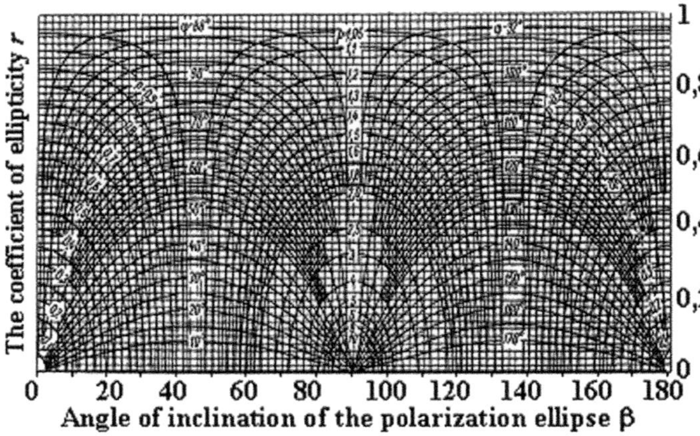

Fig. 5.9 Nomogram for determining the value of r and β

the polarization ratio with the decomposition of an elliptically polarized field in a linear basis.

5.3 Representation of Radio Waves' Polarization in Cartographic Projections of the Poincaré Sphere

The graphical representation above of the polarization on the Poincaré sphere makes it possible to clearly illustrate both the polarization state of the wave and the transformation of the polarization basis.

At the same time, the sphere is inconvenient for graphical constructions, since it is a three-dimensional surface. It is especially difficult to use methods of computer imaging of the polarization diagrams of the Poincaré sphere. Nonetheless, modern computer technologies make such constructions easy, accompanied with spectacular rotations reflecting the polarization state of the sphere. Now polarization meters are combined with a computer that displays the current polarization of the analyzed wave. Such technologies became of particular importance in organizing virtual controls of radiated signals characteristics and procedures for processing signals received by a radio polarimeter. Even more sophisticated technologies are designed to map the scattering properties of observed objects on the Poincaré sphere, both concentrated and surface or volume distributed.

However, very often, the use of computer technology included, many practical calculations are organized and conducted with the use of the method of cartographic projection. Using the cartographic projection, one can obtain a two-dimensional representation of the surface of the Poincaré sphere. In computer representation of polarized radar data, all the initial coordinates are spherical:

- radius of the sphere R;
- scale factor S;
- latitude 2ε, $2\varepsilon_0$ (coordinate of the center);
- longitude 2τ, $2\tau_0$ (coordinate of the center);
- $0 \leq \varepsilon \leq 45°$—upper hemisphere;
- $-45° \leq \varepsilon \leq 0$—lower hemisphere.

5.3.1 Mercator Cylindrical Conformal Projection

The projection of Mercator, proposed in 1569 by Gerardus Mercator, was used for navigation, both during the exploration of the world's oceans, and in the era of space exploration. The cylindrical projection of Mercator marked the beginning of modern cartographic projections.

The mathematical transformation of the mapping process involves the transformation of spherical coordinates into rectangular coordinates. Suppose that the coordinate x is arbitrarily defined as the origin of the transformation, the coordinate y is determined by the conformality condition. The coordinate transformation in this case will look like this:

$$\begin{pmatrix} x \\ y \end{pmatrix} = \begin{pmatrix} R \cdot S \cdot (2\tau - 2\tau_0) \\ R \cdot S \cdot \ln \tan(\pi/4 + \varepsilon) \end{pmatrix}. \tag{5.7}$$

In Fig. 5.10, the x-axis is along the equator, and the $\varepsilon = \pi/4$ axis is along the central meridian. All other meridians are equally spaced straight lines, and the parallels are straight lines that are unevenly distributed.

The center of the projection is a point with coordinates ($2\tau_0 = 0°, 2\varepsilon_0 = 0°$) that corresponds to a horizontal polarization basis. Hence, the vertical projection corresponds to the two extreme points on the left and right of the projection. The equator, which is the projection of the x-axis, corresponds to bases of linear polarization.

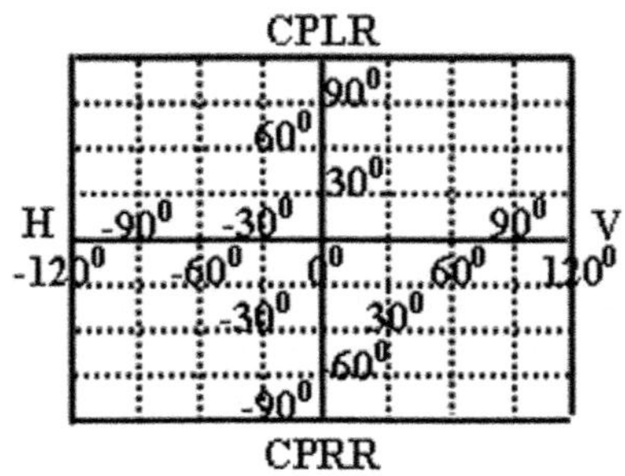

Fig. 5.10 Mercator cylindrical conformal projection: V is the vertical polarization, H is the horizontal polarization, CPLR—circular polarization of the left rotation, CPRR—circular polarization of the right rotation

The first advantage of the Mercator cylindrical conformal projection is that the meridians and parallels are the common vertical and horizontal lines. The second advantage of this type of cartographic projection is easy readability.

It follows from the basic nature of the cartographic projection that each rather narrow latitude zone anywhere on the Poincaré sphere, with the exception of the poles, widens or shrinks to the desired size and represents a real, usable image of a point on the Poincaré sphere. Such a view was widely used by Hoinen in his studies.

It should be noted that when the map is stretched beyond 60° north latitude, the scale factor will constantly increase. In the event that the upper horizontal line (90°) corresponds to the circular polarization of the left rotation, and the lower line to the circular polarization of the right rotation, the present cartographic representation is highly distorted. From this, we can conclude that the peculiarity of the Mercator projection is its improbability (in the sense of mapping the area).

5.3.2 τ–ε Diagrams

In addition to the Mercator cylindrical conformal projection, there is another convenient way of representing a polarization basis from the Poincaré sphere onto the plane of Cartesian coordinates. In the coherent case, the Stokes parameters g_i of each polarization basis can be uniquely calculated through the ellipticity angle ε and the orientation angle τ:

$$\vec{g} = \begin{bmatrix} g_1 \\ g_2 \\ g_3 \\ g_4 \end{bmatrix} = \begin{bmatrix} |E_H|^2 + |E_V|^2 \\ |E_H|^2 - |E_V|^2 \\ 2|E_H||E_V|\cos\phi \\ 2|E_H||E_V|\sin\phi \end{bmatrix} = A^2 \begin{bmatrix} 1 \\ \cos 2\varepsilon \cos 2\tau \\ \cos 2\varepsilon \sin 2\tau \\ \sin 2\varepsilon \end{bmatrix}, \tag{5.8}$$

where $A^2 = a^2 + b^2 = |E_H|^2 + |E_V|^2$ is the “size” of the polarization ellipse; a and b, respectively, the large and small semiaxes of the polarization ellipse; ϕ—relative phase (the phase difference between two orthogonal components of the electric field).

Let the domains of definition ε and τ be $(-\pi/4; \pi/4)$ and $(-\pi/2; \pi/2)$, respectively. Then, the expression for the coordinate transformation will look like this:

$$\begin{pmatrix} x \\ y \end{pmatrix} = \begin{pmatrix} \tau \\ \varepsilon \end{pmatrix}, \tag{5.9}$$

where x and y are the Cartesian coordinates.

Any polarization basis is determined by a point on the plane with coordinates (τ, ε) (Fig. 5.11). The points on the axis x $(\varepsilon = 0)$ correspond to linear polarizations, and the circular polarizations of the left and right rotation are represented by the points $\varepsilon = \pi/4$ and $\varepsilon = -\pi/4$. When conjugated, symmetric and orthogonal

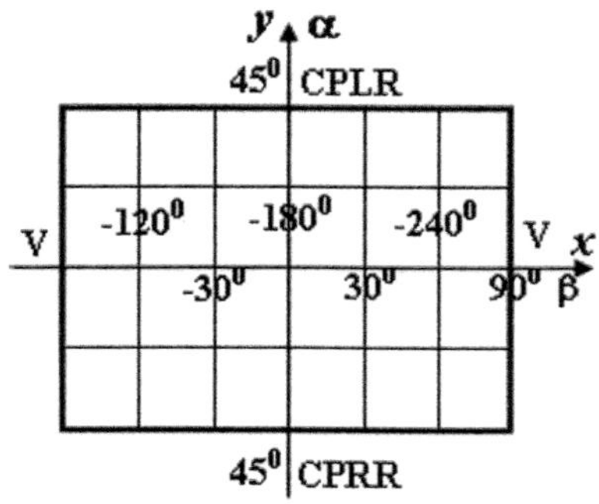

Fig. 5.11 Cartographic projection of the Poincaré sphere in the coordinates: α—ellipticity angle and β—orientation angle

ellipses will be represented, respectively $(x \to x, y \to -y)$, $(x \to -x, y \to -y)$ and $(x \to x - \pi/2, y \to -y)$.

5.3.3 *The Lambert Polar Projection*

This type of cartographic projection is suitable for cluster analysis of complex polarization states, since the Lambert polar projection has the property of conformality.

Mathematical transformation using this projection involves the transition from a spherical coordinate system to a polar coordinate system. In particular, the conversion process consists of determining the parameter by solving the following sinusoidal function:

$$c = \cos^{-1}\{(\sin(2\varepsilon)\sin(2\varepsilon_0) + \cos(2\varepsilon_0)\cos(2\tau - 2\tau_0))\}. \tag{5.10}$$

The conversion operation is as follows:

$$\begin{pmatrix} \gamma \\ \theta \end{pmatrix} = \begin{pmatrix} 2\sin(c/2) \\ 2\tau - 2\tau_0 \end{pmatrix}, \tag{5.11}$$

where γ and θ are the radius and the azimuth angle, respectively (resulting transformation coordinates).

The Lambert polar projection can be obtained directly from the projection of the cone. Lines of equal longitude (meridians) converge to the pole (the center of the projection), and lines of equal latitude (parallel) are concentric circles with respect to the center of the projection.

The main property of this projection is that the area of the projected surface is proportional to the surface area of the sphere. In fact, this means that the ratio of the areas of any part of the projection is equal to the ratio of the areas of the entire projection. In the cluster analysis of complex polarization states, the average value

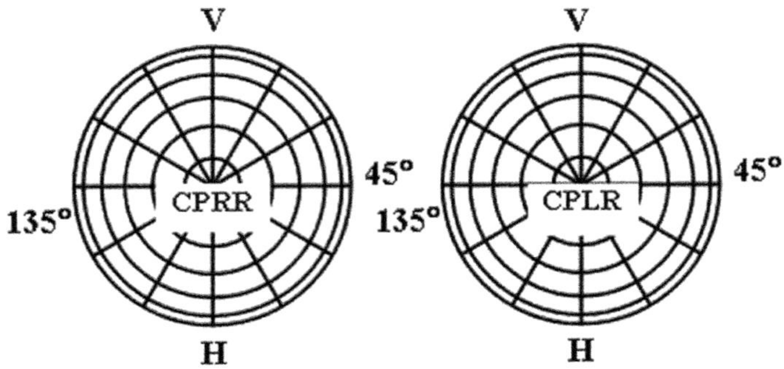

Fig. 5.12 Azimuthal polar projection of equal areas of Lambert

and arrangement of the polarization zeros do indeed appear on the projection, because such a representation of the area is plausible.

Due to its circular nature, the Lambert map projection is suitable for displaying regions of very lengthy in all directions. The polar projection of Lambert is proposed for clustering (partitioning into groups) of polarization states located near the pole (the projection center), that is, groups of closely spaced circular polarization bases.

The disadvantage of this projection is that for the upper and lower hemispheres, it needs its own separate projection, that is, one projection for the left-handed polarization basis and one for the right-handed polarization basis (Fig. 5.12).

5.3.4 *Elliptical Projection of Aitoff–Hammer*

The projection of Aitoff–Hammer is another projection of equal areas, which is based on a spherical model.

The process of mathematical transformation for this case consists of the transition from a spherical coordinate system to a rectangular coordinate system. It consists of two stages. First, the calculation of the three parameters is used in the transformation, and the second is the implementation of the transformation itself.

Expression for intermediate parameters:

$$A_1 = \sqrt{2\left(1 - \cos 2\varepsilon \cos \frac{2\tau - 2\tau_0}{2}\right)}, \quad A_2 = \tan^{-1}\left(\frac{\sin \frac{2\tau - 2\tau_0}{2}}{\tan 2\varepsilon}\right). \tag{5.12}$$

Transformation operation:

$$\begin{pmatrix} x \\ y \end{pmatrix} = \begin{bmatrix} 2RS\,A_1 \sin(A_2) \\ RS\,A_1 \cos(A_2) \end{bmatrix}, \tag{5.13}$$

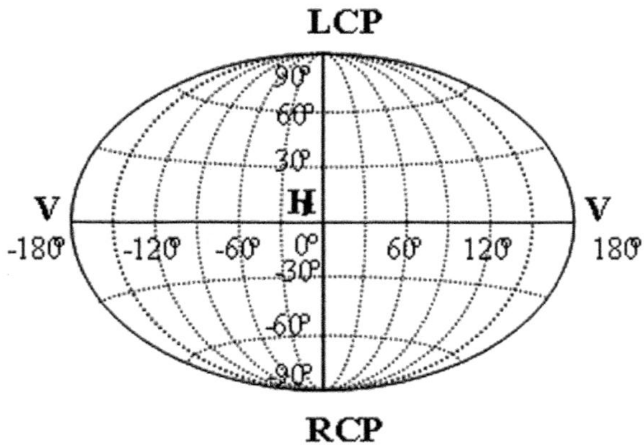

Fig. 5.13 Elliptical projection of equal areas of Aitoff–Hammer

where x and y are the resulting transformation coordinates (longitude and latitude lines, respectively).

Figure 5.13 shows the projection of Aitoff–Hammer. The projection center, which corresponds to the basis of horizontal polarization on the Poincaré sphere, has the coordinates $\tau_0 = 0$ and $\varepsilon_0 = 0$.

In the image, only the equator and the central meridian are straight lines, and the remaining parallels and meridians are curved lines. This projection of equal areas provides an opportunity to see the entire surface of the sphere simultaneously. This makes the projection of Aitoff–Hammer convenient for displaying data on unknown polarization states. This cartographic representation was first used by Rayvan's radar polarimetry to analyze the characteristics of the polarization states of noise clusters. The cartographic projection of Aitoff–Hammer is also a convenient means of representing distributions on the Poincaré sphere similar to the Lambert cylindrical projection of equal areas.

In addition, the elliptical projection of Aitoff–Hammer can give a more accurate representation for regions close to the poles, since it has a circular shape in the region of the poles. This projection is also plausible, which means the equality of the ratio of the areas in any part of the projection and the ratio of the areas of the entire projection. Nonetheless, the projection of Molveyd, which is very close to the projection of Aitoff–Hammer, is more preferable.

5.3.5 Molveyd Projection of Equal Areas

The Molveyd projection is based on the spherical model of the Earth, and therefore, it is often used to obtain geographic maps of the globe (Fig. 5.14).

The main task in constructing this projection is to arrange the parallels in such a way that the property of the area equivalence is preserved. The process of

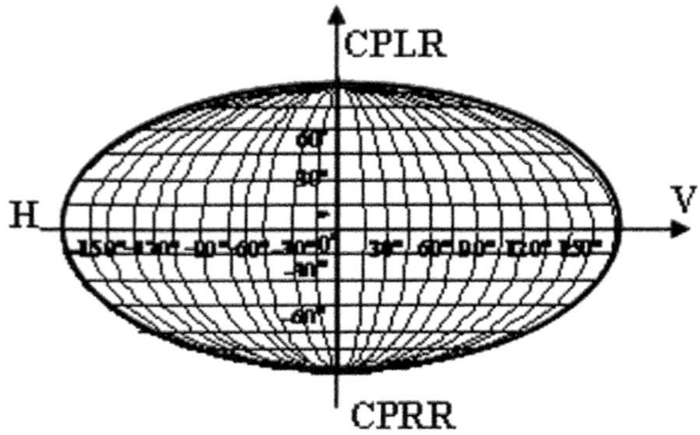

Fig. 5.14 Molveiyd projection

transforming this projection includes the transformation of a spherical coordinate system into a rectangular coordinate system:

$$\begin{bmatrix} x \\ y \end{bmatrix} = \begin{bmatrix} \frac{\sqrt{2}}{90^{\circ}}(2\tau - 2\tau_0)RS\cos(2\varepsilon) \\ \sqrt{2}RS\sin(2\varepsilon) \end{bmatrix}. \tag{5.14}$$

The coordinate grid for this projection is shown in Fig. 5.14, where the center of the projection corresponds to the point with the coordinates $\tau = 0$ and $\varepsilon = 0$. Equator and parallels on the projection are straight horizontal lines. All meridians are ellipses except for the central meridian, which is a rectilinear ellipse (a straight line perpendicular to the equator). Two extreme points on the central meridian correspond to bases of circular polarization of the left and right rotation, respectively.

5.4 Representation of Thermal Radio Emission Polarization

Let us consider a geometric interpretation of the thermal radio emission polarization. We will align the Poincaré sphere so that the polarizations corresponding to the unit vectors of the zero polarization basis are represented by the points F and G at the north and south poles (Fig. 1.4). Then the polarizations, for which $\gamma = \text{const}$, and $\Phi = 2\delta$ is arbitrarily, will correspond to the small circle on the Poincaré sphere. In this case, on the polarizations displayed on the poles, the radio brightness temperature is extreme, and on polarizations displayed on the equator ($\gamma = 45°$), it has an average value $\left(\chi_1^2 + \chi_2^2\right)/2T_0$. Polarizations in which $\Phi = \text{const}$, and γ—arbitrarily, are mapped onto the meridian. When the polarization is changed along the meridian, the rate of change in the radio brightness temperature is maximal. From the geometrical representation considered, the practically important conclusion follows: If the polarization properties of the receiving antenna change so that its polarization is mapped to a circle ($\gamma = \text{const}$), then the radiation source is represented as a polarization isotropic (homogeneous); that is, its radio brightness temperature does not depend on the kind of polarization; if the mapping takes place onto the meridian ($\delta = \text{const}$), then the emitter has a maximum polarization contrast.

In practice, it is often required to find out the maximum and minimum emissivity of a body, provided that measurement of radio brightness temperature is possible only on a small number of some standard types of polarization. Usually, these can be various linear polarizations, which are changed by the standard rotation of receiving antennas in space, as well as circular polarizations. Let us consider the problem of determining the extreme values of the radio brightness temperature with the minimum number of measurements on these types of polarizations.

Figure 5.15 shows the Poincaré sphere, aligned so that the extreme polarizations are represented by the points *A* and *B* (the north and south poles). Let the linear polarizations correspond to the points of the large circle $R_1SMR_2NR_1$. The angle between its diameter R_1R_2 and its axis *OA* will be denoted as $2\gamma_0$, i.e., $\angle R_1OA = 2\gamma_0$. The circular polarization for a given orientation of the sphere will be represented by a point *T*, lying on the perpendicular to the plane of this circle, i.e., $OT \perp R_1OR_2$. As evident from the figure $\angle AOT = 90° - 2\gamma_0$. Let the measurement be performed on some linear polarization (provisionally to be called horizontal) and correspond to the point *S*. Denote $\angle R_1OS$ with 2ψ and $\angle SOA$ with 2γ. Note that the rotation of the receiving antenna in space by the angle ψ_1 will correspond to the moving of the point *S* along the circle $R_1SMR_2NR_1$, to the point S_1, so that the angle $\angle S_1OS = 2\psi_1$. Consideration of the trihedral angle OAR_1S yields the following relation:

$$\cos 2\gamma = \cos 2\gamma_0 \cos 2\psi. \tag{5.15}$$

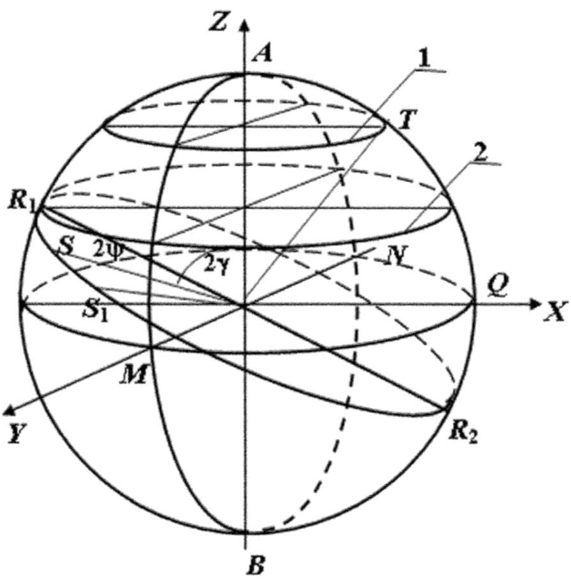

Fig. 5.15 On representation of the thermal radio emission polarization on the Poincaré sphere

On the horizontal polarization (point S), the emissivity $\Pi_{\Gamma\Pi}$ can be found using equality (5.15):

$$\Pi_{\Gamma\Pi} = \Pi_{\Sigma} + \frac{\chi_1^2 + \chi_2^2}{2} \cos 2\gamma_0 \cos 2\psi. \tag{5.16}$$

To determine Π_{Σ}, it suffices to carry out another measurement with vertical polarization, for which

$$\Pi_{B\Pi} = \Pi_{\Sigma} - \frac{\chi_1^2 - \chi_2^2}{2} \cos 2\gamma_0 \cos \psi. \tag{5.17}$$

Adding the equalities (5.16) and (5.17), we obtain: $2\Pi_{\Sigma} = \chi_1^2 + \chi_2^2$. It is evident from these relations that by measuring the emissivity only on linear polarizations (only the angle ψ changes), it is impossible to determine the difference $\left(\chi_1^2 - \chi_2^2\right)$, since it enters into these expressions with an unknown factor $\cos 2\gamma_0$. Therefore, a measurement is to be made on such a polarization at which the angle between the OA axis and the radius drawn to the polarization image point would differ fromγ_0. For example, it can be circular polarization (point T), for which we can write:

$$\Pi_{K\Pi} = \Pi_{\Sigma} + \frac{\chi_1^2 + \chi_2^2}{2} \sin 2\gamma_0. \tag{5.18}$$

The relations (5.17) and (5.18) are a system of three equations with four unknowns: Π_{Σ}; $\left(\chi_1^2 - \chi_2^2\right)$; $2\gamma_0$ and 2ψ.

Therefore, another measurement is necessary. For this purpose, we will use a linear polarization obtained by rotating the receiving (conditionally horizontally polarized) antenna by an angle of 45° ($\angle SOS_1 = 90°$) that yields the missing equation:

$$\Pi_{45} = \Pi_{\Sigma} - \frac{\chi_1^2 - \chi_2^2}{2} \cos 2\gamma_0 \sin 2\psi. \tag{5.19}$$

The relations (5.16) and (5.19) make it possible to find Π_{Σ}:

$$\Pi_{\Sigma} = \frac{\Pi_{\Gamma\Pi} + \Pi_{B\Pi}}{2} = \frac{\chi_1^2 + \chi_2^2}{2} = \frac{T_{\Gamma\Pi} + T_{B\Pi}}{2T_0}. \tag{5.20}$$

It follows from expressions (5.17) and (5.18) that:

$$tg2\psi = \frac{\Pi_{\Sigma} - \Pi_{45}}{\Pi_{\Sigma} - \Pi_{B\Pi}} = \frac{T_{\Sigma} + T_{45^\circ}}{T_{\Sigma} + T_{B\Pi}} = \frac{T_{\Gamma\Pi} + T_{B\Pi} - 2T_{45^\circ}}{T_{\Gamma\Pi} - T_{B\Pi}}, \tag{5.21}$$

where T_{Σ}—is the radio brightness temperature measured in the total channel; $T_{\Gamma\Pi}$—radio brightness temperature, measured in one of the channels; $T_{B\Pi}$—radio brightness temperature measured in the orthogonal channel; T_{45°—the radio brightness temperature measured in the first channel, provided that the antenna is rotated in space by 45°.

It can also be found from expressions (5.17) and (5.18) that:

$$\tan 2\gamma_0 = \frac{(\Pi_{\text{КП}} - \Pi_\Sigma)\cos 2\psi}{\Pi_\Sigma - T_{\text{ВП}}} = \frac{2T_{\text{КП}} - T_{\text{ГП}} - T_{\text{ВП}}}{\sqrt{(T_{\text{ГП}} - T_{\text{ВП}} - 2T_{45^\circ})^2 + (T_{\text{ГП}} - T_{\text{ВП}})^2}}, \tag{5.22}$$

where $T_{\text{КП}}$ is the radio brightness temperature measured on the circular polarization.

It follows from relation (5.18):

$$\chi_1^2 - \chi_2^2 = 2 \cdot \frac{T_\Sigma}{T_0}\sqrt{\left(1 - \frac{T_{\text{КП}}}{T_\Sigma}\right)^2 + \left(1 - \frac{T_{45^\circ}}{T_\Sigma}\right)^2 + \left(1 - \frac{T_{\text{ГП}}}{T_\Sigma}\right)^2}. \tag{5.23}$$

Relations (5.20) and (5.23) are a system of two equations with two unknowns, the solution of which allows one to determine the extreme values of the emissivity:

$$\left.\begin{aligned} \frac{T_{\max}}{T_0} &= \Pi_{\max} = \chi_1^2 = \frac{T_\Sigma}{T_0}\left[1 + \sqrt{\left(1 - \frac{T_{\text{ГП}}}{T_\Sigma}\right)^2 + \left(1 - \frac{T_{\text{КП}}}{T_\Sigma}\right)^2 + \left(1 - \frac{T_{45^\circ}}{T_\Sigma}\right)^2}\right] \\ \frac{T_{\min}}{T_0} &= \Pi_{\min} = \chi_2^2 = \frac{T_\Sigma}{T_0}\left[1 - \sqrt{\left(1 - \frac{T_{\text{ГП}}}{T_\Sigma}\right)^2 + \left(1 - \frac{T_{\text{КП}}}{T_\Sigma}\right)^2 + \left(1 - \frac{T_{45^\circ}}{T_\Sigma}\right)^2}\right] \end{aligned}\right\} \tag{5.24}$$

and degree of polarization

$$m = \frac{1}{2}\sqrt{\left(1 - \frac{T_{\text{ГП}}}{T_\Sigma}\right)^2 + \left(1 - \frac{T_{\text{КП}}}{T_\Sigma}\right)^2 + \left(1 - \frac{T_{45^\circ}}{T_\Sigma}\right)^2}. \tag{5.25}$$

If the measurements make it possible to find a linear polarization in which the radio brightness temperature (in the class of linear polarizations) is extreme (point R_1 or R_2 in Fig. 5.15), then the calculated formulas are simplified, since in this case $\angle R_1OS = 2\psi = 0$, and therefore $\gamma = \gamma_0$, as follows from the Formula (5.21):

$$T_{45^\circ} = 0.5\,(T_{\text{ГП}} + T_{\text{ВП}}). \tag{5.26}$$

Thus, the need for measurement on a linear polarization with an angle of 45° disappears. In this case:

$$\left.\begin{aligned}\frac{T_{\max}}{T_0}&=\chi_1^2=\frac{T_\Sigma}{T_0}\left[1+\sqrt{\left(1-\frac{T_{\text{ТП}}}{T_\Sigma}\right)^2+\left(1-\frac{T_{\text{КП}}}{T_\Sigma}\right)^2}\right]\\ \frac{T_{\min}}{T_0}&=\chi_2^2=\frac{T_\Sigma}{T_0}\left[1-\sqrt{\left(1-\frac{T_{\text{ТП}}}{T_\Sigma}\right)^2+\left(1-\frac{T_{\text{КП}}}{T_\Sigma}\right)^2}\right]\end{aligned}\right\}. \tag{5.27}$$

Consequently, in order to obtain the required values, only three measurements are sufficient in this case.

The relations (5.24) make it possible to obtain necessary and sufficient conditions for the polarization isotropy of the radiator, which takes place, as follows from the definition for $\chi_1 = \chi_2$, i.e., when $T_{\max} = T_{\min}$, which is possible if and only if

$$T_{\text{ТП}} = T_{\text{ВП}} = T_{\text{КП}} + T_{45^\circ}. \tag{5.28}$$

It should be borne in mind that under horizontal, any linear polarization is understood, and under the vertical one—orthogonal to it. The worst case, corresponding to condition (5.28), arises if the horizontal polarization is displayed in Fig. 5.15 at point M, the vertical one at point N, the circular one at point Q. Then, the point R_1 will transit to point A, and R_2—to point B. It is quite clear that instead of linear polarization with an angle of 45°, any other linear polarization can be used. The fulfillment of the condition $m = 0$ on any two pairs of orthogonal polarizations will indicate that a large circle of linear polarizations coincides with the equator, and then, the circular polarizations of right and left rotation are extreme, or that the radiator is polarization isotropic. The degeneracy condition of the radiator can be obtained from expression (5.24). Since $T_{\min}$ needs to be equals to zero, it is required that

$$(1 - T_{\text{ТП}}/T_\Sigma)^2 + (1 - T_{\text{КП}}/T_\Sigma)^2 + (1 - T_{45^\circ}/T_\Sigma)^2 = 1, \tag{5.29}$$

which predetermines the need for four measurements.

5.5 Representation of the Polarization State on the Plane

The problem of infinite values of polarization characteristics in radio polarimetry
In radio polarimetry, as well as electrodynamics, there are often cases when it is required to find some averaged characteristics of the polarization state for the analyzed wave. Most often this problem arises when considering the guiding structures, for example, waveguides and light guides, especially when these structures are filled with chiral media. For the polarization state, the phasor state is rather universal and convenient; it easily "gets rid" of such insignificant from the point of view of polarization characteristics as wave energy and the initial phase. The description of the polarization state on the complex plane $\underline{W} = W_1 + iW_2$—the Poincaré plane also

has the same advantages, since the points of the Poincaré plane are in one-to-one correspondence with the phasor by the fractional linear transformation.

However, complex variables for the phasor and complex variables of the Poincaré plane turn to infinity, and for the most popular states in radio polarimetry—linear or circular polarizations. Consequently, this circumstance completely excludes the possibility of averaging the polarization characteristics along a contour or a cross section. We need to look for such a transformation of the complex planes of the phasor or Poincaré so that the obtained characteristics in the form of a certain complex number (the complex numbers here are due to the requirement to combine two numbers necessary for the representation of the polarization state into a single organic algebraic object) never lead to infinity. Note that one of the most popular ways of such a transformation is the Poincaré sphere, corresponding to the stereographic projection of the Poincaré plane.

An effective "planar" (and not the just mentioned "spherical") way of overcoming infinity in the description of the polarization state was proposed by Shevchenko.

<u>The complex plane of F. I. Fedorov</u> Shevchenko's approach to the formation of a complex plane for the representation of the polarization state characteristics is the development of the method of description proposed by F. I. Fedorov for describing gyrotropic media, where he used the transformation of the complex plane of a phasor according to the rule:

$$\underline{\Phi} = \arctan\, \underline{p}, \tag{5.30}$$

where the complex analog of the arctangent function, if we consider only its main value, transforms the complex plane of the phasor into a vertical strip symmetric with respect to the imaginary axis in the complex plane $\underline{\Phi}$, of width π (Fig. 5.10).

The complex coefficient of polarization $\underline{\Phi}$ depends on the geometric parameters of the polarization ellipse as follows:

$$\mathrm{Re}\,\underline{\Phi} = \beta, \quad \left|\tan\left(\mathrm{Im}\,\underline{\Phi}\right)\right| = \alpha. \tag{5.31}$$

Thus, it turns out that $\underline{\Phi} = \infty$ corresponds to a wave of right circular polarization and $\underline{\Phi} = -\infty$ to a left one. In the area near to the zero of the complex plane, $\underline{\Phi} \approx \underline{p}$.

<u>The complex plane of V. V. Shevchenko.</u> The complex plane of V. V. Shevchenko $\underline{Ш}$ is formed from the complex plane of F. I. Fedorov according to the rule:

$$\mathrm{Re}\,\underline{Ш} = \tan(\mathrm{Re}\,\underline{\Phi}),\ \mathrm{Im}\,\underline{Ш} = th(\mathrm{Im}\underline{\Phi}). \tag{5.32}$$

Its points are connected with the phasor by the following relations:

$$\underline{p} = \frac{4\underline{Ш}}{4 - \underline{Ш}^2 + \underline{Ш}^*}, \tag{5.33}$$

and

$$\underline{\text{Ш}} = 2\underline{p}\frac{(\underline{p}^*)^2 + 1 - \left|\underline{p}^2 + 1\right|}{(\underline{p}^*)^2 - \underline{p}^2}. \tag{5.34}$$

The last relation can also be written as a pair for real and imaginary parts $\underline{\text{Ш}}$:

$$\operatorname{Re}\underline{\text{Ш}} = \frac{\left|\underline{p}\right|^2 - 1 - \left|\underline{p}^2 + 1\right|}{2\operatorname{Re}\underline{p}}, \quad \operatorname{Im}\underline{\text{Ш}} = \frac{\left|\underline{p}\right|^2 + 1 - \left|\underline{p}^2 + 1\right|}{2\operatorname{Im}\underline{p}}. \tag{5.35}$$

For small modulus phasors $\left|\underline{p}\right| \ll 1$, the relations (5.34) and (5.35) allow the following transformation:

$$\underline{\text{Ш}} = \underline{p}\left(1 + \frac{(\underline{p}^*)^2 - \underline{p}^2}{4}\right), \tag{5.36}$$

and

$$\operatorname{Re}\underline{\text{Ш}} = \operatorname{Re}\underline{p}(1 + (\operatorname{Im}\underline{p})^2), \quad \operatorname{Im}\underline{\text{Ш}} = \operatorname{Im}\underline{p}(1 + (\operatorname{Re}\underline{p})^2). \tag{5.37}$$

Figure 5.16 shows the complex plane of Shevchenko, from which it can be seen that here the phasor plane transits on the complex plane $\underline{\text{Ш}}$ into a horizontal strip symmetrically aligned about the imaginary axis of the width $2i$; that is, again this complex plane contains infinity.

The plane of Fedorov–Shevchenko To exclude the infinities on the planes $\underline{\text{Ш}}$ and $\underline{\Phi}$, V. V. Shevchenko proposed to unite them into a new plane (Fig. 5.17), which can be considered as both complex $\underline{\Phi\text{Ш}}$ and Cartesian $\Phi\text{Ш}$. The fact is that the pair of

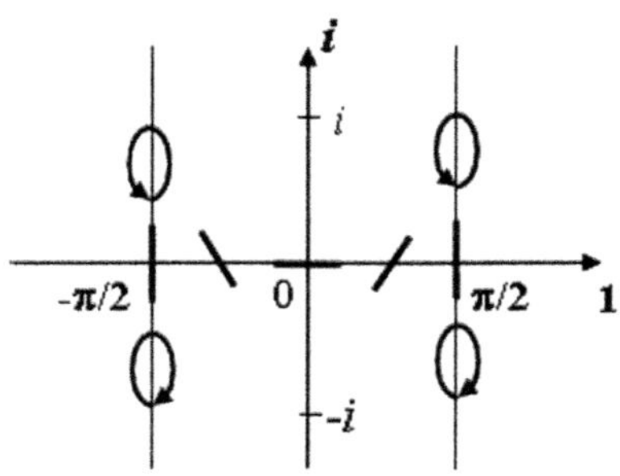

Fig. 5.16 Complex plane V. V. Shevchenko

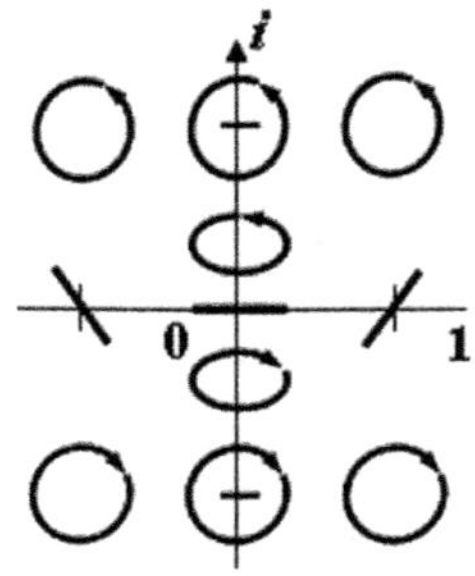

Fig. 5.17 Complex plane of F. I. Fedorov–V. V. Shevchenko

numbers formed as a result of this combination does not allow to form an analytic function, and therefore, it is advisable to tell of the plane precisely ΦШ.

The abscissa axis of the plane ΦШ corresponds to the real axis of the complex plane $\underline{\Phi}$ and the ordinate axis to the imaginary axis of the complex plane $\underline{Ш}$. In this plane, it is now advisable to take into account only a rectangular region, horizontally of width π, and vertically 2. The coordinates of this Cartesian plane in this rectangular region are related to the phasor as follows:

$$\Phi = \arctan\left(\frac{\left|\underline{p}\right|^2 - 1 - \left|\underline{p}^2 + 1\right|}{2\mathrm{Re}\,\underline{p}}\right), \tag{5.38}$$

and

$$Ш = \frac{\left|\underline{p}\right|^2 + 1 - \left|\underline{p}^2 + 1\right|}{2\,\mathrm{Im}\,\underline{p}}. \tag{5.39}$$

The rectangular region of the plane ΦШ thus formed does not contain infinities and allows us to describe all Φ states and hence all possible transformations.

Part II
Transformation of the Polarization Structure of the Scattered and Eigen Emission of Navigation Observation Objects

Greater part of the most important practical applications, ideas and results of the radiophysical theory are related to the analysis of the scattered field. Here it is necessary to specify, first of all, radar detection and ranging, as well as the use of radar methods, when performing the tasks of glaciology, monitoring, pathology, non-destructive testing, trajectory measurements, planetology, remote research of geophysical and hydrological objects, climatology and meteorology.

There is a constant growth in both the arsenal of scattered field research means, and the nomenclature of the characteristics of the probed objects. If, at an early stage of the development of scattered field analysis methods, the main attention was paid to the detection of anomalies in the objects under research and to the determination of the kinematic (i.e. motion-related) characteristics of the monitored objects: range, angular coordinates, speed, then recently there has been a noticeable shift to the development of methods and algorithms for obtaining information on the type, class of the object, its dimensions, electrophysical properties of the surface, and structural properties. Special hopes are placed on time (non-stationary) methods using wideband and ultra wideband signals such as small low-range pulses and Walsh waves.

The following three groups of methods and ways to describe and research a scattered field, discussed in the literature, may be distinguished: methods of diffraction theory, methods of scattering theory, and phenomenological representations.

The diffraction theory combines the methods of theoretical determination of electromagnetic fields that arise when various objects are placed in the field of given sources. Phenomenological representations come from the approach to the scattered field as a carrier of information about the source, which allows us to treat the scattered field as a corresponding signal, usually—a spatio-temporal one.

Intermediate position between these two groups of methods is occupied by scattering theory, the subject of which is the research and formation of field structures that reflect the specificity of the process of interaction of the electromagnetic field and the matter, and allow the parametrization of the quantities observed in practice.

The main attention in this part is given to the phenomenological representation of the scattering process, since the field is considered here as an element of an information channel comprising, in addition to the field under research, devices that extract information necessary for solving practical issues. It is clear that this view is closely related to practical activities. The investigation of this interconnection determines the realizable conditions for observing the scattered field, the set of parameters to be recorded, and the required class of scatterers. Further, the requirements identified are "transmitted on a chain basis", imposing certain restrictions on the structures used in the scattering theory, and forming problems that are to be solved in the diffraction theory.

Proceeding from the above, it becomes clear that the phenomenological methods shall be related to the cybernetic "black box" method and, as applied to the scattering problems under consideration, are intended to fix the influence of the incident field (radio wave which irradiates the object) $\vec{E}_u$ on the scattered field $\vec{E}_p$ through the scatterer. The methods of diffraction theory shall explain the influence of the fields on each other.

Another consequence of the methodology of the phenomenological level of describing the scattered field is that the scattering process described at the level of the above-mentioned "black box" can almost always be modeled with a sufficiently simple multi-pole device, whose "inputs" and "outputs" are determined by the conditions for observing the scatterer.

A complete description of the scattered field, as a system with an infinite number of degrees of freedom, requires the formation of an appropriate complex continual spatio-temporal object—a mathematical model (see clause 6.2). This is one of the problems of the scattering theory. However, this circumstance also has a negative side. The simplicity of the proposed descriptions for various aspects of the scattering process at the phenomenological level leads to the fact that researchers have many private methods of research and analysis and, as a consequence, quantitative characteristics and parameters. In this connection, it is necessary to identify the interconnections between these characteristics, and therefore also to classify them, and that is why it is hardly possible in this approach to do without the scattering theory, which is the necessary basis for interpreting the selected and observed phenomenological characteristics.

Chapter 6
Scattering Matrix and Its Basic Properties

6.1 Scattered Field and Its Characteristics

The structural representations for the field existing outside the scatterer are based on the division of the observed total field $\vec{E}_\Sigma$ into the incident $\vec{E}_u$ and scattered $\vec{E}_p$, i.e., $\vec{E}_\Sigma = \vec{E}_u + \vec{E}_p$.

Thus, the field scattered by the object is defined as the vector difference between the total field and the field that would be in space in the absence of the scatterer. The distinguishing of the scattered field is often a rather complicated task, and this problem is essentially absent only for backscattering, since the incident and scattered fields are automatically separated in this case, as they propagate in opposite directions.

An original reference task is to study the scattering process when the object is probed with a plane wave.

If the scatterer has finite dimensions, and observations are carried out at a sufficient distance, the scattered field can be approximated by a spherical wave.

Figure 6.1 shows the results of the calculation of the scattered field for balls of different radii a. The shadow-forming back lobe of the scattered field is clearly visible, which, when superimposed on the incident field, ensures its "suppression". As can be seen, the magnitude of the lobe and the degree of "suppression" increase as parameter $ka = 2\pi a/\lambda$ increases. The same figure shows the effect of the type of polarization of a sphere-irradiating plane wave.

Let us define a spherical coordinate system centered at a certain point of scattering. The energy flux of the scattered wave in a solid angle directed along $\vec{r}$ and equal to $\mathrm{d}\Omega = \sin\theta\,\mathrm{d}\theta\,\mathrm{d}\varphi$ is proportional to the value $\vec{E}_p^{0+}(\theta, \varphi)\vec{E}_p^{0}(\theta, \varphi)\sin\theta\mathrm{d}\theta\mathrm{d}\varphi$, where $\vec{E}_p^{0}$—three-dimensional complex vector—is determined by its components in a given coordinate system.

To estimate the scattering power of the scatterer, let us discuss the following. Let in the direction of the scatterer, located at a distance R from the transmitting antenna, the latter emits an electromagnetic wave of power P_Σ. In this case, the modulus of the Umov–Poynting vector of this wave near the scatterer will be equal

Kozlov A. I. et al., *Introduction to the Theory of Radiopolarimetric Navigation Systems*, Springer Aerospace Technology, https://doi.org/10.1007/978-981-13-8395-3_6

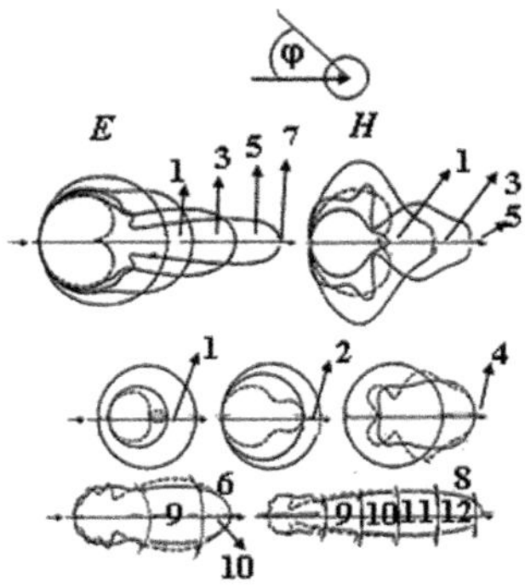

Fig. 6.1 On diffraction on the ball 1—$ka = 0.5$; 2—$ka = 1.0$; 3—$ka = 1.5$; 4—$ka = 2.0$; 5—$ka = 4.0$; 6—$ka = 5.0$; 7—$ka = 8.0$; 8—$ka = 10$; 9—$F = 1$; 10—$F = 2$, 11—$F = 3$; 12—$F = 4$

to $\Pi_u = P_\Sigma / 4\pi R^2$, and the modulus of its electric vector $E_u = \sqrt{30 P_\Sigma}/R$. After the interaction of the emitted wave with the scatterer, the latter itself becomes the source of electromagnetic radiation (a scattered wave), which in the immediate vicinity of the scatterer will be characterized by its electric vector $\vec{E}_p$ and its Umov–Poynting vector $\vec{\Pi}_p$.

When determining general characteristics of the scatterer based on the above, two approaches are possible. The first one will rely on comparing the electric vectors of the incident and scattered radio waves, and the second will rely on the comparison of the Umov–Poynting vector for the wave irradiating the scatterer with the power of the scattered wave.

Let us consider each of the approaches separately.

Since this is a linear scattering process (nonlinear scattering will be considered in clause 6.3), it is obvious that the complex electric vector of the radio wave irradiating the scatterer $\vec{E}_u$ will be associated with an analogous vector of the scattered radio wave $\vec{E}_p$ by means of some linear operator S through the use of the obvious relation:

$$\vec{E}_p = S\,\vec{E}_p \tag{6.1}$$

In this case, the operator S, called the scattering matrix, is a square matrix of 3×3 in size, the elements of which $s_{mn},\;\; m, n = \overline{1,3}$ are *dimensionless* complex numbers. If the vectors $\vec{E}_u$ and $\vec{E}_p$ are considered at distances where the spherical wave can be regarded as flat with a given accuracy, then the dimension of the scattering matrix S decreases and becomes equal to 2×2.

As seen from above, the scattering matrix is a direct characteristic of the scatterer itself.

If we take into account that the scattering of waves occurs in all directions (θ_p, φ_p), and their irradiation is also possible from all directions θ_u, φ_u , then it is obvious that the elements of the scattering matrix depend on these angles, i.e., $s_{mn} \equiv s_{mn}(\theta_u, \varphi_u; \theta_p, \varphi_p)$. This gives grounds for discussing the scattering diagram (SD). If we confine ourselves to the consideration of scattering in the direction of the transmitting antenna only, then we speak of a backscattering diagram (BSD).

Thus, the scatterer is characterized by a scattering matrix having the following form:

$$S\left(\vec{k}_u;\theta,\varphi\right)=\begin{pmatrix} s_{11}\left(\vec{k}_u;\theta,\varphi\right) & s_{12}\left(\vec{k}_u;\theta,\varphi\right) \\ s_{21}\left(\vec{k}_u;\theta,\varphi\right) & s_{22}\left(\vec{k}_u;\theta,\varphi\right)\end{pmatrix}, \tag{6.2}$$

where $\vec{k}_u$—wave vector of the radiated plane wave, and the angles $\left(\theta_p,\varphi_p\right)$ are accordingly replaced by (θ,φ).

The second approach, related to the determination of the scatterer's characteristics, as mentioned above, is based on a comparison of the energy characteristics of the radiated and scattered fields.

The modulus of the Umov–Poynting vector of the incident wave near the scatterer will be equal to $\Pi_u = P_\Sigma/4\pi R^2$. It is obvious that the scattered radiation power P_p will be proportional to the value Π_u, while the coefficient of proportionality $\sigma^2\left(\theta,\varphi;\vec{k}_u\right)$, having the area dimension and called the effective scattering area (ESA) or, more accurately, the ESA diagram, will characterize the scattering properties of the scatterer. Since the radiation energy is proportional to the squared modulus of the electric vector, this allows to write the ESA expression through the electric vector of the radiated wave, measured directly at the transmitting antenna—$\vec{E}_u^0\left(\vec{k}_u\right)$, and the electric vector of the scattered wave, measured directly at the scatterer—$\vec{E}_p^0(\theta,\varphi)$, with the help of an obvious equation:

$$\sigma^2\left(\theta,\varphi;\vec{k}_u\right)=4\pi R^2\frac{\vec{E}_p^{0+}(\theta,\varphi)\vec{E}_p^0(\theta,\varphi)}{\vec{E}_u^{0+}\left(\vec{k}_u\right)\vec{E}_u^0\left(\vec{k}_u\right)}. \tag{6.3a}$$

Note that the product $4\pi R^2\vec{E}_p^{0+}(\theta,\varphi)\vec{E}_p^0(\theta,\varphi)$ *is numerically* equal to $\vec{E}_p^{0+}(\theta,\varphi)\vec{E}_p^0(\theta,\varphi)$, if $\vec{E}_p^0(\theta,\varphi)$ means the value of the electric vector of the scattered wave measured directly at the receiving antenna.

In view of the last remark, we further use a somewhat different formula instead of formula (6.3a):

$$\sigma^2\left(\theta,\varphi;\vec{k}_u\right)=\frac{\vec{E}_p^{0+}(\theta,\varphi)\vec{E}_p^0(\theta,\varphi)}{\vec{E}_u^{0+}\left(\vec{k}_u\right)\vec{E}_u^0\left(\vec{k}_u\right)}. \tag{6.3b}$$

Adopting *formally* Eq. (6.3b), it is also necessary to *formally* assume that the electric vectors of the radiated and scattered fields have, as it were, *different dimensions.*

When integrating (6.3b) over all directions of scattering, a characteristic of the scattering power is formed, called the integral ESA:

$$\sigma_{инт}\left(\vec{k}_u\right)=\int_0^{\pi}\int_0^{2\pi}\frac{\vec{E}_p^{0+}(\theta,\varphi)\vec{E}_p^0(\theta,\varphi)}{\vec{E}_u^{0+}\left(\vec{k}_u\right)\vec{E}_u^0\left(\vec{k}_u\right)}\sin\theta d\theta d\varphi. \tag{6.4}$$

The form of relations (6.3b) and (6.4) suggests that we should single out a characteristic corresponding to ratio $\vec{E}_p^0 \Big/ \vec{E}_u^0$. Since the dividend and the divisor are vectors, this ratio will be determined for a linear scatterer by some complex matrix of 3×3 in size, which, when the irradiation and scattering directions are varied, determines the scattering amplitude $S\left(\vec{k}_u; \theta, \varphi\right)$:

$$\vec{E}_p^0(\theta, \varphi) = S\left(\vec{k}_u; \theta, \varphi\right)\vec{E}_u^0\left(\vec{k}_u\right). \tag{6.5}$$

In Eq. (6.5), if we assume that the values $\vec{E}_u^0\left(\vec{k}_u\right)$ and $\vec{E}_p^0(\theta, \varphi)$ are the electric vectors of the radiated and scattered fields measured directly at the antenna, it is necessary to assume that the *dimension of the scattering amplitude corresponds to the length.*

The incident flat and scattered spherical waves at a large distance from the scatterer can always be written in two-component symbols. This will lead to the fact that the scattering process will be characterized by a square complex matrix of 2×2 in size (6.2).

The difference between the two considered approaches is that for the same formula of the scattering matrix in form (6.1), the elements of this matrix are dimensionless in the first case and have the length dimension in the second case. However, since both formulas are formally identical, all subsequent calculations refer to both representations in the same plan.

One of the main theorems of the scattering theory asserts that the scattering amplitude makes it possible to reconstruct a field for a finite scatterer in the near zone, up to a radial component. The uniqueness theorem establishes a one-to-one relationship between the scattered field and the scatterer, provided that the latter does not contain zero-volume singularities (pins, planes) as surface elements. Thus, the scattering amplitude is uniquely related to the scatterer, and therefore is a complete characteristic of its scattering power.

Between the ESA diagram and the scattering amplitude, the following relationship can be established on the basis of ratios (6.2) and (6.4):

$$\sigma^2(\theta, \varphi, k_u) = S_p\left(S^+\left(\vec{k}_u; \theta, \varphi\right)S\left(\vec{k}_u; \theta, \varphi\right)\right). \tag{6.6}$$

Let us give some information about the scattered field and its characteristics, which follow from the theories of diffraction and scattering.

Relying on differential geometry in scattering theory, a number of statements can be formulated regarding the polarization structure of the scattered field. Let a field of tangent vectors be given on the sphere. Then the Poincaré–Hopf theorem is correct for such a field, which asserts that the sum of the indices of all singular points of the vector field is equal to 2. We successively define the concepts that make up the statement of the theorem.

A tangent vector is associated with each point of the surface, the direction of which varies continuously from point to point, which represents the field of tangent vectors

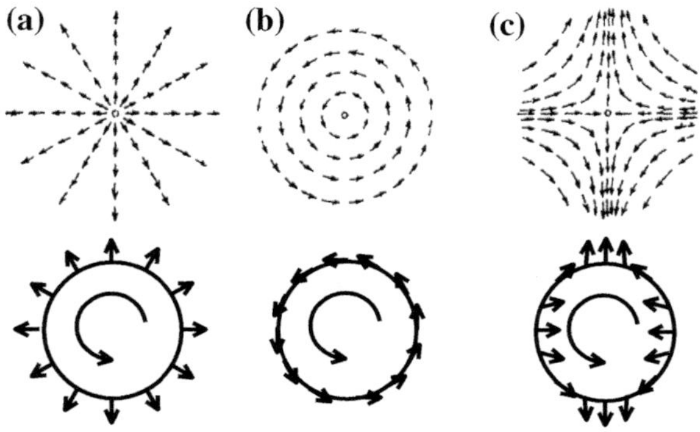

Fig. 6.2 Three kinds of special points

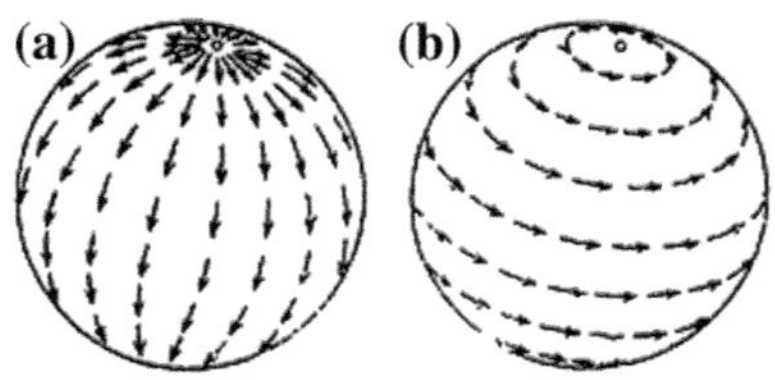

Fig. 6.3 Fields of tangent vectors

on the surface (in this case—on the sphere). Discontinuity occurs in the singular points of the vector field. As shown in differential geometry, singular points of three kinds are possible for the sphere, as indicated in Fig. 6.2 with the letters *a*, *b*, and *c*. When traversing singular points "*a*" and "*b*", vectors undergo a rotation in the direction of traversal, and for a singular point "*c*"—in the opposite direction. Index $+1$ is assigned to special points "*a*" and "*b*", and index -1 is assigned to a singular point "*c*".

We can offer two fields of tangent vectors that preserve their direction. As can be seen from Fig. 6.3, each of them has two singular points with the index $+1$, which corresponds to the conditions of the theorem under discussion. Other types of vector fields may be also offered, but the sum of the indices of singular points for all of them shall be equal to 2.

This essentially proves the assertion that it is impossible to create an isotropically scattering scatterer of a linearly polarized field. One can also speak of the impossibility of maintaining a constant elliptic polarization of a scattered wave on the sphere. In this case, components with linear polarizations will appear at least for two directions or the scattered field will be equal to zero. By combining two tangent vector fields and setting singular points for each of them, we can discard the unrealizable combinations of the polarization structure of the scattered field using the Poincaré–Hopf theorem.

For the stationary case, when the field components perform oscillations of type $e^{i\omega t}$, the diffraction theory sets the natural scale of the characteristics of the scattering task and provides the corresponding physical representations for the formation of the field structure in the scattering theory and for the formation of parameters observable under the phenomenological approach. The main parameter of this scale is the ratio of the wavelength λ to the characteristic size of inhomogeneities and scatterers a. Three groups of scattering problems can be distinguished for objects and inhomogeneities of finite dimensions: Rayleigh, resonance, and quasi-optical.

For Rayleigh scattering ($ka \ll 1$), it is characteristic that, first, the amplitude of the scattered field is proportional to the volume of the object and slightly depends on its shape, and, second, the phases of the scattered signals do not depend on the scattering angle and the target shape. The latter circumstance leads to the maintenance of the phase difference in the scattering between the polarization components, and hence, the linear polarization for Rayleigh scattering is an invariant—only its slope changes.

The scattering power of the Rayleigh reflector is satisfactorily described by the ESA diagram:

$$b^2 = \left[2\left|\vec{k}_u\right|^2 V\left(1 + \frac{e^{-y}}{\pi y}\right)\right]^2 \cdot \frac{1}{\pi}, \tag{6.7}$$

where V—volume of the scatterer, y—length/width ratio of the object. Due to the similarity of the EPR diagram to the isotropic diagram, arguments in (6.7) at the ESA value are omitted.

The criterion of the resonance zone is the approximate equation $ka \approx 1$. This area is most difficult for research, because here, to some extent, it is necessary to use rigorous methods of solving wave equations. The resonance region occupies an intermediate position between the Rayleigh and quasi-optical regions, and therefore, at least in the case of qualitative consideration, methods corresponding to neighboring regions can be applied in this region.

To consider the mechanism of resonant phenomena in scattering, we shall proceed from the Rayleigh region. In the case of Rayleigh scattering, the object was essentially "inert" to the incident wave—all points of the object are in-phase excited by the incident wave. As the size of the object increases, the wave, figuratively speaking, can "enter" the object (unless it is metal), excite currents there that generate their own electromagnetic waves, which, together with the "entered" wave and after interacting with the electrical inhomogeneities inside the object (in particular with its lower boundary), "exit" from the object and is combined with the wave reflected from the outer surface of the object. Usually, the polarization of the "resonated" wave can be arbitrary. It is reasonable that such a wave in this case carries information about the internal structure of the object.

In addition to the described "internal" resonance, an "external" resonance is possible. In this case, the main role is played by "creeping" waves, the existence of which is predicted by the diffraction theory, and detected experimentally with the help of nano- and picosecond sounding method. These diffracted waves arising at the edge of the zone of geometric shadow propagate around the object and contribute to the resulting scattered signal.

"Creeping" waves preserve the perpendicularity of the electric field vector of the surface of the object, and therefore, the polarization of the wave at the point directly behind the object directly depends on the geometry of the object and on the path that this diffracted wave "has passed" to this point. Thus, here again the polarization structure of the signal is determined by the shape of the object, not only by the optically observable side, but also by the shadow side.

Even such a purely qualitative consideration is sufficient to judge the complexity of the phenomenon of resonance scattering, leading to an unstable situation. Small changes in the causes of resonance scattering lead to a radical change in the observed phenomena. In this sense, the problem of resonance scattering is incorrect.

The high-frequency region is characterized by a criterion $ka \gg 1$. This is the region of scattering most widely considered in the literature. The regularities of this region used in the next articles will be discussed as they are described.

6.2 Stable Objects of Navigational Observation

6.2.1 Complete Scattering Matrix

If linear boundary conditions exist on the surface of the object, then its scattering power can be characterized by the relationship between the coefficients of the expansion of the incident and scattered fields over a certain system of basis functions. We will search for the specified representation if two requirements are fulfilled: First, the system of functions shall preserve all the information necessary to identify boundary conditions on the surface of the object and, second, this system of functions shall be of the same type, and also suitable for describing the incident and scattered waves for arbitrary polarized and spatiotemporal structures.

The first requirement makes it necessary in the representation of the scattered field to use the scattering amplitude (6.4), which uniquely determines the scatterer. The incident field shall be represented by a plane wave, bearing in mind that the spatiotemporal structure of the incident wave as its main element contains a plane wave—all the remaining contribution of the incident wave can be estimated by expansion in the spatial spectrum of plane waves' incident on the scatterer from various directions. Thus, the "primary" wave for an incident wave is a plane wave, and the "primary" wave for a scattered wave is a spherical wave. If some system of basis functions is chosen, then the incident and scattered fields are, in general, represented by infinite-dimensional column vectors of the expansion coefficients. In turn, the scattering properties of the object are described in this case by an infinite-dimensional square matrix of the coefficients of proportionality between the indicated expansion coefficients. The infinite-dimensional matrix being considered will be denoted by S_∞, where the subscript indicates the dimension of the matrix, and is called the complete scattering matrix.

If the scatterer does not absorb the energy incident on it, then the complete scattering matrix shall be unitary: $S_\infty S_\infty^+ = I$, where I—in this case, a single infinite-dimensional matrix.

Since the complete scattering matrix satisfies the conditions of the uniqueness theorem (it is uniquely related to the scattering diagram), a change in the irradiation direction of the scatterer or the polarization state of the incident wave leads to a unitary transformation of the complete scattering matrix.

As a system of basis functions in scattering theory, a complete orthonormal system of vector spherical functions $\vec{Y}_{LN}^M(\theta, \varphi)$ is chosen. It is possible to factorize the spatial and polarization structures in these functions. The spatial structure is determined by spherical harmonics, which in the spherical coordinate system (r, θ, φ) are periodic for φ and for θ. The product of these functions determines the spatial structure of the vector spherical harmonics.

Serial expansion of a plane incident wave in vector spherical harmonics is provided below:

$$\vec{E}^{\text{in}} \mathrm{e}^{-j\vec{k}^{\text{in}} r} = \frac{2\pi}{kr} \sum_{LMN} (-i)^{N+1} \vec{Y}_{LN}^M \left[\mathrm{e}^{i\pi N} w_N^{(-)}(k, r) + w_N^{(+)}(k, r) \right] C_{LN}^M, \quad (6.8)$$

where $w_N^{(+)}(k, r)$ and $w_N^{(-)}(k, r)$—the Ricatti–Hankel function—determine the radial dependence of divergent and convergent waves.

As can be seen from Eq. (6.8), a plane wave can be represented as a sum of converging and diverging spherical waves of equal amplitude. This can also be shown by simple reasoning. A plane wave with respect to the origin (the scatterer) can be represented as two plane waves, one of which falls on the scatterer, and the other goes away from it. When these plane waves are expanded in any complete basis system of functions, this relation between the waves converging in and diverging from the given point shall be preserved.

The scattering diagram affects the contribution to the total field of the diverging waves, i.e., allows the following expansion:

$$\vec{E}_0^{SC} \left(\vec{k}^{\text{in}}, \theta_S, \varphi_S \right) = \frac{2\pi}{kr} \sum_{LMN} \left[(-i)^{N+1} \vec{Y}_{LN}^M w_{(L)}^+(k, r) C_{LN}^M \right] S_{LN\infty}, \quad (6.9)$$

where $S_{LN\infty}$—elements of the complete scattering matrix.

The use of vector spherical harmonics allows us to consider the problem of the influence of the eigenfunctions of the scattering object, describing the scatterer's response as a whole to the effect of the irradiating field, which is especially important in the resonance region.

Spherical wave vector functions describe oscillations corresponding to characteristic types, i.e., resistant to the distributions of the current charge and the field, related primarily to the shape of the body. This explains the unambiguous relationship of the matrix S_∞ with currents flowing through the scatterer, which leads to the need for experimental current studies, which are usually carried out under the conditions of anechoic chambers.

6.2.2 Single-Position Scattering Matrix

In scattering theory, if information on the mechanism of the formation of a scattered field is not involved, the main method for describing the scattering power of an object is the ESA diagram, the scattering amplitude, and the complete scattering matrix. This is connected with the main problem for the scattering theory: clarifying the conditions for obtaining information about the object of observation on the basis of an investigation of the field scattered by it. In practice, the conditions for obtaining these characteristics of scattering power cannot usually be achieved, and therefore, they deal with observing the object under certain angles or with a fixed position of the "irradiation sections" and analyzing the scattered field. Phenomenological ideas already come into effect here, when the characteristics of the scattering power of the body are rigidly bound to the possible conditions for its observation.

When the conditions for observing the object are fixed, the ESA diagram and the scattering amplitude degenerate into some set of ESAs or scattering matrices (the expressions for them coincide with (6.2)–(6.6), while the arguments of the function do not run through the entire set of observation conditions any more, but determine the realizable observational conditions), for which we retain the same notations as for the ESA diagram and the scattering amplitude, respectively.

Elements of the scattering matrix $S\left(\vec{k}_u; \theta, \varphi\right)$, which will be further denoted as $s_{ij}\left(\vec{k}_u; \theta, \varphi\right)$, $(i, j = 1, 2)$, represent some complex numbers depending on the angle of incidence of the wave on the object and the angle of observation of the scattere d field. As was shown in clause 6.1, the ESA and the moduli of the elements of the scattering matrix are invariant to the distance to the scatterer.

The scattering matrix $S\left(\vec{k}_u; \theta, \varphi\right)$ is a complete characteristic of the scattering properties of a stable object, but under fixed observation conditions. The completeness of the description here lies in the fact that the amplitude, phase, and polarization of each spectral component of the scattered wave for any given parameters of the irradiating field and the observation conditions are determined with the aid of this matrix. The complex character of the elements of the scattering matrix directly indicates the allowance for phase progressions caused by the dispersion of the orthogonally polarized components of the irradiating wave. Absolute values of the arguments of all elements of the matrix $s_{ij}\left(\vec{k}_u; \theta, \varphi\right)$ are also determined by the distance to the scatterer. When this distance is changed, the arguments of all s_{ij} are changed the same way, and such a phase factor can be taken out of the matrix. The first one includes the absolute phase, which depends on the movement of the object along the line of sight, whereas the second one contains information only about the difference between the arguments of all elements, while it is convenient to choose one of them as real.

From the reciprocity theorem, it follows that replacing the direction of irradiation with the direction of observation and vice versa entails transposition of the scattering matrix, i.e., $S\left(\vec{k}_u, \vec{r}\right) = S^{\mathrm{T}}\left(-\vec{r}, -\vec{k}_u\right)$, wherein $\vec{r}$ determines the direction of the scattered wave.

Elements of the scattering matrix can be found from the scattering amplitude or, more commonly, experimentally. Despite the huge variety of scatterers, nevertheless, there are fairly general patterns for the elements of their scattering matrices.

We begin the investigation with the most simple and, at the same time, the most frequently encountered case of backscattering. In this case, $S\left(\vec{k}_u, -\vec{k}_u\right)$ and it corresponds to single-position (or monostatic) scattering, when the direction of irradiation of the object and the reception of the scattered field coincide.

In this case, we get: $S\left(\vec{k}_u, \vec{r}\right) = S^{\mathrm{T}}\left(\vec{k}_u, \vec{r}\right)$ (with antiparallelism $\vec{k}_u$ and $\vec{r}$). In this way, $s_{12}\left(\vec{k}_u, \vec{r}\right) = s_{21}\left(\vec{k}_u, \vec{r}\right)$, which indicates the symmetry of the scattering matrix.

It follows that in case of single-position scattering an object is characterized by three complex numbers: s_{11}, s_{12}, s_{22}. The main consequence of the symmetry of the scattering matrix follows from consideration of those transformations to which the matrix S is subjected when the polarization basis is changed.

For a polarization basis formed by two polarized waves with unit moduli of complex vectors $\vec{e}_1$ and $\vec{e}_2$, propagating in the same direction as the analyzed wave, we introduce the following notation:

$$(\vec{e}_1, \vec{e}_2)^{\mathrm{T}} = \hat{\chi}. \tag{6.10}$$

We draw the reader's attention to the fact that such a record is purely formal in character—parameter $\hat{\chi}$ on the right-hand side of formula (6.10) is not an ordinary complex vector, but only denotes the set of two complex vectors (this circumstance is underlined by the sign "∧" above the basis symbol). Nevertheless, when dealing with such a record in the same way as with a complex vector, we can represent the expansion of an arbitrary elliptically polarized wave $\vec{E}$ in the basis $\hat{\chi}$ as a scalar product:

$$\left(\vec{E} \cdot \hat{\chi}\right) = (E_1, E_2)(\vec{e}_1, \vec{e}_2)^{\mathrm{T}} = E_1\vec{e}_1 + E_2\vec{e}_2 = \left(\vec{E}\vec{e}_1^{\,*}\right)\vec{e}_1 + \left(E_2\vec{e}_2^{\,*}\right)\vec{e}_2 \tag{6.11}$$

(in clauses 6.2.2 and 6.2.3 it is assumed that $\vec{e}_1 \perp \vec{e}_2$, i.e., polarization basis is orthogonal).

We recall that if the polarization basis changes $\hat{\chi}_C \to \hat{\chi}_H$ ("o" and "H" denote the old and new bases) in accordance with rule (1.38):

$$\hat{\chi}_H = Q(\hat{\chi}_H, \hat{\chi}_C)\hat{\chi}_C, \tag{6.12}$$

then the components of vector $\vec{E}$ undergo the following transformation:

$$E(\hat{\chi}_H) = Q^*(\hat{\chi}_H, \hat{\chi}_C)E(\hat{\chi}_C), \tag{6.13}$$

where $E(\hat{\chi}_H) = \left(E_1(\hat{\chi}_H),\ E_2(\hat{\chi}_H)\right)^{\mathrm{T}}$ and $E(\hat{\chi}_C) = \left(E_1(\hat{\chi}_C),\ E_2(\hat{\chi}_C)\right)^{\mathrm{T}}$—vectors formed by the components of vector $\vec{E}$ in the new and old bases, $Q(\hat{\chi}_H,\ \hat{\chi}_C)$—unitary matrix of the transition from the old basis to the new one.

Conditions (6.12) and (6.13) ensure the invariance of the wave energy characteristics in any polarization basis.

Since the incident and scattered waves can be analyzed in any basis in determining the scattering matrix (6.4), the fixation of the notation for the polarization bases will affect the notation of the initial relations for the scattering matrix as follows:

$$E_p(\hat{\chi}^p) = S(\hat{\chi}^p, \hat{\chi}^u)E_u(\hat{\chi}^u). \tag{6.14}$$

We note that since $\hat{\chi}^u \neq \hat{\chi}^p$ (in the general case), the scattering matrix $S(\hat{\chi}^p, \chi^u)$ is no longer symmetric—the symmetry is restored for a single-position matrix with the coincidence of the bases $\hat{\chi}^p = \hat{\chi}^u$.

Since we are dealing with the transformation of the scattering matrix with a change in the polarization bases, it is therefore necessary to assume that Eq. (6.14) is written in old bases $\hat{\chi}_C^u$ and $\hat{\chi}_C^p$.

For an incident wave, in the transition to a new basis $\hat{\chi}_H^u$, formulas (6.12) and (6.13) can be used, since the polarization properties of this wave are estimated in a direction that coincides with the direction of its propagation. In reflection, the opposite situation occurs: The wave propagates toward the observer. At this position, the direction of rotation of the electric field vector changes. In order to take this fact into account for the scattered wave, (6.12) shall be represented as follows:

$$\hat{\chi}_H^{*p} = Q^*(\hat{\chi}_H^p, \hat{\chi}_C^p)\hat{\chi}_C^p, \tag{6.15}$$

whence instead of Eq. (6.13) for the coefficients of expansion in such complex-conjugate bases, we obtain:

$$E_p(\hat{\chi}_H^p) = Q(\hat{\chi}_H^p,\ \hat{\chi}_C^p)E_p(\hat{\chi}_C^p). \tag{6.16}$$

Substituting Eq. (6.13) for the incident wave and (6.16) for the scattered wave into (6.14) and taking into account the unitary property of the matrix $Q(\hat{\chi}_H^u, \hat{\chi}_C^u)$, expression (6.14) can be transformed to the form:

$$\begin{aligned} E_p(\hat{\chi}_H^p) &= Q(\hat{\chi}_H^p, \hat{\chi}_C^p)S(\hat{\chi}_C^p, \hat{\chi}_C^u)Q^*(\hat{\chi}_C^u, \hat{\chi}_H^u)E_u(\hat{\chi}_H^u) \\ &= Q(\hat{\chi}_H^p, \hat{\chi}_C^p)S(\hat{\chi}_C^p, \hat{\chi}_C^u)Q^{\mathrm{T}}(\hat{\chi}_H^u, \hat{\chi}_C^u)E_u(\hat{\chi}_H^u) = S(\hat{\chi}_H^p, \hat{\chi}_C^u)E_u(\hat{\chi}_H^u). \end{aligned} \tag{6.17}$$

It follows that the monostatic scattering matrix with the polarization basis changes according to the rule:

$$S(\hat{\chi}_H^p, \hat{\chi}_H^u) = Q(\hat{\chi}_H^p, \hat{\chi}_C^p) S(\hat{\chi}_C^p, \hat{\chi}_C^u) Q^{\mathrm{T}}(\hat{\chi}_H^u, \hat{\chi}_C^u), \tag{6.18}$$

which corresponds to the so-called generalized congruent transformation.

When the polarization bases $\hat{\chi}_C^p$ and $\hat{\chi}_C^u$ coincide, as well as $\hat{\chi}_H^p$ and $\hat{\chi}_H^u$, the relation (6.18) is simplified:

$$S(\hat{\chi}_H) = Q(\hat{\chi}_H, \hat{\chi}_C) S(\hat{\chi}_C) Q^{\mathrm{T}}(\hat{\chi}_H, \hat{\chi}_C). \tag{6.19}$$

Transformation (6.19) is called congruent.

The congruent transformation preserves the symmetry of the scattering matrix in the new polarization basis. Let us show this:

$$\begin{aligned} S^{\mathrm{T}}(\hat{\chi}_H) &= \left[Q(\hat{\chi}_H, \hat{\chi}_C) S(\hat{\chi}_C) Q^{\mathrm{T}}(\hat{\chi}_H, \hat{\chi}_C)\right]^{\mathrm{T}} = Q(\hat{\chi}_H, \hat{\chi}_C) S^{\mathrm{T}}(\hat{\chi}_C) Q^{\mathrm{T}}(\hat{\chi}_H, \hat{\chi}_C) \\ &= Q(\hat{\chi}_H, \hat{\chi}_C) S(\hat{\chi}_C) Q^{\mathrm{T}}(\hat{\chi}_H, \hat{\chi}_C) = S(\hat{\chi}_H), \end{aligned} \tag{6.20}$$

i.e., $s_{12}(\hat{\chi}_H) = s_{21}(\hat{\chi}_H)$.

The determinant of matrix S in the general case is a complex number. When the basis is changed, the phase of the determinant changes, which follows from (1.43) and (6.19):

$$\det S(\hat{\chi}_H) = \mathrm{e}^{2\mathrm{i}\varphi} \det S(\hat{\chi}_C). \tag{6.21}$$

As is known, any square matrix S can be reduced to the diagonal form S_D using two unitary matrices Q_1 and Q_2 using the relation:

$$Q_1 S Q_2 = S_D. \tag{6.22}$$

If we consider that S is a symmetric matrix, then it follows from (6.22) that:

$$S_g^{\mathrm{T}} = (Q_1 S Q_2)^{\mathrm{T}} = Q_2^{\mathrm{T}} S^{\mathrm{T}} Q_1^{\mathrm{T}} = Q_2^{\mathrm{T}} S Q_1^{\mathrm{T}} = S_D. \tag{6.23}$$

Comparing expressions (6.22) and (6.23), we find $Q_2 = Q_1^{\mathrm{T}}$, i.e., taking into account Eq. (6.19) we find that matrix $S(\hat{\chi}_C)$ can be reduced to a diagonal form by some congruent transformation $\left(s_{12}(\hat{\chi}_H) = s_{21}(\hat{\chi}_H) = 0\right)$.

Polarization basis $\hat{\chi}_0$, in which the scattering matrix has diagonal shape (in this case the scattering matrix will be denoted by $S(\hat{\chi}_0)$), is called the eigenpolarization basis of the scatterer, and the off-diagonal elements of the scattering matrix, which will be further denoted by $\lambda_1(\hat{\chi}_0)$ and $\lambda_2(\hat{\chi}_0)$—the eigenvalues of matrix S. This basis determines two orthogonal polarizations $\vec{e}_{10}$ and $\vec{e}_{20}$, for which the cross-component in the scattered wave is absent. Let us call the polarizations of the waves, which coincide with the unitary vectors of polarization basis $\hat{\chi}_0 = (\vec{e}_{10}, \vec{e}_{20})^{\mathrm{T}}$, the eigenpolarizations of the scatterer.

Physically, the foregoing means that when the scatterer is irradiated with a wave whose polarization coincides with one of the eigenpolarizations, the shape and spatial orientation of the polarization ellipse of the scattered wave will coincide with the corresponding parameters of the irradiating wave. Directions of rotation of vectors will be opposite when observing both waves from the side of the scatterer or the source of the irradiating wave, and coinciding when viewed along the normals to the front of each wave.

If the scattering matrix by means of some unitary matrix $Q(\hat{\chi}_H, \hat{\chi}_C)$ is reduced to the diagonal shape $S_g(\hat{\chi}_0)$, then after applying to $S_g(\hat{\chi}_0)$ congruent transformation (6.9) with the matrix

$$Q(\hat{\chi}_0^1, \hat{\chi}_0) = \begin{pmatrix} \mathrm{e}^{-0.5\alpha} & 0 \\ 0 & \mathrm{e}^{-0.5\beta} \end{pmatrix} \tag{6.24}$$

the diagonality of the desired matrix $S_g(\hat{\chi}_0^1)$, as well as the modules of its eigenvalues, will not change. Only the arguments of the eigenvalues will be changed: For the first of the eigenvalues, it will decrease by α, while for the second—by β. Consequently, there is always such a polarization basis $\hat{\chi}_0$, in which the scattering matrix is diagonal, and its eigenvalues λ_1 and λ_2 are real non-negative numbers. The scattering matrices in this basis will be denoted by $S_0(\hat{\chi}_0)$.

Using transformation (6.19) and Eq. (1.43), we can obtain a representation of the elements of the scattering matrix in an arbitrary basis. The initial one will be the basis in which the scattering matrix becomes $S(\hat{\chi}_0)$. Any other basis with respect to it will be determined using parameters γ, δ, η and φ of unitary matrix (1.43).

Taking into account the explicit form of matrix $Q(\hat{\chi}_H, \hat{\chi}_0)$, we obtain the following expression for the scattering matrix in an arbitrary basis:

$$\left.\begin{aligned} s_{11}(\hat{\chi}_H) &= \mathrm{e}^{2\mathrm{i}\varphi}\mathrm{e}^{2\mathrm{i}(\eta+\delta)}\left(\lambda_1\mathrm{e}^{-2\mathrm{i}\delta}\cos^2\gamma + \lambda_2\mathrm{e}^{2\mathrm{i}\delta}\sin^2\gamma\right) \\ s_{12}(\hat{\chi}_H) &= \mathrm{e}^{2\mathrm{i}\varphi}\left(-\lambda_1\mathrm{e}^{-2\mathrm{i}\delta} + \lambda_2\mathrm{e}^{2\mathrm{i}\delta}\right)\sin\gamma\cos\gamma \\ s_{22}(\hat{\chi}_H) &= \mathrm{e}^{2\mathrm{i}\varphi}\mathrm{e}^{-2\mathrm{i}(\eta+\delta)}\left(\lambda_1\mathrm{e}^{-2\mathrm{i}\delta}\sin^2\gamma + \lambda_2\mathrm{e}^{2\mathrm{i}\delta}\cos^2\gamma\right) \end{aligned}\right\} \tag{6.25}$$

If the initial basis is not eigenbasis, but some arbitrary basis $\hat{\chi}_C$, then the elements of the scattering matrix in another basis $\hat{\chi}_H$, determined with respect to the initial one with the help of parameters γ, δ, η and φ, will be as follows:

$$\left.\begin{aligned} s_{11}(\hat{\chi}_H) &= \mathrm{e}^{2\mathrm{i}\varphi}\mathrm{e}^{2\mathrm{i}(\eta+\delta)}\left(s_{11}(\hat{\chi}_C)\mathrm{e}^{-2\mathrm{i}\delta}\cos^2\gamma + s_{22}(\hat{\chi}_C)\mathrm{e}^{2\mathrm{i}\delta}\sin^2\gamma + s_{12}(\hat{\chi}_C)\sin 2\gamma\right) \\ s_{12}(\hat{\chi}_H) &= \mathrm{e}^{2\mathrm{i}\varphi}\left[\left(-s_{11}(\hat{\chi}_C)\mathrm{e}^{-2\mathrm{i}\delta} + s_{22}(\hat{\chi}_C)\mathrm{e}^{2\mathrm{i}\delta}\right)\sin\gamma\cos\gamma + s_{12}(\hat{\chi}_C)\cos 2\gamma\right] \\ s_{22}(\hat{\chi}_H) &= \mathrm{e}^{2\mathrm{i}\varphi}\mathrm{e}^{-2\mathrm{i}(\eta+\delta)}\left(s_{11}(\hat{\chi}_C)\mathrm{e}^{-2\mathrm{i}\delta}\sin^2\gamma + s_{22}(\hat{\chi}_C)\mathrm{e}^{2\mathrm{i}\delta}\cos^2\gamma - s_{12}(\hat{\chi}_C)\sin 2\gamma\right) \end{aligned}\right\} \tag{6.26}$$

Scattering matrix S, as well as coherence matrix K, has two invariants. Let us consider them.

A direct check shows that for the scattering matrix, when the polarization basis changes, the sum of the squares of the moduli of its elements remains unchanged, which is the first invariant:

$$A = |s_{11}|^2 + |s_{22}|^2 + 2|s_{12}|^2 = \sigma_{11}^2 + \sigma_{22}^2 + 2\sigma_{12}^2 = \lambda_1^2 + \lambda_2^2 = \sigma_\Sigma^2, \tag{6.27}$$

where σ_{ij}—particular values of single-position ESA, obtained from (6.2) when $\vec{E}_p^0 \equiv E_{ip}\vec{e}_i$, and $\vec{E}_u^0 \equiv E_{ju}\vec{e}_j$ (compared to (6.11)) $(i, j = 1, 2)$, with the observation condition corresponding to the single-position case.

Value A is the sum of all the particular ESAs of the object; we will call it the complete ESA of the target σ_Σ^2.

The second invariant of the scattering matrix is the modulus of its determinant:

$$B = |\det S| = \left|s_{11}s_{22} - s_{12}^2\right| = |\lambda_1\lambda_2|. \tag{6.28}$$

However, as the second invariant, it is more expedient to choose not the value B, and some combination of numbers A and B :

$$q = \sqrt{1 - 4\frac{B^2}{A^2}} = \frac{|\lambda_1|^2 - |\lambda_2|^2}{|\lambda_1|^2 + |\lambda_2|^2}, \tag{6.29}$$

where, as in the future, $|\lambda_1| \geq |\lambda_2|$ is taken, which is a formal analog of the degree of polarization of a partially polarized wave (3.10). Let us call the parameter q the polarization anisotropy degree of the object, or simply the anisotropy degree.

Note that sometimes, especially in radar metrology tasks, the anisotropy degree is determined differently:

$$q' = \frac{|\lambda_1| - |\lambda_2|}{|\lambda_1| + |\lambda_2|} = \frac{\sqrt{1+q} - \sqrt{1-q}}{\sqrt{1+q} - \sqrt{1-q}}. \tag{6.30}$$

so we must carefully “decipher” the formulas and algorithms available in the literature. It follows from (6.30) that:

$$q = \frac{2q'}{\left(1 + q'^2\right)}. \tag{6.31}$$

The relations (6.27)–(6.29) allow us, by means of the scattering matrix known in some basis, to find the moduli of its eigenvalues:

$$\left.\begin{aligned} |\lambda_1| &= \sqrt{0.5\left(A + \sqrt{A^2 - 4B^2}\right)} = \sigma_\Sigma\sqrt{0.5(1+q)} \\ |\lambda_2| &= \sqrt{0.5\left(A - \sqrt{A^2 - 4B^2}\right)} = \sigma_\Sigma\sqrt{0.5(1-q)} \end{aligned}\right\}. \tag{6.32}$$

We also note one interesting relation obtained by subtracting the squares of the moduli of the diagonal elements in formulas (6.25):

$$\sigma_{11}^2 - \sigma_{22}^2 = q\sigma_{\Sigma}^2 \cos 2\gamma, \tag{6.33}$$

where 2γ is the angle between the points on the Poincaré sphere corresponding to the polarization basis $\hat{\chi}_0$ and the basis where measurements are taken.

If we analyze formulas (6.25), then we can establish that in the case of general transmission and reception of the scattering power meter antenna, there are always two bases in which an arbitrary scattering matrix has the following form:

$$S = \frac{e^{2i\varphi}}{\sqrt{2}}\sigma_{\Sigma}\begin{pmatrix} 0 & (-1)^n i\sqrt[4]{1-q^2} \\ (-1)^n i\sqrt[4]{1-q^2} & e^{-2i\eta}\left(\sqrt{1-q}-\sqrt{1+q}\right) \end{pmatrix} \tag{6.34}$$

or similar:

$$S = \frac{e^{2i\varphi}}{\sqrt{2}}\sigma_{\Sigma}\begin{pmatrix} e^{-2i\eta}\left(\sqrt{1+q}-\sqrt{1-q}\right) & (-1)^n i\sqrt[4]{1-q^2} \\ (-i)^n i\sqrt[4]{1-q^2} & 0 \end{pmatrix} \tag{6.35}$$

The first unitary vector of each of the two bases under consideration determines such polarization of the irradiating wave at which the signal scattered by the object is polarized orthogonally and cannot be received by an antenna common to irradiation and reception. As a result, despite the presence of a scattered wave, the scattering profile meter cannot observe the object, so it seems logical to call such polarizations as zero-signal polarizations.

Thus, the condition, under which the scattered signal from the object of observation does not fall into the receiving path, in the eigenbasis $\hat{\chi}_0$ has the following form:

$$(E_{1u}, E_{2u})\begin{pmatrix} \lambda_1 & 0 \\ 0 & \lambda_2 \end{pmatrix}\begin{pmatrix} E_{1u} \\ E_{2u} \end{pmatrix} = \lambda_1 (E_{1u})^2 + \lambda_2 (E_{2u})^2 = 0. \tag{6.36}$$

Hence, it follows that for the zero-signal polarization:

$$E_{2u} \Big/ E_{1u} = \pm i\sqrt{\lambda_2 / \lambda_1} = \pm i\sqrt[4]{(1-q)(1+q)}. \tag{6.37}$$

Note that, the zero-signal polarizations are not orthogonal for an arbitrary target. For the orthogonality of these polarizations, it is necessary that condition $|\lambda_1| = |\lambda_2|$, i.e., $q = 0$, is met.

Comparing a pair of eigenpolarizations and a pair of zero-signal polarizations, it should be noted that specifying the last polarizations is a more complete description of the polarization profile of the scatterer. The eigenpolarizations determine only the eigenbasis, while the zero-signal polarizations determine, in addition, the ratio of the reflection coefficients for both eigenpolarizations.

Thus, the zero-signal polarizations are such invariants of the scattering matrix, which completely describe the polarization profile of a stable object of observation in a given direction.

In the polarization basis, determined with respect to the original eigenbasis by parameters $2\gamma = \frac{1}{2}\pi(1+2n)$, $\eta + \delta = 0.5\pi n$, equation $s_{11} = s_{22}$ is always met.

Let us return to formulas (6.25). Considering the first and third equations together, it is not difficult to express $\lambda_1 e^{-2i\delta}$ and $\lambda_2 e^{2i\delta}$ from them. After substituting the formulas found into the second equation, it is possible to establish a direct connection between the cross- and diagonal elements of the scattering matrix:

$$2s_{12}\cos 2\gamma = \left[s_{22}e^{2i(\eta+\delta)} - s_{11}e^{-2i(\eta+\delta)}\right]\sin 2\gamma. \tag{6.38}$$

Equation (6.38) can be called the basic equation of the elements of the scattering matrix.

Using formulas (6.25) and (6.33), we can obtain relations connecting the squares of the moduli of the elements of the scattering matrix, i.e., partial ESA, with its invariants and angles δ and γ :

$$\left.\begin{aligned} |s_{11}|^2 &= 0.5\sigma_\Sigma^2(1 + q\cos 2\gamma) - |s_{12}|^2 \\ |s_{22}|^2 &= 0.5\sigma_\Sigma^2(1 - q\cos 2\gamma) - |s_{12}|^2 \\ |s_{12}|^2 &= 0.25\sigma_\Sigma^2\left(1 - \sqrt{1-q^2}\cos 4\delta\right)\sin^2 2\gamma \end{aligned}\right\} \tag{6.39}$$

Let us analyze the results obtained.

The third of the equations written allows us to determine the boundaries of possible changes in the module of the cross-element:

$$0 \le |s_{12}|^2 \le 0.25\sigma_\Sigma^2\left(1 + \sqrt{1-q^2}\right). \tag{6.40}$$

As can be seen, the maximum achievable value $|s_{12}|^2$ is realized in scatterers, for which $q = 0$. In this case, $|s_{12}|^2_{\max}$ will be equal to half the full ESA; for scatterers, for which $q = 1$, the corresponding value will be two times less.

Maximization of $|s_{12}|^2$ occurs in the polarization basis, where $2\gamma = 0.5\pi$ and $4\delta = \pi$; in this same basis, $|s_{11}|^2$ and $|s_{22}|^2$ are equal to each other:

$$|s_{11}|^2 = |s_{22}|^2 = \frac{\sigma_\Sigma^2}{4}\left(1 - \sqrt{1-q^2}\right). \tag{6.41}$$

Thus, if it is the polarization basis, where $|s_{11}| = |s_{22}|$ is known, then it is easy to find the invariants of the scatterer by means of formulas (6.40) and (6.41):

$$\left.\begin{aligned} q &= 2|s_{11}s_{22}|/\sqrt{|s_{11}|^2 + |s_{22}|^2} \\ \sigma_\Sigma^2 &= 2\left(|s_{11}|^2 + |s_{22}|^2\right) \end{aligned}\right\} \tag{6.42}$$

The obtained relations make it possible to obtain a formula for the arguments of the elements of the scattering matrix. Using Eqs. (6.28), (6.33), and (6.39) after cumbersome calculations, the following expression can be found for the generalized phase $\Phi = \psi_{11} + \psi_{22} - 2\psi_{12}$, where $\psi_{ij} = \arg s_{ij}$:

$$\cos \Phi = \frac{z^2 \sin^2 2\gamma - 2z + 2q^2}{z\sqrt{\left(2 - z \sin^2 2\gamma\right)^2 - 4q^2 \cos 2\gamma}}, \tag{6.43}$$

where $z = 1 - \sqrt{1 - q^2} \cos 4\delta$.

Let us consider two particular cases: $q = 0$ and $q = 1$. For the first of them, formula (6.43) gives $\Phi = \pi$, and for the second—$\Phi = 0$.

The above relations allow us to make a primary classification of scatterers in terms of the anisotropy degree q.

By analogy with the introduction of the concepts of polarized, partially polarized, and non-polarized waves, let us introduce a similar classification for scatterers. Objects with an anisotropy degree $q = 0$ will be called polarization isotropic; if $q = 1$, we will speak of a degenerate object (the meaning of such a determination will be explained below).

It follows from relation (6.29) that a necessary and sufficient condition for the polarization isotropy is the equality of the moduli of the eigenvalues of the scattering matrix: $|\lambda_1| = |\lambda_2|$. In accordance with relations (6.25), this allows us to represent the elements of the scattering matrix in the polarization basis, whose symbol relative to the eigen one is as follows: $\chi_0 = (\gamma, \eta, \delta)$, in the following form (for $\varphi = 0$):

$$\left.\begin{aligned} s_{11} &= s_{22}^* = \lambda_1 (\cos 2\delta - i \cos 2\gamma \sin 2\delta) \mathrm{e}^{2\mathrm{i}(\eta+\delta)} \\ s_{12} &= i\lambda_1 \sin 2\delta \sin 2\gamma \end{aligned}\right\} \tag{6.44}$$

or

$$\left.\begin{aligned} s_{11} &= -s_{22}^* = \lambda_1 (\cos 2\delta \cos 2\gamma - i \sin 2\delta) \mathrm{e}^{2\mathrm{i}(\eta+\delta)} \\ s_{12} &= -\lambda_1 \cos 2\delta \sin 2\gamma \end{aligned}\right\}. \tag{6.45}$$

As follows from formulas (6.27)–(6.29), the requirement of isotropy ($q = 0$) leads to equality $|s_{11}|^2 + |s_{12}|^2 = \left|s_{11}s_{22} - s_{12}^2\right|$, which, after squaring, leads to a rigid condition for the generalized phase Φ , where $\psi_{mn} = \arg s_{mn}$:

$$\Phi = \psi_{11} + \psi_{22} - 2\psi_{12} = \pi(1 + 2n). \tag{6.46}$$

Thus, for a polarization-isotropic object, there always exist both the reduced requirement for the arguments of the elements of the scattering matrix, and the equality $|s_{11}| = |s_{22}|$.

If the scatterer has both zero-signal polarizations, then it is always polarization isotropic. It follows from (6.33) that in polarization bases with $2\gamma = \pi/4$, the moduli of the diagonal elements of the scattering matrix are equal. Analysis of the geometric meaning of quantities γ and δ shows that the equality of modules under consideration takes place in all polarization bases whose unitary vectors are mapped onto a large circle of the Poincaré sphere that is equidistant from points corresponding to eigenpolarizations. Such a circle is formed by the intersection of the Poincaré sphere and the plane perpendicular to the diameter passing through points corresponding to eigenpolarizations. On one side of the plane of the great circle—$|s_{11}| > |s_{22}|$, and on the other side—on the contrary, $|s_{22}| > |s_{11}|$. We call polarization bases with $|s_{11}| = |s_{22}|$ the class B bases (analog of the first singular basis for the coherence matrix). It follows from (6.33) that if the moduli of the diagonal elements of the scattering matrix in some basis are equal (with the exception of class B bases), then they will be equal in every other basis. If Eq. (6.33) is satisfied, then for the further analysis it is necessary to verify the fulfillment of Eq. (6.29).

For scatterers with $q = 1$, i.e., less (modulo) of eigenvalues $\lambda_2 = 0$, the determinant of the scattering matrix is zero; the scattering matrix is degenerate. This served as the basis for calling such scatterers degenerate ones. The general form of the scattering matrix of the degenerate target, as follows from (6.25), will be:

$$S = \lambda e^{2i(\varphi-\delta)}\begin{pmatrix} e^{2i(\eta+\delta)}\cos^2\gamma & -\sin\gamma\cos\gamma \\ -\sin\gamma\cos\gamma & e^{-2i(\eta+\delta)}\sin^2\gamma \end{pmatrix}. \tag{6.47}$$

The condition that the determinant of the scattering matrix be zero (the degeneracy condition of the target) establishes a unique relationship between the arguments of the elements s_{mn} :

$$\Phi = \psi_{11} + \psi_{22} - 2\psi_{12} = 2\pi n, \quad n = 0, \pm 1, \ldots \tag{6.48}$$

Comparison of equalities (6.46) and (6.48) shows that with a decrease in the anisotropy degree from 1 to 0, the generalized phase of the scattering matrix increases from 0 to π.

The above expression makes it possible to establish a fairly simple criterion for determining the degeneracy of the target from the results of amplitude measurements only (e.g., effective scattering areas). If, instead of complex elements s_{mn}, we substitute their modules proportional to the square root of the ESA—σ^2_{mn}, then the determinant of the matrix $S_{ЭПР}$ for degenerate targets:

$$S_{ЭПР} = \begin{pmatrix} \sigma_{11} & \sigma_{12} \\ \sigma_{12} & \sigma_{22} \end{pmatrix}. \tag{6.49}$$

will be zero.

As an exception to this rule, as direct calculations show, only the class B polarization bases ($2\gamma = \pi(n + 0.5)$), for which, in addition, condition $2\eta = 0.25\pi(1 + 2n)$ is met. In these bases and for non-degenerate targets, $\det S_{ЭПР} = 0$.

In conclusion of the consideration of the monostatic scattering matrix, we present its representation in terms of such parameters of the eigenbasis as the ellipticity of one of the unit vectors α and its slope angle β. In this case, the parameters of the second unitary vector will be $-\alpha$ and $\beta + \pi/2$. This representation, which is a particular case of relation (6.25), contains, as its parameters, quantities that do not depend on the observation conditions, which is very convenient in solving the problem of recognizing scatterers.

For a linearly polarized basis $\hat{\chi}_л$ (horizontal unitary vector—vertical unitary vector), the expression for the scattering matrix in terms of the parameters of the eigenbasis and the eigenvalues has the following form:

$$S(\hat{\chi}_л) = \frac{\lambda_1 + \lambda_2}{2}\begin{pmatrix} \cos 2\alpha - q' \cos 2\beta + i \sin 2\alpha \sin 2\beta & -q' \sin 2\beta - i \sin 2\alpha \cos 2\beta \\ -q' \sin 2\beta - i \sin 2\alpha \cos 2\beta & \cos 2\alpha + q' \cos 2\beta - i \sin 2\alpha \sin 2\beta \end{pmatrix}, \quad (6.50)$$

where q'—degree of polarization anisotropy in the sense of (6.30).

If the eigenbasis is linear, i.e., $\alpha = 0$, expression (6.50) is transformed to the Shupyatskiy formula [1]:

$$S(\hat{\chi}_л) = \lambda_2 \begin{pmatrix} 1 & 0 \\ 0 & 1 \end{pmatrix} + (\lambda_1 - \lambda_2)\begin{pmatrix} \cos^2\beta & \sin\beta\cos\beta \\ \sin\beta\cos\beta & \sin^2\beta \end{pmatrix}, \quad (6.51)$$

widely used in radar metrology due to the fact that the eigenbasis of most objects is linear.

The problem of representing the scattering matrix of two objects in the same basis with the help of the invariants (6.27) and (6.29) mentioned in 6.3 is of great importance for practical application. Suppose that on a Poincaré sphere, the pair of diametrically opposite points (1-1) maps the eigenpolarization basis of the first object $\hat{\chi}_C^1$, and the pair (2-2)—the second object $\hat{\chi}_0^2$. Let us denote the angle between them through $2\gamma_{12}$. One of the scattering matrices in each of these bases is diagonal:

$$S_D^1(\hat{\chi}_0^1) = \begin{pmatrix} \lambda_1^{(1)} & 0 \\ 0 & \lambda_2^{(1)} \end{pmatrix}, \quad S_D^2(\hat{\chi}_0) = \begin{pmatrix} \lambda_1^{(2)} & 0 \\ 0 & \lambda_2^{(2)} \end{pmatrix}, \quad (6.52)$$

where the complex numbers $\lambda_j^{(i)}$ $(i, j = 1, 2)$ denote the j-th eigenvalue of the scattering matrix of the *i-th* object. For definiteness, we assume $\left|\lambda_1^{(i)}\right| \geq \left|\lambda_2^{(i)}\right|$, $(i = 1, 2)$. When moving to a new basis, matrices $S_D^1(\hat{\chi}_0^1)$ and $S_D^2(\hat{\chi}_0^2)$ are subject to congruent transformation (6.19).

Points of large circles whose planes are perpendicular to the diameters (1-1) and (2-2) (corresponding angles γ of the matrix are equal to $\pi/4$). The basis represented by the points of intersection of the large circles under consideration will be assigned the index (0-0); it is called a zero basis and denoted as $\hat{\delta}_z$ (z—zero). The scattering matrices of the two objects under consideration in the zero basis will have the following form:

$$\left.\begin{array}{l} S^1(\hat{\chi}_z) = Q(\hat{\chi}_z, \hat{\chi}_0^1) S_D^1(\hat{\chi}_0^1) Q^{\mathrm{T}}(\hat{\chi}_z, \hat{\chi}_0^1) \\ S^2(\hat{\chi}_z) = Q(\hat{\chi}_z, \hat{\chi}_0^2) S_D^2(\hat{\chi}_0^2) Q^{\mathrm{T}}(\hat{\chi}_z, \hat{\chi}_0^2) \end{array}\right\} \tag{6.53}$$

The last equations imply inverse formulas:

$$\left.\begin{array}{l} S_D^1(\hat{\chi}_0^1) = Q^+(\hat{\chi}_z, \hat{\chi}_0^1) S^1(\hat{\chi}_z) Q^*(\hat{\chi}_z, \hat{\chi}_0^1) \\ S_D^2(\hat{\chi}_0^2) = Q^+(\hat{\chi}_z, \hat{\chi}_0^2) S^2(\hat{\chi}_z) Q^*(\hat{\chi}_z, \hat{\chi}_0^2) \end{array}\right\} \tag{6.54}$$

On the other hand, in accordance with transformations of type (6.53), we have:

$$\left.\begin{array}{l} S_D^1(\hat{\chi}_0^1) = Q(\hat{\chi}_0^1, \hat{\chi}_z) S^1(\hat{\chi}_z) Q^{\mathrm{T}}(\hat{\chi}_0^1, \hat{\chi}_z) \\ S_D^2(\hat{\chi}_0^2) = Q(\hat{\chi}_0^2, \hat{\chi}_z) S^2(\hat{\chi}_z) Q^{\mathrm{T}}(\hat{\chi}_0^2, \hat{\chi}_z) \end{array}\right\} \tag{6.55}$$

Comparison of expressions (6.54) and (6.55) shows that

$$Q^+(\hat{\chi}_z, \hat{\chi}_0^1) = Q(\hat{\chi}_0^1, \hat{\chi}_z), \qquad Q^+(\hat{\chi}_z, \hat{\chi}_0^2) = Q(\hat{\chi}_0^2, \hat{\chi}_z), \tag{6.56}$$

whence $\gamma_{01} = \gamma_{10}$, $\delta_{01} = \pi + \delta_{10}$, $(\varphi + \eta)_{01}^{01} = (\varphi + \eta)_{02}^{01} = -(\varphi + \eta)_{01}^{10} = -(\varphi + \eta)_{02}^{10} = \varphi_1$, while the matrix Q is determined by Eq. (1.43).

Using the relation (6.53) and the requirement $\gamma = \pi / 4$, let us write the scattering matrix of the first object in the zero basis:

$$S^1(\hat{\chi}_z) = \frac{1}{2}\begin{pmatrix} \lambda_{11}(\hat{\chi}_z) & \lambda_{12}(\hat{\chi}_z) \\ \lambda_{21}(\hat{\chi}_z) & \lambda_{22}(\hat{\chi}_z) \end{pmatrix}, \tag{6.57}$$

where $\lambda_{11} = \lambda_{22} = \lambda_1^1 e^{-2i\varphi_1} - \lambda_2^1 e^{-2i\delta_{10}}$, $\lambda_{12} = \lambda_1^1 e^{2i(-\varphi_1 + \delta_{10})} - \lambda_2^1 e^{2i(\varphi_1 - \delta_{10})}$.

Matrix $S^2(\hat{\chi}_z)$ has a similar form.

Formula (6.57) contains low-use parameters φ_1, δ_{10}. It is important to express them in terms of the quantities connecting the bases (1-1) and (2-2), i.e., angles γ_{12} and δ_{12}.

The transition from basis (1-1) to basis (2-2) can be done directly and also according to the scheme $\hat{\chi}_0^1 \to \hat{\chi}_z \to \hat{\chi}_0^2$:

$$\begin{aligned} S^1(\hat{\chi}_0^2) &= Q(\hat{\chi}_0^2, \hat{\chi}_0^1) S_D^1(\hat{\chi}_0^1) Q^{\mathrm{T}}(\hat{\chi}_0^2, \hat{\chi}_0^1) \\ &= Q(\hat{\chi}_0^2, \hat{\chi}_z)\big(Q(\hat{\chi}_z, \hat{\chi}_0^1) S_D^1(\hat{\chi}_0^1) Q^{\mathrm{T}}(\hat{\chi}_z, \hat{\chi}_0^1)\big) Q^{\mathrm{T}}(\hat{\chi}_0^2, \hat{\chi}_z) \\ &= Q(\hat{\chi}_0^2, \hat{\chi}_z) Q(\hat{\chi}_z, \hat{\chi}_0^1) S_D^1(\hat{\chi}_0^1) Q^{\mathrm{T}}(\hat{\chi}_z, \hat{\chi}_0^1) Q^{\mathrm{T}}(\hat{\chi}_0^2, \hat{\chi}_z). \end{aligned} \tag{6.58}$$

(here we use the Eq. (6.56)). Hence, we have:

$$Q(\hat{\chi}_0^2, \hat{\chi}_0^1) = Q(\hat{\chi}_z, \hat{\chi}_0^2)^+ Q(\hat{\chi}_z, \hat{\chi}_0^1) = Q(\hat{\chi}_0^2, \hat{\chi}_z) Q^+(\hat{\chi}_0^1, \hat{\chi}_z), \tag{6.59}$$

which gives

$$\left.\begin{aligned}(\varphi_2-\varphi_1)-(\delta_{10}-\delta_{20})&=2\gamma_{12}\\(\varphi_2-\varphi_1)+(\delta_{10}-\delta_{20})&=2\varphi_{12}\\(\varphi_2+\varphi_1)+(\delta_{10}+\delta_{20})&=\pi+2\delta_{12}\end{aligned}\right\}\tag{6.60}$$

The relations (6.60) are a system of three equations with four unknowns. Let us write one of its symmetric solutions.

$$\left.\begin{aligned}2\varphi_1&=\gamma_{12}+\varphi_{12}+\delta_{12}+0.5\pi\\2\delta_{10}&=-\gamma_{12}+\varphi_{12}+\delta_{12}+0.5\pi\\2\varphi_2&=-\gamma_{12}-\varphi_{12}+\delta_{12}+0.5\pi\\2\delta_{20}&=\gamma_{12}-\varphi_{12}+\delta_{12}+0.5\pi\end{aligned}\right\}\tag{6.61}$$

Let us substitute this solution in the general form (6.57) for the scattering matrix $S^1(\hat{\chi}_z)$:

$$S^1(\hat{\chi}_z)=\frac{i}{2}\begin{pmatrix}(\mu_1-\mu_2)e^{i\gamma_{12}} & \mu_1+\mu_2\\ \mu_1+\mu_2 & (\mu_1-\mu_2)e^{-i\gamma_{12}}\end{pmatrix}\tag{6.62}$$

and for the matrix $S^2(\hat{\chi}_z)$:

$$S^2(\hat{\chi}_z)=\frac{i}{2}\begin{pmatrix}(\tau_1-\tau_2)e^{-i\gamma_{12}} & \tau_1+\tau_2\\ \tau_1+\tau_2 & (\tau_1-\tau_2)e^{i\gamma_{12}}\end{pmatrix},\tag{6.63}$$

where

$$\begin{aligned}\mu_1&=-\lambda_1^1(\hat{\chi}_0^1)e^{-2i\delta_{12}}, & \mu_2&=-\lambda_2^1(\hat{\chi}_0^1)e^{2i\delta_{12}};\\ \tau_1&=-\lambda_1^2(\hat{\chi}_0^2)e^{-2i(\varphi_{12}+\delta_{12})}, & \tau_2&=-\lambda_2^2(\hat{\chi}_0^2)e^{2i(\varphi_{12}+\delta_{12})}.\end{aligned}$$

Formulas (6.62) and (6.63) give the required representations of the scattering matrices of two objects in the same polarization basis, expressed in terms of the eigenvalues of each scattering matrix and the parameters connecting the eigenpolarizations of each scatterer.

6.2.3 Two- and Multiple-Position Scattering Matrix

When fixing different directions of irradiation and analysis of the scattered wave in the scattering amplitude (6.5), a bistatic scattering matrix is formed. It is clear that the bistatic case is the most common for the phenomenological representation: Monostatic scattering profile can be considered as a certain subset characterized by the coincidence of these directions, and the case n (positional scattering) is reduced to n (independent bistatic observations).

The main parameter of bistatic scattering is the bistatic angle Ψ, which is defined as follows. Through the origin, coinciding with some point of the scatterer, a plane parallel to the vectors $\vec{k}_u$ and $\vec{r}$ is drawn. The angle between the two vectors measured in this plane is the bistatic angle, and the plane itself is called the scattering plane.

With the introduction of the scattering plane into consideration, we can refine the coordinate system used in clause 6.1 when analyzing the scattered field. The spherical coordinate system is chosen here so that the equation of the scattering plane is represented as $\theta = \pi/2$. Now, for any point of the scattering plane, unitary vector $\overset{\circ}{\vec{\varphi}} = \frac{\text{grad}\varphi}{|\text{grad}\varphi|}$ lies in the plane, and unitary vector $\overset{\circ}{\vec{\theta}} = \frac{\text{grad}\theta}{|\text{grad}\theta|}$—perpendicular to it. Let us define the pairs of unitary vectors $\overset{\circ}{\vec{\theta}}_u, \overset{\circ}{\vec{\varphi}}_u$ and $\overset{\circ}{\vec{\theta}}_p, \overset{\circ}{\vec{\varphi}}_p$ for the irradiation directions and for the analysis of the scattered field as specified. It remains to define two more unitary vectors $\overset{\circ}{\vec{r}}_u$ and $\overset{\circ}{\vec{r}}_p$ for these pairs.

Since unitary vectors $\overset{\circ}{\vec{\theta}}_u$ and $\overset{\circ}{\vec{\theta}}_p$ are collinear, and unitary vectors $\overset{\circ}{\vec{\varphi}}_u$ and $\overset{\circ}{\vec{\varphi}}_p$ lie in one plane, it is necessary to distinguish between possible four cases (bistatic angle $\aleph$ is selected from 0 to 2π):

1. Unitary vectors $\overset{\circ}{\vec{\theta}}_u$ and $\overset{\circ}{\vec{\theta}}_p$ are parallel, and unitary vectors $\overset{\circ}{\vec{\varphi}}_u$ and $\overset{\circ}{\vec{\varphi}}_p$ are directed in the scattering plane in one direction of rotation with respect to the origin;
2. Unitary vectors $\overset{\circ}{\vec{\theta}}_u$ and $\overset{\circ}{\vec{\theta}}_p$ are antiparallel, and unitary vectors $\overset{\circ}{\vec{\varphi}}_u$ and $\overset{\circ}{\vec{\varphi}}_p$ have different directions with respect to the rotation of the plane;
3. Unitary vectors $\overset{\circ}{\vec{\theta}}_u$ and $\overset{\circ}{\vec{\theta}}_p$ are parallel, and unitary vectors $\overset{\circ}{\vec{\varphi}}_u$ and $\overset{\circ}{\vec{\varphi}}_p$ are directed, as in case 2;
4. Unitary vectors $\overset{\circ}{\vec{\theta}}_u$ and $\overset{\circ}{\vec{\theta}}_p$ are antiparallel, and unitary vectors $\overset{\circ}{\vec{\varphi}}_u$ and $\overset{\circ}{\vec{\varphi}}_p$ are directed as in case 1.

If the third unitary vectors $\overset{\circ}{\vec{r}}_u$ or $\overset{\circ}{\vec{r}}_p$ are always chosen to coincide with the direction of wave propagation, then the triples of unitary vectors in cases 1 and 2 belong to different types of coordinates (one—right, and the other—left). In cases 3 and 4, the coordinate systems are of the same type (both right or both left). Incidentally, monostatic scattering in clause 6.2.2 was described in different types of coordinate systems corresponding to case 1.

If unitary vectors $\overset{\circ}{\vec{\theta}}$ and $\overset{\circ}{\vec{\varphi}}$ coincide with the unitary vectors of the linear polarization basis, then we say that the polarization state of the wave is investigated in a natural basis. A natural basis for bistatic scattering is the main one in determining any other basis. Its exceptional role is seen from the following example.

Suppose that some scatterer has a symmetry plane and a linearly polarized wave falls on it, the polarization of which coincides with one of the unitary vectors of the natural basis. It is easy to see that the bistatic matrix, defined in the natural basis, is diagonal. The diagonality of the matrix follows from the symmetry of the currents excited at the object and the conditions for recording the scattered field. In this case, there is mutual compensation of the cross-components of the current.

Recalling monostatic scattering (6.2.2), we can state that the natural basis in this case is eigen (for both bi- and monostatic scattering and for any chosen types of coordinate systems and any bistatic angle).

For the bistatic scattering matrix, we can use the notation [compared to (6.14) $S = S(\Psi, \hat{\chi}^p, \hat{\chi}^u)$] indicating both the bistatic angle and the nature of the polarization bases. We note that, in contrast to the monostatic case, where the discrepancy of the bases $\hat{\chi}^p$ and $\hat{\chi}^u$ is the exception rather than the rule (the antenna system of the polarimeter is mutual—the characteristics of the antenna, e.g., polarization, in the transmission and reception modes are the same) in the bistatic mode, it is necessary to assume a priori the fact of difference of these bases. Establishing the fact of the coincidence of the bases, when the transmitting and receiving antenna systems are spaced from each other in space, is a rather complicated procedure.

In the case of heterogeneous polarization bases of the incident wave (input basis) and the scattered wave (output basis), the scattering matrix measured in the natural basis is transformed according to the rule:

$$S(\Psi, \hat{\chi}_H^p, \hat{\chi}_H^u) = Q_1(\hat{\chi}_H^p, \hat{\chi}_e) S(\Psi, \hat{\chi}_e) Q_2^{\mathrm{T}}(\hat{\chi}_H^u, \hat{\chi}_e). \tag{6.64}$$

If the starting situation in the study of matrix transformation is an arbitrary input $\hat{\chi}_e^u$ and output $\hat{\chi}_e^p$ bases, then relation (6.64) will have the following form:

$$S(\Psi, \hat{\chi}_H^p, \hat{\chi}_H^u) = Q_1(\hat{\chi}_H^p, \hat{\chi}_C^p) S(\Psi, \hat{\chi}_C^p, \hat{\chi}_C^u) Q_2^{\mathrm{T}}(\hat{\chi}_H^u, \hat{\chi}_C^u), \tag{6.65}$$

Where Q_1 and Q_2 are determined in relations (6.52) and (6.53)—unitary matrices of the transformation of the output and input polarization bases. Case of coincidence of $\hat{\chi}_H^p$ and $\hat{\chi}_H^u$ corresponds to a transformation from a natural basis, when the transformation matrices Q_1 and Q_2 are equal and have the form $Q(\alpha, -\alpha, \beta, \beta + \pi/2)$.

For analogous polarization bases, analogous formulas have the following form:

$$\left.\begin{aligned} S(\Psi, \hat{\chi}_H^p, \hat{\chi}_H^u) &= Q_1^*(\hat{\chi}_H^p, \hat{\chi}_C) S(\Psi, \hat{\chi}_e) Q_2^{\mathrm{T}}(\hat{\chi}_H^u, \hat{\chi}_e) \\ S(\psi, \hat{\chi}_H^p, \hat{\chi}_H^u) &= Q_1^*(\hat{\chi}_H^p, \hat{\chi}_C^p) S(\psi, \hat{\chi}_C^p, \hat{\chi}_C^u) Q_2^{\mathrm{T}}(\hat{\chi}_H^u, \hat{\chi}_C^u) \end{aligned}\right\} \tag{6.66}$$

Here the case of coincidence of polarization bases $\hat{\chi}_H^p$ and $\hat{\chi}_H^u$ corresponds to the transformation from the natural basis $\hat{\chi}_e$, when the transformation matrices Q_1 and Q_2 are equal and have the form $Q(\alpha, -\alpha, \beta, \beta + \pi/2)$.

The difference between relations (6.59)–(6.65) and representation (6.66) is that the matrix for the scattered wave in the second group of formulas is complex conjugate, since the rotation of the scattered wave vector changes after the coordinate system is changed. In this case, the following representation holds for the polarization basis $\vec{e}_H^{\,*} = Q_2^*(\hat{\chi}_H^p, \hat{\chi}_C^p)\vec{e}_C$ and, respectively, for the polarization vector $\vec{E}_H = Q(\hat{\chi}_H^p, \hat{\chi}_C^p)\vec{E}_C$. If the input and output bases coincide, transformation

(6.64)–(6.65) is called congruent, and transformations (6.66) are called similarity transformations. It is expedient to preserve these names for the case of non-coinciding bases and call them a generalized congruent transformation and generalized similarity transformation.

The generalized similarity transformation is used mainly for bistatic angles Ψ, which satisfy the inequality $\pi/2 \le \Psi \le \pi$, and the generalized congruent transformation—at angles Ψ less than $\pi/2$.

The bistatic scattering matrix is reduced by converting the similarity to the triangular form. Physically, this means that in bistatic observation there is a polarization of the irradiating wave, in the scattering of which the polarization parameters do not change. In contrast to the monostatic case, diagonalization of the bistatic scattering matrix can be carried out only for different input and output polarization bases.

In addition, if the propagation directions of the irradiating and scattered waves change, we can show with the help of the reciprocity theorem that the following equality holds:

$$S\left(\Psi, \hat{\chi}_H^p, \hat{\chi}_H^u\right) = S^{\mathrm{T}}\left(\Psi, \hat{\chi}_C^u, \hat{\chi}_C^p\right), \tag{6.67}$$

while $\hat{\chi}_H^p = \hat{\chi}_H^{p*}$ and $\hat{\chi}_H^u = \hat{\chi}_C^{u*}$.

In the monostatic case, when $\Psi = 0$, $\hat{\chi}_C^u = \hat{\chi}_C^{p*}$ and hence $\hat{\chi}_H^p = \hat{\chi}_H^{u*}$, but relation (6.67) takes the form: $S\left(\hat{\chi}^p, \hat{\chi}^u\right) = S^{\mathrm{T}}\left(\hat{\chi}^u, \hat{\chi}^p\right)$, which indicates the symmetry of the monostatic scattering matrix (see clause 6.2.2).

We will now point out a number of useful facts about the ESA, which generalize the main results of the previous sections of this chapter. Monostatic ESA will be denoted as $\sigma^2\left(\vec{k}_u\right)$, and bistatic—as $\sigma^2\left(\Psi, \vec{k}_u, \vec{r}\right)$ (note that $\sigma^2\left(\vec{k}_u\right) = \sigma^2\left(0°, \vec{k}_u, -\vec{k}_u\right)$).

At small bistatic angles $\left(\Psi \le \pi/2\right)$:

$$\sigma^2\left(\psi, \vec{k}_u, \vec{r}\right) = \sigma^2\left[\left(-\frac{\vec{k}_u}{\left|\vec{k}_u\right|} + \frac{\vec{r}}{|\vec{r}|}\right)\left|-\frac{\vec{k}_u}{\left|\vec{k}_u\right|} + \frac{\vec{r}}{|\vec{r}|}\right|^{-1}\right], \tag{6.68}$$

(the arguments in σ^2 in the first part of the equality define the bisector of the angle between $\vec{k}_u$ and $\vec{r}$).

Relation (6.68) is also valid for scattering matrices if, of course, the bistatic matrix is defined in a generalized congruent basis, and the monostatic matrix is in the same basis as the bistatic matrix.

We also introduce the following characteristics of the scattering properties of the object: average bistatic ESA

$$<\sigma^2\left(\Psi, \vec{k}_u\right)> \overset{def}{=} \frac{1}{4\pi}\int_0^{4\pi} \sigma^2\left(\Psi, \vec{k}_u, \vec{r}\right) d\vec{r} = \frac{1}{4\pi}\sigma^2_{\textit{инт}}\left(\vec{k}_u\right), \tag{6.69}$$

and average monostatic ESA:

$$< \sigma^2 > \overset{\text{def}}{=} \frac{1}{4\pi} \int\limits_0^{4\pi} \sigma^2\left(\vec{k}_u\right) \mathrm{d}\vec{k}_u. \tag{6.70}$$

In this case, the following equalities are valid:

$$\left.\begin{array}{l} \lim\limits_{\lambda \to 0} < \sigma^2\left(\vec{k}_u\right) >= 2\Sigma\left(\vec{k}_u\right) \\ \lim\limits_{\lambda \to 0} < \sigma^2\left(\vec{k}_u\right) >= \frac{1}{4}\Theta \end{array}\right\}, \tag{6.71}$$

binding $\sigma^2\left(\vec{k}_u\right)$ and σ_C^2 with the area of the projection of the object on the direction of irradiation $\Sigma\left(\vec{k}_u\right)$ and with the area of its surface Θ. For Rayleigh scattering:

$$\lim_{\lambda \to 0}\left(\sigma^2\left(\vec{k}_u\right) + \sigma^2\left(\vec{k}_u, \vec{r}\right) + < \sigma^2 > + < \sigma^2\left(\vec{k}_u\right) > \right)\lambda^4 = 0\left(V^2\right), \tag{6.72}$$

where V—volume of the scatterer (we recall that notation $f(z) = 0(\varphi(z))$ in the analysis indicates that $|f(z)| < c|\varphi(z)|$, c—some constant, or $f(z)/\varphi(z)$ is limited by the area of its determination).

The introduced averaged characteristics of the scattering power make it necessary to carefully consider the procedure for determining the scattering power. In the scattering amplitude (6.5), the order of variation in the irradiation directions of the object and the analysis of the scattered wave are not determined. It is important here that the scattering amplitude is given immediately for all the indicated directions.

Usually, the scattering profile is presented in the form of circular diagrams characterizing the dependence of the scattering properties of the object on the incidence angle of the incident wave and on the angle between the irradiation and scattering directions. The following diagrams were used: scattering diagram, backscattering diagram, and bistatic scattering diagram.

The scattering diagram determines the nature of the scattered field around the object at a fixed position of the object and the irradiation direction $\left(\vec{k}_u = \text{const}, \vec{r}/|\vec{r}| = \text{var}\right)$.

The scattering diagram (SD) is measured by moving the receiving antenna of the polarimeter around the object. The distance from the object to the antenna remains unchanged.

The backscattering diagram (BSD) shows the dependence of the scattered field on the position of the scatterer when the receiving and incident antennas of the polarimeter are aligned. The backscattering diagram is measured by rotating the object around the axis for a fixed position of the transmitting and receiving antenna system $\left(\vec{k}_u/\left|\vec{k}_u\right| = -\vec{r}/|\vec{r}| = \text{var}\right)$.

The bistatic scattering diagram (BSD) shows the dependence of the scattered field on the object view when the receiving antenna is spaced relative to the bistatic transmitting antenna Ψ.

In clause 6.1, it was noted that the scattering amplitude ensures the uniqueness of the problem of determining the object from the detected scattered field for smooth objects. If the object contains singularities (parts of the surface that have zero volume), then two different polarizations of the irradiation are needed. The only exception here is the vibrator located in the direction of irradiation. Further, this degenerate case will be excluded from consideration.

6.2.4 Scattering Matrix in Affine Polarization Bases

To consider the properties of the scattering matrix in affine polarization bases, we choose natural basis (see 6.2.2) $\hat{\chi}_e$ as the initial one. The affinity of the basis for both the incident and scattered waves will be denoted by the superscript, thereby indicating its contravariance (clause 1.4.2). Thus, the old and new bases for the incident and scattered waves (see (6.13)–(6.20)) with their affinity will be denoted by $\hat{\chi}_u^H$, $\hat{\chi}_u^C$ and $\hat{\chi}_p^H$, $\hat{\chi}_p^C$, i.e., indexes will interchange. The characteristic form of the formula for the transition to an affine basis for the scattering matrix is as follows:

$$S\left(\hat{\chi}_p^H, \hat{\chi}_C^H\right) = Q'\left(\hat{\chi}_p^H, \hat{\chi}_e^H\right) S\left(\hat{\chi}_e\right) Q''\left(\hat{\chi}_u^H, \hat{\chi}_e^u\right), \tag{6.73}$$

where Q' and Q''—matrices that determine the transition from a natural basis to an affine basis, but taking into account the properties of the matrix as an object of transformation from one coordinate system to another (as it may be, as in (6.13), that $Q'' = Q'^{\mathrm{T}}$, etc.).

The continuity of the method used for the affine basis with the notation for the orthogonal polarization basis is visible, which makes it possible to form two Tables 6.1 and 6.2. Table 6.1 shows the expressions for a monostatic scattering matrix for various bases of the incident and scattered waves.

In Table 6.1, you can select the data blocks corresponding to the elements of the transformed scattering matrix. Thus, block I with elements 1–4 determines the initial monostatic matrix in the natural basis (for the sake of generality, various designations s_{12} and s_{21} are preserved, which makes it possible to extend the results to the bistatic case, if a generalized congruential transformation takes place here). Blocks II and III correspond to a transformation of type (6.19) with a transition to a circular basis and linearly polarized basis rotated by $\pi / 4$. It can be seen that with $s_{12}\left(\hat{\chi}_C\right) = s_{21}\left(\hat{\chi}_C\right)$, the symmetry of the matrix here is also preserved in the new polarization basis.

Blocks IV–VII correspond to the transformation of the scattering matrix, when the bases of the incident $\hat{\chi}_H^u$ and scattered waves $\hat{\chi}_H^p$ are different. The remaining blocks correspond to the case of the orthogonal polarization basis of the incident wave $\hat{\chi}_H^u$ and an affine basis for the scattered wave $\hat{\chi}_p^H$. Opposite case: The wave that irradiates the scatterer is investigated in the affine basis, and the scattered wave, in the orthogonal basis, is reflected in Table 6.2.

Table 6.1 Expressions for a monostatic scattering matrix for various bases of the incident and scattered waves

Reception		Radiation					
		→	↑	↗	↘	↺	↻
1	→	S_{11}	S_{12}	$\frac{\sqrt{2}}{2}(S_{11}+S_{12})$	$\frac{\sqrt{2}}{2}(S_{11}-S_{12})$	$\frac{\sqrt{2}}{2}(S_{11}+iS_{12})$	$\frac{\sqrt{2}}{2}(iS_{11}+S_{12})$
	↑	S_{21}	S_{22}	$\frac{\sqrt{2}}{2}(S_{21}+S_{22})$	$\frac{\sqrt{2}}{2}(S_{21}-S_{22})$	$\frac{\sqrt{2}}{2}(S_{21}+iS_{22})$	$\frac{\sqrt{2}}{2}(iS_{11}+S_{22})$
2	↺	$\frac{\sqrt{2}}{2}(S_{11}+iS_{21})$	$\frac{\sqrt{2}}{2}(S_{12}-iS_{22})$	$\frac{1}{2}(S_{11}+S_{12}+iS_{21}+iS_{22})$	$\frac{1}{2}(S_{11}+S_{12}+iS_{21}-iS_{22})$	$\frac{1}{2}(S_{11}+iS_{12}+iS_{21}-S_{22})$	$\frac{1}{2}(iS_{11}+S_{12}-S_{21}+iS_{22})$
	↻	$\frac{\sqrt{2}}{2}(iS_{11}+S_{21})$	$\frac{\sqrt{2}}{2}(iS_{12}+S_{22})$	$\frac{1}{2}(iS_{11}+iS_{12}+S_{21}+S_{22})$	$\frac{1}{2}(iS_{11}-iS_{12}+S_{21}-S_{22})$	$\frac{1}{2}(iS_{11}-S_{12}+S_{21}+iS_{22})$	$\frac{1}{2}(-S_{11}+iS_{12}+iS_{21}+S_{22})$
3	↗	$\frac{\sqrt{2}}{2}(S_{11}+S_{21})$	$\frac{\sqrt{2}}{2}(S_{11}-S_{22})$	$\frac{1}{2}(S_{11}-S_{12}+S_{21}+S_{22})$	$\frac{1}{2}(S_{11}-S_{12}+S_{21}-S_{22})$	$\frac{1}{2}(S_{11}+iS_{12}+S_{21}+iS_{22})$	$\frac{1}{2}(iS_{11}+S_{12}+iS_{21}+S_{22})$
	↘	$\frac{\sqrt{2}}{2}(S_{11}-S_{21})$	$\frac{\sqrt{2}}{2}(S_{12}-S_{22})$	$\frac{1}{2}(S_{11}+S_{12}-S_{21}-S_{22})$	$\frac{1}{2}(S_{11}-S_{12}-S_{21}+S_{22})$	$\frac{1}{2}(S_{11}+iS_{12}-S_{21}-iS_{22})$	$\frac{1}{2}(iS_{11}+S_{12}-iS_{21}-S_{22})$
4	→	$S_{11}-iS_{21}$	$S_{12}-iS_{22}$	$\frac{\sqrt{2}}{2}(S_{11}+S_{12}-iS_{21}-iS_{22})$	$\frac{\sqrt{2}}{2}(S_{11}-S_{12}-iS_{21}+iS_{22})$	$\frac{\sqrt{2}}{2}(S_{11}+iS_{12}-iS_{21}+S_{22})$	$\frac{\sqrt{2}}{2}(iS_{11}-S_{12}+S_{21}-iS_{22})$
	↺	$\sqrt{2}S_{21}$	$\sqrt{2}S_{22}$	$S_{21}+S_{22}$	$S_{21}-S_{22}$	$S_{21}+iS_{22}$	$iS_{21}+S_{22}$
5	↑	$iS_{11}+S_{21}$	$iS_{21}+S_{22}$	$\frac{\sqrt{2}}{2}(iS_{11}+iS_{12}+S_{21}-S_{22})$	$\frac{\sqrt{2}}{2}(iS_{11}-iS_{12}+S_{21}-S_{22})$	$\frac{\sqrt{2}}{2}(iS_{11}-S_{12}-S_{21}+iS_{22})$	$\frac{\sqrt{2}}{2}(-S_{11}+iS_{12}+iS_{21}+S_{22})$
	↺	$-i\sqrt{2}S_{11}$	$-i\sqrt{2}S_{12}$	$-i(S_{11}+S_{12})$	$-i(S_{11}-S_{12})$	$-iS_{11}+S_{12}$	$S_{11}-iS_{12}$
6	↗	$e^{i\frac{\pi}{4}}(S_{11}-iS_{21})$	$e^{i\frac{\pi}{4}}(S_{12}-iS_{22})$	$\frac{\sqrt{2}}{2}(S_{11}+S_{12}-iS_{21}+iS_{22})e^{i\frac{\pi}{4}}$	$\frac{\sqrt{2}}{2}(S_{11}-S_{12}-iS_{21}+iS_{22})e^{i\frac{\pi}{4}}$	$\frac{\sqrt{2}}{2}(S_{11}+iS_{12}-iS_{21}+S_{22})e^{j\frac{\pi}{4}}$	$\frac{\sqrt{2}}{2}(iS_{11}+S_{12}+S_{21}-iS_{22})e^{j\frac{\pi}{4}}$
	↺	$e^{i\frac{\pi}{4}}(-S_{11}+S_{21})$	$e^{i\frac{\pi}{4}}(-S_{12}+S_{22})$	$\frac{\sqrt{2}}{2}(-S_{11}-S_{12}+S_{21}+S_{22})e^{i\frac{\pi}{4}}$	$\frac{\sqrt{2}}{2}(-S_{11}+S_{12}+S_{21}-S_{22})e^{i\frac{\pi}{4}}$	$\frac{\sqrt{2}}{2}(-S_{11}-iS_{12}+S_{21}+iS_{22})e^{i\frac{\pi}{4}}$	$\frac{\sqrt{2}}{2}(-iS_{11}-S_{12}+iS_{21}+S_{22})e^{i\frac{\pi}{4}}$
7	↘	$e^{-i\frac{\pi}{4}}(S_{11}-iS_{21})$	$e^{-i\frac{\pi}{4}}(S_{12}-iS_{22})$	$\frac{\sqrt{2}}{2}(S_{11}+S_{12}-iS_{21}-iS_{22})e^{-i\frac{\pi}{4}}$	$\frac{\sqrt{2}}{2}(S_{11}-S_{12}-iS_{21}+iS_{22})e^{-i\frac{\pi}{4}}$	$\frac{\sqrt{2}}{2}(S_{11}+iS_{12}-iS_{21}+S_{22})e^{-i\frac{\pi}{4}}$	$\frac{\sqrt{2}}{2}(iS_{11}+S_{12}+S_{21}-iS_{22})e^{-i\frac{\pi}{4}}$
	↺	$e^{-i\frac{\pi}{4}}(S_{11}-S_{21})$	$e^{-i\frac{\pi}{4}}(S_{12}+S_{22})$	$\frac{\sqrt{2}}{2}(S_{11}+S_{12}-S_{21}+S_{22})e^{-i\frac{\pi}{4}}$	$\frac{\sqrt{2}}{2}(S_{11}-S_{12}+S_{21}-S_{22})e^{-i\frac{\pi}{4}}$	$\frac{\sqrt{2}}{2}(S_{11}+iS_{12}+S_{21}+iS_{22})e^{-i\frac{\pi}{4}}$	$\frac{\sqrt{2}}{2}(iS_{11}+S_{12}+iS_{21}-S_{22})e^{-i\frac{\pi}{4}}$

Table 6.2 Expressions for a monostatic scattering matrix for various bases of the incident and scattered waves

Reception		Radiation					
		→	↑	↗	↘	↺	↻
8	→	$S_{11}+iS_{21}$	$S_{12}+iS_{22}$	$\frac{\sqrt{2}}{2}(S_{11}+S_{12}+iS_{21}+iS_{22})$	$\frac{\sqrt{2}}{2}(S_{11}-S_{12}+iS_{21}-iS_{22})$	$\frac{\sqrt{2}}{2}(S_{11}+iS_{12}+iS_{21}-S_{22})$	$\frac{\sqrt{2}}{2}(iS_{11}+S_{12}-S_{21}+iS_{22})$
	↻	$\sqrt{2}S_{21}$	$\sqrt{2}S_{22}$	$S_{21}+S_{22}$	$S_{21}-S_{22}$	$S_{21}+iS_{22}$	$iS_{21}+S_{22}$
9	↑	$-iS_{11}+S_{21}$	$-iS_{12}-S_{22}$	$\frac{\sqrt{2}}{2}(-iS_{11}-iS_{12}+S_{21}+S_{22})$	$\frac{\sqrt{2}}{2}(iS_{11}+iS_{12}+S_{21}-S_{22})$	$\frac{\sqrt{2}}{2}(-iS_{11}+S_{12}+S_{21}-iS_{22})$	$\frac{\sqrt{2}}{2}(S_{11}-iS_{12}+iS_{21}+S_{22})$
	↻	$i\sqrt{2}S_{11}$	$i\sqrt{2}S_{12}$	$i(S_{11}+S_{12})$	$i(S_{11}-S_{12})$	$iS_{21}-S_{22}$	$-S_{21}+iS_{22}$
10	↗	$e^{-i\frac{\pi}{4}}(S_{11}+iS_{21})$	$e^{-i\frac{\pi}{4}}(S_{12}+iS_{22})$	$\frac{\sqrt{2}}{2}(S_{11}+S_{12}+iS_{21}+iS_{22})e^{-i\frac{\pi}{4}}$	$\frac{\sqrt{2}}{2}(S_{11}-S_{12}+iS_{21}-iS_{22})e^{-i\frac{\pi}{4}}$	$\frac{\sqrt{2}}{2}(S_{11}+iS_{12}+iS_{21}-S_{22})e^{-i\frac{\pi}{4}}$	$\frac{\sqrt{2}}{2}(iS_{11}+S_{12}-S_{21}+iS_{22})e^{-i\frac{\pi}{4}}$
	↻	$e^{-i\frac{\pi}{4}}(-S_{11}-S_{21})$	$e^{-i\frac{\pi}{4}}(-S_{12}+S_{22})$	$\frac{\sqrt{2}}{2}(-S_{11}-S_{12}+S_{21}-S_{22})e^{-i\frac{\pi}{4}}$	$\frac{\sqrt{2}}{2}(-S_{11}+S_{12}+S_{21}-S_{22})e^{-i\frac{\pi}{4}}$	$\frac{\sqrt{2}}{2}(-S_{11}-iS_{12}+S_{21}+iS_{22})e^{-i\frac{\pi}{4}}$	$\frac{\sqrt{2}}{2}(-iS_{11}-S_{12}+iS_{21}+S_{22})e^{-i\frac{\pi}{4}}$
11	↘	$e^{i\frac{\pi}{4}}(S_{11}+iS_{21})$	$e^{i\frac{\pi}{4}}(S_{12}+iS_{22})$	$\frac{\sqrt{2}}{2}(S_{11}+S_{12}+iS_{21}+iS_{22})e^{i\frac{\pi}{4}}$	$\frac{\sqrt{2}}{2}(S_{11}-S_{12}+iS_{21}-iS_{22})e^{i\frac{\pi}{4}}$	$\frac{\sqrt{2}}{2}(S_{11}+iS_{12}+iS_{21}-S_{22})e^{i\frac{\pi}{4}}$	$\frac{\sqrt{2}}{2}(iS_{11}+S_{12}-S_{21}+iS_{22})e^{i\frac{\pi}{4}}$
	↻	$e^{i\frac{\pi}{4}}(S_{11}+S_{21})$	$e^{i\frac{\pi}{4}}(S_{12}+S_{22})$	$\frac{\sqrt{2}}{2}(S_{11}+S_{12}+S_{21}+S_{22})e^{i\frac{\pi}{4}}$	$\frac{\sqrt{2}}{2}(S_{11}-S_{12}+S_{21}-S_{22})e^{i\frac{\pi}{4}}$	$\frac{\sqrt{2}}{2}(S_{11}+iS_{12}+S_{21}-iS_{22})e^{i\frac{\pi}{4}}$	$\frac{\sqrt{2}}{2}(iS_{11}+S_{12}+iS_{21}+S_{22})e^{i\frac{\pi}{4}}$
12	→	$S_{11}-S_{21}$	$S_{12}-S_{22}$	$\frac{\sqrt{2}}{2}(S_{11}+S_{12}-S_{21}-S_{22})$	$\frac{\sqrt{2}}{2}(S_{11}-S_{12}-S_{21}+S_{22})$	$\frac{\sqrt{2}}{2}(S_{11}+iS_{12}-S_{21}-iS_{22})$	$\frac{\sqrt{2}}{2}(S_{11}+S_{12}-iS_{21}-S_{22})$
	↗	$\sqrt{2}S_{21}$	$\sqrt{2}S_{22}$	$S_{21}-S_{22}$	$S_{21}-S_{22}$	$S_{21}+iS_{22}$	$iS_{21}+S_{22}$
13	→	$S_{11}+S_{21}$	$S_{12}+S_{22}$	$\frac{\sqrt{2}}{2}(S_{11}+S_{12}+S_{21}+S_{22})$	$\frac{\sqrt{2}}{2}(S_{11}-S_{12}+S_{21}-S_{22})$	$\frac{\sqrt{2}}{2}(S_{11}+iS_{12}+S_{21}+iS_{22})$	$\frac{\sqrt{2}}{2}(iS_{11}+S_{12}+iS_{21}+S_{22})$
	↘	$-\sqrt{2}S_{21}$	$-\sqrt{2}S_{22}$	$-(S_{21}-S_{22})$	$-S_{21}+S_{22}$	$-(S_{21}+iS_{22})$	$-(iS_{21}+S_{22})$
14	↑	$-S_{11}+S_{21}$	$-S_{12}+S_{22}$	$\frac{\sqrt{2}}{2}(-S_{11}-S_{12}+S_{21}+S_{22})$	$\frac{\sqrt{2}}{2}(-S_{11}+S_{12}+S_{21}-S_{22})$	$\frac{\sqrt{2}}{2}(-S_{11}-iS_{12}+S_{21}-iS_{22})$	$\frac{\sqrt{2}}{2}(iS_{11}-S_{12}+iS_{21}+S_{22})$
	↗	$\sqrt{2}S_{21}$	$\sqrt{2}S_{12}$	$S_{21}+S_{22}$	$S_{11}-S_{12}$	$S_{11}+iS_{12}$	$iS_{11}+S_{12}$
15	↑	$S_{11}+S_{21}$	$S_{12}+S_{22}$	$\frac{\sqrt{2}}{2}(S_{11}+S_{12}-S_{21}-S_{22})$	$\frac{\sqrt{2}}{2}(S_{11}-S_{12}+S_{21}-S_{22})$	$\frac{\sqrt{2}}{2}(S_{11}+iS_{12}+S_{21}+iS_{22})$	$\frac{\sqrt{2}}{2}(iS_{11}+S_{12}+iS_{21}+S_{22})$
	↘	$\sqrt{2}S_{11}$	$\sqrt{2}S_{12}$	$S_{21}+S_{12}$	$S_{11}-S_{12}$	$S_{11}+iS_{12}$	$iS_{11}+S_{12}$

Table 6.3 can be obtained from Table 6.1, retaining the orthogonality of the basis when receiving the signal and specifying various combinations of polarization vectors during irradiation of the object, since the polarization-irradiating signals are independent in this case. These tables do not take into account the phase shifts caused by the difference in the instantaneous positions of the electric field vector in the polarization ellipses. However, assuming that Tables 6.1 and 6.2 give the initial stock of information on the affine transformation, we first consider the full theory of transformations with allowance for the above phase shifts, and then we investigate the affine transformations in the "geometric language" (see clause 1.4.2).

Let us define the unitary vectors of the affine basis with respect to the natural parameters $(\gamma_1, \gamma_2, \delta_1, \delta_2, \varphi_{01}, \varphi_{02})$. Then, in the monostatic case, instead of (6.19) and taking into account equality (1.59), we obtain:

$$S(\hat{\chi}^H) = \left[Q^+(\hat{\chi}^H, \hat{\chi}_e)\right]^{-1} S(\hat{\chi}_e) Q^{\mathrm{T}}(\hat{\chi}^H, \hat{\chi}_e). \tag{6.74}$$

With $\gamma_1 = \gamma_2, \delta_1 = \delta_2$, formula (6.74) transforms into (6.19).

Inverse matrix $\left[Q(\hat{\chi}^H, \hat{\chi}^e)\right]^{-1}$ is determined by means of expression (1.62); it exists for any parameters except those that satisfy condition (1.65).

In the affine basis, as can be seen from Table 6.1 and 6.2,

there is usually no equality between elements $s_{12}(\hat{\chi}^H)$ and $s_{21}(\hat{\chi}^H)$.

Let us illustrate the basic formula (6.74) with reference to the definition of the polarization basis using geometric parameters—the elliptic angle and the slope angle of the major axis of the ellipse. In this case, matrix $Q(\hat{\chi}^H, \hat{\chi}_e)$ has the form (1.79), which allows to write the following expression for $\left[Q^+(\hat{\chi}^H, \hat{\chi}_e)\right]^{-1}$:

$$\begin{aligned} \left[Q^+(\hat{\chi}^H, \hat{\chi}_e)\right]^{-1} &= \frac{1}{\sin(\beta_2 - \beta_1)\cos(\alpha_2 + \alpha_1) - i\cos(\beta_2 - \beta_1)\sin(\alpha_2 - \alpha_1)} \\ &\times \begin{pmatrix} \cos\alpha_2 \sin\beta_2 - i\sin\alpha_2\cos\beta_2 & -\cos\alpha_2\cos\beta_2 - i\sin\alpha_2\sin\beta_2 \\ -\cos\alpha_1\sin\beta_1 + i\sin\alpha_1\cos\beta_1 & \cos\alpha_1\cos\beta_1 + i\sin\alpha_1\sin\beta_1 \end{pmatrix}. \end{aligned} \tag{6.75}$$

It can be directly verified that with $\alpha_2 = -\alpha_1, \beta_2 = \beta_1 + \pi/2$ for an orthogonal basis:

$$\begin{aligned} Q(\hat{\chi}_H, \hat{\chi}_e) &= \left[Q^+(\hat{\chi}^H, \hat{\chi}_e)\right]^{-1} \\ &= \begin{pmatrix} \cos\alpha_2 \sin\beta_2 - i\sin\alpha_2\cos\beta_2 & -\cos\alpha_2\cos\beta_2 - i\sin\alpha_2\sin\beta_2 \\ -\cos\alpha_1\sin\beta_1 + i\sin\alpha_1\cos\beta_1 & \cos\alpha_1\cos\beta_1 + i\sin\alpha_1\sin\beta_1 \end{pmatrix}. \end{aligned} \tag{6.76}$$

On the basis of relations (1.78), (6.74), and (6.75), the affine blocks in Tables 6.1 and 6.2 are calculated. Let us give an example of such a calculation. Let the polarization basis $\hat{\chi}_p^H$ correspond to a linear polarization with a slope angle $+\pi/4$ and a right circular polarization, and the basis $\hat{\chi}_u^H$ will vary, remaining orthogonal.

Table 6.3 Expressions for a monostatic scattering matrix for various bases of the incident and scattered waves

Reception		Radiation					
		→	↑	↗	↘	↺	↻
1	→	S_{11}	S_{21}	$\frac{\sqrt{2}}{2}(S_{11}+S_{21})$	$\frac{\sqrt{2}}{2}(S_{11}-S_{21})$	$\frac{\sqrt{2}}{2}(S_{11}+iS_{21})$	$\frac{\sqrt{2}}{2}(iS_{11}+S_{21})$
	↺	$\frac{\sqrt{2}}{2}(S_{11}+jS_{12})$	$\frac{\sqrt{2}}{2}(S_{21}-jS_{22})$	$\frac{1}{2}(S_{11}+iS_{12}-S_{21}+iS_{22})$	$\frac{1}{2}(S_{11}+iS_{12}-S_{21}-iS_{22})$	$\frac{1}{2}(S_{11}+iS_{12}+iS_{21}-S_{22})$	$\frac{1}{2}(iS_{11}-S_{12}+S_{21}+iS_{22})$
2	↑	S_{12}	S_{22}	$\frac{\sqrt{2}}{2}(S_{12}+S_{22})$	$\frac{\sqrt{2}}{2}(S_{12}-S_{22})$	$\frac{\sqrt{2}}{2}(S_{12}+iS_{22})$	$\frac{\sqrt{2}}{2}(iS_{12}+S_{22})$
	↺	$\frac{\sqrt{2}}{2}(S_{11}-iS_{12})$	$\frac{\sqrt{2}}{2}(S_{21}+iS_{22})$	$\frac{1}{2}(S_{11}+iS_{12}+S_{21}+iS_{22})$	$\frac{1}{2}(S_{11}+iS_{12}-S_{21}-iS_{22})$	$\frac{1}{2}(S_{11}+iS_{12}+iS_{21}-S_{22})$	$\frac{1}{2}(iS_{11}-S_{12}+S_{21}+iS_{22})$
3	↗	$\frac{\sqrt{2}}{2}(S_{11}+S_{12})$	$\frac{\sqrt{2}}{2}(S_{21}+S_{22})$	$\frac{1}{2}(S_{11}-S_{12}-S_{21}+S_{22})$	$\frac{1}{2}(S_{11}+S_{12}-S_{21}-S_{22})$	$\frac{1}{2}(S_{11}+S_{12}+iS_{21}+iS_{22})$	$\frac{1}{2}(iS_{11}+iS_{12}+S_{21}+S_{22})$
	↺	$\frac{\sqrt{2}}{2}(S_{11}+iS_{12})$	$\frac{\sqrt{2}}{2}(S_{21}+iS_{22})$	$\frac{1}{2}(S_{11}+iS_{12}+S_{21}+jS_{22})$	$\frac{1}{2}(S_{11}+iS_{12}-S_{21}-iS_{22})$	$\frac{1}{2}(S_{11}+iS_{12}+iS_{21}-S_{22})$	$\frac{1}{2}(iS_{11}-S_{12}+S_{21}+iS_{22})$
4	↘	$\frac{\sqrt{2}}{2}(S_{11}-S_{12})$	$\frac{\sqrt{2}}{2}(S_{21}-S_{22})$	$\frac{1}{2}(S_{11}-S_{12}-S_{21}-S_{22})$	$\frac{1}{2}(S_{11}-S_{12}-S_{21}+S_{22})$	$\frac{1}{2}(S_{11}-S_{12}+iS_{21}-iS_{22})$	$\frac{1}{2}(iS_{11}-iS_{12}+S_{21}-S_{22})$
	↺	$\frac{\sqrt{2}}{2}(S_{11}-iS_{12})$	$\frac{\sqrt{2}}{2}(S_{21}+S_{22})$	$\frac{1}{2}(S_{11}+iS_{12}+S_{21}+iS_{22})$	$\frac{1}{2}(S_{11}+iS_{12}-S_{21}-iS_{22})$	$\frac{1}{2}(S_{11}+iS_{12}+iS_{21}-S_{22})$	$\frac{1}{2}(iS_{11}-S_{12}+S_{21}+iS_{22})$
5	↑	S_{11}	S_{21}	$\frac{\sqrt{2}}{2}(S_{11}+S_{21})$	$\frac{\sqrt{2}}{2}(S_{11}-S_{21})$	$\frac{\sqrt{2}}{2}(S_{11}+iS_{21})$	$\frac{\sqrt{2}}{2}(iS_{11}+S_{21})$
	↺	$\frac{\sqrt{2}}{2}(iS_{11}+S_{12})$	$\frac{\sqrt{2}}{2}(iS_{21}+S_{22})$	$\frac{1}{2}(iS_{11}+S_{12}+iS_{21}+S_{22})$	$\frac{1}{2}(iS_{11}+S_{12}-iS_{21}-S_{22})$	$\frac{1}{2}(iS_{11}+S_{12}-iS_{21}+iS_{22})$	$\frac{1}{2}(-S_{11}+iS_{12}+iS_{21}+S_{22})$
6	↗	S_{12}	S_{22}	$\frac{\sqrt{2}}{2}(S_{12}+S_{22})$	$\frac{\sqrt{2}}{2}(S_{12}-S_{22})$	$\frac{\sqrt{2}}{2}(S_{12}+iS_{22})$	$\frac{\sqrt{2}}{2}(iS_{12}+S_{22})$
	↺	$\frac{\sqrt{2}}{2}(iS_{11}+S_{22})$	$\frac{\sqrt{2}}{2}(iS_{21}-S_{22})$	$\frac{\sqrt{2}}{2}(iS_{11}+S_{12}+iS_{21}+S_{22})$	$\frac{\sqrt{2}}{2}(iS_{11}+S_{12}-iS_{21}+iS_{22})$	$\frac{\sqrt{2}}{2}(iS_{11}+S_{12}-S_{21}+iS_{22})$	$\frac{\sqrt{2}}{2}(-S_{11}+iS_{12}+iS_{21}+S_{22})$
7	↘	$\frac{\sqrt{2}}{2}(S_{11}+S_{12})$	$\frac{\sqrt{2}}{2}(S_{21}-S_{22})$	$\frac{1}{2}(S_{11}+S_{12}-S_{21}-S_{22})$	$\frac{1}{2}(S_{11}+S_{12}-S_{21}-S_{22})$	$\frac{1}{2}(S_{11}+S_{12}+iS_{21}-iS_{22})$	$\frac{1}{2}(iS_{11}+iS_{12}+S_{21}+S_{22})$
	↺	$\frac{\sqrt{2}}{2}(iS_{11}+S_{12})$	$\frac{\sqrt{2}}{2}(iS_{11}+S_{22})$	$\frac{1}{2}(iS_{11}+S_{12}+iS_{21}+S_{22})$	$\frac{1}{2}(iS_{11}+S_{12}-iS_{21}-S_{22})$	$\frac{1}{2}(iS_{11}+S_{12}-S_{21}-jS_{22})$	$\frac{1}{2}(-S_{11}+iS_{12}+iS_{21}+S_{22})$
8		$\frac{\sqrt{2}}{2}(S_{11}-S_{12})$	$\frac{\sqrt{2}}{2}(S_{21}-S_{22})$	$\frac{1}{2}(S_{11}-S_{12}+S_{21}-S_{22})$	$\frac{1}{2}(S_{11}-S_{12}-S_{21}+S_{22})$	$\frac{1}{2}(S_{11}-S_{12}+iS_{21}-iS_{22})$	$\frac{1}{2}(iS_{11}-iS_{12}+S_{21}-S_{22})$
		$\frac{\sqrt{2}}{2}(iS_{11}+S_{12})$	$\frac{\sqrt{2}}{2}(iS_{21}-S_{22})$	$\frac{1}{2}(iS_{11}+S_{12}+iS_{21}+S_{22})$	$\frac{1}{2}(iS_{11}+S_{12}+iS_{21}-S_{22})$	$\frac{1}{2}(iS_{11}+S_{12}-S_{21}+iS_{22})$	$\frac{1}{2}(-S_{11}+iS_{12}+iS_{21}+S_{22})$

The parameters of the receiving polarization basis are obtained equal to $\alpha_1 = 0, \beta_1 = \pi/4$, $\alpha_2 = \pi/4\, \beta_2 = \pi/2$, which will determine the form of matrix $[Q^+(\hat{\chi}^H, \hat{\chi}_e)]^{-1}$:

$$[Q^+(\hat{\chi}_p^H, \hat{\chi}_e^p)]^{-1} = \mathrm{e}^{\mathrm{i}\frac{\pi}{4}} \begin{pmatrix} 1 & -i \\ -1 & 1 \end{pmatrix}. \tag{6.77}$$

If the basis of the incident wave has parameters $\alpha_1 = 0\, \beta_1 = 0, \alpha_2 = 0, \beta_2 = \pi/2$, then the affine scattering matrix is obtained equal to:

$$S(\hat{\chi}_p^H, \hat{\chi}_e^u) = \mathrm{e}^{\mathrm{i}\frac{\pi}{4}} \begin{pmatrix} 1 & -i \\ -1 & 1 \end{pmatrix} \begin{pmatrix} s_{11} & s_{12} \\ s_{21} & s_{22} \end{pmatrix} \begin{pmatrix} 1 & 0 \\ 0 & 1 \end{pmatrix} = \mathrm{e}^{\mathrm{i}\frac{\pi}{4}} \begin{pmatrix} s_{11} - is_{21} & s_{12} - is_{22} \\ -s_{11} + s_{21} & -s_{12} + s_{22} \end{pmatrix}. \tag{6.78}$$

When the basis of the irradiating wave is rotated by $+\pi/4 : \alpha_1 = 0$, $\beta_1 = \pi/4$, $\alpha_2 = 0$, $\beta_2 = -\pi/4$, we will receive:

$$\begin{aligned} S(\hat{\chi}_p^H, \hat{\chi}_H^u) &= \mathrm{e}^{\mathrm{i}\frac{\pi}{4}} \frac{\sqrt{2}}{2} \begin{pmatrix} 1 & -i \\ -1 & 1 \end{pmatrix} \begin{pmatrix} s_{11} & s_{12} \\ s_{21} & s_{22} \end{pmatrix} \begin{pmatrix} 1 & 1 \\ 1 & -1 \end{pmatrix} \\ &= \mathrm{e}^{\mathrm{i}\frac{\pi}{4}} \frac{\sqrt{2}}{2} \begin{pmatrix} s_{11} + s_{12} - i(s_{21} + s_{22}) & s_{11} - s_{12} - i(s_{21} - s_{22}) \\ -s_{11} - s_{12} + s_{21} + s_{22} & -s_{11} + s_{12} + s_{21} - s_{22} \end{pmatrix}. \end{aligned} \tag{6.79}$$

Finally, the transition to the circular basis of the incident wave determines the following set of parameters: $\alpha_1 = \pi/4$, $\beta_1 = 0$, $\alpha_2 = -\pi/4$, $\beta_2 = \pi/2$, what gives:

$$\begin{aligned} S(\hat{\chi}_p^H, \hat{\chi}_H^u) &= \mathrm{e}^{\mathrm{i}\frac{\pi}{4}} \frac{\sqrt{2}}{2} \begin{pmatrix} 1 & -i \\ -1 & 1 \end{pmatrix} \begin{pmatrix} s_{11} & s_{12} \\ s_{21} & s_{22} \end{pmatrix} \begin{pmatrix} 1 & i \\ i & 1 \end{pmatrix} \\ &= \mathrm{e}^{\mathrm{i}\frac{\pi}{4}} \frac{\sqrt{2}}{2} \begin{pmatrix} s_{11} + s_{22} + i(s_{12} - s_{21}) & i(s_{11} - s_{22}) + s_{12} + s_{21} \\ -s_{11} + s_{21} - i(s_{12} - s_{22}) & -i(s_{11} - s_{21}) - s_{12} + s_{22} \end{pmatrix}. \end{aligned} \tag{6.80}$$

Clause 6.2.4 has been written with the participation of Ye.V. Kusakina.

6.3 Objects of Navigational Observation with Nonlinear Characteristics

6.3.1 Nonlinear Scattering

As shown by experimental studies, in a number of cases, when electromagnetic waves are reflected from targets, there is a so-called nonlinear scattering, when the connection between $\vec{E}_{omp}$ and $\vec{E}_{nao}$ can no longer be represented in the form of Eq. (6.1), since the spectrum of the reflected wave differs from the corresponding spectrum of the incident wave, primarily, by the presence of additional components. This indicates that the components of the reflected wave are obtained by some nonlinear transformation of the components of the incident wave.

Examples of nonlinear processes are effects of nonlinear radar detection, and nonlinear phenomena arising when ionosphere is exposed to irradiation or occurring during propagation of radio waves (primarily the Luxembourg–Gorky effect). In connection with the creation of a variety of artificial scatterers with nonlinear characteristics, the processes occurring in ferromagnetic and ferroelectric environments, various effects in plasma and magnetohydrodynamic generators, phenomena of ferromagnetic resonance, etc., shall be subjected to description and investigation. We note that almost all nonlinear effects in the propagation of radio waves create the ionosphere.

Let us consider some examples of phenomena responsible for the appearance of nonlinear properties in the scattering of radio waves.

The linearity of the problems in the scattering and diffraction theory is due to the harmonic character of the charge oscillations in the scatterer matter when they are harmonic-excited. The harmonic character of the oscillations determines the monochromaticity of the waves emitted by the charges.

The equation of motion of a non-harmonic oscillator has the following form:

$$m_e \frac{\mathrm{d}^2 x(t)}{\mathrm{d}t^2} + \gamma \frac{\mathrm{d}x(t)}{\mathrm{d}t} + m_e \omega_0^2 x(t) + \alpha x^2(t) = q_e E \cos\left(\omega t - \vec{k}\vec{r}\right), \tag{6.81}$$

where m_e and q_e—mass and charge of the oscillator, α—nonlinearity-determining coefficient, γ—attenuation coefficient, $x(t)$—charge displacement.

If the atom is located at the center of symmetry, then a nonlinearity of a higher order appears due to the presence of not only the component $\alpha x(t)$, but also $\beta x^2(t)$, as well as higher multipole components, in the last differential equation. All this leads to the establishment of induced dipole moments at frequencies 2ω , 3ω, and higher. The ratio of the corresponding dipole moments of the oscillator system under consideration at frequencies 2ω and ω for the quantum-mechanical description is as follows:

$$\vec{P}_e(\vec{r}, 2\omega)/\vec{P}_e(\vec{r}, \omega) = q_e \vec{E} x / |\omega - \omega_0 + i\gamma|, \tag{6.82}$$

where x—amplitude of oscillations, $h\omega_0$—difference between energy levels.

When approaching resonance $\omega - \omega_0$, the nonlinearity grows rapidly in contrast to the case when the basic frequency and its harmonics are far from ω_0, where the nonlinearity is negligibly small.

It is seen from Eq. (6.82) that for the appearance of nontrivial polarization effects, it is necessary to investigate the case when ω_0^2, or α, or γ are tensors, and direction $x(t)$ in this case will not coincide with direction $\vec{E}$. For this reason, it is necessary to consider a method for representing the solution of a differential equation when the coefficients are tensors or, in general, matrices. Such a representation is possible when using multidimensional Voltaire-Wiener series.

The Voltaire-Wiener functional series are successfully used in the analysis of nonlinear systems. For one-dimensional input $U(t)$ and output signals $y(t)$, these series have the following form:

$$y(t) = W_0(t) + \int_{-\infty}^{\infty} W_1(t, S_1)U(S_1)dS_1 + \int\int_{-\infty}^{\infty} W_2(t, S_1, S_2)U(S_1)U(S_2)dS_1dS_2$$

$$+ \cdots + \int \cdots \int_{-\infty}^{\infty} W_n(t, S_1, S_2, \ldots, S_n)U(S_1)U(S_2)\cdots U(S_n)dS_1dS_2\cdots dS_n, \quad (6.83)$$

where $W_1(\xi)$, $W_2(\xi)$, ...—kernel of the functional series.

Prospectivity of the Voltaire-Wiener series is determined by their following properties:

- explicit relationship of response and effect;
- simultaneous accounting of inertial and non-inertial properties of the signal conversion channel;
- generalized description of the channel properties, which transforms into the description using the convolution integral (Duhamel's integral)—for linear systems, and into a power-law series—for non-inertial nonlinearities;
- representation of nonlinearities by kernels $W_1(\xi)$, $W_2(\xi)$, ..., $W_n(\xi)$ independent of the input signal;
- possibility of using well-known impulse and frequency transfer functions.

Using the Voltaire-Wiener series, the problem of separating the types of nonlinearity for subsequent classification is easily solved, and a relatively simple connection of the nonlinearity criteria applied in practice with the characteristics of the kernels $0 \le |s_{12}|^2 \le 0.25\sigma_\Sigma^2\left(1 + \sqrt{1 - q^2}\right)$ of the functional series is established.

The transition to multidimensional Voltaire-Wiener series is connected with tensors, or more precisely with tensor notation, to the characterization of which we now proceed.

For ordinary matrices (A), the tensor notation A_α^β can be used, if we agree that the lower index (subscript) defines the rows of the matrix, and the upper index (superscript)—its columns. The vector in this case is denoted by A_α, and the transposed vector—A^α. As in tensor analysis, indices can be denoted arbitrarily—it is not the index value that matters, but its location (mute indices).

Probably, the notation for the product of the matrix $C_\alpha^\gamma = A_\alpha^\beta B_\beta^\gamma$, where the action of the Einstein rule is assumed—summation over repeated indices (in this case over β) will not cause difficulties.

Accordingly, the product of the matrix by the vector will be written as follows: $C_\alpha = A_\alpha^\beta B_\beta$, i.e., we obtain a vector. Naturally, the subscripts shall have the same dimension.

Later, when generalizing to higher-dimensional matrices, it is no longer possible to store the upper and lower positions of the indices. Henceforth, the indices will be written below, preserving the first index to denote the rows. The above relations are written as $C_{\alpha\gamma} = A_{\alpha\beta} B_{\beta\gamma}$ and $C_\alpha = A_{\alpha\beta} B_\beta$.

Now, let us define matrices of dimension three or more: $A_{\alpha\beta\gamma}$ and $C_{\alpha\beta\gamma\delta}$. They can be represented as three-dimensional and n-dimensional matrices consisting of two-dimensional, three-dimensional, etc., layers. To select the layers, let us consider the procedure for fixing the indices. By definition, the notation C_α represents a record of vector $(C_1, C_2, C_3)^{\mathrm{T}}$ (only two- and three-dimensional cases are important to us). Accordingly, the record $A_{2\beta}$ determines the second side of the matrix, i.e., (A_{21}, A_{22}, A_{23}). Fixing the index in a three-dimensional matrix, for example, the third one, results in a two-dimensional matrix $C_{\alpha\beta 2}$. A set of two-dimensional matrices (layers) $A_{\alpha\beta 1}$, $A_{\alpha\beta 2}$, $A_{\alpha\beta 3}$ characterizes the entire three-dimensional matrix $A_{\alpha\beta\gamma}$, as well as the two-dimensional matrix is determined by one-dimensional row matrices or column matrices, depending on which index is fixed in it—the first or second. Similarly, you can specify three-dimensional layers for a four-dimensional matrix, etc.

With the help of layers, it is possible to define the operation of multiplication of multidimensional matrices. Let us consider first the multiplication of a three-dimensional matrix $A_{\alpha\beta\gamma}$ by column B_γ. Similarly to the two-dimensional case, we divide the matrix into layers, fixing the last index: $A_{\alpha\beta 1}$, $A_{\alpha\beta 2}$, $A_{\alpha\beta 3}$. Now the multiplication operation of is carried out as $A_{\alpha\beta 1} B_1 + A_{\alpha\beta 2} B_2 + A_{\alpha\beta 3} B_3 = C_{\alpha\beta}$ (we recall that B_1 , B_2 , B_3 are now numbers), i.e., weighted summation of the matrix layers $A_{\alpha\beta\gamma}$.

The multiplication of a three-dimensional matrix by a two-dimensional matrix is organized according to a similar rule: $A_{\alpha 11} B_{11} + A_{\alpha 12} B_{12} + A_{\alpha 13} B_{13} + \ldots + A_{\alpha 32} B_{32} + A_{\alpha 33} B_{33}$, which leads to the appearance of a vector C_α, i.e., $A_{\alpha\beta\gamma} B_{\beta\gamma} = C_\alpha$. From what has been said, the record of the multiplication procedure $A_{\alpha\beta\gamma} B_\gamma = C_{\alpha\beta\delta}$ becomes understandable and, in general, the dimension of the product matrix is equal to the sum of the dimensions of the initial matrices minus the doubled number of those indices by which convolution is carried out during multiplication.

In view of the above rules for multiplication of matrices, the functional series (6.83) for nonlinear scattering can be represented in the following form:

$$\vec{E}_\alpha^{(p)}(\vec{r}, t) = \int_{-\infty}^{\infty} S_{\alpha\beta}(\vec{r}, \tau_1) \vec{E}_\beta^{(u)}(t - \tau_1, \vec{r}) \mathrm{d}\tau_1 + \int \int_{-\infty}^{\infty} S_{\alpha\beta\gamma}(\vec{r}, \tau_1, \tau_2) \vec{E}_\beta^{(u)}(t - \tau_1, \vec{r}) \times \vec{E}_\gamma^{(u)}(t - \tau_2, \vec{r}) d\tau_1 \mathrm{d}\tau_2 + \ldots$$

$$+\int \dots \int_{-\infty}^{\infty} S_{\alpha\beta\gamma\dots\lambda}(\vec{r}, \tau_1, \tau_2, \dots, \tau_n) E_p^{(u)}(t - \tau_n) \mathrm{d}\tau_1 \dots \mathrm{d}\tau_n. \tag{6.84}$$

Here it is assumed that all the incident waves $\vec{E}_\beta^{(u)}(t, \vec{r}), \vec{E}_\gamma^u(t, \vec{r}), \dots, \vec{E}_\lambda^u(t, \vec{r})$ affect a nonlinear object from one direction (otherwise the dependence of the incident waves on the wave vector should be reflected). Of course, in relation (6.84), all the incident waves can represent the same wave, but determined for different time delays. To avoid confusion with mute indices, the indices characterizing the incident and scattered waves are enclosed in parentheses.

Expression (6.84) represents a multidimensional Voltaire-Wiener series in the absence of spatial dispersion (integration is carried out only in a time variable); summation is proposed for repeated indices.

Functional series (6.84) represents identity—the results of all integrations are indeed vectors. This representation is given in a form valid only for stationary systems (the scattering properties of the object do not change during observation), which will substantially simplify further reasoning and allow us to introduce spectral procedures. As in the original Voltaire-Wiener series (6.83), the kernels $S_{\alpha\beta}$, $S_{\alpha\beta\gamma}$, $S_{\alpha\beta\dots\lambda}$ shall satisfy the following requirements: the symmetry, i.e., the values of the kernels shall not change when the arguments are interchanged, and the continuity for each of its variables individually. Since the integrals in series (6.84) ensure linearity for each signal $\vec{E}_\alpha^u$, $\vec{E}_\gamma^u$ individually, it can be verified that for each component $E_1^{(p)}(t, \vec{r})$, $E_2^{(p)}(t, \vec{r})$, $E_3^{(p)}(t, \vec{r})$ the conditions of the one-dimensional Voltaire theorem on the possibility of representing the functional by series (6.83) are satisfied. Together, these conditions ensure the representation of a multidimensional nonlinear scatterer by functional series (6.84).

It is necessary to pay attention that multidimensional kernels $S_{\alpha\beta}(\vec{r}, \tau)$, $S_{\alpha\beta\gamma}(\vec{r}, \tau_1, \tau_2)$, etc., now represent the properties of the scatterer, completely independent of the magnitude of the exciting field, or of its temporal structure. If there is no temporal variance, then $S_{\alpha\beta}(\vec{r}, \tau) = S_{\alpha\beta}(\vec{r})\delta(\tau)$, $S_{\alpha\beta\gamma}(\vec{r}, \tau_1, \tau_2) = S_{\alpha\beta\gamma}(\vec{r})\delta(\tau_1)\delta(\tau_2)$, $S_{\alpha\beta\gamma\delta}(\vec{r})\delta(\tau_1)\delta(\tau_2)\delta(\tau_3)$ relation (6.84) becomes an algebraic one, describing a non-inertial nonlinear scatterer:

$$E_\alpha^{(p)}(\vec{r}, t) = S_{\alpha\beta}(\vec{r})\vec{E}_\beta^{(u)}(\vec{r}, t) + S_{\alpha\beta\gamma}(\vec{r})\vec{E}_\beta^{(u)}(\vec{r}, t)E_\gamma^{(u)}(\vec{r}, t) + \dots \tag{6.85}$$

Dimension $S_{\alpha\beta}$—meter, $S_{\alpha\beta\gamma}$—square meter, and in general for n-dimensional kernel $S_{\alpha\beta\dots\lambda}$—meter to the power of $n - 1$. It is precisely these dimensions that will ensure the independence of the introduced characteristics of the scattering power from the distance to the scatterer and, in general, from the polarization of the incident wave, in the sense that these kernels, like the scattering matrix, allow us to determine the scattered wave for any polarization of the incident wave.

The consideration of the stationary case makes us consider the problem of changing the representation of the scattering kernels $S_{\alpha\beta\gamma}(\omega, \omega')$, $S_{\alpha\beta\gamma\delta}(\omega, \omega', \omega'')$, etc., depending on the polarization basis used.

In accordance with the notation used in clause 6.2, the transformation of the polarization basis of the incident wave will be denoted as follows:

$$\begin{aligned}\vec{E}_{\varepsilon}^{(u)}\left(\omega,\vec{k},\hat{\chi}_H\right) &= Q_{\varepsilon\beta}\left(\hat{\chi}_H,\hat{\chi}_C\right)\vec{E}_{\beta}^{(u)}\left(\omega,\vec{k},\hat{\chi}_C\right);\\ \vec{E}_{\nu}^{(u)}\left(\omega,\vec{k},\hat{\chi}_H\right) &= Q_{\nu\gamma}\left(\hat{\chi}_H,\hat{\chi}_C\right)\vec{E}_{\gamma}^{(u)}\left(\omega,\vec{k},\hat{\chi}_C\right) \quad \text{etc}\end{aligned} \tag{6.86}$$

The inverse transformations are written as follows:

$$\vec{E}_{\beta}^{(u)}\left(\omega,\vec{k},\hat{\chi}_C\right) = Q_{\varepsilon\beta}^{-1}\left(\hat{\chi}_H,\hat{\chi}_C\right)\vec{E}_{\varepsilon}^{(u)}\left(\omega,\vec{k},\hat{\chi}_H\right). \tag{6.87}$$

In this case, for products typical for individual members of the multidimensional Voltaire-Wiener series, we can define transformations to the new basis as follows:

$$\left.\begin{aligned}&S_{\alpha\beta}\left(\hat{\chi}_C\right)\vec{E}_{\beta}^{(u)}\left(\hat{\chi}_C\right) = S_{\alpha\beta}\left(\hat{\chi}_C\right)Q_{\varepsilon\beta}^{-1}\left(\hat{\chi}_H,\hat{\chi}_C\right)E_{\varepsilon}^{(u)}\left(\hat{\chi}_H\right)\\ &S_{\alpha\beta\gamma}\left(\hat{\chi}_C\right)\vec{E}_{\beta}^{(u)}\left(\hat{\chi}_C\right)\vec{E}_{\gamma}^{(u)}\left(\hat{\chi}_C\right) = S_{\alpha\beta\gamma}\left(\hat{\chi}_C\right)Q_{\varepsilon\beta}^{-1}\left(\hat{\chi}_H,\hat{\chi}_C\right)Q_{\nu\gamma}^{-1}\left(\hat{\chi}_H,\hat{\chi}_C\right)\\ &\quad\times\vec{E}_{\varepsilon}^{(u)}\left(\hat{\chi}_H\right)\vec{E}_{\nu}\left(\hat{\chi}_H\right)\end{aligned}\right\} \tag{6.88}$$

(all arguments that do not reflect the specificity of the transformation of the polarization basis are omitted here).

If we take into account that the result of the transformation for the scattered field is obtained in the old basis $\hat{\chi}_C$, then product (6.88) shall be supplemented by multiplication by matrix $Q_{\mu\alpha}\left(\hat{\chi}_H,\hat{\chi}_C\right)$ (again we draw attention to the fact that a mute index can be denoted by any letter, so long as they are not confused during multiplication). All this leads to the following record for scattering kernels in a new polarization basis:

$$\begin{aligned}S_{\mu\varepsilon E}\left(\hat{\chi}_H\right) &= S_{\alpha\beta}\left(\hat{\chi}_C\right)Q_{\mu\alpha}\left(\hat{\chi}_H,\hat{\chi}_C\right)Q_{\varepsilon\beta}^{-1}\left(\hat{\chi}_H,\hat{\chi}_C\right);\\ S_{\nu\varepsilon\mu}\left(\hat{\chi}_H\right) &= S_{\alpha\beta\gamma}\left(\hat{\chi}_C\right)Q_{\ \alpha}\left(\hat{\chi}_H,\hat{\chi}_C\right)Q_{\varepsilon\beta}^{-1}\left(\hat{\chi}_H,\hat{\chi}_C\right)Q_{\mu\nu}^{-1}\left(\hat{\chi}_H,\hat{\chi}_C\right) \quad \text{etc}\end{aligned} \tag{6.89}$$

Here the rule of indices is also observed during multiplication.

6.3.2 Scattering Matrix of Nonlinear Reflectors

Let us consider, in a somewhat different way, the transformation of the polarization basis for non-inertial scatterers, when the nonlinearity is described by the dependence:

$$E_x^p = \sum_{m=1}^{N} A_m\left(E_x^u\right)^m. \tag{6.90}$$

It is required to determine the response of such scatterers in the case when the incident wave is elliptically polarized, which corresponds to the most general situation $\vec{E} = \left(E_x, E_y\right)^{\mathrm{T}}$. We confine ourselves to the case when all A_m, except A_2, are equal to zero.

For the scatterers under investigation, first of all, we need to find the appropriate form of transformation of the incident wave components. The desired transformation shall be invariant to different representations of the complex vector describing the incident wave, i.e., its form (but not the coefficients) shall remain unchanged.

It is not difficult to verify that the general form of such a transformation should be written as follows:

$$\left.\begin{array}{l} E_x^p = a_{11}\left|E_x^u\right|^2 + a_{12}\left|E_x^u E_y^{u*}\right| + a_{21}\left|E_x^{u*} E_y^u\right| + a_{22}\left|E_y^u\right|^2; \\ E_y^p = b_{11}\left|E_x^u\right|^2 + b_{12}\left|E_x^u E_y^{u*}\right| + b_{21}\left|E_x^{u*} E_y^u\right| + b_{22}\left|E_y^u\right|^2 \end{array}\right\} \tag{6.91}$$

or in the matrix form:

$$\begin{pmatrix} E_x^p \\ E_y^p \end{pmatrix} = \begin{pmatrix} a_{11}\; a_{12}\; a_{21}\; a_{22} \\ b_{11}\; b_{12}\; b_{21}\; b_{22} \end{pmatrix} \left(\left|E_x^u\right|^2, E_x^u E_y^{u*}, E_x^{u*} E_y^u, \left|E_y^u\right|^2\right)^{\mathrm{T}}. \tag{6.92}$$

Let us write the Eq. (6.92) in the complex form:

$$\vec{E}^p = A_2 \vec{E}^u_{(2)}. \tag{6.93}$$

The meaning of the notations introduced is obvious. Thus, in order to describe the scattering properties of a nonlinear "purely square" scatterer, in the general case, eight complex numbers are required, i.e., 16 valid ones. These numbers form a rectangular matrix of 4×2. The decomposition of the complex vector describing the incident wave into orthogonal components can be carried out in countless ways. Naturally, in this case E_x and E_y means not Cartesian projections, but some other projections corresponding to the chosen polarization basis. In different bases, the elements of the matrix A_2 will be also different. Let us consider how they will change when the polarization basis is varied.

As follows from (2.30), the old components of the electric vector $\vec{E}\left(\hat{\chi}_C\right)$ will be associated with new ones $\vec{E}\left(\hat{\chi}_H\right)$ using a linear transformation:

$$\vec{E}\left(\hat{\chi}_C\right) = Q\left(\hat{\chi}_C, \hat{\chi}_H\right)\vec{E}_H, \tag{6.94}$$

where $Q\left(\hat{\chi}_C, \hat{\chi}_H\right)$—unitary matrix of the form (1.43).

Direct calculations give the following relation:

$$\vec{E}^u_{(2)}\left(\hat{\chi}_C\right) = Q_{(2)}\left(\hat{\chi}_C, \hat{\chi}_H\right)\vec{E}^u_{(2)}\left(\hat{\chi}_H\right), \tag{6.95}$$

where

$$Q_{(2)}(\hat{\chi}_C, \hat{\chi}_H) = \mathrm{e}^{2\mathrm{i}\varphi}$$
$$\begin{pmatrix} \mathrm{e}^{2\mathrm{i}\eta}\cos^2\gamma & 0.5\mathrm{e}^{2\mathrm{i}(\eta+\delta)}\sin 2\gamma & 0.5\mathrm{e}^{2\mathrm{i}(\eta+\delta)}\sin 2\gamma & \mathrm{e}^{2\mathrm{i}(\eta+2\delta)}\sin^2\gamma \\ -0.5\mathrm{e}^{-2\mathrm{i}\delta}\sin 2\gamma & \cos^2\gamma & -\sin^2\gamma & 0.5\mathrm{e}^{2\mathrm{i}\delta}\sin 2\gamma \\ -0.5\mathrm{e}^{-2\mathrm{i}\delta}\sin 2\gamma & -\sin^2\gamma & \cos^2\gamma & 0.5\mathrm{e}^{2\mathrm{i}\delta}\sin 2\gamma \\ \mathrm{e}^{-2\mathrm{i}(\eta+2\delta)}\sin^2\gamma & -0.5\mathrm{e}^{-2\mathrm{i}(\eta+\delta)}\sin 2\gamma & -0.5\mathrm{e}^{-2\mathrm{i}(\eta+\delta)}\sin 2\gamma & \mathrm{e}^{-2\mathrm{i}\eta}\cos^2\gamma \end{pmatrix}. \tag{6.96}$$

A direct check shows that the matrix $Q_{(2)}(\hat{\chi}_C, \hat{\chi}_H)$ is also unitary, since $Q_{(2)}(\hat{\chi}_C, \hat{\chi}_H)Q^{+}_{(2)}(\hat{\chi}_C, \hat{\chi}_H) = I$.

Thus, when the polarization basis is changed, the column matrix of the quadratic elements shall be multiplied by unitary matrix $Q_{(2)}(\hat{\chi}_C, \hat{\chi}_H)$, which coincides with matrix $L(\hat{\chi}_C, \hat{\chi}_H)$ ((7.118) of clause 7.1), which transforms the covariance matrix during linear scattering.

In order to find out how matrix A_2 changes when the polarization basis changes, we shall substitute relations (6.94) and (6.95) into (6.93), and take the complex-conjugate quantity as its right-hand side:

$$\vec{E}_p(\hat{\chi}_H) = Q^{\mathrm{T}}(\hat{\chi}_H, \hat{\chi}_C)A_2(\hat{\chi}_C)Q_{(2)}(\hat{\chi}_C, \hat{\chi}_H)\vec{E}^{(u)}(\hat{\chi}_H). \tag{6.97}$$

Thus, when the polarization basis is changed, matrix A_2 is subjected to the following transformation:

$$A_2(\hat{\chi}_H) = Q^{\mathrm{T}}(\hat{\chi}_H, \hat{\chi}_C)A_2(\hat{\chi}_C)Q_{(2)}(\hat{\chi}_C, \hat{\chi}_H), \tag{6.98}$$

wherein $Q(\hat{\chi}_H, \hat{\chi}_C)$ and $Q_{(2)}(\hat{\chi}_C, \hat{\chi}_H)$ are unitary matrices.

In order to estimate the power of the scattered backward wave, it is necessary to perform operation $\vec{E}^p + \vec{E}^{p*}$:

$$\begin{aligned} \Pi = \vec{E}^p + \vec{E}^{p*} &= \left[Q^{\mathrm{T}}(\hat{\chi}_H, \hat{\chi}_C)A_2(\hat{\chi}_C)Q_{(2)}(\hat{\chi}_C, \hat{\chi}_H)E^n_{(2)}(\hat{\chi}_H)\right]^{\mathrm{T}} \\ &\quad \times \left[Q^*(\hat{\chi}_H, \hat{\chi}_C)A^*_2(\hat{\chi}_C)Q^*_2(\hat{\chi}_C, \hat{\chi}_H)E^n_2(\hat{\chi}_H)\right] \\ &= E^n_{(2)}(\hat{\chi}_H)Q^{\mathrm{T}}_{(2)}(\hat{\chi}_C, \hat{\chi}_H)A^{\mathrm{T}}_2(\hat{\chi}_C)A^*_2(\hat{\chi}_C)Q^*_2(\hat{\chi}_C, \hat{\chi}_H)E^{n*}_{(2)}(\hat{\chi}_H). \end{aligned} \tag{6.99}$$

Let us introduce the notations:

$$G_{(2)}(\hat{\chi}_C) = A^{\mathrm{T}}_2(\hat{\chi}_C)A^*_2(\hat{\chi}_C). \tag{6.100}$$

Matrix $G_{(2)}(\hat{\chi}_C)$ is a square Hermitian matrix with the size of 4×4.

As can be seen from (6.100), matrix $G_{(2)}(\hat{\chi}_C)$, when the polarization basis is changed, undergoes a similarity transformation:

$$G_{(2)}(\hat{\chi}_C) = Q^{\mathrm{T}}_2(\hat{\chi}_C, \hat{\chi}_H)G_{(2)}(\hat{\chi}_C)Q^*_{(2)}(\hat{\chi}_C, \hat{\chi}_H). \tag{6.101}$$

Equation (6.92) allows us to give recommendations on the experimental determination of the parameters of the scatterer—elements of matrix A_2. For this, it is necessary to perform measurements on four polarizations of the irradiating wave, for example, horizontal—$E_1^n = \begin{pmatrix} 1 & 0 \end{pmatrix}^T$, vertical—$E_2^n = \begin{pmatrix} 0 & 1 \end{pmatrix}^T$, linear angular—$\pi/4$ $E_3^n = 0.5\begin{pmatrix} 1 & 1 \end{pmatrix}^T$, and circular—$E_4^n = 0.5\sqrt{2}\begin{pmatrix} 1 & i \end{pmatrix}^T$. Knowing the amplitude and phase of the scattered wave upon irradiation by the waves of these polarizations makes it possible to uniquely determine the elements of matrix A_2—a_{mn}, b_{mn}.

In the case when in Eq. (6.90) all $m \neq 0$, we will have the following instead of the relation (6.18):

$$\vec{E}^p = \sum_{m=1}^{N} A_m \left(\vec{E}^n\right)^m \tag{6.102}$$

.

It can be shown here that with a change in the polarization basis, matrices A_m will be transformed as follows:

$$A_m(\hat{\chi}_H) = \sum_{m=1}^{N} Q^T_{(m)}(\hat{\chi}_C, \hat{\chi}_H) A_m(\hat{\chi}_C) Q_{(m)}(\hat{\chi}_C, \hat{\chi}_H), \tag{6.103}$$

where $Q_{(m)}(\hat{\chi}_C, \hat{\chi}_H)$—unitary matrix, which is formed similar to $Q_{(2)}(\hat{\chi}_C, \hat{\chi}_H)$.

The presence of nonlinearity in the reflective characteristics of radar targets fundamentally changes the basic radar equation. However, before proceeding to the derivation and analysis of this equation, we shall consider in some detail a number of aspects related to the "purely" radar characteristic for nonlinear reflectors, which is the effective scattering area of reflectors with nonlinear properties.

6.3.3 The Effective Scattering Area of Nonlinear Reflectors

When determining the ESA with respect to reflectors with "nonlinear" properties (hereinafter referred to as "nonlinear targets"), several different approaches are possible. Let us consider each of them separately.

Let Π_u—Umov–Poynting vector of the electromagnetic wave irradiating nonlinear target. If the frequency of the sounding signal is f, then the electromagnetic wave reflected from the target will contain, in addition to the fundamental frequency f, a wide spectrum of harmonics of the sounding signal at frequencies $2f$, $3f$, $4f$, … Consequently, the total power $P_{\Sigma p}$ of the electromagnetic wave scattered by such a target in the direction of the receiving antenna will include both the power dissipated at the frequency of the sounding signal and the powers scattered by the harmonics of the sounding signal:

$$P_{\Sigma p}(C) = \sum_{n=1}^{\infty} P_{n\,p}(C), \tag{6.104}$$

where $P_{n\,p}(C)$—power of an electromagnetic wave scattered by the nonlinear target on the n-th harmonic in the direction of the receiving antenna.

In parentheses, the letter "C" indicates that the named values are measured in the immediate vicinity of the target. Since this kind of explanation is required in the future throughout the text of this section, we shall indicate in brackets the place to which this or that parameter belongs. If we are talking about the parameters ξ, measured in the immediate vicinity of the reflector (nonlinear target), we will write $\xi(C)$ then for such quantities; if these quantities are measured in the immediate vicinity of the receiving antenna, we will write $\xi(A)$.

The power scattered by the reflector is proportional to the Umov–Poynting vector $\Pi_u(C)$ of the electromagnetic field incident on it:

$$P_{\Sigma p}(C) = \sigma_{\Sigma}^{\prime 2} \Pi_u(C), \tag{6.105}$$

where proportionality coefficient $\sigma_{\Sigma}^{\prime 2}$ is called the total ESA of the reflector (in the considered case of the total ESA of the nonlinear target).

Let us represent the total ESA as the sum of the ESAs—$\sigma_n^{\prime 2}$ on each of the harmonics:

$$\sigma_{\Sigma}^{\prime 2} = \sum_{n=1}^{\infty} \sigma_n^{\prime 2}. \tag{6.106}$$

After substituting equalities (6.105) and (6.106) into formula (6.104), we will obtain:

$$P_{\Sigma p}(C) = \Pi_u(C) \sum_{n=1}^{\infty} \sigma_n^{\prime 2}. \tag{6.107}$$

If the reception of a reflected electromagnetic wave from a nonlinear target is carried out only on the n-th harmonic of the sounding signal, the interest is the only the n-th component of the total ESA target, which in this case according to Eqs. (6.105)–(6.107) will be determined as follows:

$$\sigma_n^{\prime 2} = \frac{P_{n\,p}(C)}{\Pi_u(C)} = 4\pi R^2 \frac{P_{n\,p}(C)}{P_u(A)} = 4\pi R^2 \left[\frac{E_{n\,p}(C)}{E_u(A)}\right]^2, \tag{6.108}$$

where $P_u(A)$—power of the electromagnetic wave emitted by the transmitter at frequency f in the direction of the nonlinear target, determined in the immediate vicinity of the transmitting antenna, $E_u(A)$—electric vector of this wave, $E_{n\,p}(C)$—

electric vector of the electromagnetic wave scattered by the reflector at frequency nf, R—distance between the antenna and the reflector.

Formula (6.108) is inherently one of the definitions of the ESA of a nonlinear target. The dimension of the introduced ESA coincides with the dimension of the ESA in classical radar detection and represents m^2.

Let us consider a somewhat different approach to the determination of the ESA of a nonlinear target.

In classical radar detection, the value of the electric vector of an irradiated electromagnetic wave $E_u(A)$, determined directly near the receiving antenna, and the electric vector of the electromagnetic wave scattered by the reflector—$E_p(C)$—are related to each other by the known relation:

$$E_p(C) = \frac{\sigma}{\sqrt{4\pi}R} E_u(A), \tag{6.109}$$

where σ^2—ESA of the reflector.

Assume the nonlinear properties of the reflector are described by power dependence of the n-th degree. In this case, by analogy with expression (6.109), we can write:

$$E_{np}(C) = \frac{\sigma_n}{\sqrt{4\pi}R} E_u^n(A). \tag{6.110}$$

The coefficient of proportionality in formula (6.110) σ_n^2 can also be called the ESA of a nonlinear target on the n-th harmonic. In this case, the ESA can be represented as follows:

$$\sigma_n^2 = 4\pi R^2 \left[\frac{E_{np}(C)}{E_u^n(A)}\right]^2. \tag{6.111}$$

It can be seen that the ESA of the nonlinear target thus introduced differs from the ESA determined by formula (6.108). First of all, we must talk about the differences in the dimensions of the quantities considered. If $\sigma_n'^2$ found by formula (6.108), as already mentioned, had the dimension m^2 on all harmonics, then σ_n^2 calculated through relation (6.111) will have the dimension $m^{2n}Vt^{2(1-n)}$. This, naturally, creates certain difficulties, first, in the interpretation of the ESA, and secondly, in comparing the ESA on different harmonics (quantity $\sigma_n'^2$ is apparently free of this lack). Note that for $n = 1$ (classical radar detection), both ESA coincide.

Let us find the relationship between quantities $\sigma_n'^2$ and σ_n^2, for which we transform formula (6.111):

$$\sigma_n^2 = 4\pi R^2 \left[\frac{E_{np}(C)}{E_u^n(A)}\right]^2 = 4\pi R^2 \left[\frac{E_{np}(C)}{E_u(A)}\right]^2 E_u^{2(1-n)}(C) = \sigma_n'^2 E_u^{2(1-n)}(C). \tag{6.112}$$

Formula (6.112) can be expressed in terms of more natural characteristics such as radiation power $P_u(A)$ and antenna gain at frequency $f - G_1$. Assuming that the wave incident on the target is plane, we can write the equality for the Umov–Poynting vector:

$$\Pi_u(C) = \frac{P_u(A)G_1}{4\pi R^2} = \frac{E_u^2(C)}{120\pi}. \tag{6.113}$$

Equalities (6.112) and (1.113) lead to the following relation connecting the nonlinear target defined by different ESA methods:

$$\sigma_n^2 = \sigma_n'^2 \left(\frac{30 P_u(A) G_1}{R^2} \right)^{1-n}. \tag{6.114}$$

One more definition of the ESA of a nonlinear target similar to definition (6.111) is possible:

$$\sigma_n''^2 = \frac{P_{n\,p}(C)}{\Pi_u^n(C)}, \tag{6.115}$$

ESA $\sigma_n''^2$ introduced using formula (6.115) has the same dimension as σ_n^2, i.e., $m^{2n} W t^{(1-n)}$.

Let us find the relationship between $\sigma_n''^2$ and $\sigma_n'^2$, as well as between $\sigma_n''^2$ and σ_n^2. Let us transform formula (6.115) taking into account relations (6.108), (6.112), and (6.113):

$$\begin{aligned} \sigma_n''^2 &= \frac{P_{n\,p}(C)}{\Pi_u^n(C)} = \frac{P_{n\,p}(C)}{\Pi_u(C)} \cdot \frac{1}{\Pi_u^{n-1}(C)} = \frac{\sigma_n'^2}{\Pi_u^{n-1}(C)} = \frac{(120\pi)^{n-1}\sigma_n'^2}{E_u^{2(n-1)}(C)} \\ &= (120\pi)^{n-1} \frac{E_u^{2(n-1)}(C)}{E_u^{2(n-1)}(C)} \sigma_n^2 = (120\pi)^{n-1} \sigma_n^2. \end{aligned} \tag{6.116}$$

Relationship between $\sigma_n''^2$ and $\sigma_n'^2$ can also be represented in the following form, where relations (6.113) and (6.116) are used:

$$\sigma_n'^2 = \left(\frac{P_\Sigma G_1}{4\pi R^2} \right)^{n-1} \sigma_n''^2. \tag{6.117}$$

Formulas (6.116) and (6.117) make it possible to carry out the required recalculation. For $n = 1$, all three ESA definitions obviously coincide.

Theoretical calculation of the total ESA or even ESA on the harmonic of a real nonlinear target is a very difficult task. As the simplest model of such a target, it is most convenient to use a half-wave vibrator, in the middle of which a contact with nonlinear properties is included.

The theoretical conclusion and experimental verification of the "nonlinear" ESA of a half-wave vibrator for metal-oxide-metal contacts is given in the literature. The ESA of a half-wave vibrator with a cubic volt-ampere characteristic when receiving a reflected signal at the third harmonic is represented in [1].

For "very strong" signals, the ESA expression takes the following form:

$$\sigma_3'^2 \cong \frac{1}{100}\frac{G_3}{R_A}\sqrt[3]{\frac{\lambda^2 W_0}{\beta \Pi}\frac{R_0}{R_A}}, \tag{6.118}$$

where R_A—resistance of the vibrator at the point where the contact is switched on, R_0—initial (at zero voltage) contact resistance, $W_0 = 120$ O—wave impedance of free space, β—coefficient characterizing the nonlinearity of the current–voltage characteristic ($1/V^2$).

Of greatest interest is the consideration of the ESA of a half-wave vibrator having "nonlinear" properties when weak signals act on it. The formula for the "nonlinear" effective scattering area of the vibrator for weak signals has the form:

$$\sigma_3'^2 = \frac{\beta^2 G_3 W_0 \lambda^6 R_A \Pi^2}{4\pi^6 R_0^2\left(1+\frac{R_A}{R_0}\right)^8\left[1+\left(\frac{f}{f_A}\right)^2\right]^3\left[1+9\left(\frac{f}{f_A}\right)^2\right]}, \tag{6.119}$$

where f—frequency of the electromagnetic field acting on the vibrator; f_A—boundary frequency of the actual contact, taking into account the radiation resistance.

The analysis of expressions (6.112) and (6.113) shows that for "very strong" signals, the "nonlinear" ESA of the vibrator decreases with increasing irradiation power flux (for some parameters of the contact, its breakdown can occur before the condition of "very strong" signals is reached). For weak signals, there is a quadratic dependence of the signal power reflected on the third harmonic of the sounding signal on the irradiation power flux. When vibrations of two frequencies act on the vibrator, the ESA formula has a similar form and differs only in a numerical coefficient of the unity order.

As calculations show, the ESA value of a half-wave vibrator on the third harmonic is two orders of magnitude smaller than the ESA of the vibrator at the frequency of the sounding signal.

Formulas (6.118) and (6.119) contain the initial resistance and the boundary frequency of the contact as parameters R_0 and f_A, respectively. R_0 and f_A depend on the thickness of the oxide layer in the metal-oxide-metal contact and on the contact area. With an increase in the thickness of the oxide layer, the initial resistance of the contact increases, and the boundary frequency decreases; hence, the ESA of the target having such a contact decreases for weak signals and increases for strong signals. As the contact area increases, R_o decreases and f_A increases; hence, the ESA of the nonlinear target increases for the case of weak signals and decreases for strong signals. This dependence of the nonlinear target ESA on the physical properties of the contact leads to the conclusion on the great difficulties related to the accurate measurement of the ESA of the nonlinear real targets on harmonics, since any movement of the contacting surfaces leads to a change in the ESA for such a target on the n-th harmonic. Moreover, two identical targets having equal ESA, when detected at the sounding frequency, can have different ESAs when detected on the harmonics of the sounding signal.

6.3.4 Basic Equality of Nonlinear Radiolocation

The basic radar equation (range equation) applied to free space (i.e., without taking into account the effect of ground and absorption in the atmosphere) for radar reflectors establishes a relationship between the radar station characteristics, the reflector parameters, and the range of this radar station.

Let us derive the basic radar equation (range equation) when a point nonlinear target is detected. As an ESA of a nonlinear target, we will use σ_n^2 defined by formula (6.111).

The density of the power flux of the electromagnetic wave incident on the target $\Pi_u(C)$, taking into account the directional properties of the transmitting antenna, is described by expression (6.113).

If the reflector has nonlinear properties, the power of the signal reflected from it on the n-th harmonic with the account of formulas (6.108) and (6.112) may be represented as follows:

$$P_{np} = \frac{P_u(A)G_1}{4\pi R^2}\sigma_n'^2 = \frac{P_u(A)G_1}{4\pi R^2}\sigma_n^2 E_p^{2(n-1)}(C). \tag{6.120}$$

For monostatic radar ranging, the power flux density of a wave reflected from a nonlinear target at the antenna can be written as:

$$\Pi_n(A) = \frac{P_{np}(C)}{4\pi R^2}. \tag{6.121}$$

After substituting the expression for P_{np} from (6.120) into (6.121), we obtain an expression for the power flux density of the electromagnetic wave reflected from the nonlinear target at the receiving antenna:

$$\Pi_n(A) = \frac{P_u(A)G_1}{\left(4\pi R^2\right)^2} E_u^{2(n-1)}(C)\sigma_n^2. \tag{6.122}$$

Output power of the receiving antenna $P_n(A)$ and power flux density at the receiving antenna $\Pi_n(A)$ are related by the following relationship:

$$P_n(A) = S_n(A)\Pi_n(A), \tag{6.123}$$

while between the gain of the receiving antenna on the n-th harmonic of the signal—G_n and the effective area of the antenna on the same harmonic—$S_n(A)$, there is a well-known relationship:

$$S_n(A) = \frac{\lambda_n^2 G_n}{4\pi}, \tag{6.124}$$

where λ_n—wavelength on the n-th harmonic of the sounding signal, wherein it is natural that $\lambda_n = \lambda/n$.

Using Eqs. (6.123) and (6.124), we obtain an expression for the output signal power of the receiving antenna:

$$P_{n\,p}(A) = \frac{\lambda^2 G_n}{4\pi} \Pi_n(A). \tag{6.125}$$

After substituting the value of power flux density $\Pi_n(A)$ from Eq. (6.122) into formula (6.123), we obtain:

$$P_{n\,p}(A) = \frac{P_u(A) G_1 G_n \lambda_n^2}{(4\pi)^3 R^4} E_p^{2(n-1)}(C) \sigma_n^2. \tag{6.126}$$

The strength of the electric field of an electromagnetic wave incident on a nonlinear target E_u is related to the power of the radiated signal, as can be seen from formula (6.113), by the following relation:

$$E_u(C) = \frac{\sqrt{30 P_u(A) G_1}}{R}. \tag{6.127}$$

After substituting the expression for $E_u(C)$ from relation (6.127) into Eq. (6.126), the formula for the output power of the receiving antenna will take the following form:

$$P_{n\,p}(A) = \frac{P_u(A) G_1 G_n \lambda_n^2}{(4\pi)^3 R^4} \cdot \frac{[30 P_u(A) G_1]^{n-1}}{R^{2(n-1)}} \sigma_n^2. \tag{6.128}$$

As is known, detection of radar targets (irrespective of whether reflectors have nonlinear properties or not) is based on the principles of storage of energy of a useful signal. In this connection, let us move in Eq. (6.128) from power to energy characteristics.

Since relation (6.128) remains valid at any time t, the powers that appear in it can also be considered as a function of time, i.e., $P_u(A) \equiv P_u(A; t)$ and $P_{n\,p}(A) \equiv P_{n\,p}(A; t)$. In the future, since we are talking about the characteristics measured in the immediate vicinity of the antennas, the letter A in the argument will be omitted, and instead of (6.128) we will have:

$$P_{n\,p}(t) = \frac{P_u(t) G_1 G_n \lambda_n^2}{(4\pi)^3 R^4} \cdot \frac{[30 P_u(t) G_1]^{n-1}}{R^{2(n-1)}} \sigma_n^2 = k P_u^n(t). \tag{6.129}$$

The meaning of the parameter k introduced into formula (6.129) is obvious.

When determining the maximum range of the radar station, it is necessary to establish a relationship between the parameters characterizing the effectiveness of target detection (in particular, the probability of correct detection of target D and false alarm F) and the value of the received energy of the signal reflected from a nonlinear target.

The energy of the received signal $Э_{np}$ is connected with the output power of the receiving antenna by the following relation:

$$Э_{np} = \int_0^{t_s} P_{np}(t)dt, \tag{6.130}$$

where t_s—time of observation of the reflected signal.

By substituting expression (6.130) into formula (6.129), we express the energy of the received signal through the mean value of quantity $P_u^n(t)$, which we denote by $\left(P_u^n\right)_m$:

$$Э_{np} = k\int_0^{t_s} P_u^n(t)dt = k\left(P_u^n\right)_m t_s. \tag{6.131}$$

In view of this, we finally get:

$$Э_{np} = \frac{30^{n-1}\left(P_u^n\right)_m G_1^n G_n \lambda^2 \sigma_n^2 t_s}{(4\pi)^3 R^{2(n+1)}}. \tag{6.132}$$

The probability of correct detection of a radar target depends on the ratio of the energy of the received signal (in this case on the n-th harmonic of the sounding signal) to noise power density $N_0/2$. Let us find this ratio:

$$a_n = \frac{2Э_{np}}{N_0} = \frac{2}{N_0}\cdot\frac{30^{n-1}\left(P_u^n\right)_m G_1^n G_n \lambda^2 \sigma_n^2 t_s}{(4\pi)^3 R^{2(n+1)}}. \tag{6.133}$$

Expression (6.133) is valid for any range "radar station–nonlinear target" (for reflectors without nonlinear properties, $n = 1$) for radar observation in free space. However, the most important and often used radar parameter is the maximum range $R_{n\,\max}$, on which the target will be detected. This maximum range corresponds to the lowest value of the received energy of the wave reflected from the target $Э_{n\min}$, at which the presence of the target is fixed in the end device with a given probability.

This enables us, using formula (6.132), to write:

$$Э_{n\min} = \frac{\left(P_u^n\right)_m t_s G_1 G_n \lambda_n^2}{(4\pi)^3} \cdot \frac{(30G_1)^{n-1}}{R_{n\max}^{2(n+1)}} \sigma_n^2. \tag{6.134}$$

From expression (6.134), we find the maximum range of the radar station $R_{n\max}$ for the case when the point radar reflector has nonlinear properties of the n-th power, and the reception of the reflected signal occurs on the n-th harmonic of the sounding signal:

$$R_{n\max} = \sqrt[2(n+1)]{\frac{30^{n-1}\left(P_u^n\right)_m G_1^n G_n \lambda_n^2 \sigma_n^2 t_s}{(4\pi)^3 Э_{n\min}}}. \tag{6.135}$$

Relation (6.135) *is the equation of the radar detection range for nonlinear targets with a power-law character of nonlinearity.*

Unlike classical radar detection, where the maximum range and detection characteristics do not depend on the type of sounding signal, there is such dependence with nonlinear radar detection. Let us illustrate what has been said.

In the case of radiation with continuous constant-power oscillation P_u there will be equality $\left(P_u^n\right)_m = P_u^n$, and therefore, Eq. (6.135) takes the following form:

$$\left(R_{n\max}\right)_{непр} = \sqrt[2(n+1)]{\frac{P_u t_s G_1 G_n \lambda_n^2 \sigma_n^2}{(4\pi)^3 Э_{n\min}} (30P_u G_1)^{n-1}}. \tag{6.136}$$

Let us consider the case when the target is irradiated by a sequence of N pulses, each of which has power $P_{uм}$ and duration τ, wherein T is the repetition period. For such a situation, formula (6.135) takes the following form:

$$\left(R_{n\max}\right)_{uмn} = \sqrt[2(n+1)]{\frac{30^{n-1} P_{uм}^n \tau N G_1^n G_n \lambda_n^2 \sigma_n^2}{(4\pi)^3 Э_{n\min}}} \tag{6.137}$$

Let us use the elementary transformation:

$$P_{uм}^n \frac{\tau}{T} = P_0 P_{uм}^{n-1}, \tag{6.138}$$

where P_0—average power over a period T.

Taking into account equality (6.138), formula (6.137) takes the following form:

$$\left(R_{n\max}\right)_{\text{имп}} = 2(n+1)\sqrt{\frac{P_0 G_1 G_n \lambda_n^2 \sigma_n^2 NT}{(4\pi)^3 Э_{n\min}}(30P_{\text{им}}G_1)^{n-1}}. \tag{6.139}$$

In order to make a correct comparison of the continuous and pulsed regimes, it is necessary to require the equality of the radiated energy for the same time interval, which is equivalent to the equality of the average signal powers over the same time interval, i.e., it is necessary to consider that. Substitution of this equality into formula (6.139) leads to the following expression:

$$\left(R_{n\max}\right)_{\text{имп}} = \left(R_{n\max}\right)_{\text{непр}} \cdot q^{\frac{n-1}{2(n+1)}}, \tag{6.140}$$

where $q = T/\tau$—duty cycle, i.e., ratio of the pulse repetition period to pulse duration.

As can be seen from the obtained formula, the range of detection of a nonlinear target in a pulsed regime can turn out to be substantially greater than for a continuous one. So for $n = 2$ and duty cycle $q = 1000$, this growth is 3.16 times, and for $n = 3$ and the same duty cycle—5.6 times.

Let us illustrate the expressions obtained using the example of a half-wave vibrator. To this end, we use expressions (6.119), (6.112), and (6.136). After simple transformations for the case of weak signals on the third harmonic, we get:

$$R\text{max} = \sqrt[8]{\frac{(P_u G_1)}{(4\pi)^5} \cdot \frac{G_3 \lambda_3^2 \gamma}{P\text{min}}}, \tag{6.141}$$

where $\gamma = \frac{\alpha^2 \beta^3 R_B G_3 \lambda_3^6}{4\pi^6 R_0^6\left(1+\frac{R_B}{R_0}\right)^8\left[1+\left(\frac{\omega}{\omega_0}\right)^2\right]\left[1+9\left(\frac{\omega}{\omega_0}\right)\right]^2}$.

In case of strong signals:

$$R\text{max} = \left[\frac{(P_u G_1)^{1/3}}{(4\pi)^{7/3}} \cdot \frac{G_3 \lambda_3^2 \delta}{P\text{min}}\right]^{\frac{3}{14}}, \tag{6.142}$$

where $\delta = \frac{1}{100}\frac{G_3}{R_A} \cdot \sqrt[3]{\lambda^2\left(\frac{\beta R_A}{R_0}\right)^{-2}}$.

After substituting γ in the formula with the values of parameters taken from [2]: $\alpha = 1/V^2$; $W_0 = 120\,Om$; $G_3 = 15$; $R_0 = R_A = 73\,Om$; $\Delta\omega/\omega_0 \approx 0$ we obtain: For a VHF band— $R \approx \sqrt[8]{P_0 P_{\text{им}}^2 N \sigma_3^2}, km$; for an S band— $R \approx 8{,}8\sqrt[8]{P_0 P_{\text{им}}^2 N \sigma_3^2}, km$.

When deriving the given formulas, it was assumed that in the VHF band $G_1 = 3$; $G_3 = 15$, and in S band $G_1 = G_3 = 5000$, the receiver noise factor is 0.2; $N_0 = 60 \cdot 1.38 \cdot 10^{-23}$ W/Hz, N—number of pulses.

Calculation by formula (6.119) gave $\sigma_3^2 = 0,028m^6/V^{-4}$.

When choosing $P_0 = 20$ W, $P_{im} = 10$ kW in the VHF band, we get $R = 2\sqrt[8]{N}$ km When choosing $P_0 = 50$ W, $P_{uM} = 60\,kW$ in S band, we get $R = 50\sqrt[8]{N}$ km.

6.4 Fluctuating Objects

The scatterers considered in the previous sections belonged to a class of so-called stable scatterers, and elements of their scattering matrix remained unchanged during the observation process. Although this situation is mainly characteristic of controlled laboratory studies, for example, in the physical modeling of the scattering channel, the results obtained are very useful, since, on the one hand, they allow us to understand the specific nature of the scattering process and possible ways of representing it, and, on the other hand, they can serve as a basis for the analysis of the dynamic case, when the elements of the scattering matrix vary in time, including randomly. Let us clarify the last statement.

Such sources of change in reflecting properties like motion, swing, rotations of the scatterer in the process of observation are quite obvious. Most of the real scatterers can be considered as objects consisting of a set of scattering elements and creating interfering signals in the direction of observation. This "granularity" of the scatterer leads to fluctuations in amplitude, phase, frequency, and polarization of the scattered signal, and, as applied to radar observation, manifests itself in the existence of specific sources of errors, such as angular, range, and velocity noise, when measuring target coordinates. It is for this reason that objects whose scattering capacity varies in time are classified as fluctuating ones.

The most complete characteristic of the scattering properties of fluctuating objects is the scattering matrix, in which the elements are time functions. Generally speaking, in this case formula $\vec{E}_p = S(t)\vec{E}_u$ becomes problematic. The point is that for the stable target in relation (6.4), it was assumed that $\vec{E}_u$ and $\vec{E}_p$ are complex vectors corresponding to time dependence $e^{i\omega t}$. It can be seen that this assumption is not fair at least for $\vec{E}_p$, which makes us take a closer look at the possibility of using complex vectors in analyzing the apparatus.

At the same time, the concept of an analytic polarization signal makes it possible to circumvent this difficulty. For this purpose, the time-domain signals are compared with the help of the Hilbert transform to the components of the incident and scattered fields. These components form, respectively, the vectors of analytical signals $\vec{E}_u(t)$ and $\vec{E}_p(t)$. This makes it possible to use the following property of the Hilbert transform: $\Gamma(M(t)\sigma(t)) = M(t)\Gamma(\sigma(t))$. A necessary condition for observing the last relation is the disjointness of the spectra of the slow multiplier $M(t)$ and the rapid multiplier $\sigma(t)$.

Thus, if the fluctuations of the elements of the scattering matrix $S(t)$ are low frequency and their spectrum does not intersect the spectrum of vector $\vec{E}_u(t)e^{i\omega_0 t}$ (ω_0—carrier frequency), then the following is true:

$$\vec{E}_p(t) = S(t)\vec{E}_u(t), \tag{6.143}$$

Thus, the following step is taken in the formation of the hierarchy of scattering characteristics: scattering amplitude (clause 6.1) → static scattering characteristics (scattering diagrams) (clause 6.2) → dynamic scattering characteristics, just considered and described by matrix $S(t)$. Dynamic characteristics in the foreign literature are called object signatures and represent the dependence of statistical characteristics $S(t)$, $\sigma^2(t)$ on time.

As applied to radar detection, dynamic scattering characteristics are transformed into dynamic radar characteristics. These characteristics are determined for given viewing angles of the object and for the known motion dynamics of its elements. In this case, the time structure of the received radar signal is similar to the time structure of the radar characteristics.

From the above stochasticity of both the nature of the object and the observation conditions, the individual dynamic scattering characteristics can be regarded as realizations of some random process. If such a hypothesis is confirmed when checking the stability of statistical observable properties, then it should be asserted that there are corresponding statistical characteristics of the scattering.

The statistical scattering characteristics can thus be used to implement the random process, distribution laws, and certain average parameters that determine the basic properties of the original random process. The statistical scattering characteristics complete the hierarchy of the scattering characteristics used.

Let us return again to the description of the scattering by means of formula (6.143).

As in the case of regular scattering, when the polarization basis is varied, monostatic matrix $S(t)$ will be subjected to a congruent transformation (6.19), and for each fixed moment of time there will always exist such a polarization basis, where the scattering matrix is diagonalized. However, the coordinates of this basis on the Poincaré sphere $(2\gamma, 2\delta)$ will not remain constant and will change continuously over time. As for the definition of the scattering matrix, the degree of anisotropy of the object, then they will also become time functions remaining invariant to a change in the polarization basis. It should be emphasized that all the relations obtained in Sect. 6.2 remain valid for fluctuating scatterers with the proviso that their parameters depend on time. Here we should pay attention to the fact that in the future, we will have to deal with the probability distribution densities determined for real numbers. In this connection, it seems expedient to write the elements of the scattering matrix, which are complex numbers, in the form $s_{mn}(t) = \operatorname{Re} s_{mn}(t) + i \operatorname{Im} s_{mn}(t) \equiv s_{mnR}(t) + i s_{mnI}(t)$.

In case of accidental change in $s_{mn}(t)$, the most complete characteristic of a fluctuating scatterer is the six-dimensional probability distribution law for the real and imaginary parts of the elements of the scattering matrix $W_6(s_{mn})$.

To analyze the properties of a fluctuating object, it is required to know the explicit form of function $W_6(s_{mn})$, which entails the need to build statistical models of scattering capacity. This, in turn, requires different classifications of scatterers and sharply increases the scope of analysis.

Despite this, as will be shown below, there are general patterns for the numerical characteristics of these laws (mathematical expectations, variance, correlation coefficient), irrespective of their explicit form.

Let us begin our consideration of fluctuating objects by revealing such properties.

By averaging the elements of the scattering matrix, we can speak of an average scattering matrix $\overline{S}$ with elements $\overline{s_{mn}}$. Such a matrix does not differ from the scattering matrix of stable objects; therefore, all formulas and conclusions, presented in Sects. 6.1–6.2, remain valid for it. In particular, the invariant to a change in the polarization basis is the sum:

$$\overline{|s_{11}|}^2 + \overline{|s_{22}|}^2 + 2\overline{|s_{12}|}^2 = \sigma_\Sigma^2, \tag{6.144}$$

where σ_Σ^2 shall mean some analog of the total ESA of the scatterer.

Let us now turn to the analysis of second-order moments.

First of all, we note that formulas (6.27) remain valid for an arbitrary fluctuating basis at any fixed time. By performing a statistical averaging of equality (6.27) and performing the corresponding addition, we obtain:

$$\overline{|s_{11}|^2} + \overline{|s_{22}|^2} + 2\overline{|s_{12}|^2} = \overline{\sigma_\Sigma^2} \equiv \sigma_{\Sigma\Phi}^2. \tag{6.145}$$

Thus, the mean sum of the squares of the moduli of the elements of the scattering matrix does not depend on the choice of the polarization basis. By subtracting (6.145) from equality (6.145), we obtain an important conclusion that the variance total of the elements of the scattering matrix is also its invariant.

As already mentioned, in order to describe the properties of fluctuating targets, an apparatus based on the use of a four-dimensional complex vector $\vec{S}(t)$, leading to the so-called 4 × 4-order covariance matrix D_4, is preferable. The off-diagonal elements of this matrix are complex correlation coefficients, which will be used in the future in one of three representations, the meaning and the notations of which are sufficiently clear $\overline{s_{mn}s_{pk}^*} = \sigma_{mn}\sigma_{pk}r_{mnpk}e^{iq_{mnpk}} = \sigma_{mn}\sigma_{pk}r_{mnpk}\left(\cos q_{mnpk} + i\sin q_{mnpk\ I}\right) \equiv \left(\overline{s_{mn}s_{pk}^*}\right)_R + i\left(\overline{s_{mn}s_{pk}^*}\right)_I$, while the diagonal elements are the mean square values $\overline{|s_{mm}|^2} = \sigma_{mm}^2$, $(m, n, p, k = 1, 2)$:

$$D_4 = \begin{pmatrix} \sigma_{11}^2 & \sigma_{11}\sigma_{12}r_{1112}\mathrm{e}^{\mathrm{i}q_{1112}} & \sigma_{11}\sigma_{21}r_{1121}\mathrm{e}^{\mathrm{i}q_{1121}} & \sigma_{11}\sigma_{22}r_{1122}\mathrm{e}^{\mathrm{i}q_{1122}} \\ \sigma_{11}\sigma_{12}r_{1112}\mathrm{e}^{-\mathrm{i}q_{1112}} & \sigma_{12}^2 & \sigma_{12}\sigma_{21}r_{1221}\mathrm{e}^{\mathrm{i}q_{1221}} & \sigma_{12}\sigma_{22}r_{1222}\mathrm{e}^{\mathrm{i}q_{1222}} \\ \sigma_{1121}\sigma_{21}r_{1121}\mathrm{e}^{-\mathrm{i}q_{1121}} & \sigma_{12}\sigma_{21}r_{1221}\mathrm{e}^{-\mathrm{i}q_{1221}} & \sigma_{21}^2 & \sigma_{21}\sigma_{22}r_{2122}\mathrm{e}^{\mathrm{i}q_{2122}} \\ \sigma_{11}\sigma_{22}r_{1122}\mathrm{e}^{-\mathrm{i}q_{1122}} & \sigma_{12}\sigma_{22}r_{1222}\mathrm{e}^{-\mathrm{i}q_{1222}} & \sigma_{21}\sigma_{22}r_{2122}\mathrm{e}^{-\mathrm{i}q_{2122}} & \sigma_{22}^2 \end{pmatrix} \tag{6.146}$$

Matrix D_4 can be represented as the sum of its real and imaginary parts $D_4 = \mathrm{Re}D_4 + i\mathrm{Im}D_4 \equiv D_{4R} + iD_{4I}$. It can also be formed by a column matrix N, whose elements are elements of the scattering matrix:

$$D_4 = \overline{NN^+}, \tag{6.147}$$

where $N^{\mathrm{T}} = \left(s_{11}(t)\ s_{12}(t)\ s_{21}(t)\ s_{22}(t)\right)^{\mathrm{T}}$.

The column matrix N, in turn, can also be represented as the sum of two column matrices: $N = \mathrm{Re}N + i\mathrm{Im}N \equiv N_R + iN_I$.

For the case of monostatic radar detection—$s_{12} = s_{21}$; therefore, the covariance matrix is degenerate, its determinant is zero, and it is described by nine parameters (if $s_{12} \neq s_{21}$, then the number of parameters increases to 16). If we use expression (6.18) for the transformation of the scattering matrix, in which s_{12} and s_{21} are written instead of the off-diagonal elements, and also the general form of the 2×2-order unitary matrix Q (1.42), then instead of (6.26), which establishes a connection between the elements of the scattering matrix in the old and new bases, we obtain:

$$N_H = L\ N_C, \tag{6.148}$$

where transition matrix L is defined by the equality:

$$L = \mathrm{e}^{2\mathrm{i}\varphi} \begin{pmatrix} \mathrm{e}^{2\mathrm{i}\eta}\cos^2\gamma & 0.5\mathrm{e}^{2\mathrm{i}(\eta+\delta)}\sin 2\gamma & 0.5\mathrm{e}^{2\mathrm{i}(\eta+\delta)}\sin 2\gamma & \mathrm{e}^{2\mathrm{i}(\eta+2\delta)}\sin^2\gamma \\ -0.5\mathrm{e}^{-2\mathrm{i}\delta}\sin 2\gamma & \cos^2\gamma & -\sin^2\gamma & 0.5\mathrm{e}^{2\mathrm{i}\delta}\sin 2\gamma \\ -0.5\mathrm{e}^{-2\mathrm{i}\delta}\sin 2\gamma & -\sin^2\gamma & \cos^2\gamma & 0.5\mathrm{e}^{2\mathrm{i}\delta}\sin 2\gamma \\ \mathrm{e}^{-2\mathrm{i}(\eta+2\delta)}\sin^2\gamma & -0.5\mathrm{e}^{-2\mathrm{i}(\eta+\delta)}\sin 2\gamma & -0.5\mathrm{e}^{-2\mathrm{i}(\eta+\delta)}\sin 2\gamma & \mathrm{e}^{-2\mathrm{i}\eta}\cos^2\gamma \end{pmatrix} \tag{6.149}$$

The expression for L can be obtained using the Kronecker multiplication symbol $L = Q \otimes Q^*$.

A direct check proves the unitarity of matrix L, i.e., $LL^+ = I$.

Let us find how the covariance matrix changes D_4 (in the old basis D_{4C}) in the transition to a new polarization basis (in a new basis D_{4H}):

$$(D_4)_H = \overline{N_H N_H^+} = \overline{LN_C(LN_C)^+} = LD_{4C}L^+. \tag{6.150}$$

Equation (6.150) is a similarity transformation over a degenerate Hermitian matrix D_{4C}.

The invariants of the covariance matrix are its trace equal to the total ESA, and the determinant (for monostatic scattering, as mentioned above, it is simply zero).

A number of general relations are also valid for elements of the covariance matrix. Let us get some of them.

Let us determine whether coefficient r_{1122} be arbitrary or be subject to any restrictions. For this, we must require that relations $\overline{|s_{mn}|} > 0$ are met in any basis. It follows from (6.150) that:

$$\begin{aligned}\overline{|s_{12}|^2_H} &= 0.25\left[\sigma_{11}^2 + \sigma_{22}^2 - 2\sigma_{11}\sigma_{22}\cos(q_{1122} + 4\delta)\right]\sin^2 2\gamma \\ &\quad + \sigma_{12}^2\cos^2 2\gamma + 0.5\sigma_{11}\sigma_{22}r_{1112}\sin\gamma\cos^2\gamma\cos(q_{1112} - 2\delta) \\ &\quad - 0.5\sigma_{12}\sigma_{22}r_{1222}\sin\gamma\cos^2\gamma\cos(q_{1222} - 2\delta) \qquad (6.151)\end{aligned}$$

.

The transformations of (6.151) show that $\overline{|s_{12}|^2} > 0$ for any γ, δ, if the following inequality holds:

$$\begin{aligned}&R_{12}^2 + R_{12}\left[R_1^2\cos 2\Phi_1 + R_2^2\cos 2\Phi_2 - 2R_1R_2\cos(\Phi_1 + \Phi_2)\right] \\ &\quad + R_1^2 + R_2^2 - 2R_1R_2\cos(\Phi_1 - \Phi_2) \le 1, \qquad (6.152)\end{aligned}$$

where

$$R_1 = \sigma_{11}^2 r_{1112}/\sigma^2;\quad R_2 = \sigma_{22}^2 r_{1222}/\sigma^2;\quad R_{12} = 2\sigma_{11}^2\sigma_{22}^2 r_{1122}/\sigma^4;$$

$$2\Phi_1 = q_{1122} - 2q_{1112};\quad 2\Phi_2 = q_{1122} - 2q_{1222};\quad \sigma^2 = \sqrt{\sigma_{11}^4 + \sigma_{22}^4}.$$

As it can be seen, R_1, R_2, R_{12} are positive and less than one. Formula (6.152) is the square inequality with respect to R_{12}. If we denote the coefficient for R_{12} through $2X$, and the free term—through Y, then

$$-X - \sqrt{X^2 - Y} \le R_{12} \le -X + \sqrt{X^2 - Y}, \qquad (6.153)$$

while $|X| \le 1$; $|Y| \le 1$.

Analysis of formula (6.153) shows that the range of admissible values X and Y ($|X| \le 1$, $|Y| \le 1$) can be divided into five subdomains (Fig. 6.4):

1. The first quadrant and part of the second quadrant above the parabola enveloping the family of straight lines is a subdomain containing such values X and Y, which cannot exist simultaneously.
2. Trapezium in the third quadrant with vertices at the points $(-1; 0)$, $(-1; -1)$, $(0; -1)$, and $(0.5; 0)$—subdomain within which any values of X and Y are admissible; no additional restrictions are imposed on the values of coefficient R_{12}, i.e., $0 \le R_{12} \le 1$.

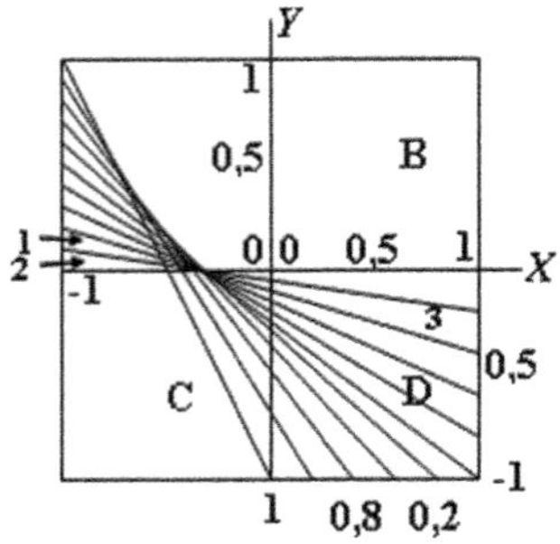

Fig. 6.4 The range of admissible values for the correlation coefficient between elements of the scattering matrix. **A**—no solutions: **1**—restrictions from above and from below, **2**—restrictions from below. **B**—there are no solutions. **C**—there are no restrictions. **D**—restrictions from above 3

3. The fourth quadrant and the triangle in the third quadrant with vertices at the points (0; 0); (0.5; 0), (0; −1) is a subdomain within which coefficient R_{12} cannot exceed the values of the parameter indicated on the line passing through the point (X, Y)—an additional upper limit (e.g., for $X = 0.5$; $Y = 0.5$ we have $0 \leq R_{12} \leq 0,4)$.
4. Triangle in the second quadrant with vertices at the points (−1; −1); (−1; 1); (−0.5; 0)—subdomain, where there is a similar additional limitation, but lower one (e.g., for $X = -0.8$; $Y = 0.25$ we have $0.2 \leq R_{12} \leq 1)$.
5. Part of the second quadrant bounded from above by a parabola enveloping the family of straight lines, from the bottom—by the abscissa axis, from the left—by a straight line passing through the points (−1; 1) and (0.5; 0),—subdomain where the values of coefficient R_{12} are further limited both from above and from below (e.g., for $X = -0.5$; $Y = 0.15$ we have $0.2 \leq R_{12} \leq 0.8)$.

In order to fully assess the limitations on coefficient R_{12}, it is necessary to consider inequalities $\overline{\left|s_{jj}\right|^2} > 0$ in formulas (6.151) for any $\gamma, \delta \quad (j = 1, 2)$. Unfortunately, it is not possible to obtain the final relation in a compact form; however, our estimates show that condition $\overline{|s_{12}|^2} > 0$ is more stringent.

In all the above relations, only mean square values of the elements of the scattering matrix appear, irrespective of the specific form of the multidimensional probability distribution law s_{mn}. To obtain more detailed links, knowledge of the explicit form of these laws is already necessary.

Let us now consider some of their properties.

Let the scattering vector of form $\vec{N}(t_1, t_2) = \vec{N}$ corresponds to an ensemble of statistical characteristics sufficient to describe its properties within the framework of the correlation theory under the condition of stationarity and stationary connection of the vector components. In view of what has been said, we introduce instead of two time moments t_1 and t_2—current time t and time shift τ. We denote as follows: $W\left(\vec{N}, \tau\right)$—probability distribution density, $\vec{N}_0$—vector of mathematical expectation, $K(\tau)$—correlation matrix.

It follows from (6.148) that $\vec{N}_{0H}(t) = L\vec{N}_{0C}(t)$.

In view of what has been said, we have:

$$K(\tau) = \left[\vec{N}(t) - \vec{N}_0\right]\left[\vec{N}(t+\tau) - \vec{N}_0\right]^+, \tag{6.154}$$

i.e., when passing to a new basis, the correlation matrix, like covariance matrix (6.147), is transformed in accordance with law (6.148):

$$K_H(\tau) = LK_C(\tau)L^+. \tag{6.155}$$

Let us determine the regularities of transformation $W\left(\vec{N}, \tau\right)$. It is known that under functional transformations of vector random processes $\vec{X}$ on this $\vec{Y} = F\left(\vec{X}\right)$, where F—transformation functional, the following equality holds:

$$W\left(\vec{Y}\right) = |J|W\left(\vec{X}\right), \tag{6.156}$$

where $W\left(\vec{Y}\right)$, $W\left(\vec{X}\right)$—corresponding probability distribution density of vector processes $\vec{Y}, \vec{X}$, $|J| = \left|\frac{\partial \vec{X}}{\partial \vec{Y}}\right|$—Jacobian of transformation.

The key to solving the problem of establishing the patterns of density variation $W\left(\vec{N}, \tau\right)$ when the scattering vector is transformed with matrix L is the occurrence of the Jacobian of this transformation.

In accordance with the basis transforming rules, we have $S_H = QS_CQ^{\mathrm{T}}$. In this case, the desired Jacobian has the following form: $|J| = \left|\frac{\partial S_{mnH}(t)}{\partial S_{mnC}(t)}\right| = \det L = 1$.

This leads to the conclusion, important for practical applications, about the invariance of the law of probability distribution of the statistical scattering matrix with respect to the change in the polarization basis.

If, for example, the multidimensional law in some basis is the Gaussian law, then it remains the Gaussian law in any other basis. The same applies to one-dimensional Gaussian probability distribution laws.

When solving a number of problems, knowing only the vector of mathematical expectation of the process and its matrix of second initial moments $D_4(t_1, t_2)$, statistically binding elements of matrix $S(t)$ at different instants of time (t_1 , t_2), is not enough. In this connection, it becomes necessary to obtain an explicit form of the probability distribution law for the scattering matrix in two or several time instants $W\left(\vec{N}, t_1, \ldots, t_N\right)$. In the general case, this law describes the properties of scattering vector $\vec{N}(t)$ in $(4 \times n)$-dimensional complex space. Within the framework of the correlation theory, it is sufficient to study the distribution law of the statistical scattering matrix at two fixed time instants $W\left(\vec{N}, t_1, t_2\right)$. With this restriction, the properties of the matrix are described by the joint eight-dimensional distribution law of the components of vector $\vec{N}(t_1, t_2)$. In the general case, it is impossible to predict the nature of the element distribution law of matrix $S(t)$, while probability density $W\left(\vec{N}, t_1, t_2\right)$ can be established based on some general physical considerations.

The following model of the fluctuating object became widespread.

Let there be an object in the resolving volume of the polarimeter consisting of a large number of reflectors with small and close values of the efficiency of the scattering surfaces. Suppose, in addition, there is one reflector among them commensurable with the total surface of the reflector of the rest of the reflector set. If we assume that the random nature of the changes in the elements of the scattering matrix of such an object is due to the wander of the reflectors relative to the center of mass and relative to each other, and also to the displacement of the object as a whole, then it is natural to assume that the arguments of the elements of scattering matrix $s_{mn}(t)$ will be uniformly distributed in interval $[0, 2\pi]$. Then, assuming that the nature of the displacement of elementary reflectors remains unchanged during the observation and the reflecting properties of the element with the dominant effective scattering area are also constant over the observation interval, it can be assumed that the fluctuations are stationary, and the elements of its scattering matrix are stationary and stationary connected random time functions.

If the number of elementary reflectors with commensurable effective scattering areas for elements $s_{mn}(t)$ is large, then the central limit theorem of probability theory holds, and therefore, the distribution of the elements of the scattering matrix can be considered normal. The parameters of this distribution are the vector of mean values $\vec{N}_0$ and the correlation matrix $K(\tau = t_1 - t_2)$ (6.154):

$$W\left(\vec{N}, \tau\right) = \frac{\exp\left\{-0.5\left[\vec{N}(\tau) - \vec{N}_0\right]K^{-1}(\tau)\left[\vec{N}(t) - \vec{N}_0\right]^+\right\}}{(2\pi)^2\sqrt{\det K(\tau)}}. \tag{6.157}$$

The resulting expression indicates that all two-dimensional laws $W_2(s_{mn}, \tau)$ will also be Gaussian laws in which the real and imaginary parts of elements will appear as variables s_{mn}. The above uniformity of the law of distribution of arguments s_{mn} provides statistical independence among themselves $\mathrm{Re}s_{mn}$, $\mathrm{Im}s_{mn}$ and the equality of the variances of these quantities. It can be shown elementary that the distribution of the moduli of the elements of the scattering matrix within the framework of the model in question will in general be subject to the Rice law:

$$W(|s_{mn}|) = \frac{|s_{mn}|}{d^2}\exp\left(-\frac{a^2 + |s_{mn}|^2}{2d^2}\right)I_0\left(\frac{a|s_{mn}|}{d^2}\right), \tag{6.158}$$

where d^2—variance of random variable $\mathrm{Re}s_{mn}$, equal to the variance of random variable $\mathrm{Im}s_{mn}$, a parameters—mean values of these variables.

In the absence of elements that give a stable reflection $a = 0$, Rice's law transits into Rayleigh's law:

$$W(|s_{mn}|) = \frac{|s_{mn}|}{d^2} \exp\left(-\frac{|s_{mn}|^2}{2d^2} \right). \tag{6.159}$$

Let us return to formulas (6.25). If we write expressions for real and imaginary parts of elements s_{mn}, then direct calculations show that assuming independence of all the random variables $\lambda_1, \lambda_2, \gamma, \delta$, the uniform probability distribution law for $\varphi, \eta, \gamma, \delta$, and the Rayleigh law for λ_1 and λ_2, then $\mathrm{Re} s_{mn}$ and $\mathrm{Im} s_{mn}$ in any polarization basis will have a Gaussian probability distribution law.

This statistical independence, as follows from direct calculations, will lead to statistical independence s_{11}, s_{12}, s_{22}. Thus, for scatterers modeled by a system of independent reflectors, the following equality holds:

$$\overline{s_{11}s_{22}^*} = \overline{s_{11}s_{12}^*} = \overline{s_{12}s_{22}^*} = 0. \tag{6.160}$$

The restriction on the statistical independence of random variables λ_1 and λ_2 will lead to the same result.

This allows us to describe the investigated class of scatterers on the basis of the following multidimensional probability distribution law:

$$\begin{aligned} W(\lambda_1, \lambda_2, \gamma, \delta, \varphi_{01}, \varphi_{02}) &= W(\lambda_1, \lambda_2) W(\gamma) W(\delta) W(\eta) W(\delta) \\ &= \frac{4\lambda_1\lambda_2}{(2\pi)^4 (1 - R^2) d_1^2 d_2^2} \exp \\ &\left\{ -\frac{1}{(1 - R^2)} \left(\frac{\lambda_1^2}{d_1^2} + \frac{\lambda_2^2}{d_2^2} \right) \right\} I_0 \left(\frac{2R}{1 - R^2} \frac{\lambda_1\lambda_2}{d_1 d_2} \right), \end{aligned} \tag{6.161}$$

where d_1^2, d_2^2—variance of random variables $\lambda_1 \equiv |\lambda_1|$ and $\lambda_2 \equiv |\lambda_2|$, respectively, R—coefficient of correlation between them.

Expression (6.161) makes it possible to find the probability distribution density of the total ESA of the target $\sigma_\Sigma^2 = \lambda_1^2 + \lambda_2^2$.

Cumbersome calculations lead to the following result:

$$W\left(\sigma_\Sigma^2\right) = \frac{sh\left(\sigma_\Sigma^2 \sqrt{\alpha^2 + \beta^2}\right)}{\sqrt{\alpha^2 + \beta^2}} \cdot \frac{\exp\left\{ -\left(1 - R^2\right)^{-1} \left(R_1^{-2} + R_2^{-2}\right) \sigma_\Sigma^2 \right\}}{\left(1 - R^2\right) d_1^2 d_2^2}, \tag{6.162}$$

where $\alpha = \left(1 - R^2\right)^{-1} \left(d_2^{-2} - d_1^2\right)$; $\beta = R\left[\left(1 - R^2\right) d_1 d_2\right]$.

When the variances are equal $d_1^2 = d_2^2 = d^2$ and in the absence of correlation $(R = 0)$, formula (6.162) is transformed to the following form:

$$W\left(\sigma_\Sigma^2\right) = \frac{\sigma_\Sigma^2}{d^4} \exp\left\{ -\frac{\sigma_\Sigma^2}{d^2} \right\}. \tag{6.163}$$

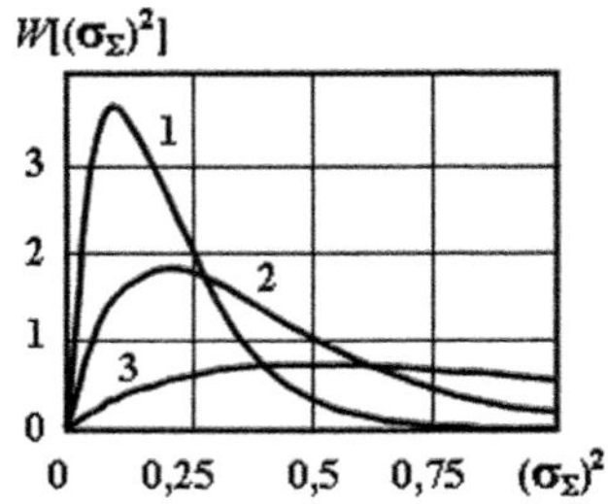

Fig. 6.5 The density of the distribution of the probabilities of total ESR. 1 – $d^2 = 0.1$; 2 – $d^2 = 0.2$; 3 – $d^2 = 0.5$. $d_1 = d = d$; $R = 0$

The corresponding distribution laws are shown in Fig. 6.5. From Eq. (6.162), the mean value and the variance of the total ESA can be easily found:

$$\left.\begin{aligned} \overline{\sigma_\Sigma^2} &= d_1^2 + d_2^2 \\ \overline{\sigma_\Sigma^4} - \overline{\sigma_\Sigma^2}^2 &= d_1^4 + d_2^2 + 2Rd_1^2 d_2^2 \end{aligned}\right\} \qquad (6.164)$$

Of great practical interest is the law of probability distribution of the anisotropy degree of objects modeled by a system of independent reflectors. If we use definition $q = \left||\lambda_1|^2 - |\lambda_2|^2\right| / \sigma_\Sigma^2$ and transformation rules of distribution laws, it can be shown that in this case:

$$W(q) = 2\Omega\left(1 - R^2\right)\frac{(1+\Omega) - q(1-\Omega)}{\left\{[(1+\Omega) - q(1-\Omega)]^2 - 4\Omega R^2\left(1 - q^2\right)\right\}^{1.5}} + \left.\frac{(1+\Omega) + q(1-\Omega)}{\left\{[(1+\Omega) + q(1-\Omega)]^2 - 4\Omega R^2\left(1 - q^2\right)\right\}^{1.5}}\right\}, \qquad (6.165)$$

where $\Omega = d_1^2 / d_2^2$—degree of asymmetry of the target.

In the particular case of a polarization-isotropic scatterer ($\Omega = 1$), the average value of the anisotropy degree will be determined by the following equality:

$$\overline{q} = \frac{\sqrt{1 - R^2}}{R^2}\left(1 - \sqrt{1 - R^2}\right), \qquad (6.166)$$

and the variance

$$\overline{q^2} - (\overline{q})^2 = \frac{1 - R^2}{R^2}\left[\frac{1}{2R}\ln\frac{1+R}{1-R} - \left(1 - \sqrt{1 - R^2}\right)^2 - 1\right]. \qquad (6.167)$$

The corresponding curves are shown in Fig. 6.6.

Formulas (6.148) make it possible to obtain a number of inequalities for estimating variances D_{mn}^2 in an arbitrary polarization basis (in the formulas given below, d_{mn}^2 means the value of the variances in the polarization basis in which the measurement is made):

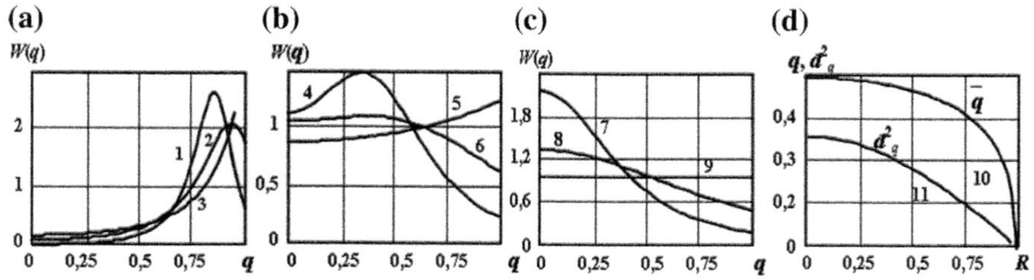

Fig. 6.6 Probability distribution density $W(q)$ at different values of asymmetry coefficient Ω and correlation R (fig. a, b and c) and dependence of the mean value q (curve 10 on fig. d) and its dispersion (curve 11 on fig. d) on the correlation coefficient R. The curve on fig. a, b, and c correspond the following values of Ω and R

$$\begin{cases} \frac{D_\Sigma^2}{4} + \frac{d_{12}^2}{2} \le D_{11}^2 \le \frac{D_\Sigma^2}{2} + \frac{d_{11}+d_{22}}{2}\sqrt{(d_{11}-d_{22})^2 + 4d_{12}^2} \\ \begin{cases} \frac{d_{11}^2+d_{22}^2}{4} \le D_{12}^2 \le \frac{D_\Sigma^2}{2} - \frac{(d_{11}-d_{22})^2}{4}, & \text{if } d_{11}^2 + d_{22}^2 > 4d_{12}^2 \\ d_{12}^2 \le D_{12}^2 \le \frac{D_\Sigma^2}{2} - \frac{(d_{11}-d_{22})^2}{4}, & \text{if } d_{11}^2 + d_{22}^2 < 4d_{12}^2 \end{cases} \\ \frac{D_\Sigma^2}{4} + \frac{d_{12}^2}{2} \le D_{22}^2 \le \frac{D_\Sigma^2}{2} + \frac{d_{11}+d_{22}}{2}\sqrt{(d_{11}-d_{22})^2 + 4d_{12}^2} \end{cases} . \tag{6.168}$$

The above set of inequalities can be very useful in practical radio polarimetry.

Despite the random nature of the change in the elements of the scattering matrix, it is possible to carry out some classification of the fluctuating reflectors. If it turns out that in one of the polarization bases - $s_{12}(t) \equiv 0$, then the scattering matrix in this basis has a diagonal form, and therefore, reflectors having a scattering matrix of the following form

$$\Lambda_{0-0}(t) = \begin{pmatrix} \lambda_1(t) & 0 \\ 0 & \lambda_2(t) \end{pmatrix} \tag{6.169}$$

in one of the bases will be called diagonal targets. Relations (6.25) remain valid for them. A diagonal target always has a pair of non-orthogonal zero polarizations. Time dependencies $\lambda_1(t)$ and $\lambda_2(t)$ can be obtained from equalities (6.32). By analogy with the case of stable targets, we can introduce the notion of a "degenerate target", for which $\lambda_2(t) \equiv 0$. This goal has an invariant determinant of the scattering matrix, which is always zero.

The scattering matrix of such targets has the general form (6.47). The generalized phase requirement (6.48) for the arguments of the elements of the scattering matrix s_{mn} also remains in force.

As can be seen, the immediate possibilities of classifying fluctuating targets by their scattering matrices are much more modest than for stable targets. If we move from temporary random processes $s_{mn}(t)$ to their energy characteristics, it is possible to obtain classification of fluctuating targets similarly to the classification of stable targets by means of a sufficiently small set of constant numbers. The energy characteristics of the fluctuating targets are the Graves matrix and the covariance (correlation) scattering matrix considered in Sect. 7.

Similarly, fluctuating scatterers can be investigated for bistatic observation conditions. The description of the scattering channel is identical in many respects to the description used in the monostatic case, and therefore is not given here.

References

1. Kozlov AI, Logvin AI, Sarychev VA (2005) Polarization of radio waves. Book 1. Polarization structure of radar signals. Moscow. Radio engineering
2. Kozlov AI, Logvin AI, Sarychev VA (2008) Polarization of radio waves. Book 3. Radiopolarimetry of structurally complex signals. Radio engineering, Moscow

Chapter 7
Own Radio Emission and Scattering of Radio Waves

7.1 Reflection from the Stable Reflecting Objects

When analyzing the energy characteristics of the reflected signal, the so-called Graves matrix is most convenient. The density of power flux Π reflected by the signal target is proportional to the parameter $E_{1p}^2 + E_{2p}^2$. After introducing the column matrix $\vec{E} = (E_1 E_2)^{\mathrm{T}}$, the expression for Π in the matrix form can be written as follows:

$$\Pi = \vec{E}_p^+ \vec{E}_p. \tag{7.1}$$

If we move to the radiated waves, then

$$\Pi = \left(S\vec{E}_u\right)^+ S\vec{E}_u = \vec{E}_u^+ S^+ S\vec{E}_u = \vec{E}_u^+ G\vec{E}_u \tag{7.2}$$

while in (7.2) we have introduced notation

$$G = S^+ S = S^* S. \tag{7.3}$$

Hermitian matrix G is called the Graves matrix. In its expanded form, it has the following form:

$$G = \begin{pmatrix} |s_{11}|^2 + |s_{12}|^2 & s_{11}^* s_{12} + s_{12}^* s_{22} \\ s_{11}s_{12}^* + s_{12}s_{22}^* & |s_{12}|^2 + |s_{22}|^2 \end{pmatrix}. \tag{7.4}$$

Let us consider how the Graves matrix changes as the polarization basis changes. Using Formula (6.19), we obtain

$$G_H = S_H^+ S = \left(QS_C Q^{\mathrm{T}}\right)^+ \left(QS_C Q^{\mathrm{T}}\right) = Q^* G_C Q^{\mathrm{T}}. \tag{7.5}$$

Kozlov A. I. et al., *Introduction to the Theory of Radiopolarimetric Navigation Systems*, Springer Aerospace Technology, https://doi.org/10.1007/978-981-13-8395-3_7

Thus, in the transition to a new polarization basis, the Graves matrix undergoes a similarity transformation using the same unitary matrix as matrix S. Therefore, if some unitary matrix Q leads matrix S to diagonal form, then this matrix also diagonalizes the matrix G.

Let us choose eigen target basis as the initial basis, then the Graves matrix in this basis will have the following form:

$$G_0 = \begin{pmatrix} |\lambda_1|^2 & 0 \\ 0 & |\lambda_2|^2 \end{pmatrix}. \tag{7.6}$$

Using Formulas (7.5) and (1.42), we can write the Graves matrix in an arbitrary basis determined by the parameters $(\gamma, \delta, \varphi)$ in relation to the initial one:

$$G = \frac{\sigma_\Sigma^2}{2}\begin{pmatrix} 1 + q\cos 2\gamma & -q\mathrm{e}^{-2i(\eta+\delta)}\sin 2\gamma \\ -q\mathrm{e}^{2i(\eta+\delta)}\sin 2\gamma & 1 - q\cos 2\gamma \end{pmatrix}. \tag{7.7}$$

Formula (7.7) formally coincides with the analogous representation for coherence matrix (3.11). It is easy to see that the invariants of the Graves matrix are its determinant and trace:

$$\begin{aligned} \det G &= |\det S|^2 = \left|s_{11}s_{22} - s_{12}^2\right|^2 = |\lambda_1\lambda_2|^2 = 0.25\sigma_\Sigma^2\left(1 - q^2\right); \\ S_P G &= |s_{11}|^2 + 2|s_{12}|^2 + |s_{22}|^2 = |\lambda_1|^2 + |\lambda_2|^2 = \sigma_\Sigma^2. \end{aligned} \tag{7.8}$$

For polarization-isotropic targets $(\lambda_1 = \lambda_2, q = 0)$, the Graves matrix is always an identity matrix, and for a degenerate target $(\lambda_2 = 0, q = 1)$, it has the following form:

$$G_{B\text{Ц}} = \frac{|\lambda|^2}{2}\begin{pmatrix} 2\cos^2\gamma & -e^{-2i(\eta+\delta)}\sin 2\gamma \\ -e^{2i(\eta+\delta)}\sin 2\gamma & 2\sin^2\gamma \end{pmatrix}. \tag{7.9}$$

In the class B polarization basis, $(2\gamma = 0.5\pi(1 + 2n), n = 0, \pm 1, \ldots)$
The Graves matrix is written as follows:

$$G_B = \frac{\sigma_\Sigma^2}{2}\begin{pmatrix} 1 & -q\mathrm{e}^{-2\mathrm{i}(\eta+\delta)} \\ -q\mathrm{e}^{2\mathrm{i}(\eta+\delta)} & 1 \end{pmatrix} \tag{7.10}$$

The obtained relations make it possible to proceed to an estimate of the power of the wave reflected by the target as a function of the type of polarization of the incident wave (it seems that when it comes to scattering or reflecting radio waves, it is more convenient to use the term radar target or simply the target instead of the concepts of a scatterer or a reflector). Since the energy of the reflected wave does not depend on the choice of the polarization basis, the desired relations are best sought in the eigenbasis of the target.

Let the power of the incident wave be unity, in this case vector $\vec{E}_u$, describing the radiated wave in its eigenbasis of the target, can always be represented in the following form:

$$\vec{E}_u = \begin{pmatrix} E_1 & E_2 \end{pmatrix}^{\mathrm{T}} = \begin{pmatrix} \mathrm{e}^{i\varphi}\cos\gamma & \mathrm{e}^{-i\varphi}\sin\gamma \end{pmatrix}^{\mathrm{T}}. \tag{7.11}$$

If we use relations (7.2) and (7.6), then

$$\Pi = |\lambda_1|^2|E_1|^2 + |\lambda_2|^2|E_2|^2 = |\lambda_1|^2\cos^2\gamma + |\lambda_1|^2\sin^2\gamma = \frac{\sigma_\Sigma^2}{2}(1 + q\cos 2\gamma). \tag{7.12}$$

Assuming for definiteness that $|\lambda_1|^2 \geq |\lambda_2|^2$, we obtain:

$$\begin{cases} \Pi_{\max} = |\lambda_1|^2 = 0.5\sigma_\Sigma^2(1+q) \\ \Pi_{\min} = |\lambda_2|^2 = 0.5\sigma_\Sigma^2(1-q) \end{cases}. \tag{7.13}$$

Extreme values of the reflected power take place in those cases when the polarization of the incident wave coincides with the eigen polarizations of the target.

If the analysis is carried out not in the eigen, but in an arbitrary basis, then the angle 2γ in Formula (7.12) will no longer coincide with angle 2γ in Formula (7.11). It can be expressed in terms of the elements of the scattering matrix using relations (3.27) and (3.30):

$$\cos 2\gamma = \frac{\sigma_{11}^2 - \sigma_{22}^2}{q\sigma_\Sigma^2} = \frac{\sigma_{11}^2 - \sigma_{22}^2}{\sqrt{\left(\sigma_{11}^2 - \sigma_{22}^2\right)^2 + 4\sigma_{12}^2\left(\sigma_{11}^2 + \sigma_{22}^2 + 2\sigma_{11}\sigma_{22}\cos\Phi\right)}}, \tag{7.14}$$

where the generalized phase $\Phi = \psi_{11} + \psi_{22} - 2\psi_{12}$.

Let us orient the Poincaré sphere so that the polarization of the target corresponds to the north and south poles (Fig. 7.1). Since the reflected power does not depend on angle φ, when the imaging point moves along the DC parallel, the power of the reflected signal will not change. When moving along the meridian, on the contrary, the reflected power decreases from the value $|\lambda_1|^2$ at the north pole (point B) to $|\lambda_2|^2$ at the south pole (point B'). On a parallel defined by angle 2γ, it will be equal to the value calculated by Formula (7.12).

Relation (7.12) shows that the reflected power does not depend on the type of polarization of the irradiating wave if and only if $|\lambda_1| = |\lambda_2|$ $(q = 0)$, i.e., when the target is polarization isotropic.

Let us pay attention that in case of change of kinds of polarization by such law at which $\gamma = \text{const}$, any goal will be perceived as polarization-isotropic one.

Let us now turn to the solution of an important practical problem for determining the power and degree of polarization of an electromagnetic wave reflected from a

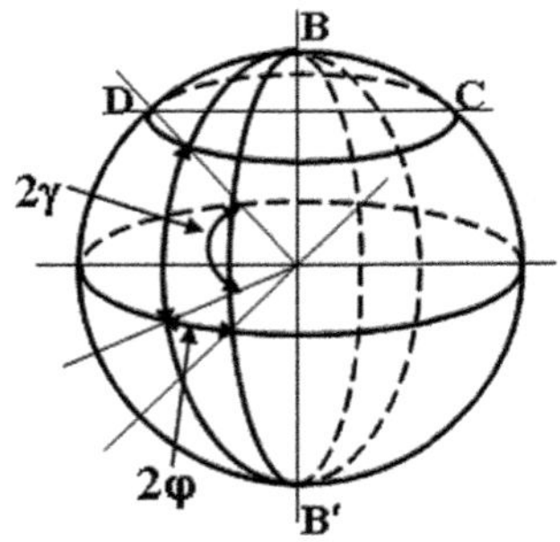

Fig. 7.1 Poincare sphere [illustration of the Formula (7.12)]

stable target when it is irradiated by a partially polarized wave whose degree of polarization is m_0.

The solution of the required problem begins with the recording of Graves matrices of two targets, expressed in terms of their invariants, in a single polarization basis. For this, we use Formulas (6.62) and (6.63), which give the required representations:

$$\left.\begin{aligned} G_1 &= \frac{\sigma_{\Sigma 1}^2}{2}\begin{pmatrix} 1 & q_1 e^{i\gamma_{12}} \\ q_1 e^{-i\gamma_{12}} & 1 \end{pmatrix} \\ G_2 &= \frac{\sigma_{\Sigma 2}^2}{2}\begin{pmatrix} 1 & q_2 e^{i\gamma_{12}} \\ q_2 e^{-i\gamma_{12}} & 1 \end{pmatrix} \end{aligned}\right\} \tag{7.15}$$

where $\sigma_{\Sigma 1}^2$, $\sigma_{\Sigma 2}^2$—total ESAs of the first and second targets, respectively, and q_1, q_2—degree of their anisotropy.

It is quite obvious that if we consider the Graves matrix referring to some target, and coherence matrix K_0 irradiating this target of the wave, then, acting in a completely analogous way, we can obtain representations of these matrices in terms of their invariants in the same basis. In this case, the first equality in Formula (7.15) remains unchanged, and instead of the second one we get:

$$K_0 = \frac{J_0}{2}\begin{pmatrix} 1 & m_0 e^{i\gamma_{12}} \\ m_0 e^{-i\gamma_{12}} & 1 \end{pmatrix}, \tag{7.16}$$

where J_0—the power of the incident wave and angle $2\gamma_{12}$—central angle on the Poincaré sphere between the images of the second singular basis of the coherence matrix and the eigenbasis of the target.

Using the definition of the coherence matrix, we can also associate the coherence matrices for the incident and reflected waves:

$$K = SK_0S^*, \tag{7.17}$$

where K—coherence matrix of the reflected wave.

Direct calculations based on relations (7.8), (7.16), (6.62), and (6.63) give the following expressions for the power of the reflected wave J_{omp} and its degree of polarization m:

$$\left.\begin{array}{l} J_{\text{omp}} = 0.5 J_0 \sigma_\Sigma^2 (1 + m_0 q \cos 2\gamma_{12}) \\ m^2 = 1 - \frac{(1-q^2)(1-m_0^2)}{(1+m_0 q \cos 2\gamma_{12})^2} \end{array}\right\}. \tag{7.18}$$

As can be seen from Eq. (7.18), the maximum reflected power is $0.5 J_0 \sigma_\Sigma^2 (1 + m_0 q)$, and therefore to increase it, it is necessary to use polarized waves ($m_0 = 1$)—if $0 \le 2\gamma_{12} \le 0.5\pi$, and unpolarized ($m_0 = 0$)—if $0.5\pi < 2\gamma_{12} < \pi$.

As for the degree of polarization of the reflected wave, it can be either more or less than quantity m_0. When the target is irradiated with an unpolarized wave ($m_0 = 0$), the degree of polarization of the reflected wave turns out to be equal to the degree of anisotropy of target ($m = q$). If reception of a partially polarized wave is carried out on some fixed polarization, then the output signal of the receiving device $J_{\Pi P}$ is proportional to one of the diagonal elements of the coherence matrix written in the corresponding basis (3.11) $J_{\Pi P} = 0.5 J_P \sigma_\Sigma^2 (1 + m_0 \cos 2\gamma_{34})$, where angle $2\gamma_{34}$ is the central angle between the images on the Poincaré sphere of the second singular basis and the basis in which the wave is received.

For the case under consideration, as follows from (7.18), we have:

$$J_{\Pi P} = 0.25 J_0 \sigma_\Sigma^2 (1 + m_0 q \cos 2\gamma_{12})(1 + m_0 \cos 2\gamma_{34}). \tag{7.19}$$

Figure 7.2 shows the dependences of the degree of polarization of the reflected wave on the degree of polarization of the incident wave at two limiting values of angle $2\gamma_{12}$, equal to 0 and π. As can be seen from Formula (7.12) and figures, with $2\gamma_{12} < 0.5\pi$, $m > m_0$, while at $0.5\pi \le 2\gamma_{12} \le \pi$ value m can be either more or less than m_0.

Figure 7.3 shows the dependence of the "reflection coefficient" $R = J/(\sigma_\Sigma^2 J_0)$, calculated by Formula (7.19), on the degree of polarization of the incident wave m_0 also for two values of angle $2\gamma_{12}$. The upper curves correspond to $2\gamma_{34} = 0$, lower to $2\gamma_{34} = \pi$.

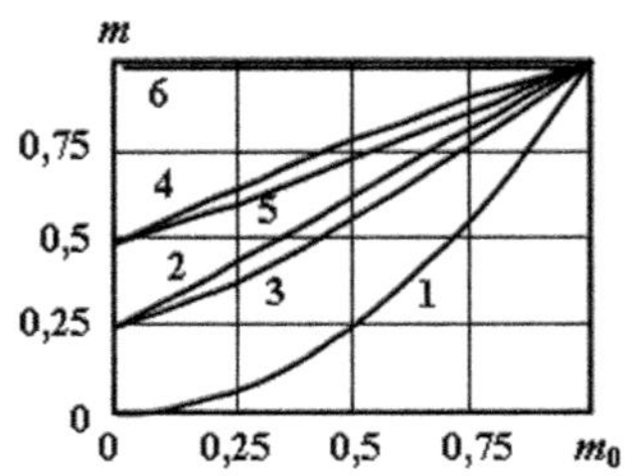

Fig. 7.2 Dependence of the polarization degree of reflected wave on the polarization degree of incident wave m_0 at: 1—$q = 0$. 2—$q = 0.5$; $2\gamma_{12} = \pi/3$. 3—$q = 0.5$; $2\gamma_{12} = \pi/6$. 4—$q = 0.7$; $2\gamma_{12} = \pi/3$. 5—$q = 0.7$; $2\gamma_{12} = \pi/6$

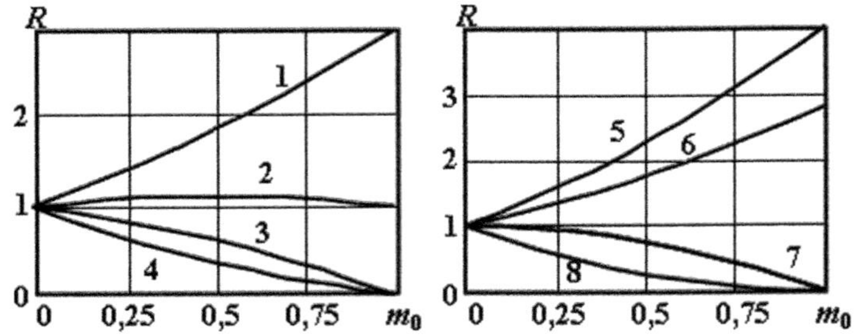

Fig. 7.3 Dependence of the generalized reflection coefficient on the incident wave polarization degree at the following parameters values: 1—$q = 1$; $2\gamma_{12} = 0$; $2\gamma_{34} = 0$. 2—$q = 1$; $2\gamma_{12} = \pi$; $2\gamma_{34} = 0$. 3—$q = 1$; $2\gamma_{12} = 0$; $2\gamma_{34} = \pi$. 4—$q = 1$; $2\gamma_{12} = \pi$; $2\gamma_{34} = \pi$. 5—$q = 0.5$; $2\gamma_{12} = 0$; $2\gamma_{34} = 0$. 6—$q = 0.5$; $2\gamma_{12} = \pi$; $2\gamma_{34} = 0$. 7—$q = 0.5$; $2\gamma_{12} = 0$; $2\gamma_{34} = \pi$. 8—$q = 0.5$; $2\gamma_{12} = \pi$; $2\gamma_{34} = \pi$

The obtained relations clearly demonstrate that a change in the polarization characteristics in a number of cases can lead to appreciable energy gains.

In practice, there is often a problem when it is necessary to achieve not only an increase in the reflected power but an increase in the contrast of one target against the other.

For this, let us compare the power flux density of signals reflected from two targets (λ-target and P-target) with different scattering matrices, when they are irradiated by a completely polarized wave. If we use relation (7.15), then the desired ratio (radar contrast) will be:

$$g = \frac{\Pi_\Lambda}{\Pi_P} = \frac{\sigma^2_{\Sigma\Lambda}}{\sigma^2_{\Sigma P}} \cdot \frac{1 + q_\lambda \sin 2\gamma \cos(2\delta + \gamma_{12})}{1 + q_P \sin 2\gamma \cos(2\delta - \gamma_{12})} = \frac{\sigma^2_{\Sigma\Lambda}}{\sigma^2_{\Sigma P}} \cdot g', \tag{7.20}$$

where $\sigma^2_{\Sigma\Lambda}$ and q_Λ, $\sigma^2_{\Sigma P}$ and q_P—total ESAs and degrees of anisotropy of the first and second targets, respectively, γ_{12}—central angle on the Poincaré sphere between images of eigenbases of targets, 2γ and 2δ—parameters that determine the polarization of the wave irradiating these targets.

It can be seen from (7.20) that the ratio g is the product of two factors, the first of which does not depend on the type of polarization of the incident wave (the ratio of total ESAs of targets), and the second which takes into account their polarization anisotropy and depends on the type of polarization of the irradiating wave (γ, 2δ).

The rather cumbersome transformations of Formula (7.20) make it possible to find an explicit expression for the extreme values of radar contrast:

$$g_{\max,\min} = \frac{\sigma^2_{\Sigma\Lambda}}{\sigma^2_{\Sigma P}} \cdot \frac{\sqrt{q_\Lambda^2 + q_P^2 - 2q_\Lambda q_P \cos 2\gamma_{12} - q_\Lambda^2 q_P^2 \sin^2 2\gamma_{12}} \mp q_\Lambda q_P \cos 2\gamma_{12} \pm q_\Lambda^2}{\sqrt{q_\Lambda^2 + q_P^2 - 2q_\Lambda q_P \cos 2\gamma_{12} - q_\Lambda^2 q_P^2 \sin^2 2\gamma_{12}} \pm q_\Lambda q_P \cos 2\gamma_{12} \mp q_P^2}, \tag{7.21}$$

while extreme values are reached when:

$$\begin{cases} 2\delta = \arctan\left(\frac{q_P + q_\Lambda}{q_P - q_\Lambda} \tan \gamma_{12}\right) + (-1)^n \arcsin \frac{q_P q_\Lambda \sin 2\gamma_{12}}{\sqrt{q_\Lambda^2 + q_P^2 - 2q_P q_\Lambda \cos 2\gamma_{12}}}. \\ 2\gamma = \pi m, \quad n = 0; 1 \end{cases} \tag{7.22}$$

Figure 7.4 shows the dependence of the extreme values of the second factor (g') on g_Λ at various q_P. The upper curves of the figure correspond to the maximum value of the factor, the lower ones to the minimum value. In other words, when the type of polarization of the incident wave changes, the ratio of the power flux density will vary from $g'_{\max}(\sigma^2_{\Sigma\Lambda}/\sigma^2_{\Sigma P})$ to $g'_{\min}(\sigma^2_{\Sigma\Lambda}/\sigma^2_{\Sigma P})$.

Another particular problem, which is also of great practical importance, is the problem of the best allocation of the useful component in the received electromagnetic wave against the background of interfering unpolarized and partially polarized components. (Such a situation involves the reception of a useful signal against the background of the receiver's intrinsic noise and interfering partially polarized wave).

So, let the antenna receive a useful partially polarized wave with a power of J_C and a polarization degree of m_C, an interfering partially polarized wave with a power of J_Π and a polarization degree of m_Π and an unpolarized wave with a power of $J_{H\Pi}$. It is required to estimate the maximum possible ratio of the polarized component of the useful signal to the sum of all other components.

Let point A on the Poincaré sphere (Fig. 7.5) correspond to the completely polarized component of the useful signal. The angular coordinates of this point will be

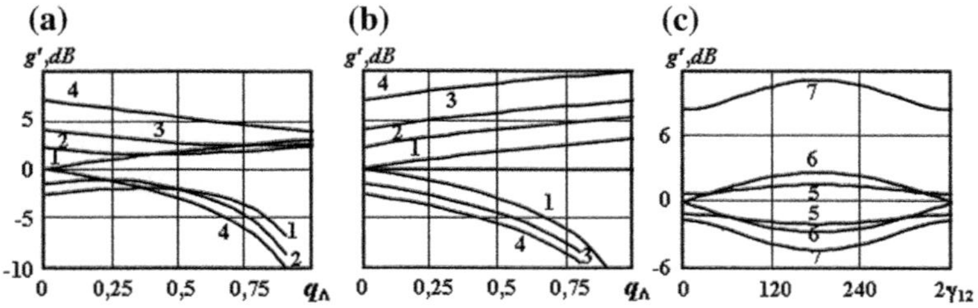

Fig. 7.4 Dependence of extreme values g', dB on q_Λ (a and b) for different values of q_P and γ_{12} and from γ_{12} (в) for different values of q_P and q_Λ. In figure **a** $\gamma_{12} = 45°$, in figure **b** $\gamma_{12} = 180°$. 1—$q_P = 0$; 2—$q_P = 0.4$; 3—$q_P = 0.6$; 4—$q_P = 0.8$; 5—$q_\Lambda = 0.3$; $q_P = 0.1$; 6—$q_\Lambda = 0.3$; $q_P = 0.3$; 7—$q_\Lambda = 0.3$; $q_P = 0.9$

denoted by ($2\alpha_1$ and $2\beta_1$). To the analogous component for the interfering wave, we assign point *B* with coordinates ($2\alpha_2$ and $2\beta_2$). To point *C* with coordinates (2α, 2β), we assign the polarization of the receiving antenna.

It follows from (7.18) that the power emitted at the antenna output from the reception of the polarized component of the useful signal will be $m_C J_C \cos 2\gamma$, from the interfering wave—$0.5 J_\Pi (1 + m_\Pi \cos 2\gamma_2)$, and from unpolarized wave—$0.5 J_{H\Pi}$. In addition, an unpolarized component of the useful signal with a power of $0.5(1 - m_C) J_C$ is received. The angles used here are shown in Fig. 7.5 ($\angle AOC = 2\gamma_1, \angle BOC = 2\gamma_2$). The desired ratio, subject to maximization, will be as follows:

$$z = \frac{m_C J_C \cos 2\gamma_1}{J_{m\Sigma} + m_\Pi J_\Pi \cos 2\gamma_2}, \tag{7.23}$$

where $J_{m\Sigma} = 0.5[J_m + (1 - m_C) J_C + (1 - m_\Pi) J_\Pi]$.

In accordance with the formulas for spherical trigonometry, we have:

$$\cos 2\gamma_i = \cos 2\alpha \cos 2\alpha_i + \sin 2\alpha \sin 2\alpha_i \cos(\beta - \beta_i). \tag{7.24}$$

By substituting (7.24) into (7.23), we finally obtain the following relation for the required z:

$$z = \frac{m_C J_C [1 + \cos 2\alpha \cos 2\alpha_1 + \sin 2\alpha \sin 2\alpha_1 \cos 2(\beta - \beta_1)]}{J_{m\Sigma} + m_\Pi J_\Pi [1 + \cos 2\alpha \cos 2\alpha_2 + \sin 2\alpha \sin 2\alpha_2 \cos 2(\beta - \beta_2)]}. \tag{7.25}$$

To find α and β that maximize z, we proceed as follows. Let us assume that $\beta = \beta_1$. Then it is evident that extreme values z are achieved with $\beta = \beta + \pi n$, i.e., point *C* must lie on the meridian passing through points *A* and *B*. Because of the arbitrary choice of the *XYZ* axes in Fig. 7.5, this means that the maximization of z

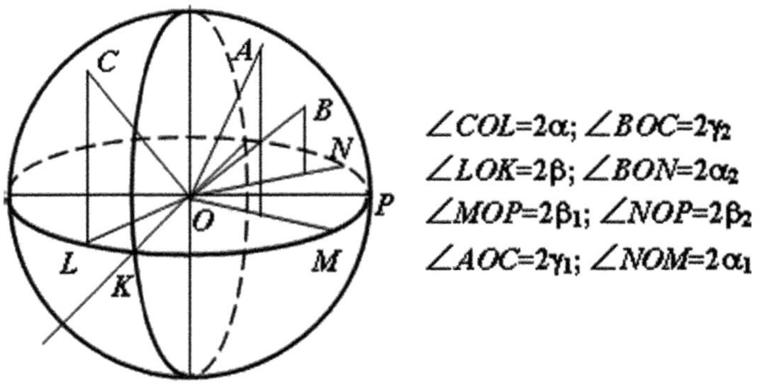

Fig. 7.5 The Poincaré sphere (illustration of receiving the useful and interference partially polarized radio waves)

for arbitrary β_1 and β_2 holds if and only if points A, B, and C lie on a large circle of the Poincaré sphere.

We orient the sphere so that points A, B, and C lie on the meridian. The formula of the desired relation (7.25) does not change in this case. However, it can now be asserted that point C will lie on the ABC meridian, and therefore the maximization of z by parameter β will occur when $\beta = \beta_1 = \beta_2$. This allows us to write Eq. (7.25) in the following form:

$$z_{\max}(\alpha) = \frac{m_C J_C[1 + \cos 2(\alpha - \alpha_1)]}{J_{m\Sigma} + m_\Pi J_\Pi[1 + \cos 2(\alpha - \alpha_2)]}. \tag{7.26}$$

It should be emphasized that the differences of $(\alpha - \alpha_1)$ and $(\alpha - \alpha_2)$ in Formula (7.26) do not depend on the initial orientation of the sphere since they are invariants to any of its rotations.

When calculating the corresponding derivatives, we find that the extremum $z(\alpha)$ takes place when

$$2\alpha = \arctan \frac{K^2 \cos 2\alpha_1 - \cos 2\alpha_2}{\sin 2\alpha_2 - K^2 \sin 2\alpha_1} \pm \arcsin \frac{\sin 2\alpha_0}{1 - K^2 \cos 2\alpha_0 - K^4}, \tag{7.27}$$

where $\alpha = (\alpha_1 - \alpha_2)$ is invariant to the rotation of the sphere, $K^2 = 1 + J_{m\Sigma}/(m_\Pi J_\Pi)$.

The substitution of (7.27) into Formula (7.26) determines the required maximum value:

$$z_{\max} = \frac{m_C J_C}{m_\Pi J_\Pi + J_{m\Sigma}} \left(1 + \frac{m_\Pi J_\Pi}{J_{m\Sigma}} \sin^2 \alpha_0\right). \tag{7.28}$$

It is clear from (7.28) that the greatest gain is realized when $\alpha_0 = 0.5\pi$, i.e., in the case when the polarization of the received useful wave is orthogonal to the polarization of the interference wave. In this case—$z_{\max} = m_C J_C / J_{m\Sigma}$. If, however, the named polarizations coincide, then $z_{\max} = m_C J_C / (J_{m\Sigma} + m_\Pi J_\Pi)$.

7.2 Reflection from Fluctuating Reflectors

7.2.1 The Graves Matrix and the Covariance Matrix

A convenient and informative device for describing the properties of fluctuating targets is the Graves average energy matrix $\overline{G}$, which is obtained as a result of averaging the formula $\overline{G} = \overline{S^* S}$:

$$\overline{G} = \overline{S^* S} = \begin{pmatrix} \overline{|s_{11}|^2} + \overline{|s_{12}|^2} & \overline{s_{11}^* s_{12}} + \overline{s_{12}^* s_{22}} \\ \overline{s_{11} s_{12}^*} + \overline{s_{12} s_{22}^*} & \overline{|s_{12}|^2} + \overline{|s_{22}|^2} \end{pmatrix}, \tag{7.29}$$

The formula for the average power flux density in accordance with Formula (7.2) in this case will have the form—$\Pi = \vec{E}_u^+ \overline{G} \vec{E}_u$. When we move to a new basis, matrix $\overline{G}$, as well as matrix G (7.5), undergoes a similarity transformation: $\overline{G_H} = Q^* \overline{G_C} Q^{\mathrm{T}}$.

Consequently, there is always such a polarization basis in which the mean Graves matrix has a diagonal form. For fluctuating purposes, this basis plays the same role as its eigenbasis for stable targets. We call it a zero basis, or a class *A* polarization basis, and denote the corresponding Graves matrix by $\overline{G_0}$, which can be obtained from Formula (7.4) by averaging its elements, while the off-diagonal elements must be assumed to be zero:

$$\overline{G_0} = \begin{pmatrix} \sigma_{011}^2 + \sigma_{012}^2 & 0 \\ 0 & \sigma_{012}^2 + \sigma_{022}^2 \end{pmatrix}, \tag{7.30}$$

where $\overline{s_{ij}^2} = \sigma_{0ij}^2$, $(i, j = 1, 2)$—mean values of the squares of the moduli of the elements of the scattering matrix in the zero polarization basis.

If the zero basis is assumed to be initial, then by means of equality $\overline{G_H} = Q^* \overline{G_0} Q^{\mathrm{T}}$ we can write down the general form of the averaged Graves matrix in an arbitrary polarization basis:

$$\overline{G} = \frac{\sigma_{\Sigma\Phi}^2}{2} \begin{pmatrix} 1 + q_\Phi \cos 2\gamma & -q_\Phi \cdot e^{-2i(\eta+\delta)} \sin 2\gamma \\ -q_\Phi \cdot e^{2i(\eta+\delta)} \sin 2\gamma & 1 - q_\Phi \cos 2\gamma \end{pmatrix}. \tag{7.31}$$

In Formula (7.31), the following notations are introduced:

$\sigma_{\Sigma\Phi}^2 = \sigma_{011}^2 + \sigma_{022}^2 + 2\sigma_{012}^2$—total ESA of the fluctuating target;

$q_\Phi = \left(\sigma_{011}^2 - \sigma_{022}^2\right) / \sigma_{\Sigma\Phi}^2$—degree of anisotropy of the fluctuating target, while this definition can be considered as a generalization of the concept of the anisotropy degree, introduced for stable targets, while it is obvious that $|q_\Phi| \le 1$.

It follows from (7.31) that, first, as already shown, the trace of the averaged Graves matrix, equal to $\sigma_{\Sigma\Phi}^2$, and its determinant equal to $\overline{G} = 0{,}25\sigma_{\Sigma\Phi}^4 \left(1 - q_\Phi^2\right)$, are invariant to a change in the polarization basis.

If we write down the difference between the diagonal elements in Formula (7.31), we will obtain:

$$\sigma_{11}^2 - \sigma_{22}^2 = \left(\sigma_{011}^2 - \sigma_{022}^2\right) \cos 2\gamma = q_\Phi \sigma_{\Sigma\Phi}^2 \cos 2\gamma, \tag{7.32}$$

where $\overline{|s_{mn}|^2} = \sigma_{mn}^2 (m, n = 1, 2)$—mean values of the squares of the moduli of the elements of the scattering matrix in an arbitrary polarization basis.

This equality shows that if the mean square values of the elements of the scattering matrix are equal to each other in any basis (excluding those where $2\gamma = 0.5\pi$), then they will be equal in any other basis.

Let us assume, by analogy (as was done for stable targets), that a polarization-isotropic fluctuating target is a target in which $q_{\Phi} = 0$, i.e., $\sigma_{011} = \sigma_{022}$, while a polarization degenerate fluctuation target is a target in which $q_{\Phi} = 1$, i.e., $\sigma_{012} = \sigma_{022} = 0$.

The above reasoning shows that the mean values of the squared moduli of the diagonal elements of the scattering matrix for an isotropic fluctuating target in any PB are equal to each other, and consequently the Graves matrix of such targets in any PB is a unitary matrix:

$$\overline{G_{nu\!u}} = \frac{\sigma_{\Sigma\Phi}^2}{2}\begin{pmatrix} 1 & 0 \\ 0 & 1 \end{pmatrix} = \frac{\sigma_{\Sigma\Phi}^2}{2} I. \tag{7.33}$$

For a degenerate target, the eigenbasis coincides with the zero basis; matrix $\bar{G}$ in this case, in an arbitrary basis, assumes a general form, which can be easily obtained from Formulas (6.47) and (7.29), where $\overline{|\lambda^2|}$ means σ_{011}^2.

By introducing the complex coefficient of covariance $r_{mn\ pt} \cdot e^{iq_{mn\ pt}}$, defined by means of equality $\overline{s_{mn}^* s_{pt}} = \sigma_{mn}\sigma_{pt} r_{mn\ pt} e^{iq_{mn\ pt}}$, the averaged Graves matrix can be written in the following form:

$$\overline{G} = \begin{pmatrix} \sigma_{11}^2 + \sigma_{12}^2 & \sigma_{12}\left(\sigma_{11} r_{1112} e^{iq_{1112}} + \sigma_{22} r_{1222} e^{iq_{1222}}\right) \\ \sigma_{12}\left(\sigma_{11} r_{1112} e^{-iq_{1112}} + \sigma_{22} r_{1222} e^{-iq_{1222}}\right) & \sigma_{12}^2 + \sigma_{22}^2 \end{pmatrix} \tag{7.34}$$

Let us now consider the properties of mixed second-order moments to which the following products refer in this case: $\overline{s_{12}s_{22}^*}$, $\overline{s_{11}s_{12}^*}$, $\overline{s_{11}^*s_{22}}$.

Using Formulas (7.31) and (7.29), let us write the following equation:

$$\sigma_{12}\left(\sigma_{11} r_{1112} e^{iq_{1112}} + \sigma_{22} r_{1222} e^{iq_{1222}}\right) = -0{,}5\sigma_{\Sigma\Phi}^2 q_{\Phi} e^{-2i(\eta+\delta)} \sin 2\gamma. \tag{7.35}$$

A number of useful relations follow from (7.35).

First of all, in the zero basis ($2\gamma = 0$), for all targets, except those with $\sigma_{12} = 0$, we have the following equality:

$$\left.\begin{aligned} r_{01112}\sigma_{011} &= \sigma_{022} r_{01222} \\ q_{01112} &= q_{01222} + (2n+1)\pi, \end{aligned}\right\} \quad n = 0, \pm 1, \pm 2. \tag{7.36}$$

for polarization-isotropic targets $\left(q_{\Phi}=0\right)$, Formulas (7.32) and (7.35) give the following equality (valid in any basis):

$$\left.\begin{aligned} r_{1112} &= r_{1222} \\ q_{1112} &= q_{1222} + (2n+1)\pi \end{aligned}\right\}. \tag{7.37}$$

Let us pay attention to the fact that simultaneous vanishing of r_{1112} and r_{1222} is possible only for polarization-isotropic targets.

The complex covariance coefficients that take into account the statistical relationships between the elements of the scattering matrix vary with the polarization basis, and the mean Graves matrix becomes diagonal if conditions (7.36) are satisfied, which shows that in the zero polarization basis, the complex correlation coefficients between the off-diagonal and diagonal elements of the scattering matrix and their mean square values are dependent quantities. The foregoing means that in an arbitrary polarization basis, in the general case, a certain functional connection must exist between the named parameters.

Using the zero polarization basis as the initial one, we can write the expressions for the mean square values of the elements of the scattering matrix in an arbitrary polarization basis (parameters γ, η, δ with respect to the zero basis, in this case φ—arbitrarily).

The required relations can be obtained on the basis of Formulas (6.26) and (7.36):

$$\left.\begin{aligned} \sigma^2_{11(22)} = {}& \sigma^2_{011(022)} \cos^4\gamma + \sigma^2_{022(011)} \sin^4\gamma + \sigma^2_{012} \sin^2 2\gamma \\ & +0.5\sigma_{011}\sigma_{022} r_{01122} \cos(4\delta + q_{01122}) \sin^2 2\gamma \\ & +\sigma_{011}\sigma_{012} r_{01112} \cos(2\delta - q_{01112}) \sin 4\gamma \\ \sigma^2_{012} = {}& \left(\sigma^2_{011} + \sigma^2_{012}\right) \sin^2\gamma \cos^2\gamma + \sigma^2_{012} \cos^2 2\gamma \\ & -0.5\sigma_{011}\sigma_{022} r_{01122} \cos(4\delta - q_{01122}) \sin^2 2\gamma \\ & -\sigma_{011}\sigma_{012} r_{01112} \cos(2\delta - q_{01112}) \sin 4\gamma \end{aligned}\right\} \tag{7.38}$$

Relations (7.38) are of great importance for practice, since the values given are proportional to the corresponding ESA values.

Expressions for complex covariance coefficients have the same structure and can be obtained from the same relations. In view of their cumbersomeness, they are not explicitly presented.

The left-hand side of Eq. (7.35) is also zero if $\sigma_{12} = 0$. This is possible only if $s_{12}(t) \equiv 0$. Under such a situation, we are dealing with a target whose fluctuation scattering matrix is reduced to a diagonal form in some polarization basis. Such targets (objects) will be called diagonal targets. Using Formula (6.26), we can obtain the root-mean-square values of the squares of the moduli of the elements of the scattering matrix of diagonal targets:

$$\left.\begin{aligned} & \sigma_{Д11(22)} = \sigma^2_{1(2)} \cos^4\gamma + \sigma^2_{2(1)} \sin^4\gamma + 0.5\sigma_1\sigma_2 r_{12} \cos(4\delta - q_{12}) \sin^2 2\gamma \\ & \sigma^2_{Д12} = \sigma_1^2 + \sigma_2^2 - 0.5\sigma_1\sigma_2 r_{12} \cos(4\delta - q_{12}) \sin^2 2\gamma \end{aligned}\right\}, \tag{7.39}$$

where σ_1^2, σ_2^2—rms values of the diagonal elements of the corresponding scattering matrix, r_{12}, q_{12}—module and phase of the complex correlation coefficient between them, respectively.

We return to the inequalities related to the estimate of r_{1122}. If the class A polarization basis, in which the mean Graves matrix is diagonal, is chosen as the initial one, then inequality (6.152) can be substantially simplified, since equalities (7.36) are simplified. By using them, we get:

$$\begin{aligned} &r_{011122}^2 + 2r_{01222}r_{01122}\cos(q_{01122} - q_{01112}) \\ &\quad - \left(r_{01112}^2 + r_{01222}^2\right)^2 (2r_{01112}r_{01222})^{-2} + \left(r_{01222}^2 + r_{01112}^2\right)^2 \le 0. \end{aligned} \quad (7.40)$$

The solution of this inequality can be represented as the following restriction on coefficient r_{01122}:

$$r_{01122} \le -X\cos\varphi_0 + \frac{X}{Y}\sqrt{Y^2\cos\varphi_0 - 2Y + 1}, \quad (7.41)$$

In this case, quantities X, Y and φ_0 introduced in Formula (7.41) have the following meanings:

$$\left.\begin{aligned} X &= r_{01112}r_{01222} \\ Y &= \tfrac{2r_{01112}^2 r_{01222}^2}{r_{01112}^2 + r_{01222}^2} \\ \varphi_0 &= q_{01122} - 2q_{01112} \end{aligned}\right\}. \quad (7.42)$$

Since coefficient r_{01122} always "exists," then the expression under the radical in relation (7.41) should not be negative. This requirement allows us to impose some restrictions on the phase angle

$$|\sin\varphi_0| \le 1 - Y. \quad (7.43)$$

Inequalities (7.41) and (7.43) impose a restriction on the phase and the modulus of the correlation coefficient of the off-diagonal elements of the scattering matrix. A number of interesting analogous relations for special targets were obtained in the literature.

Let us consider the properties of the target, in which the elements of the scattering matrix are statistically independent quantities. Equality (7.35) shows that in the zero polarization basis ($\gamma = 0$), in the absence of a statistical relationship between the elements of the scattering matrix s_{11} and s_{12}, i.e., at $r_{01112} = 0$, there are three possible situations: either $\sigma_{012} = 0$, or $\sigma_{022} = 0$, or $r_{01222} = 0$. In the first case, the target should be assigned to the diagonal class, because condition $\sigma_{012} = 0$ unequivocally requires that $s_{012}(t) \equiv 0$, the second condition leads to degenerate targets. In the third case, condition $r_{01112} = r_{01222} = 0$ removes the restrictions imposed on r_{01122} and q_{01122} by inequalities (7.41) and (7.43). In this case, in the zero polarization basis, the complex covariance coefficient $r_{01122}e^{iq_{01122}}$, as well as σ_{011}, σ_{022}, and σ_{012}, can be arbitrary.

If, as a result of measurements in a certain polarization basis, it turns out that $r_{1112} = r_{1222} = 0$, then in accordance with Formula (7.35), this will mean that either the basis in which the measurement is carried out belongs to class A ($\gamma = 0$), or $\sigma_{011} = \sigma_{022}$. The latter, in accordance with relations (7.38), leads to equality $\sigma_{11} = \sigma_{22}$ in any polarization basis, i.e., to a statistically isotropic target.

Let us consider one more particular case when in a zero polarization basis (class A basis) for some target, there is no statistical relationship between each pair of elements of the scattering matrix ($r_{01112} = r_{01222} = r_{01122} = 0$).

For this target, in the polarization basis, which is determined with respect to the initial basis by means of parameters γ, η, δ, it follows from relations (6.26) that:

$$\left.\begin{array}{l}
\sigma^2_{11(22)} = \sigma^2_{011(022)}\cos^4\gamma + \sigma^2_{022(011)}\sin^4\gamma + \sigma^2_{012}\sin^2 2\gamma \\
\sigma^2_{012} = \left(\sigma^2_{011} + \sigma^2_{012}\right)\sin^2\gamma\cos^2\gamma + \sigma^2_{012}\cos^2 2\gamma \\
\sigma_{11}\sigma_{12}r_{1112}e^{iq_{1112}} = -\frac{\sin 2\gamma}{4}\left[\left(\sigma^2_{011} - \sigma^2_{022}\right) + \left(\sigma^2_{\Sigma} - 6\sigma^2_{012}\right)\cos 2\gamma\right]e^{2i(\eta+\delta)} \\
\sigma_{12}\sigma_{22}r_{1112}e^{iq_{1222}} = -\frac{\sin 2\gamma}{4}\left[\left(\sigma^2_{011} - \sigma^2_{022}\right) - \left(\sigma^2_{\Sigma} - 6\sigma^2_{012}\right)\cos 2\gamma\right]e^{2i(\eta+\delta)} \\
\sigma_{11}\sigma_{22}r_{1112}e^{iq_{1122}} = \frac{\sin 2\gamma}{4}\left(\sigma^2_{\Sigma} - 6\sigma^2_{012}\right)e^{4i(\eta+\delta)}
\end{array}\right\}. \quad (7.44)$$

From the above relations, it is clear that the following equality necessarily takes place for the target in question:

$$\arg\overline{s_{11}s^*_{22}} = 2\arg\overline{s_{11}s^*_{12}} + n\pi = 2\arg\overline{s_{12}s^*_{22}} + m\pi, \quad n, m = 0, 1, 2, \ldots \quad (7.45)$$

Condition (7.45) is necessary, but by no means sufficient for the statistical independence of the elements of the scattering matrix.

We note that when the polarization basis is changed, the elements of the scattering matrix become statistically dependent quantities. As can be seen from relations (7.44), if $\sigma_{011} \neq \sigma_{022}$—then there is no other polarization basis other than zero, wherever r_{1112} and r_{1222} would simultaneously vanish. Moreover, if $\sigma_{022} < 2\sigma_{012} < \sigma_{011}$, then r_{1112} and r_{1222} are zero only in the zero basis. In B class ($2\gamma = \pi/2 + n\pi$), it follows from relations (7.44) that:

$$\left.\begin{array}{l}
r_{1112} = r_{1222} = \frac{\sigma^2_{011} - \sigma^2_{022}}{\sqrt{\sigma^2_{011} + \sigma^2_{022}}\sqrt{\sigma^2_{\Sigma} + 2\sigma^2_{012}}} \\
r_{1122} = \frac{\sigma^2_{\Sigma} - 6\sigma^2_{012} + 2\sigma_{011}\sigma_{022}r_{01122}\cos(2\delta - q_{01122})}{\sigma^2_{\Sigma} + 2\sigma^2_{012}} \\
2q_{1112} = 2q_{1222} = q_{1122} = 2(\eta + \delta) + 2\pi
\end{array}\right\} \quad (7.46)$$

Suppose that for some target in some polarization basis, but not in the zero one, equality $r_{1112} = r_{1222} = 0$ exists. This means that the mean Graves matrix is diagonal in this basis. It follows from (7.35) that if $\gamma \neq 0$, then $\sigma_{011} = \sigma_{022}$, and therefore, as already mentioned above, equality $\sigma_{11} = \sigma_{22}$ holds in any polarization basis. Thus, the statistical independence between the elements of the scattering matrix s_{11} and s^*_{12},

s_{12} and s_{22}^* not in the zero basis occurs only for polarization-isotropic targets. When the polarization basis is changed between the elements of the scattering matrix, a statistical relationship appears.

It follows from relations (7.44) that for targets, for which the mean square values of the elements of the scattering matrix are related by

$$\sigma_{11}^2 = \sigma_{22}^2 = 2\sigma_{12}^2 \tag{7.47}$$

there is no statistical dependence between all elements of the scattering matrix when the polarization basis is changed. In this case, $\sigma_{11}^2 = \sigma_{22}^2 = 2\sigma_{12}^2 = \sigma_{011}^2 = \sigma_{022}^2 = 2\sigma_{012}^2$ and $\left|\overline{s_{11}s_{12}^*}\right| = \left|\overline{s_{12}s_{22}^*}\right| = \left|\overline{s_{11}s_{22}^*}\right| = 0$.

Equation (7.47) is a necessary and sufficient condition for the statistical independence of the elements of the scattering matrix in any polarization basis. In other words, if relation (7.47) is satisfied in some arbitrary basis, then it will be satisfied in any other basis. We note that the assertions formulated are analogous to the corresponding condition for a completely unpolarized wave.

7.2.2 *Polarization Expansion of the Fluctuating Object*

In the analysis or processing of the signal reflected by the target, various equivalent representations of the real object in the form of a set of some targets whose statistical characteristics have known properties play an important role. As a rule, with various expansions, it is desirable to insure the statistical independence of the signals from the elements of the said set. In this case, the covariance scattering matrix and the Graves matrix of the original object will simply be equal to the sum of the corresponding matrices of the constituent parts. Let us show that an arbitrary fluctuating target can be uniquely represented as a sum of polarization-isotropic and diagonal purposes.

Let us consider relations (7.38), representing a general formula of the mean square values of the elements of the scattering matrix in an arbitrary polarization basis. If we put $\sigma_{012} = 0$ and $r_{01122} = 0$ in these formulas, the remaining formulas will be the corresponding variances for the diagonal target, which does not have a statistical relationship between the elements $\lambda_1(t)$ and $\lambda_2(t)$ of the scattering matrices ($\overline{\lambda_1\lambda_2^*} = \sigma_1\sigma_2 r_{12}\mathrm{e}^{\mathrm{i}q_{12}} = 0$). Let us denote the mean square values of these elements in accordance with relations (7.39) in terms of $\sigma_{\text{Д}}^2$, i.e., $\sigma_{\text{Д}}^2 \equiv \sigma_0^2\ (r_{12} = 0)$ ($r_{12} = 0$). The "zero" index will be referred to the initial zero basis, therefore $\sigma_1^2 = \sigma_{011}^2, \sigma_2^2 = \sigma_{022}^2$. In this case, we have:

$$\left.\begin{aligned} \sigma_{11}^2 &= \sigma_{\text{Д}11}^2 + \sigma_{\text{пиц}}^2 \\ \sigma_{22}^2 &= \sigma_{\text{Д}22}^2 + \sigma_{\text{пиц}}^2 \\ \sigma_{12}^2 &= \sigma_{\text{Д}12}^2 + \sigma_{012}^2 - \sigma_{\text{пиц}}^2 \end{aligned}\right\}, \tag{7.48}$$

where

$$\sigma_{nuu}^2 = \left[\sigma_{012}^2 + 0{,}5\sigma_{011}\sigma_{022}r_{01122}\cos(4\delta - q_{01122})\right]\sin^2 2\gamma +$$
$$+\sigma_{011}\sigma_{012}r_{01112}\cos(2\delta - q_{01112})\sin 4\gamma$$

Equation (7.48) gives the required expansion into polarization-isotropic and diagonal targets. The eigenbasis of the diagonal target coincides with the zero basis of the fluctuating target, and the corresponding eigenvalues will be equal to σ_{011}^2 and σ_{022}^2. In other words, for the resulting diagonal target in the zero basis, the scattering matrix has the following form:

$$S_{Д} = \begin{pmatrix} \lambda_1(t) & 0 \\ 0 & \lambda_2(t) \end{pmatrix}, \tag{7.49}$$

while $\overline{\lambda_1\lambda_2} = 0$.

The resulting expansion leads to the representation of the zero basis mean Graves matrix in the following form:

$$\bar{G} = \begin{pmatrix} \sigma_{011}^2 & 0 \\ 0 & \sigma_{022}^2 \end{pmatrix} + \sigma_{012}^2 \begin{pmatrix} 1 & 0 \\ 0 & 1 \end{pmatrix}. \tag{7.50}$$

For the diagonal target, it is possible to calculate the complex coefficient of covariance between the elements s_{11} and s_{12}^* using relations (7.39):

$$\left. \begin{array}{l} r_{1112} = \left| \dfrac{(\sigma_{011}^2 - \sigma_{022}^2) + (\sigma_{011}^2 + \sigma_{022}^2)\cos 2\gamma}{4\sqrt{\sigma_{011}^2 + \sigma_{022}^2}\sqrt{\sigma_{011}^2 \cos^4\gamma + \sigma_{022}^2 \sin^4\gamma}} \sin 2\gamma \right| \\ q_{1112} = 2(\eta + \delta) + (n+1)\pi \end{array} \right\}. \tag{7.51}$$

Figure 7.6 shows the dependence of the modulus of the covariance coefficient r_{1112} from angle γ and relation $q = \sigma_{022}/\sigma_{011}$. Since equality $r_{1112}(q, \gamma) = r_{1112}(1/q, 90° - \gamma)$ takes place, then values r_{1112} for $q >$ 1n can be also found from the constructed curves. For this, we must take quantity $1/q$ instead of q, and instead of angle γ—angle $(90° - \gamma)$.

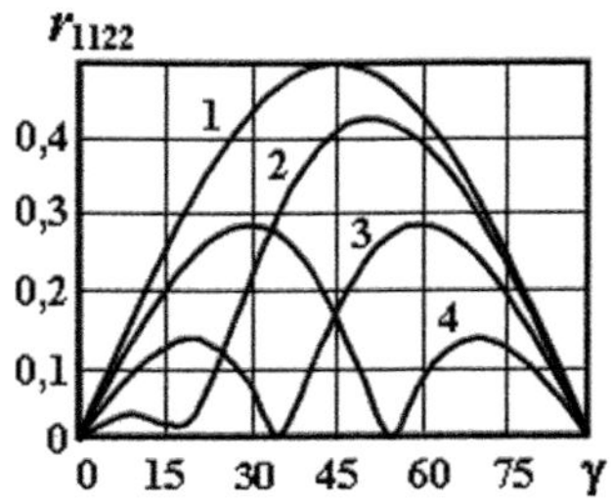

Fig. 7.6 Dependence of the covariance coefficient r_{1122} of the diagonal target, for which $\overline{\lambda_1\lambda_2} = 0$, on the angle γ for different values of the parameter q. 1—$q = 0$. 2—$q = 0.5$. 3—$q = 2$. 4—$q = 10$

Formula (7.39) also makes it possible to calculate the complex coefficient of covariance between elements $s_{11}(t)$ and $s_{22}(t)$ of the scattering matrix of the constructed diagonal target:

$$r_{1122} = \frac{\left(\sigma_{011}^2 + \sigma_{022}^2\right)\sin^2 2\gamma}{4\sqrt{\sigma_{011}^2\cos^4\gamma + \sigma_{022}^2\sin^4\gamma}\sqrt{\sigma_{011}^2\sin^4\gamma + \sigma_{022}^2\cos^4\gamma}}. \tag{7.52}$$

The curves in Fig. 7.7 establish the relationship between the quantity r_{1122} and the ratio of the eigen elements of the diagonal target in different polarization bases. As can be seen, with $\gamma = 45°$, i.e., in class *B* basis, the statistical relationship between the elements $s_{11}(t)$ and $s_{22}(t)$ of the scattering matrix of the diagonal target is maximal ($r_{1122} = 1$). If $\sigma_{011}^2 \sim \sigma_{022}^2$, then there are fairly broad classes of polarization bases, where values r_{1122} practically do not change.

As for the phase of the complex covariance coefficient, then, as follows from Formula (7.39), with $r_{12} = 0$:

$$q_{1122} = 4(\eta + \delta) + 2\pi = 2(q_{1112} + n\pi). \tag{7.53}$$

The expansion of an arbitrary fluctuating target considered above is an important generalization of the known expansion of an arbitrary stable target into a sum of polarization-isotropic and degenerate targets. With polarization selection of targets, knowledge of the relative contribution of each of the component parts makes it possible to estimate the relative efficiency of use, for example, of the most common circular or some linear polarization in practice.

7.2.3 *Reflection of Radio Waves*

We now turn to an analysis of the reflection of radio waves from fluctuating targets. Let us consider the problem of determining the characteristics of a partially polarized wave reflected from the fluctuating target, having a degree of polarization equal to m.

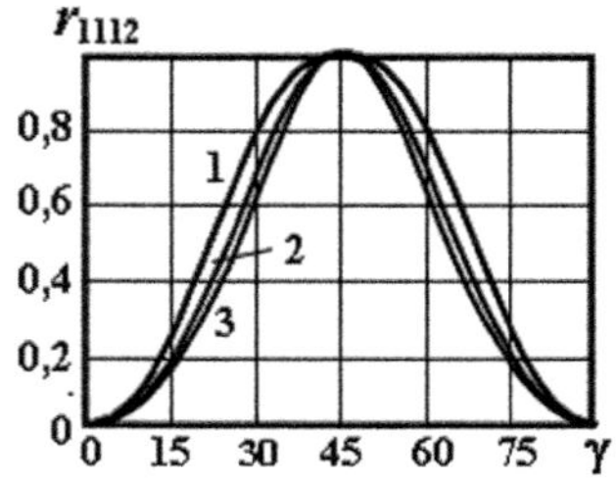

Fig. 7.7 Dependence of the covariance coefficient r_{1112} of the diagonal target, for which $\overline{\lambda_1\lambda_2} = 0$, on the angle γ for different values of the parameter q. 1—$q = 10$. 2—$q = 4$. 3—$q = 0.1$

First of all, let us determine the magnitude of the reflected power, for which we find the trace of the averaged coherence matrix of the reflected wave. In the polarization basis, where the matrix of the coherence of the incident wave is diagonal (the second singular basis), and with allowance for relations (3.11) and (7.17), we can obtain:

$$J_{\text{omp}} = 0.5\sigma_{\Sigma\Phi}^2 J_0\left(1 + mq_\Phi \cos 2\gamma\right), \tag{7.54}$$

where the angle of 2γ in this case means the angle between the images of the class A polarization basis of the fluctuating target (its Graves matrix is diagonal) and the second singular basis of the incident wave.

Let us find the degree of polarization of the reflected wave. For fluctuating purposes, it is fair to speak of the possibility of an "instantaneous" reduction of the scattering matrix to the diagonal form if only the correlation time of its elements is much greater than the period of high frequency. In this case, the eigenvalues of the scattering matrix $\lambda_1(t)$ and $\lambda_2(t)$ will be random functions of time, and the eigen polarization basis will also be random. The latter means that on the Poincaré sphere used for the geometric representation of the polarization state of a wave, diametrically opposite points $(1\ -1)$ corresponding to an eigen polarization basis will describe a certain random curve with time. In other words, the spherical coordinates of the pair $(1\ -1)$ will also be random functions of time. So, the scattering matrix of the fluctuating target at the moment of time t in the eigenbasis with symbol $(1\ -1)$ has the following form:

$$S_{1-1}(t) = \begin{pmatrix} \lambda_1(t) & 0 \\ 0 & \lambda_2(t) \end{pmatrix}. \tag{7.55}$$

Let us introduce the standard notations: $\overline{|\lambda_1|^2} = \sigma_1^2$; $\overline{|\lambda_2|^2} = \sigma_2^2$; $\overline{\lambda_1\lambda_2^*} = \sigma_1\sigma_2\rho e^{iq_{12}}$.

Matrix of coherence of the reflected wave K_{omp} is related to the coherence matrix of the incident wave by relation (7.17), in which statistical averaging is performed. Let us assign symbol $(0\ -0)$ to the second singular basis of the incident wave (matrix $\overline{K_{na\partial}}$ is diagonal). In this basis, it is obvious that the following formulas are valid: for the instantaneous coherence matrix

$$\left(K_{0-0}(t)\right)_{na\partial} = \begin{pmatrix} f_1^2(t) & f_{12}(t) \\ f_{12}^*(t) & f_2^2(t) \end{pmatrix}, \tag{7.56}$$

for the mean coherence matrix:

$$\overline{\left(K_{0-0}(t)\right)_{na\partial}} = \begin{pmatrix} r_1^2 & 0 \\ 0 & r_2^2 \end{pmatrix}. \tag{7.57}$$

In order to write coherence matrix $(K_{0-0}(t))_{nad}$ in the basis (1 −1), it must be subjected to the similarity transformation:

$$(K_{1-1}(t))_{nad} = \overset{*}{Q}_{01}(K_{0-0}(t))_{nad}Q_{01}^{T}, \tag{7.58}$$

where Q_{01} is a unitary matrix insuring a transition from the basis (0 −0) to the basis (1 −1).

By substituting relations (7.56) and (7.58) into Eq. (3.4) and performing the corresponding averaging under the assumption of statistical independence of matrices $S(t)$, $K_{nad}(t)$, $Q_{01}(t)$, taking into account Formulas (7.17) and (7.57), we obtain:

$$\overline{(K_{1-1}(t))_{nad}} = \overline{S_{1-1}\overset{*}{Q}_{01}(K_{0-0}(t))_{nad}Q_{01}\overset{*}{S}_{1-1}} \tag{7.59}$$

After the corresponding transformations, the coherence matrix of the reflected wave in some averaging basis will have the following Formula:

$$K_{\text{omp}} = \frac{J_0\sigma_\Sigma^2}{4}\begin{pmatrix} (1+q_\Phi)(1+m\overline{\cos 2\gamma_1}) & -mrb\sqrt{1-q^2}\,\overline{\sin 2\gamma_1 e^{iq_{12}}} \\ -mrb\sqrt{1-q^2}\,\overline{\sin 2\gamma_1 e^{-iq_{12}}} & (1-q_\Phi)(1-m\overline{\cos 2\gamma_1}) \end{pmatrix}, \tag{7.60}$$

where $b = \overline{\exp[-2i(\eta+\delta)]}$;

$2\gamma_1$—angle between the second singular basis of the incident wave and the image of the instantaneous position of the polarization basis (1 −1).

Since the invariants of the mean coherence matrix do not depend on the choice of basis, Formula (7.60) allows us to find a number of interesting relations. The total power of the reflected wave will be equal to the trace of the coherence matrix, i.e.,

$$J_{\text{omp}} = S_P\overline{K_{\text{omp}}} = 0{,}5\sigma_{\Sigma\Phi}^2 J_0\left(1+mq_\Phi\overline{\cos 2\gamma_1}\right). \tag{7.61}$$

Comparison of the obtained relation with equality (7.54) shows that

$$\langle\cos 2\gamma_1\rangle = \cos 2\gamma, \tag{7.62}$$

where, be reminded, 2γ—angle between the images of the A polarization basis of the fluctuating target and the second singular basis of the incident wave. Direct calculations for the degree of polarization of the reflected wave give the following relation:

$$m_{\text{omp}}^2 = 1 - \frac{\left(1 - q_{\Phi}^2\right)\left\{1 - m^2\left[\cos^2 2\gamma + \rho^2 \overline{|b|^2} \overline{\sin 2\gamma_1}^2\right]\right\}}{\left(1 + m q_{\Phi} \cos 2\gamma\right)^2}. \tag{7.63}$$

Let us consider a number of special cases.

When a fluctuating target is irradiated with an unpolarized wave ($m = 0$), it follows from Formula (7.63) that the degree of polarization of the reflected wave is equal to the degree of anisotropy of the target, i.e.,

$$m_{\text{omp}} = q. \tag{7.64}$$

In this case, the energy reflection coefficient $z_{\text{omp}} = J_{\text{omp}}/J_{na\partial}$ is equal to half of the total ESA of the target.

For a stable target, when $\gamma_1 = \text{const}$, $\rho = 1$, $|b|^2 = 1$, relation (7.63) is transformed into the following form:

$$m_{\text{omp}} = 1 - \frac{\left(1 - q_{\Phi}^2\right)\left(1 - m^2\right)}{\left(1 + m q_{\Phi} \cos 2\gamma\right)}. \tag{7.65}$$

If the target is polarization isotropic $\left(q_{\Phi} = 0\right)$, then

$$m_{\text{omp}} = m. \tag{7.66}$$

The last equation shows that, within the assumptions under which Formula (7.63) is obtained, there is a rigid connection between the elements of the scattering matrices of the polarization-isotropic target, while either

$$r_{1112} = r_{1222} = 1 \tag{7.67}$$

or this target is diagonal ($s_{12}(t) \equiv 0$), i.e., for it—$\sigma_{12} = 0$.

For an arbitrary diagonal target, $\gamma_1 = \text{const}$, $\langle |b|^2 \rangle = 1$ is to be put.

An important role in the polarization analysis belongs to the extreme values of the polarization degree of the reflected wave, the knowledge of which allows us to perform a series of limit estimates. Quantity m_{omp} for an arbitrary target of the minimum possible value, as follows from Formula (7.63), can achieve $\rho = 0$:

$$m_{\text{omp}_{\min}} = \frac{\left|q_{\Phi} + m \cos 2\gamma\right|}{1 + m q_{\Phi} \cos 2\gamma}. \tag{7.68}$$

Minimum minimore for the case $m \geq q_{\phi}$ will occur when $\cos 2\gamma = -q_{\phi}/m$, and therefore

$$m_{\text{omp}_{\text{minmin}}} = 0. \tag{7.69}$$

For the case $m \leq q_{\phi}$, this minimum will occur when $\cos 2\gamma = -1$

$$m_{\text{omp}_{\text{min min}}} = \frac{q_{\phi} - m}{1 - mq_{\phi}}. \tag{7.70}$$

m_{omp} takes the maximum value with $\rho = |b| = 1, \cos 2\gamma_1 = 1, \overline{\sin 2\gamma_1} = 0$. It will be equal to:

$$m_{\text{omp}_{\max}} = \frac{q_{\phi} + m}{1 + mq_{\phi}}. \tag{7.71}$$

Consequently, for the degree of polarization of a wave reflected from an arbitrary fluctuating target, the following inequality holds upon irradiation of a partially polarized wave:

$$\frac{q_{\phi} + m}{1 + mq_{\phi}} \geq m_{\text{omp}} \geq \begin{cases} \dfrac{q_{\phi} - m}{1 - mq_{\phi}}, & if\ m \leq q_{\phi} \\ 0, & if\ m > q_{\phi} \end{cases}. \tag{7.72}$$

Figure 7.8 shows the curves corresponding to the dependence of the extreme values m_{omp} on the degree of polarization of the incident wave m at different degrees of anisotropy q_{ϕ} of the target under study.

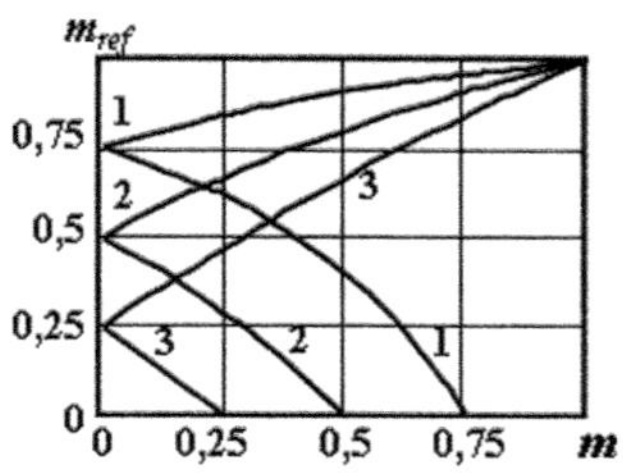

Fig. 7.8 Dependence of the extreme values of m_{ref} on the degree of polarization of the incident wave m for various degrees of anisotropy q_a of the irradiated target: 1—$q_a = 0.75$; 2—$q_a = 0.5$; 3—$q_a = 0.25$

7.3 Formation of Polarized Emission by Inhomogeneous Media

Let us consider the process of formation of polarized radiation in inhomogeneous media. We will use the calculation technique based on determining the loss of an auxiliary plane electromagnetic wave incident on the surface of the medium in the direction opposite to that for which the radiative properties of the medium are determined. The main points of this technique were described in Sect. 3.3. It is based on the reciprocity theorem, in accordance with which the polarization properties of radiothermal radiation are determined by the laws of absorption of the polarized auxiliary wave and its reflection.

As is known from the general theory, the reflection coefficient of an electromagnetic wave depends not only on the properties of the medium and the angle of incidence, but also on the nature of wave polarization. Therefore, the general case can be reduced to the consideration of two particular cases: reflection of a wave with an electric vector that is perpendicular to the plane of incidence (conventionally called horizontally polarized wave) and a wave for which vector $\vec{E}$ is parallel to the plane of incidence (vertically polarized wave).

The intrinsic radiothermal radiation of the medium is an accidental δ—correlated electromagnetic field, as a result of which it can be represented as the imposition of two independent waves (components) having mutually orthogonal polarizations. As such components, it is advisable to select waves with horizontal and vertical polarization, since these polarizations, as was shown in Sect. 3.3, often form an eigen polarization basis of the task. Coherence matrix K in this basis is diagonal and the degree of polarization of radiation m can be expressed in terms of its diagonal elements:

$$\overline{K} = \begin{pmatrix} \overline{|E_{HP}|^2} & 0 \\ 0 & \overline{|E_{VP}|^2} \end{pmatrix}, \quad m = \frac{\left|\overline{|E_{VP}|^2} - \overline{|E_{HP}|^2}\right|}{\overline{|E_{VP}|^2} + \overline{|E_{HP}|^2}}. \tag{7.73}$$

Thus, the analysis of the polarization properties of the intrinsic radiothermal radiation of a layered inhomogeneous medium can be reduced to solving the task of determining the power losses of auxiliary waves with "horizontal" and "vertical" polarization.

Let us consider the medium that fills the half-space $z \geq 0$, the electrophysical properties (complex dielectric permittivity) of which depend only on the distance to the medium surface: $\varepsilon = \varepsilon(z)$.

As an auxiliary wave, let us consider an electromagnetic wave whose electric vector is perpendicular to the plane of incidence. The electric vector of the refracted wave shall satisfy a differential equation of the following form: $\Delta E + k^2\varepsilon(z)E = 0$, which by substitution $E(x, z) = f(z)\exp\{-ik\sin\theta\}$ is reduced, in turn, to the following equation:

$$\frac{d^2 f(z)}{dz^2} + k^2\left[\varepsilon(z) - \sin^2\theta\right] f(z) = 0, \tag{7.74}$$

where —wave vector, $f(z)$—solution of Eq. (7.74) convergent to zero with $z \to 0$.

The expression for the reflection coefficient R_{HP} of the wave with horizontal polarization can be obtained using the usual conditions of continuity of the tangential components of vectors $\vec{E}$ and $\vec{H}$:

$$R_{HP} = \left.\frac{f(z,\theta) - \frac{i}{k}\sec\theta \frac{\partial f(z,\theta)}{\partial z}}{f(z,\theta) + \frac{i}{k}\sec\theta \frac{\partial f(z,\theta)}{\partial z}}\right|_{z=0}. \tag{7.75}$$

Heat loss power $d\Pi$ of the auxiliary wave in volume dV is given by the expression:

$$d\Pi(X, Y, Z) = P_H dV = \frac{\sigma}{2}|E|^2 dV = \frac{\sigma(z)}{2}|A|^2|f(z)|^2 dV, \tag{7.76}$$

where A—constant coefficient determined together with R_{HP} from the boundary conditions, P_H—volume density of heat losses.

It is seen from expression (7.76) that the volume density of thermal losses on the horizontal polarization P_H is a function of coordinate z only, which allows, as an element of volume dV, to take a layer with a thickness of dz and a unit cross-section, located at a distance z from the medium surface, and write the expression for $d\Pi(X, Y, Z)$ as follows:

$$d\Pi(z) = P_H(z)dz = 0.5\sigma(z)|A|^2|f(z)|^2 dz. \tag{7.77}$$

For a detailed analysis of the process of formation of radiation by the medium, let us turn to the WKB approximation, since in this case, it is possible to obtain fairly simple and, at the same time, quite general relations. In this case, the electric field strength of the refracted wave can be written as follows:

$$E(x, z) = \frac{1}{\sqrt[4]{\varepsilon(z) - \sin^2\theta}} \exp\left\{-ik\left[x\sin\theta + \int_0^z \sqrt{\varepsilon(z) - \sin^2\theta}\right]\right\}. \tag{7.78}$$

Let us transform the radicals into expressions (7.78). For a radical under the exponential sign, we get:

$$\sqrt{\varepsilon(z) - \sin^2\theta} = r(z) - is(z), \tag{7.79}$$

where $r(z) = g(z)\cos\left(\frac{\varphi(z)}{2}\right)$; $s(z) = g(z)\sin\left(\frac{\psi(z)}{2}\right)$, and $\psi(z)$ and $g(z)$ are related to the real and imaginary parts of the complex dielectric permittivity by the following relations:

$$\psi(z) = \arctan\frac{\varepsilon_2(z)}{\phi_1(z) - \sin^2\theta}, \quad \varepsilon(z) = \varepsilon_1(z) - \mathrm{i}\varepsilon_2(z);$$
$$g(z) = \left\{\left[\varepsilon_1(z) - \sin^2\theta\right]^2 + [\varepsilon_2(z)]^2\right\}. \tag{7.80}$$

When $\theta = 0$, the expression $\tan\psi(z)$ is the tangent of the angle of losses of the medium, and $r(z)$ and $s(z)$, respectively, are the real and imaginary components of the complex refractive index of the medium.

Let us write analogous representations for the radical in the denominator of expression (7.78):

$$\begin{array}{l}\left[\varepsilon(z) - \sin^2\theta\right]^{0.25} = r_1(z) - \mathrm{i}s_1(z)\\ r_1(z) = \sqrt{g(z)}\cos(0.25\psi(z)); \quad s_1(z) = \sqrt{g(z)}\sin(0.25\psi(z))\end{array}. \tag{7.81}$$

By substituting the expressions for the radicals from (7.80) and (7.81) into (7.78), we obtain for $E(x, z)$:

$$E(x, z) = \frac{A}{\sqrt{g(z)}}\exp\left\{-ik\left[\int_0^z g(z)\mathrm{d}z + \frac{\psi(z)}{4k} + x\sin\theta + \int_0^z r(z)\mathrm{d}z\right]\right\}. \tag{7.82}$$

The wave described by expression (7.78) is inhomogeneous: surfaces of equal amplitude are planes parallel to the boundary of the medium and surfaces of equal phase satisfy the following equation:

$$\frac{\psi(z)}{4} - kx\sin\theta - k\int_0^z r(z)\mathrm{d}z = 0. \tag{7.83}$$

In the general case, (7.83) is an equation for a complex cylindrical surface, and only for a homogeneous conducting medium it moves to the equation of the plane $x\sin\theta + zr(z) = 0$, the direction of the normal to which coincides with the direction of motion of the wave front, and angle ψ between the normal and the axis z can be found from $\psi = \arctan\frac{\sin\theta}{r(z)}$.

When $\sigma \to 0$ is inhomogeneous, the wave transforms into a homogeneous wave, and angle ψ becomes equal to an angle θ_1 determined from Snell's law.

By substituting the expression for $E(z)$ from (7.82) into (7.76), we obtain the relation for the volume power density of the thermal losses of the auxiliary wave, which coincides with the differential emissivity accurate within the constant coefficient:

$$p_H(z) = \frac{\sigma(z)}{2g(z)}\exp\left[-2k\int_0^z s(z)\mathrm{d}z\right]. \tag{7.84}$$

For a wave with vertical polarization, the electric vector $\vec{E}$ is parallel to the plane of incidence, and the vector $\vec{H}$ is perpendicular to it. In this case, it is possible to obtain a relatively simple differential equation only for function $\Phi(x, z)$:

$$\Phi(x, z) = H(x, z)/n(z). \tag{7.85}$$

By applying the variable separation method, we can obtain the following expression for $\Phi(x, z)$:

$$\Phi(x, z) = Af(z)\mathrm{e}^{-ikx\sin\theta}, \tag{7.86}$$

where function $f(z)$ is a solution of the differential equation that converges to zero at infinity:

$$f''(z) + \left[k^2\left(n^2(z) - \sin^2\theta\right) + \frac{n''(z)}{n(z)} - 2\left(\frac{n'(z)}{n(z)}\right)^2\right]f(z) = 0. \tag{7.87}$$

Thus, for $H(x, z)$, we have:

$$H(x, z) = An(z)f(z)\mathrm{e}^{-ikx\sin\theta}. \tag{7.88}$$

From the solution of the equations obtained by using the boundary conditions for the tangential components of vectors $\vec{E}$ and $\vec{H}$, we obtain the following expression for the reflection coefficient R_V on vertical polarization:

$$R_V = \frac{\frac{\mathrm{d}}{\mathrm{d}z}[n(z)f(z)]_{z=0} + \mathrm{i}kn^3(0)f(0)\cos\theta}{\frac{\mathrm{d}}{\mathrm{d}z}[n(z)f(z)]_{z=0} - \mathrm{i}kn^3(0)f(0)\cos\theta}. \tag{7.89}$$

Using relation (7.88) and performing the calculations similar to the previous ones, we obtain the expression for the volume power density of the thermal losses of the auxiliary wave with vertical polarization:

$$p_V(z) = \frac{|A|^2\sigma(z)}{2k^2|\varepsilon(z)|^2}\left\{\left|\frac{\mathrm{d}[n(z)f(z)]}{\mathrm{d}z}\right|^2 + k^2|n(z)f(z)|^2\sin\theta\right\}. \tag{7.90}$$

To obtain the ratio for $p_V(z)$, more accessible for analysis, we again turn to the WKB approximation, which makes it possible to write $H(x, z)$ as follows:

$$H(x, z) = \frac{An(z)}{\sqrt{N(z)}}\exp\left\{-\mathrm{i}k\left[x\sin\theta + \int_0^z N(z)\mathrm{d}z\right]\right\}, \tag{7.91}$$

where $N(z) = \left\{n^2(z) - \sin^2\theta + \frac{1}{k^2}\left[\frac{n''(z)}{n(z)} - 2\left(\frac{n'(z)}{n(z)}\right)^2\right]\right\}^{0.5} \approx \sqrt{n^2(z) - \sin^2\theta}$.

Thus, accurate to the constant factor, we have for the differential emissivity:

$$p_V(z) \cong \frac{\sigma(z)}{2}|A|^2\left[\frac{g(z)}{|n(z)|^2} + \frac{\sin^2\theta}{g(z)}\right]\exp\left[-2k\int\limits_0^z s(z)\mathrm{d}z\right]. \tag{7.92}$$

The results obtained can be combined by presenting them in a matrix form. This combination is not purely formal but allows for certain generalizations.

Let us introduce a matrix, which will be further called the differential loss matrix $P_{Д1}(z)$:

$$P_{Д1}(z) = \begin{pmatrix} p_H(z) & 0 \\ 0 & p_V(z) \end{pmatrix}. \tag{7.93}$$

In accordance with the foregoing, accurate to the constant factor, the nonzero coefficients of this matrix represent a contribution to the resulting radiation on both polarizations of the layer dz, located at a distance z from the boundary of the medium. By multiplying Eq. (7.93) by this factor, we obtain matrix $P_Д(z)$, which we call the differential emissivity matrix:

$$P_Д(z) = b\begin{pmatrix} p_H(z) & 0 \\ 0 & p_V(z) \end{pmatrix}. \tag{7.94}$$

As can be seen from expression (7.94), the off-diagonal elements of this matrix are zero, since it is written in its own polarization basis. When moving to another basis (from the experimental point of view when measuring on two other orthogonal polarizations; in the general case—elliptic ones), they represent a pair of complex conjugate numbers, and thus matrix $P_Д(z)$ is Hermitian conjugate.

Contribution to the resulting radiation from the near-surface layer $(0, z)$ can be expressed by matrix $Q(z)$:

$$Q(z) = b\begin{pmatrix} Q_H(z) & 0 \\ 0 & Q_V(z) \end{pmatrix}, \tag{7.95}$$

where $Q_H(z) = \int_0^z p_H(z)\mathrm{d}z$; $Q_V(z) = \int_0^z p_V(z)\mathrm{d}z$.

With $z \to 0$, matrix $Q(z)$ obviously transits to the coherence matrix:

$$K = \lim_{z\to 0} Q(z). \tag{7.96}$$

By normalizing the radiation of an absolutely black body to a single matrix, we can introduce matrices of relative emissivity (loss matrix) and radio brightness temperature:

$$\left.\begin{aligned} \Pi &= \begin{pmatrix} \chi_H & 0 \\ 0 & \chi_V \end{pmatrix} = \begin{pmatrix} 1-|R_{HP}|^2 & 0 \\ 0 & 1-|R_{VP}|^2 \end{pmatrix} \\ T &= \begin{pmatrix} T_H & 0 \\ 0 & T_V \end{pmatrix} = \begin{pmatrix} \chi_H T_0 & 0 \\ 0 & \chi_V T_0 \end{pmatrix} \end{aligned}\right\}. \tag{7.97}$$

Let us consider the differential and integral polarization characteristics of radiation of certain classes of terrestrial coverings. We start with an isothermal homogeneous medium.

Reflection coefficients R_{HP} and R_{VP} in this case, as is well known, can be calculated from the Fresnel formulas, from which a well-known fact follows that the degree of polarization of the radiothermal radiation reaches a maximum value for the angle θ determined by Brewster's law.

In this case, for the elements of the differential emissivity matrix we have:

$$p_{H,V}(z) = b_{H,V}\exp(-2kgz), \tag{7.98}$$

where $g(z) = \left[\left(\varepsilon_1^2 - \sin^2\theta\right)^2 + \varepsilon_2^2\right]\sin\frac{\psi}{2}$, $\psi = \arctan\frac{\varepsilon_2}{\varepsilon_1 - \sin^2\theta}$.

It follows from expression (7.98) that the effective thickness of the radiating layer for both polarizations is the same and increases as viewing angle θ decreases.

Let us now consider media with a transition layer. The complex permittivity variation law in this case is given by the formula:

$$\varepsilon(z) = \beta \mathrm{e}^{2\alpha z} = (\beta_1 + i\beta_2)\mathrm{e}^{2\alpha z}, \ \operatorname{Im}\alpha = 0. \tag{7.99}$$

Equations (7.74) and (7.87) have the same form, differing only in the intercept when $f(z)$.

It should be emphasized that function $f(z)$ in these equations has a different physical meaning: in the first, it denotes $E(z)$ and in the second—$H(z)/[n(z)]$. For an electric vector with horizontal polarization, we can write:

$$E_{HP}(x,z) = AH_\xi^{(2)}\left[\frac{k}{\alpha}\sqrt{\beta}\mathrm{e}^{\alpha z}\right]\mathrm{e}^{-ikx\sin\theta}, \tag{7.100}$$

where $H_\xi^{(2)}$—Hankel function of the second kind of order $\xi = \frac{k}{\alpha}\sin\theta$.

Accordingly, for a magnetic vector with vertical polarization, we have:

$$H_{VP}(x,z) = A_1 n(z) H_\eta^{(2)}\left[\frac{k}{\alpha}\sqrt{\beta}\mathrm{e}^{\alpha z}\right]\mathrm{e}^{-ikz\sin\theta}, \tag{7.101}$$

where $\eta = \sqrt{1 + \left(k^2 \sin^2 \theta\right)/\alpha^2}$, A_1—constant factor, different from A.

From Formulas (7.100) and (7.101), using the usual boundary conditions, we can obtain expressions for R_{HP} and R_{VP}, and, consequently, the matrix of radio brightness temperatures, which, in turn, makes it possible to determine the desired p_H and p_V, which in the WKB approximation will have the following form:

$$\left.\begin{aligned} p_H(z) &= C_1 e^{\alpha z} \exp\left\{-\frac{k\beta_2 e^{\alpha z}}{\alpha|\beta|}\right\} \\ P_B(z) &= C_2 e^{\alpha z} \exp\left\{-\frac{k\beta_1 e^{\alpha z}}{\alpha|\beta|}\right\} \end{aligned}\right\}, \tag{7.102}$$

where C_1 and C_2—some constants.

Thus, the differences in the formation of radiation on horizontal and vertical polarizations occur only in the layers closest to the surface. With the increasing angle θ, these differences increase. The effective thickness of the radiating layer is approximately the same on both polarizations.

Let us consider a medium with a linear law of change in the refractive index $n(z)$. In this case, the complex dielectric permittivity of the medium varies according to the law:

$$\varepsilon(z) = (\alpha z + \beta)^2, \tag{7.103}$$

where $\alpha = \alpha_1 + \mathrm{i}\alpha_2$, $\beta = \beta_1 + \mathrm{i}\beta_2$.

The corresponding Eqs. (7.74) and (7.87) for horizontal and vertical polarizations differ from each other and do not have integrals expressed in terms of the known functions. However, for small α in the WKB approximations, we can write the solution of these equations in the following form:

$$f(z) = \frac{A}{\left[(\alpha z + \beta)^2 - \sin^2 \theta\right]^{0.25}} \exp\left\{-ik \int\limits_0^z \sqrt{(\alpha z + \beta)^2 - \sin^2 \theta}\, \mathrm{d}z\right\}. \tag{7.104}$$

From expressions (7.75) and (7.89), taking into account relations (7.103), (7.104), we can obtain the reflection coefficients on the horizontal and vertical polarizations:

$$\left.\begin{aligned} R_{HP} &= \frac{\cos\theta - \sqrt{\beta^2 - \sin^2\theta} + i\alpha\lambda\beta\left(2\pi\sqrt{\beta^2 - \sin^2\theta}\right)^{-1}}{\cos\theta + \sqrt{\beta^2 - \sin^2\theta} + i\alpha\lambda\beta\left(2\pi\sqrt{\beta^2 - \sin^2\theta}\right)^{-1}} \\ R_{VP} &= \frac{\beta^3 - \sqrt{\beta^2 - \sin^2\theta} + i2\pi\alpha\left[\beta^2(\beta-\alpha) + \alpha\sin^2\alpha\right]\left[\beta\lambda\left(\beta^2 - \sin^2\theta\right)\right]^{-1}}{\beta^3 + \sqrt{\beta^2 - \sin^2\theta} + i2\pi\alpha\left[\beta^2(\beta-\alpha) + \alpha\sin^2\alpha\right]\left[\beta\lambda\left(\beta^2 - \sin^2\theta\right)\right]^{-1}} \end{aligned}\right\} \tag{7.105}$$

To analyze the radiation generation process, let us define the functions $p_H(z)$ and $p_V(z)$.

Knowing R_{HP} and R_{VP}, expressions for $p_H(z)$ and $p_V(z)$ for large z can be written in the following form:

$$p_H(z) = p_V(z) = \frac{V|A|^2}{2 \cdot \sqrt[4]{\left[n_1^2(z) - n_2^2(z) - \sin^2\theta\right] + 4n_1(z)n_2(z)}}$$
$$\times \exp\left\{-2k\int_0^z \left[\left[n_1^2(z) - n_2^2(z) - \sin^2\theta\right]^2 + 4n_1^2(z)n_2^2(z)\right]^{0.25} \sin(0.5\psi(z))\mathrm{d}z\right\} \tag{7.106}$$

Let us consider the special case, when $n_1^2(z)$ and $n_2^2(z)$ increase so that their ratio remains unchanged:

$$n_2(z) = bn_1(z), \quad 0 \le b \le 1 - \frac{1}{n_{1\min}^2(z)}. \tag{7.107}$$

By substituting $n_2^2(z)$ from (7.107) into (7.106), we obtain:

$$p_H(z) = C(b)n_1(z)\exp\left\{-2k\int_0^z \gamma(z,\theta)n_1^2(z)\mathrm{d}z\right\}, \tag{7.108}$$

where the notation is introduced

$$\gamma(z,\theta) = \frac{1}{\alpha}\sqrt{\left(1 - b^2 - \sin^2\theta\right)^2 + 8b^3\left[4b^2 + \left(1 - b^2 - \frac{\sin^2\theta}{n_1^2(z)}\right)^2\right]^{-0.5}}, \tag{7.109}$$

and let us pay attention to the fact that this quantity is almost constant, which allows us to write (7.108) in the following form:

$$p_H(z) = Cn_1(z)\exp\left\{-k\gamma n_1^2(z)\right\}. \tag{7.110}$$

It follows from (7.106) that function $p_H(z)$ can have an extremum under certain conditions. The extremum study of this function leads to the following expression for z_m—the depth of occurrence of a thin layer with the maximum relative contribution:

$$z_m = \frac{1}{\alpha_1}\left(\frac{1}{2\gamma} - \beta_1\right). \tag{7.111}$$

The maximum exists under condition $\gamma < 1/(2\beta_1)$, which corresponds to a sufficiently high rate of increase $\varepsilon(z)$.

Let us consider the process of radiation formation in media with a refractive index decreasing with depth, which is of great practical importance.

One of the possible models of such media is the exponential model with $\alpha < 0$. Another possible model arises with the following approach:

Let us set the goal to find such a law of change $\varepsilon(z)$ with the depth, at which Eqs. (7.74) and (7.87) coincide. A necessary and sufficient condition for this is the compliance with the following relation:

$$\frac{1}{n(z)} \cdot \frac{d^2 n(z)}{dz^2} - 2\left[\frac{1}{n(z)} \cdot \frac{dn(z)}{dz}\right]^2 = 0. \tag{7.112}$$

A direct substitution shows that one of the solutions of this differential equation is the function:

$$n(z) = \frac{l}{\alpha z + \beta}, \quad \varepsilon(z) = \frac{l^2}{(\alpha z + \beta)^2}, \tag{7.113}$$

which describes a medium with a monotonically decreasing complex refractive index (with $\alpha > 0$).

In case of a normal incidence of an auxiliary wave on the surface of the medium, Eqs. (7.74) and (7.87) have a general solution in the following form:

$$f(z) = A\exp\left\{\left(0.5 - \mathrm{i}\sqrt{k^2 l - 0.25}\right)z\right\}. \tag{7.114}$$

However, if the auxiliary wave is inclined, and when an intercept is added to (7.113), which allows us to describe the real media more correctly, the equations acquire a form for which there is no closed solution.

Above, the features of the polarization characteristics of non-dispersive media, to which most of the earth's rocks belong, were described. As is known, in the general case, the electrical properties of the medium are anisotropic, that is, they depend on the direction of propagation of the wave and the orientation of the electric vector. For a magnetically isotropic medium with relative magnetic permeability equal to unity, its electrical properties are completely described by the dielectric constant tensor, whose components, generally speaking, are frequency functions.

Let us consider the influence of the anisotropy of the dielectric constant on the polarization characteristics of the radiothermal radiation of such media.

The intensity of the orthogonal radiation components to be summed in one plane is determined by relation (3.150), in which δ_f is not necessarily a linear frequency function. If, for example, $\delta_f = \text{const}$, then $J = 4J_0 \cos^2 \varphi$, regardless of the bandwidth and measurement time, that is, the radiation is always polarized.

Let us consider the case when the radiothermal radiation formed in an isotropic medium passes through an anisotropic medium layer with a thickness of X_1. The phase difference δ_1 between the orthogonal components of the field is as follows:

$$\delta_1(f) = 2\pi X_1 f \left[\frac{1}{V_{\perp}(f)} - \frac{1}{V_{\mathrm{II}}(f)} \right], \tag{7.115}$$

where $V_{\perp}(f)$ and $V_{\mathrm{II}}(f)$—propagation velocity of the orthogonal components of the wave in basis $\left[\vec{e}_x, \vec{e}_y\right]$.

Chapter 8
Scattering of Polarized Radio Waves from Surface Structures and Backgrounds of Navigational Observation

8.1 Smooth Electrically Homogeneous Medium

It is important to know polarization characteristics of radio waves reflected from underlying covers, when solving the issues of remote sounding. Here first and foremost, learning the scattering parameters of exposed areas on the surface plays the most important role. It is clear that if electrophysical properties of such areas (salt content, humidity, soil composition, etc.) are changed, the basic electrodynamic characteristic of the surface—its complex dielectric constant ε—would change. The change in the complex dielectric constant leads to changes in the reflective characteristics of the underlying surface, i.e., its scattering parameters.

Let us hereafter consider the issues of establishment of relationships between the electrodynamic characteristics (complex dielectric constant) and the reflective characteristics of underlying covers on different polarizations.

The radiolocation implies the overview of underlying surfaces from upside down (from the board of the craft, satellite, fixed tower, etc.). At the same time, a certain area is irradiated on the underlying surface; its dimensions are determined by the height of observation H, the viewing angle θ, the antenna beam width (in two mutually perpendicular planes $\Delta\alpha$ and $\Delta\beta$). Figure 8.1 illustrates the stated above.

The wave from the antenna induces currents on the surface D, which are sources of the scattered field. In general cases, the energy scatters in all directions and in the direction of the antenna, in particular.

The geometric optics implies that the main energy flow coming in the direction of the antenna is created by such areas on the surface that are perpendicular to this direction (mirror reflection, highlights, or lucent areas). The presence of such areas is due to various irregularities and roughness on the surface. Wavy water surface, fairly even areas of hilly terrain, deserts, etc. are typical representatives of such surfaces. In this case, it can be assumed that the coefficient of reflection from such areas is equal to the Fresnel reflection coefficient at $\theta = 0$. Meanwhile, this coefficient is the same as horizontal and vertical polarization and equals to

Kozlov A. I. et al., *Introduction to the Theory of Radiopolarimetric Navigation Systems*, Springer Aerospace Technology, https://doi.org/10.1007/978-981-13-8395-3_8

Fig. 8.1 Overview of terrain

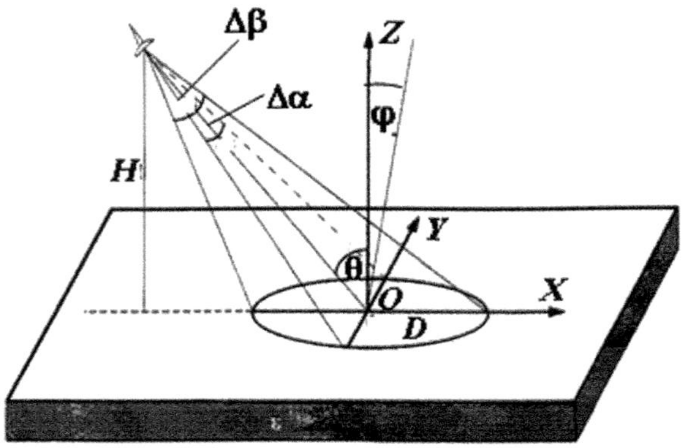

$$R_{(H,V)P} = \frac{1 - \sqrt{\varepsilon}}{1 + \sqrt{\varepsilon}}. \tag{8.1}$$

The intensity of the reflected wave is determined by the total area of such sections and their relative position on the surface. Let there be N sections. Then the resultant area would be the following total sum:

$$E_{\Sigma} = \frac{1 - \sqrt{\varepsilon}}{1 + \sqrt{\varepsilon}} \sum_{i=1}^{N} \int_{D_i} e^{i\vec{k}\vec{r}} \mathrm{d}D_i. \tag{8.2}$$

In Eq. (8.2), integration is performed point-by-point at one area, while summing up is performed by the number of areas; $\vec{r}$ is the radius vector of the antenna, and k is the wave vector.

The main thing in Eq. (8.2) is that the aggregate area may be represented by product of the Fresnel coefficient (depending only on the complex dielectric constant in this case) and some multiplier factor which is determined only by the geometry of the task and characteristics of the antenna. Therefore,

$$E_{\Sigma} = R_{(H,V)P} \Phi(\Gamma), \tag{8.3}$$

where $\Phi(\Gamma)$ is some geometric multiplier; Γ is some set of parameters characterizing the geometry of the task.

Thus, the design scattering matrix S would be as follows:

$$S = \Phi(\Gamma) \begin{pmatrix} R_{\mathrm{HP}} & 0 \\ 0 & R_{\mathrm{VP}} \end{pmatrix} = \Phi(\Gamma) \frac{1 - \sqrt{\varepsilon}}{1 + \sqrt{\varepsilon}} \begin{pmatrix} 1 & 0 \\ 0 & 1 \end{pmatrix}. \tag{8.4}$$

Let us consider another situation now. As it is well known, the direction of the reflected wave coincides with the direction of the mirrored wave, when plane radio waves reflect from smooth infinite surface. The energy does not transfer into the direction of the radiating antenna. Physically, this can be explained by the fact that there is always some elemental area ΔD on the surface that corresponds to another elemental area, the path difference from which to the antenna would equal to $\lambda/2$ (where λ is the length of the wave).

If the reflecting smooth surface is finite, then radio waves scatter in all directions and in the direction of the antenna in particular (in this case, there can be no corresponding elemental area, which was mentioned above).

So, let us consider the following task.

Let a plane wave (with the angle θ to the vertical) radiates the area D on the smooth half-space surface ($z < 0$) filled with the complex dielectric constant ε (Fig. 8.1). It is necessary to determine the scattering parameters s_{ij} $(i, j = x, y)$ of the scattering matrix S of this area.

The area in the remote zone in the direction φ—$\vec{E}^{\Sigma}(\theta, \varphi)$ can be expressed with the sum of the radiated wave $\vec{E}^{\mathrm{rad}}(\theta)$ and the wave scattered with the area D $\vec{E}^{\mathrm{sc}}(\varphi)$:

$$\vec{E}^{\Sigma}(\theta, \varphi) = \vec{E}^{\mathrm{rad}}(\theta) + \vec{E}^{\mathrm{sc}}(\varphi), \tag{8.5}$$

Let us use the method of equivalent currents to solve this task.

The main allowance would be in the fact that we shall consider currents on the surface D having the same value as in case when the whole surface of the concerned half-space is considered.

The electric vector of the radiated electromagnetic wave would be as follows:

$$\vec{E}^{\mathrm{rad}} = \left(\vec{i}, \vec{j}\right) \cdot \mathrm{e}^{\mathrm{i}\vec{k}_{\mathrm{rad}} \cdot \vec{r}}, \tag{8.6}$$

where

- the symbol $\left(\vec{i}, \vec{j}\right)$ defines conventionally $\vec{i}$ or $\vec{j}$, the component of the electric field corresponding to horizontal (HP, component $\vec{j}$) or vertical (VP, component $\vec{i}$) types of polarization of the radiated wave;
- vector $\vec{k}_{\mathrm{rad}} = \left(k \sin\theta,\ 0,\ -k \cos\theta\right)$ is a wave vector of the radiated wave (its modulus equals to $k = \frac{2\pi}{\lambda}$); $\vec{r} = (x, y, z)$—is the radius vector of the viewpoint;
- the peak of the radiating wave is considered to be equal to one.

The aggregate area in the top half-space in the case when the whole surface is radiated would be as follows:

$$\vec{E}^{\Sigma}_{(V,H)P} = \left(\vec{i}, \vec{j}\right)\mathrm{e}^{ikx \sin\theta}\left[\mathrm{e}^{-ikz \cos\theta} + R_{(V,H)P}\mathrm{e}^{ikz \cos\theta}\right], \tag{8.7}$$

where the Fresnel coefficients of reflection are of the following form:

- for horizontal polarization:

$$R_{\mathrm{HP}} = \frac{\cos\theta - \sqrt{\varepsilon - \sin^2\theta}}{\cos\theta + \sqrt{\varepsilon - \sin^2\theta}}, \tag{8.8}$$

– for vertical polarization:

$$R_{\mathrm{VP}} = -\frac{\varepsilon \cos\theta - \sqrt{\varepsilon - \sin^2\theta}}{\varepsilon \cos\theta + \sqrt{\varepsilon - \sin^2\theta}}. \tag{8.9}$$

The divider in Eq. (8.9) has a minus in order Eqs. (8.8) and (8.9) coincide with each other during vertical viewing ($\theta = 0$) when there is no difference between horizontal and vertical polarizations (mismatch of reflection coefficients at $\theta = 0$ in the Fresnel equations can be explained by the fact it is referred to the electric vector during horizontal polarization and to the magnetic vector during vertical polarization).

The following arguments can be made.

The current induced on the surface is proportional to the tangential component of the magnetic vector of the falling wave, which in its turn is proportional to the electric vector of this wave. The current directed to the smooth infinite surface makes a plane wave in the top half-space; the electric vector of this wave is proportional to the Fresnel coefficient of reflection on this polarization. This means that the peak of the current height at the finite area D (within the limits of the adopted approximation) can be considered to be proportional to the modulus of the Fresnel coefficient of reflection on the considered polarization. Since we consider the case of radiolocation and plane waves, the concerned arguments may suggest not the surface density of the current, but the linear density (in the *OX* axis) of the current, as the current in the *OY* axis may be considered constant regardless the type of polarization of the radio wave. As per the phase of the current, it fundamentally depends on the direction D of the area exposure (the angle θ).

To determine the corresponding phase correlation, let us study Fig. 8.2. It can be seen on the figure that the ray scattered in the direction φ (the ray $2O'2'$) would have altered phase in respect of the ray coming to the point O and scattered in the direction φ as well (the ray $1O1'$) by the following value: $\Delta\vartheta = kx(\sin\theta - \sin\varphi)$.

Taking into account the above stated, the expression for the electric field of the wave scattered by the elemental section in the direction φ can be as follows:

$$\Delta\vec{E}_{(V,H)P}(\theta,\varphi) = \left(\vec{i},\vec{j}\right)\cdot aR_{(V,H)P}\mathrm{e}^{ikx\Delta\vartheta}\mathrm{e}^{ikz\cos\varphi}\mathrm{e}^{-i\delta(\theta)}\Delta x, \tag{8.10}$$

where $\delta(\theta)$ is the change in the phase during reflection;

a is some proportionality coefficient to be determined.

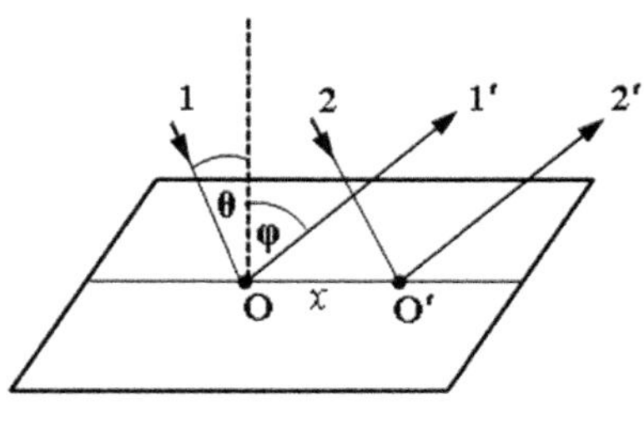

Fig. 8.2 On determining basic relations

To determine the coefficient a (within the scope of the considered approximation), let us find out the value of the electric field of the reflected wave in case of a half-space. Equation (8.10) should be integrated by x within the range of $-\infty$ and $+\infty$ for this purpose. As a result, we get the following:

$$\begin{aligned}\vec{E}_{(V,H)P}(\theta,\varphi) &= \int\limits_{-\infty}^{\infty} \Delta\vec{E}_{(V,H)P}(\theta,\varphi)\mathrm{d}x = \left(\vec{i},\vec{j}\right)\cdot\lim_{b\to\infty}\int\limits_{-b}^{b} aR_{(V,H)P}\mathrm{e}^{ikx\Delta\vartheta}\mathrm{e}^{ikz\cos\varphi}\mathrm{e}^{-\mathrm{i}\delta(\theta)}\mathrm{d}x\\ &= \left(\vec{i},\vec{j}\right)\cdot aR_{(V,H)P}\mathrm{e}^{\mathrm{i}(kz\cos\varphi-\delta(\theta))}\lim_{b\to\infty}\frac{\sin(kb\Delta\vartheta)}{k\Delta\vartheta}\\ &= \left(\vec{i},\vec{j}\right)\cdot aR_{(V,H)P}(\theta)\mathrm{e}^{\mathrm{i}(kz\cos\varphi-\delta(\theta))}\delta(\theta-\varphi)\\ &= \left(\vec{i},\vec{j}\right)\cdot R_{(V,H)P}(\theta)\mathrm{e}^{ikx\sin\theta}\mathrm{e}^{ikz\cos\theta}\delta(\theta-\varphi)\end{aligned} \tag{8.11}$$

Thus, we obtain: $a = \mathrm{e}^{ikx\sin\theta}\mathrm{e}^{\mathrm{i}\delta(\theta)}$.

The received correlations make it possible to write down an equation for the electric vector of the wave scattered by the area D and running in the direction to the radiating source ($\theta = \varphi$):

$$\vec{E}_{(V,H)P}(\theta,-\theta) = \left(\vec{i},\vec{j}\right)R_{(V,H)P}(\varepsilon,\theta)\cdot\mathrm{e}^{ikx\sin\theta}\mathrm{e}^{ikz\cos\theta}\cdot\iint\limits_{D}\mathrm{e}^{2ikx'\sin\theta}\mathrm{d}x'\mathrm{d}y'. \tag{8.12}$$

A corresponding equation for the electric vector of the exposing wave would be as follows:

$$\vec{E}_{(V,H)P}(\theta) = \left(\vec{i},\vec{j}\right)\cdot\mathrm{e}^{ikx\sin\theta}\mathrm{e}^{-ikz\cos\theta}\iint\limits_{D}\mathrm{d}x'\mathrm{d}y'. \tag{8.13}$$

To determine the target scattering parameter, it is necessary to divide Eq. (8.12) by Eq. (8.13), at $z = 0$. As a result, we shall get:

$$s_{xx} = R_{\mathrm{HP}}(\varepsilon,\theta)\frac{\iint_D \mathrm{e}^{2ikx'\sin 2\theta}\mathrm{d}x'\mathrm{d}y'}{\iint_D \mathrm{d}x'\mathrm{d}y'}. \tag{8.14}$$

The same equation we can get for the vertical polarization:

$$s_{yy} = R_{\mathrm{VP}}(\varepsilon,\theta)\frac{\iint_D \mathrm{e}^{2ikx'\sin 2\theta}\mathrm{d}x'\mathrm{d}y'}{\iint_D \mathrm{d}x'\mathrm{d}y'}. \tag{8.15}$$

The cross-scattering parameters are obviously equal to zero in the considered polarization base.

As it can be seen from Eqs. (8.14) and (8.15), the equation for the scattering parameters, as in case of highlights, represents the product of two independent multipliers. The first multiplier depends on the view angle and dielectric characteristics of the underlying surface, while the second one depends only on the geometry of the task.

Let us introduce the designation:

$$\Psi(D) = \frac{\iint\limits_{D} e^{2ikx'\sin 2\theta} dx'dy'}{\iint\limits_{D} dx'dy'}. \tag{8.16}$$

In this case, the considered scattering matrix would be:

$$S = \Psi(D) \cdot \begin{pmatrix} R_{xx} & 0 \\ 0 & R_{yy} \end{pmatrix} = -\Psi(D) \cdot \begin{pmatrix} \frac{\sqrt{\varepsilon - \sin^2\theta} - \cos\theta}{\sqrt{\varepsilon - \sin^2\theta} + \cos\theta} & 0 \\ 0 & \frac{\varepsilon\cos\theta - \sqrt{\varepsilon - \sin^2\theta}}{\varepsilon\cos\theta + \sqrt{\varepsilon - \sin^2\theta}} \end{pmatrix} \tag{8.17}$$

It is rational to reveal correlation of Fresnel coefficients on different polarizations for the further study. If is excluded from Eqs. (8.8) and (8.9), then the considered dependence would be as follows:

$$R_{\text{VP}} = R_{\text{HP}} \frac{\cos 2\theta - R_{\text{HP}}}{1 - R_{\text{HP}} \cos 2\theta}. \tag{8.18}$$

The peculiarity of this equation is that it is directly correlated Fresnel coefficients and the view angle on the observed surface. *This can at least mean that there is no principal need to make measurements on the vertical polarization. This also means that there is no need in measurements on other polarizations, since the scattering parameters of any polarization may be determined by direct calculations.*

Equations (8.17) and (8.18) allow determination of one of the basic polarization characteristics of the reflecting surface—the level of polarization anisotropy q:

$$q = \frac{S_{xx} - S_{yy}}{S_{xx} + S_{yy}} = \begin{cases} \frac{\sin^2\theta}{\cos\theta\sqrt{\varepsilon - \sin^2\theta}}, & \text{if } \theta \le \theta_{\text{Br}} \\ \frac{\cos\theta\sqrt{\varepsilon - \sin^2\theta}}{\sin^2\theta}, & \text{if } \theta > \theta_{\text{Br}} \end{cases}, \tag{8.19}$$

where θ_{Br} is the Brewster angle, equaling to $\theta_{\text{Br}} = \arctan\sqrt{\varepsilon}$ for an ideal dielectric.

An important role in the study belongs to the difference of arguments between the scattering parameters $\Delta\nu$.

Calculations by Eqs. (8.18) and (8.19) allow determination of the target difference:

$$\tan\Delta\nu = \frac{|R_{\text{HP}}|\sin\beta\sin^2 2\theta}{|R_{\text{HP}}|\left(1 + \cos^2\theta\right)\cos\beta - \left(1 + |R_{\text{HP}}|^2\right)\cos 2\theta}, \tag{8.20}$$

where $\beta = \arg R_{\text{HP}}$.

Let us introduce the designation:

$$f = \frac{R_{\mathrm{VP}}}{R_{\mathrm{HP}}}. \tag{8.21}$$

The parameter f would designate polarization ratio.

Equation (8.18) allows expressing reflection coefficients, i.e., scattering parameters via the parameter f. Corresponding transformations lead to the following equations:

$$\left.\begin{aligned} R_{\mathrm{HP}} &= \tfrac{\cos 2\theta - f}{1 - f\ \cos 2\theta} \\ R_{\mathrm{VP}} &= f\tfrac{\cos 2\theta - f}{1 - f\ \cos 2\theta} \end{aligned}\right\}. \tag{8.22}$$

The equation of the anisotropy degree becomes rather simple:

$$q = \begin{cases} \frac{1-f}{1+f}, & \text{if } \theta < \theta_{\mathrm{Br}} \\ \frac{1+f}{1-f}, & \text{if } \theta > \theta_{\mathrm{Br}} \end{cases}. \tag{8.23}$$

Equation (8.22) shows that measurement of voltage in the receiver channels (followed by calculation of their correlation and determination of phase difference) can allow determination of coefficients of reflection and, consequently, the dielectric constant of the underlying surfaces.

Thus, relative measurements of the voltage make it possible to determine the value of complex dielectric constant. Application of Eqs. (8.8), (8.9), (8.18), (8.21), and (8.22) allows obtaining of the following equation:

$$\varepsilon = \left[1 + \frac{4f}{(1-f)^2}\sin^2\theta\right]\tan^2\theta. \tag{8.24}$$

To assess practical use of Eq. (8.24), it is necessary to study relative accuracy of calculations of the complex dielectric constant. To solve this task, let us calculate logarithmic differentiation equations (8.24). After some simple transformation, we shall get:

$$\frac{\Delta\varepsilon}{\varepsilon} = 4\left|\frac{(\varepsilon - \tan^2\theta)}{\varepsilon}\right|\left[\left|\frac{1+f}{f(1-f)}\right| \cdot |\Delta f| + 2\tan\theta \cdot |\Delta\theta|\right] + 2\frac{|\Delta\theta|}{\sin\theta\cos\theta}, \tag{8.25}$$

where $|\Delta f|$ and $|\Delta\theta|$ are errors in measurements of polarization correlation and the view angle, respectively.

As it can be seen from Eq. (8.25), the accuracy of calculations of the complex dielectric constant constantly grows at $\theta \to 0$ u $\theta \to \pi/2$, i.e., during vertical and gliding view, which points on the impossibility to use these view angles for measurements with the proposed methods. The minimum measurement error occurs at viewing at the Brewster angle, when $\tan^2\theta = \varepsilon$. In this case, we shall get:

$$\left|\frac{\Delta\varepsilon}{\varepsilon}\right|_{\min} = 16\left|\frac{\varepsilon}{1+\varepsilon}\right| \cdot |\Delta f| + 2\left|\frac{1+\varepsilon}{\sqrt{\varepsilon}}\right| \cdot |\Delta\theta|. \quad (8.26)$$

Equation (8.26) states a principal perspective of the considered method.

The obtained correlations allow expressing invariants of the scattering matrix represented by Eq. (8.17).

The sum of squared modules of the scattering parameters would be: $A = \sigma_\Sigma^2 = \sum_{i,j=1}^{2} \left|s_{ij}\right|^2 = |\Psi(D)|^2\left(|R_{\text{HP}}|^2 + |R_{\text{VP}}|^2\right)$, and the module of this matrix analyzer would be $B = |\det S| = |\Psi(D)|^2 |R_{\text{HP}} R_{\text{VP}}|$.

Taking into account the set correlation between Fresnel coefficients of reflection (8.18), the last equations would be transformed as follows:

$$\begin{cases} A = \sigma_\Sigma^2 = |\Psi(D)|^2 |R_{\text{HP}}|^2 \left(1 + \left|\frac{\cos 2\theta - R_{\text{HP}}}{1 - R_{\text{HP}}\cos 2\theta}\right|^2\right) \\ B = |\det S| = |\Psi(D)|^2 |R_{\text{HP}}|^2 \left|\frac{\cos 2\theta - R_{\text{HP}}}{1 - R_{\text{HP}}\cos 2\theta}\right| \end{cases}.$$

According to Eq. (8.17), characteristic values of the matrix S would be Fresnel coefficients of reflection.

Characteristic polarization basis of the matrix S would be the pair of horizontal and vertical polarization, while its geographic coordinates on the Poincaré sphere (Fig. 1.4) would obviously be $\alpha = \beta = 0$.

8.2 Smooth Electrically Inhomogeneous Medium

8.2.1 Common Relations

Let us consider the medium filling the half-space $z < 0$, the electrophysical properties of which depend only on the distance to the medium surface: $\varepsilon = \varepsilon(z)$.

Let us consider that a plane electromagnetic wave, the electric vector of which is perpendicular to the incidence plane (horizontal polarization) falls onto the interface (the medium surface) under the angle θ. The field in the top half-space still contains falling and reflecting waves, as it was determined with Eq. (8.7). The electrical vector of the refracted wave must comply with the following differential equation:

$$\Delta E(x, z) + k^2\varepsilon(z)E(x, z) = 0. \quad (8.27)$$

After substitution of $E(x, z) = g(z)\exp\{ikx\ \sin\theta\}$, Eq. (8.27) is reduced to the following differential equation:

$$\frac{\mathrm{d}^2 g(z)}{\mathrm{d}z^2} + k^2\left[\varepsilon(z) - \sin^2\theta\right]g(z) = 0. \quad (8.28)$$

Applying the common conditions of continuity for tangential components of the vectors $\vec{E}$ and $\vec{H}$, it is possible to obtain the following equation for the coefficient of reflection on the horizontal polarization:

$$R_{\mathrm{HP}} = \frac{ikg(0,\theta)\cos\theta + \frac{\mathrm{d}g(0,\theta)}{\mathrm{d}z}}{ikg(0,\theta)\cos\theta - \frac{\mathrm{d}g(0,\theta)}{\mathrm{d}z}}, \tag{8.29}$$

where $g(z,\theta)$ is the solution of Eq. (8.28) converging to zero (at $z \to -\infty$).

The magnetic vector for the vertical polarization may be presented as follows:

$$H(x,z) = n(z)v(z)\exp(-ikx\ \sin\theta), \tag{8.30}$$

where $n(z) = \sqrt{\varepsilon(z)}$; the function $v(z)$ is the solution of the following differential equation converging to zero (at $z \to -\infty$):

$$\frac{\mathrm{d}^2[n(z)v(z)]}{\mathrm{d}z^2} + k^2\left[n^2(z) - \sin^2\theta\right]n(z)v(z) = 0, \tag{8.31}$$

As it can be seen, Eqs. (8.28) and (8.31) coincide with each other. The role of the function $g(z)$ in this case belongs to $n(z)v(z)$.

Applying the common conditions of continuity for tangential components of the vectors $\vec{E}$ and $\vec{H}$, it is possible to obtain the following equation for the coefficient of reflection on the vertical equation:

$$R_{\mathrm{VP}} = \frac{\left.\frac{\mathrm{d}g(z)}{\mathrm{d}z}\right|_{z=0} + ikg(0)\varepsilon(0)\cos\theta}{\left.\frac{\mathrm{d}g(z)}{\mathrm{d}z}\right|_{z=0} - ikg(0)\varepsilon(0)\cos\theta}. \tag{8.32}$$

Let us express the derivative from Eq. (8.29):

$$\left.\frac{\mathrm{d}g(z)}{\mathrm{d}z}\right|_{z=0} = \frac{R_{\mathrm{HP}} - 1}{R_{\mathrm{HP}} + 1} ikg(0)\cos\theta. \tag{8.33}$$

Having substituted this expression into Eq. (8.32), the direct correlation between reflection coefficients, i.e., scattering parameters, can be obtained:

$$R_{\mathrm{VP}} = \frac{(1 - R_{\mathrm{HP}}) - \varepsilon(0)(1 + R_{\mathrm{HP}})}{(1 - R_{\mathrm{HP}}) + \varepsilon(0)(1 + R_{\mathrm{HP}})}. \tag{8.34}$$

Equation (2.32) allows determination of the value of the dielectric constant at z = 0, i.e., $\varepsilon(0)$, using the known equation $g(z)$. This means that one scattering parameter can be definitely expressed with another one. Thus, there is no need in measurements of the second type of polarization, while the scattering matrix can be represented as follows:

$$S = \begin{pmatrix} R_{\mathrm{HP}} & 0 \\ 0 & \frac{(1-R_{\mathrm{HP}})-\varepsilon(0)(1+R_{\mathrm{HP}})}{(1-R_{\mathrm{HP}})+\varepsilon(0)(1+R_{\mathrm{HP}})} \end{pmatrix}. \tag{8.35}$$

The geometric multiplier $\psi(D)$ was omitted in Eq. (8.35).

Resulting from Eq. (8.35), invariants of the matrix S are:

The sum of squared modules of the scattering parameters is as follows:

$$A = \sigma_\Sigma^2 = |R_{\mathrm{HP}}|^2 + \left|\frac{(1-R_{\mathrm{HP}})-\varepsilon(0)(1+R_{\mathrm{HP}})}{(1-R_{\mathrm{HP}})+\varepsilon(0)(1+R_{\mathrm{HP}})}\right|^2,$$

and the analyzer module for this matrix is as follows:

$$B = |\det S| = \left|R_{\mathrm{HP}}\frac{(1-R_{\mathrm{HP}})-\varepsilon(0)(1+R_{\mathrm{HP}})}{(1-R_{\mathrm{HP}})+\varepsilon(0)(1+R_{\mathrm{HP}})}\right|.$$

According to Eqs. (8.34) and (8.35), characteristic values of the matrix S would be Fresnel coefficients of reflection.

Characteristic polarization basis of the matrix S would be the pair of horizontal and vertical polarization, while its geographic coordinates on the Poincaré sphere (Fig. 1.4) would obviously be $\alpha = \beta = 0$.

Let us consider several explicit examples to illustrate the obtained correlations.

8.2.2 Exponential Layer

Let the complex dielectric constant change by the exponential law:

$$\varepsilon(z) = \beta \cdot \exp\{2\alpha \cdot z\} = (\beta_1 + i\beta_2)\exp\{2\alpha \cdot z\};\ \mathrm{Im}\,\alpha = 0. \tag{8.36}$$

This medium should be called an exponential layer.

In this case, the solution of Eq. (8.28) will be the following function:

$$g(z) = H_\xi^{(2)}\left(\frac{k}{\alpha}\sqrt{\beta} \cdot \exp\{\alpha \cdot z\}\right), \tag{8.37}$$

where $H_\xi^{(2)}$ is the Hankel function of the second kind with the order $\xi = \frac{k}{\alpha}\sin\theta$.

Having done all corresponding transformations and applied the equation for the derived Hankel function, we get:

$$\left(H_v^{(2)}(x)\right)' = \frac{v}{x}H_v^{(2)}(x) - H_{v+1}^{(2)}(x), \tag{8.38}$$

we shall get:

$$R_{\mathrm{HP}} = \frac{i H_{\xi}^{(2)}\left(\frac{k}{\alpha}\sqrt{\beta}\right)\mathrm{e}^{-\mathrm{i}\theta} - \sqrt{\beta} H_{\xi+1}^{(2)}\left(\frac{k}{\alpha}\sqrt{\beta}\right)}{i H_{\xi}^{(2)}\left(\frac{k}{\alpha}\sqrt{\beta}\right)\mathrm{e}^{\mathrm{i}\theta} + \sqrt{\beta} H_{\xi+1}^{(2)}\left(\frac{k}{\alpha}\sqrt{\beta}\right)}. \tag{8.39}$$

The task of obtaining the target scattering matrix may be solved with the obtained equation.

8.2.3 *Quadratic Layer*

Let the complex dielectric constant change by the quadratic law:

$$\varepsilon(z) = (\alpha z + \beta)^2. \tag{8.40}$$

This medium should be called a quadratic layer.

It is impossible to express solution of Eq. (8.28) for such medium with the help of known functions. Though, under small α, solution can be expressed as follows:

$$g(z) = \frac{1}{\sqrt[4]{(\alpha z + \beta)^2 - \sin^2\theta}} \exp\left\{ -ik \int\limits_0^z \sqrt{(\alpha z + \beta)^2 - \sin^2 z \mathrm{d}z} \right\}. \tag{8.41}$$

Having done all necessary transformations, we get an equation of the target reflection coefficient (the parameter α is small):

$$R_{\mathrm{HP}} = \frac{\cos\theta - \sqrt{\varepsilon - \sin^2\theta}}{\cos\theta + \sqrt{\varepsilon - \sin^2\theta}}, \tag{8.42}$$

which coincides with the example of the uniform medium. Consequently, Eqs. (1.9), (1.10), and (1.17–1.26) remain correct.

8.2.4 *Polynomial Layer (Vertical Sounding)*

Vertical sounding is rather often applied to solve the tasks of remote sounding. In this case, the difference between vertical and horizontal polarization disappears during sounding of smooth surfaces, and the scattering matrix becomes ordinary. Though, the reflected wave power considerably depends on the changing characteristics of dielectric properties of the studied surfaces. We shall begin studying of such situations with a polynomial layer, where the law of the complex dielectric constant changing is described with the following equation:

$$\varepsilon(z) = (az + b)^m. \tag{8.43}$$

Solution of Eq. (8.28) at $\theta = 0$ complying with the condition of infinity would be as follows:

$$g(z) = \sqrt{az + b} \cdot J_{\frac{1}{m+2}} \left[\frac{2}{m+2} \left(\frac{k}{a} \right)^{\frac{m+2}{4}} (az + b)^{\frac{m+2}{2}} \right]. \tag{8.44}$$

8.2.5 Linear Layer (Vertical Sounding)

$m = 1$ (Eq. 8.43) corresponds to the linear layer.

Let us introduce the following designations:

$$\varepsilon(z) = az + b = (a_1 + ia_2)z + b_1 + ib_2 = \varepsilon_1(z) + i\varepsilon_2(z). \tag{8.45}$$

In this case:

$$\begin{cases} g(z) = \rho^{1/3} H_{1/3}^{(2)}(\rho) \\ \rho = \dfrac{2k}{3a}(az + b)^{1.5} \end{cases}. \tag{8.46}$$

Substitution of the obtained equations into Eq. (8.29) would lead to the following target equation:

$$R = \frac{H_{1/3}^{(2)}\left(\frac{2k}{3a}\right) - i\sqrt{b} \cdot H_{-2/3}^{(2)}\left(\frac{2k}{3a}\right)}{H_{1/3}^{(2)}\left(\frac{2k}{3a}\right) + i\sqrt{b} \cdot H_{-2/3}^{(2)}\left(\frac{2k}{3a}\right)}. \tag{8.47}$$

Analysis of Eq. (8.47) with complex values of a and b results in huge and hardly manageable expressions. Let us consider only limit cases of small and large $|a|$, as well as dependencies $|R|$ of $|a|$ for several standard and practically important cases.

For small $|a|$, when the module of the complex dielectric constant changes slowly with the depth, the inequality $|\rho| \gg 1$ takes place. In this case, we can apply asymptotic representation of the Hankel function. As a result, we shall get:

$$R = \frac{1 - b^{0.5} + ib^{-1.5}\frac{a}{k}\left(0.104 + 0.146b^{0.5}\right)}{1 + b^{0.5} + ib^{-1.5}\frac{a}{k}\left(0.104 - 0.146b^{0.5}\right)}. \tag{8.48}$$

If there is no obvious border of the section (there is no contrast at the boundary, i.e., at $b = 1$), we shall get:

$$R = 0.004|a|^2\lambda^2. \tag{8.49}$$

As it can be seen, there is almost no reflection in this case.

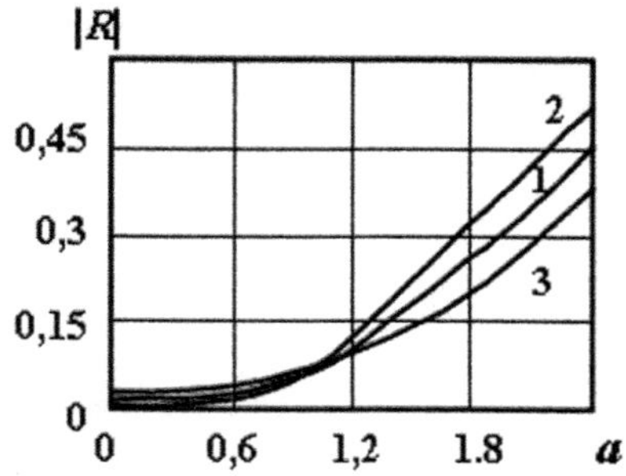

Fig. 8.3 Dependence of the reflection coefficient modulus (linear law of the complex permittivity variation) on the parameter a [1/cm], $\lambda = 3$ cm. (1) $\varepsilon(z) = 2 + iaz$; (2) $\varepsilon(z) = 1 + iaz$; (3) $\varepsilon(z) = 2 + 0.5\sqrt{2}(1 + i)az$

Another limit case can be observed at a high rate of increase in the complex dielectric constant, i.e., at a high $|a|$ and, consequently, $|\rho| \ll 1$. Here, it would be reasonable to apply the Bessel function. As a result, we shall get:

$$R = -1 + 1.58\left(\frac{2kb\sqrt{b}}{3a}\right)^{1/3}(1 - i \cdot 1.73). \qquad (8.50)$$

As it can be determined from Eq. (8.50), the coefficient of reflection is close to one at high values of $|a|$. Under the further increase in $|a|$ and λ, this coefficient would strain after one.

The stated nature of R changing remains in common for other types of polynomial dependencies. Figure 8.3 shows the dependency of the reflection coefficient $|R|^2$ and the parameter $|a|$ for the linear layer.

$$(1)\,\varepsilon(z) = 2 + iaz;\ \ (2)\,\varepsilon(z) = 1 + iaz;$$
$$(3)\,\varepsilon(z) = 2 + 0.5\sqrt{2}(1 + i)az$$

8.2.6 *Parabolic Layer (Vertical Sounding)*

$m = 2$ (Eq.8.43) corresponds to the parabolic layer.

Let us introduce the following designations:

$$\varepsilon(z) = (az + b)^2 = [(a_1 + ia_2)z + (b_1 + b_2)]^2. \qquad (8.51)$$

In this case:

$$\begin{cases} g(z) = \rho^{0.25} H_{0.25}^{(2)}(\rho) \\ \rho = \frac{k}{2a}(az + b)^2 \end{cases}. \qquad (8.52)$$

Substitution of the obtained equations into Eq. (8.29) would lead to the following target equation:

$$R = \frac{H_{0.25}^{(2)}(w) - ibH_{-0.75}^{(2)}(w)}{H_{0.25}^{(2)}(w) + ibH_{-0.75}^{(2)}(w)}, \tag{8.53}$$

where $w = \frac{kb^2}{2a}$.

In case $|a|\lambda \ll b$, i.e., $|\rho| \gg 1$, then Eq. (8.49) would be transformed into:

$$R = \frac{(1-b)kw + i(0.0094 + 0.156b)}{(1+b)kw + i(0.0094 - 0.156b)}. \tag{8.54}$$

If there is no clear boundary ($b = 1$):

$$R = 0.016|a|^2\lambda^2. \tag{8.55}$$

In another limit case, when $|\rho| \ll 1$, we shall get the following after all corresponding calculations:

$$R = -1 + \frac{5.1(1-i)}{\sqrt{|a|\lambda}}. \tag{8.56}$$

8.2.7 *Matching Layer (Vertical Sounding)*

In some cases, a thin intermediate layer appears at the interface of the medium and the atmosphere; there, the complex dielectric constant smoothly changes from the one (the atmosphere) to the value inside the medium depth; meanwhile, the curve stating this dependency has no kinks. Availability of such layer leads in some cases to a considerable reduction of the reflection coefficient due to decrease in reflections at the boundary of sections.

Thus, let us consider reflection from the following structure: The area of the space $z \geq h$ is filled with uniform medium with the complex dielectric constant ε_k. Under $0 \leq z \leq h$ the matching layer available; its complex dielectric constant ε_s changes following the law:

$$\varepsilon_s = \frac{1+\varepsilon_k}{2} + \frac{1-\varepsilon_k}{2}\cos\pi\frac{z}{h}. \tag{8.57}$$

As it can be seen from Eq. (8.57), $\varepsilon_s = 1$ under $z = 0$ (the boundary with the atmosphere), $\varepsilon_s = \varepsilon_k$ under $z = h$ (the boundary of the layer with the medium), and besides, $\varepsilon_s'(0) = \varepsilon_s'(h) = 0$, i.e., adjustment by the derivative takes place at the layer boundaries. The dependency $\varepsilon_s(z)$ for the cases of ice-water and air-ice is stated in Fig. 8.4.

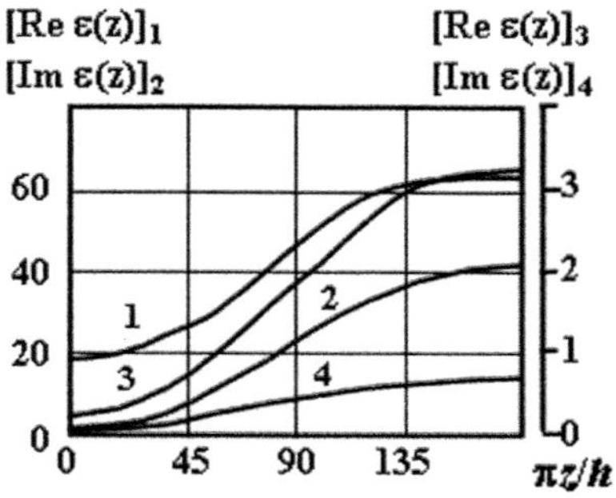

Fig. 8.4 Dependence $\varepsilon(z)$ for the coordinating layer. Curves 1 and 2 correspond to Re $[\varepsilon(z)]$ and Im $[\varepsilon(z)]$ for the system ice-water. Curves 3 and 4 are the same for the system air-ice

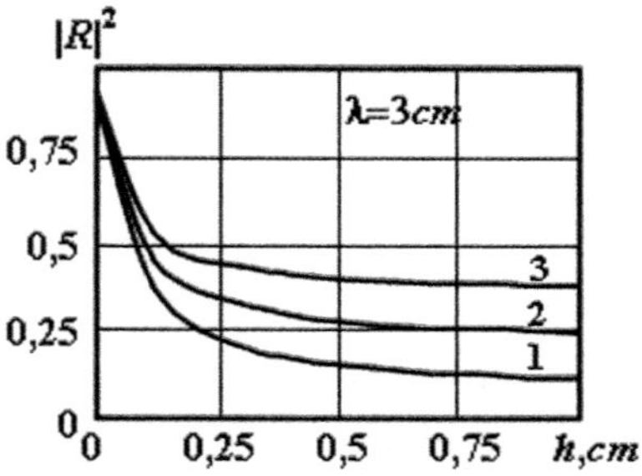

Fig. 8.5 Dependence of the reflection coefficient modulus square R on transition layer density h. In the transition layer, $\varepsilon(z)$ varies from $\varepsilon = 1$ to $\varepsilon = 65 - i40$ according to linear (1), harmonic (2) and exponential (3) laws. $\lambda = 3$ cm

The wave equation for the matching layer is reduced to the Mathieu equation by substituting of the variable.

The results of calculations of the coefficient of reflection from the thickness of the intermediate layer h, under different values of the complex dielectric constant, are stated in Fig. 8.5. It can be seen on the figure that the reflection coefficient changes from $\left|(1-\sqrt{\varepsilon})/(1+\sqrt{\varepsilon})\right|^2$, which complies with the section boundary, to zero, which corresponds to the absence of such boundary. Similar curves for the matching layer with the linear dependency (without matching with the derivative) are shown on the same figure for comparison.

As it can be seen, decrease in the reflection coefficient for the linear layer is faster than for the matching "cosinusoidal" layer. For limit cases, thick layer ($h \to \infty$) $R = 0.002|1-\varepsilon_k|^2\left(\frac{\lambda}{h}\right)^2$; thin layer ($h \to 0$) $R = \frac{20.4}{\sqrt{|1-\varepsilon_k|}}\sqrt{\frac{h}{\lambda}}$.

8.2.8 *Transition Layer (Vertical Sounding)*

The law of changing of the complex dielectric constant determined with Eq. (8.36) is applicable to the transition layer. In this case, Eqs. (8.37)–(8.39) remain correct. For the medium with a small α, Eq. (8.39) shall have the following appearance:

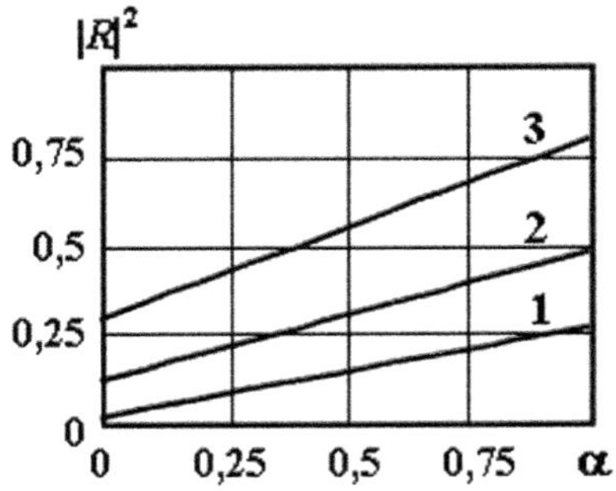

Fig. 8.6 Dependence of the reflection coefficient modulus square on parameter α for the medium with exponential law of variation ε (z). $\lambda = 3$ cm. 1—$\varepsilon(z) = \exp(1 + 0.1i\alpha z)$; 2—$\varepsilon(z) = (1 + i)\exp(\alpha z)$; 3—$\varepsilon(z) = (3.5 + 2i)\exp(\alpha z)$

$$R = \frac{1 - \sqrt{|\beta|}\mathrm{e}^{\mathrm{i}\delta/2} + 0.125\frac{i}{\rho_0}\left(1 + 3\sqrt{|\beta|}\mathrm{e}^{-\mathrm{i}\delta/2}\right)}{1 + \sqrt{|\beta|}\mathrm{e}^{-\mathrm{i}\delta/2} + 0.125\frac{i}{\rho_0}\left(1 - 3\sqrt{|\beta|}\mathrm{e}^{-\mathrm{i}\delta/2}\right)}, \tag{8.58}$$

where $\rho_0 = k\frac{\sqrt{|\beta|}}{\alpha}\mathrm{e}^{-\mathrm{i}\delta/2}$, and δ is the angle of loss.

It can be understood from Eq. (8.58) that the reflection coefficient is determined, first of all, by the break value of the complex dielectric constant at the section boundary. In particular, it is derived from Eq. (8.58) for the medium with small conductivity (freshwater):

$$R = \frac{1 - \sqrt{|\beta|}}{1 + \sqrt{|\beta|}}\left(1 - \frac{0.14\alpha\lambda}{|\beta|^2 - 1}\right). \tag{8.59}$$

In another limit case, when α is large, we shall get:

$$R \approx -1 + \frac{2N_0(\rho_0)}{N_0\left(\rho_0 + \frac{2i\sqrt{|\beta|}}{\pi\rho_0}\mathrm{e}^{\mathrm{i}\delta/2}\right)}. \tag{8.60}$$

Figure 8.6 shows a range of dependencies of the reflection coefficient from α, under different values of β and δ.

8.3 Equation for Scattering Matrix Elements

As it is stated above, calculations for the scattering parameters of smooth surfaces can be reduced to calculations of corresponding reflection coefficients. Accurate solutions of the task on reflection of radio waves from a nonuniform layer were made only for several cases, that is, why to solve these tasks we have to use different approximate methods of solution of Maxwell's equations and Eq. (8.27). In this case, it would be reasonable to get an equation directly for the reflection coefficient and solve this task in future immediately with precise and approximate methods.

Let us determine falling and reflected waves at any point of the space as follows:

– a falling wave (horizontal polarization):

$$\begin{cases} E_y = P(z)\exp\{ikxn\ \sin\theta)\} \\ H_x = -nP(z)\cos\theta\exp\{ikxn\ \sin\theta)\}\ , \\ H_z = nP(z)\sin\theta\exp\{ikxn\ \sin\theta)\} \end{cases} \tag{8.61}$$

– a reflected wave (horizontal polarization):

$$\begin{cases} E_y = R(z)\exp\{ikxn\ \sin\theta)\} \\ H_x = nR(z)\cos\theta\exp\{ikxn\ \sin\theta)\}\ . \\ H_z = nR(z)\sin\theta\exp\{ikxn\ \sin\theta)\} \end{cases} \tag{8.62}$$

The sum of the falling and the reflected waves complies with Maxwell's equations, if the functions $P(z)$ and $R(z)$ satisfy the following equation:

$$\begin{cases} \frac{\mathrm{d}P}{\mathrm{d}z} - i\beta\cdot P + \gamma(P - R) = 0 \\ \frac{\mathrm{d}R}{\mathrm{d}z} + i\beta\cdot R - \gamma(P - R) = 0 \end{cases}, \tag{8.63}$$

where $\beta = \beta(z) = kn(z)\cos\theta(z) = k\sqrt{n^2(z) - \sin^2\theta}$; $\gamma = \frac{1}{2\beta}\frac{\mathrm{d}\beta}{\mathrm{d}z}$.

Having multiplied the equation by R and divided the result by $P^2(z)$, we shall get the Riccati equation for the reflection coefficient:

$$\frac{\mathrm{d}V}{\mathrm{d}z} = -2i\beta\cdot V + \gamma\left(1 - V^2\right). \tag{8.64}$$

Having done the same transformations for the vertical polarization, we got the same Riccati equation for the reflection coefficient, where a new coefficient χ determined with the following expression plays the role of the coefficient γ in Eq. (8.64)

$$\chi = \frac{\varepsilon}{2\beta}\frac{\mathrm{d}}{\mathrm{d}z}\left(\frac{\beta}{\varepsilon}\right). \tag{8.65}$$

The following requirement should be chosen as a limit condition: $V \to 0$ at $z \to -\infty$.

Equation (8.64) can be solved with the method of step-by-step approximation.

Thus, all scattering parameters of electrically nonuniform media may be determined directly from Eq. (8.64).

8.4 Uneven Surfaces

8.4.1 Electrodynamic Models of Uneven Surfaces

Impacts of electrodynamic characteristics on the scattering parameters of the reflecting surface were studied in clauses (8.2)–(8.3). As it was stated, the impact of geometric characteristics comes down to some multiplier that depends on the geometry of the exposed area.

Totally different picture can be observed during the study of reflections from uneven surfaces. In general cases, it is impossible to divide polarization effects into those ones produced by electrodynamic characteristics of the studied medium and the ones produced due to the geometric structure.

The further analysis will be carried out with the following allowances.

The radar station with the antenna, having the radiated pattern $F(\theta, \varphi)$, is located at the height H over a statically uneven two-dimensionally rough surface $z = \xi(x, y)$ with the dielectric content $\varepsilon(x, y)$. The antenna radiates a spherical electromagnetic wave with the length that irradiates a certain area of the uneven surface. It is necessary to develop a scattering matrix of such area. Since there is a myriad of types and kinds of uneven surfaces, there would inevitably be an issue to develop a finite number of electrodynamic and statistic models covering as broad number of types and kinds of real surfaces as possible.

There is a lot of literature dedicated to electrodynamic and statistical modeling of uneven and rough surfaces, so that a separate book may be written to make a review of these issues. For further consideration, we selected the models proposed in and as the most advantageous, and we should follow these works and describe uneven surfaces using the following four models.

Model 1. A large scale $(\rho_0/\lambda \gg 1)$, smooth $(\rho_{cor}/\lambda \gg 1)$, low-pitched $\left((\nabla\xi_1)^2 < 1\right)$ surface with undefined roughness σ/λ, flat in average $\bar{\xi}_1 = 0$ (ρ_0, ρ_{cor} are the radii of correlation and curvature of surface, is the quadratic mean of roughness height).

This model can describe unfurrowed fields, hilly terrains with moderate vegetation, large waves on the water surface, etc. (see Fig. 8.7a).

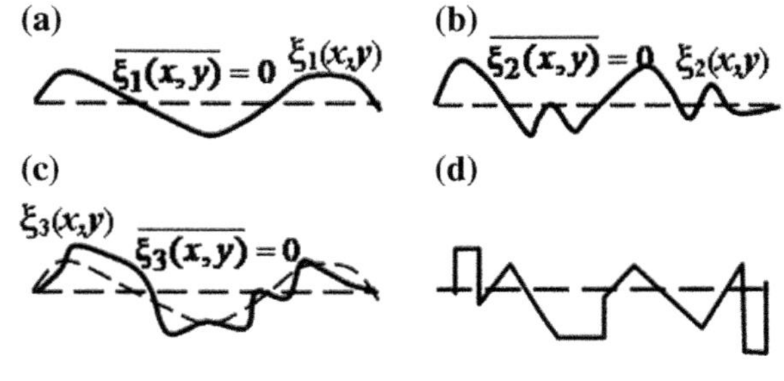

Fig. 8.7 Models of statistically uneven terrain

Model 2. Small scale ($\rho_0/\lambda \le 1$), low-pitched $\left((\nabla\xi_2)^2 \ll 1\right)$, finely roughened ($\sigma/\lambda \ll 1$) surface, flat in average $\bar{\xi}_2 = 0$.

This model can describe concrete, asphalt, sandy and gravel surfaces, calm water surface with moderate ripples, flat plains with low vegetation in centimeter and longer wave bands, etc. (see Fig. 8.7b).

Model 3. Complicated surface: continuous or discrete covering with small-scale irregularities of large scale ($\xi_3(x, y) = \xi_1(x, y) + \xi_2(x, y)$), provided that $\bar{\xi}_3 = \xi_1$.

This model can describe wave-covered water, deserts with large sand waves with sand ripples on them, plowed hilly surfaces, etc. (see Fig. 8.7c).

Model 4. The surface has different geometric shapes.

This model can describe settlements, mountains, forestlands, etc. These surfaces can be represented in centimeter and decimeter waves as a random pattern of non-coherent separate scatterers (see Fig. 8.7d).

8.4.2 The Scattering Matrix for Model 1

In compliance with the conditions imposed on Model 1, we can apply $L \gg \rho_0$; $L \gg \sigma$; $H \gg \sigma$.

The task will be solved with the following order. Falling and reflected spherical waves are differentiated into plane waves. Further, using Kirchhoff's approximations, we shall solve the task of diffraction of partial plane wave on corresponding local areas of the surface s_i which are characterized with the following random variables: height $\xi(x, y)$, gradient $\nabla\xi(x, y) = \vec{i}\frac{\partial\xi}{\partial x} + \vec{j}\frac{\partial\xi}{\partial y}$, and local Fresnel complex reflection coefficients determined with Eqs. (8.8) and (8.9). The field at the receiver point is based on the vector form of Green's theorem (Stratton-Chu's formula) with integration in illuminated areas of the surface.

The following equations are received for the scattering matrix as a result of calculations:

$$S = \begin{pmatrix} \frac{a_2 n_z^2}{1 - a_2 n_z^2}\left(\eta^2 R_{\mathrm{VP}} - a_1^2 R_{\mathrm{HP}}\right) & \frac{a_1 a_2 n_z^2 \eta}{1 - a_2 n_z^2}\left(R_{\mathrm{HP}} + R_{\mathrm{VP}}\right) \\ \frac{a_1 a_2 n_z^2 \eta}{1 - a_2 n_z^2}\left(R_{\mathrm{HP}} + R_{\mathrm{VP}}\right) & \frac{a_2 n_z^2}{1 - a_2 n_z^2}\left(a_1^2 R_{\mathrm{VP}} - \eta^2 R_{\mathrm{HP}}\right) \end{pmatrix}, \tag{8.66}$$

where

$$\begin{aligned} a_1 &= \sin\theta - \gamma\cos\theta; \quad a_2 = \cos\theta + \gamma\sin\theta; \\ n_z &= \frac{1}{\sqrt{1 + \left(\frac{\mathrm{d}\xi}{\mathrm{d}x}\right)^2 + \left(\frac{\mathrm{d}\xi}{\mathrm{d}y}\right)^2}}; \\ \eta &= \frac{\mathrm{d}\xi}{\mathrm{d}x}\sin\varphi - \frac{\mathrm{d}\xi}{\mathrm{d}y}\cos\varphi; \quad \gamma = \frac{\mathrm{d}\xi}{\mathrm{d}x}\cos\varphi + \frac{\mathrm{d}\xi}{\mathrm{d}y}\sin\varphi \end{aligned}$$

To find proper values of the matrix S, it should be multiplied and each parameter should be divided into the multiplier $\sqrt{a_1^2 + \eta^2}$. In this case,

$a_1/\sqrt{a_1^2+\eta^2} = \cos\zeta$; $\eta/\sqrt{a_1^2+\eta^2} = \sin\zeta$ may be taken. This allows presenting the scattering matrix (8.66) as follows:

$$S = -\frac{a_2 n_z^2 \sqrt{a_1^2+\eta^2}}{\left(1-a_2 n_z^2\right)} \begin{pmatrix} R_{\mathrm{HP}}\cos^2\zeta - R_{\mathrm{VP}}\sin^2\zeta & -(R_{\mathrm{HP}}+R_{\mathrm{VP}})\sin\zeta\cos\zeta \\ -(R_{\mathrm{HP}}+R_{\mathrm{VP}})\sin\zeta\cos\zeta & R_{\mathrm{HP}}\sin^2\zeta - R_{\mathrm{VP}}\cos^2\zeta \end{pmatrix}. \tag{8.67}$$

As it can be seen, Eq. (8.67) coincides with Eq. (6.25), if $\beta = \eta = \varphi = 0$ is taken in the latter.

Thus, equations for proper values of the matrix (8.66) can be written:

$$\begin{cases} \lambda_1 = -\frac{a_2 n_z^2}{1-a_2 n_z^2}\sqrt{a_1^2+\eta^2}\,R_{\mathrm{HP}} \\ \lambda_2 = \frac{a_2 n_z^2}{1-a_2 n_z^2}\sqrt{a_1^2+\eta^2}\,R_{\mathrm{VP}} \end{cases}. \tag{8.68}$$

The obtained correlations allow development of both invariants of the matrix (8.66):

$$\begin{cases} A = \sigma_\Sigma^2 = \left(\frac{a_2 n_z^2}{1-a_2 n_z^2}\sqrt{a_1^2+\eta^2}\right)^2 \left(|R_{\mathrm{HP}}|^2 + |R_{\mathrm{VP}}|^2\right) \\ B = |\det S| = \left(\frac{a_2 n_z^2}{1-a_2 n_z^2}\sqrt{a_1^2+\eta^2}\right)^2 |R_{\mathrm{HP}} R_{\mathrm{VP}}| \end{cases}. \tag{8.69}$$

To determine the coordinates of the proper basis, let us apply Eqs. (6.33) and (8.68), direct calculations of which result in the following values of the length (see Fig. 1.4) of the proper polarization basis $\cos 2\beta = \cos 2\zeta = \left|\left(\eta^2 - a_1^2\right)/\left(\eta^2 + a_1^2\right)\right|$.

As it is clear from the stated equation, one of the coordinates of the proper polarization basis on the Poincaré sphere does not depend on the complex dielectric constant and can only be determined by characteristics of the surface roughness.

As per the second geographic coordinate α (see Fig. 1.4), the same calculations can give such correlations that would reveal the fact that this coordinate also depends on the complex dielectric constant and can be determined only by characteristics of the surface roughness. Due to awkwardness of the obtained correlations, they are not represented in this text.

It is clear that horizontal and vertical polarization is not the proper polarization of considered surfaces. This leads to cross-components during reflection of horizontally and vertically polarized radio waves from such surfaces.

To identify output signals of the antenna, it is necessary to integrate all angles θ and φ according to Eq. (8.66), taking into account all radiated patterns of the antenna and the distance to it:

$$U_{ij} = \frac{1}{(2\pi)^2}\int\limits_{S_{\text{rad}}} \mathrm{d}s \int\int\limits_{-\infty}^{\infty}\int\int F(x,y)s_{ij}\left(\xi_x', \xi_y'; \theta; \varphi; \varepsilon\right)n_z$$
$$\times \exp\left\{-i\left[(k_x + k')x + \left(k_y + k_y'\right)y + (k_z + k_z')(H - \xi(x,y))\right]\right\}$$
$$\times \left(\mathrm{d}k_x \mathrm{d}k_y / k_z\right)\left(\mathrm{d}k_x' \mathrm{d}k_y' / k_z'\right) \quad (8.70)$$

Thus, Eq. (8.70) reveals that each component of the complex spherical field in the input point represents a superposition of plane waves reflected from statistically uneven surface under different angles θ and φ.

8.4.3 *The Scattering Matrix for Model 2*

Let a small-scale surface $z = \chi\ (x, y)$ be an interface of two media with the dielectric constant 1 and ε.

If it is assumed that corresponding electrical and magnetic vectors for the surface $z = 0$ have values $\vec{E}_0, \vec{H}_0$, then small-scale irregularities would result in minor changes of vectors $\vec{E}_0, \vec{H}_0$.

The surface $z = \chi\ (x, y)$ has continuous tangential components of electrical and magnetic vectors. Deviation of the surface $z = \chi(x, y)$ from the plain $z = 0$ can be considered as moderate allowance. Decomposing the boundary conditions by degrees $\chi(x, y)$, $\nabla\chi(x, y)$ into the row and confining to the first approximation of the perturbation theory, we shall get the following for the scattering parameters after some transformations:

$$S = \begin{pmatrix} (1 + R_{\text{VP}})^2 L(\theta, \varepsilon) & 0 \\ 0 & (1 + R_{\text{VP}})^2\left[(1 - R_{\text{HP}})\sin^2\theta + L(\theta, \varepsilon)\right] \end{pmatrix}, \quad (8.71)$$

where $L(\theta, \varepsilon) = \frac{\left(\sin^2\theta + \sqrt{\varepsilon - \sin^2\theta}\cos\theta\right)^2}{\varepsilon - 1}$.

The following will be invariants of this matrix:

$$\begin{cases} A = \sigma_\Sigma^2 = \left|\frac{(1-R_{\text{HP}})^2}{4R_{\text{HP}}\cos^2 2\theta}\right|^2 + \left|\frac{R_{\text{HP}}(1-R_{\text{HP}})\sin^2 2\theta - (1-R_{\text{HP}}\cos 2\theta)^2}{4R_{\text{HP}}\cos^2\theta}\right|^2 \\ B = |\det S| = \left|\frac{(1-R_{\text{HP}})^2}{4R_{\text{HP}}\cos^2 2\theta} \cdot \frac{R_{\text{HP}}(1-R_{\text{HP}})\sin^2 2\theta - (1-R_{\text{HP}}\cos 2\theta)^2}{4R_{\text{HP}}\cos^2\theta}\right| \end{cases}. \quad (8.72)$$

As it is seen from Eq. (8.71), in this case horizontal and vertical polarization is the proper polarization of the considered surfaces. This explains the absence of cross-components during reflection of horizontally and vertically polarized radio waves from such surfaces.

This result shows that the first approximation of the perturbation theory does not reveal depolarization of the field even from the surface of the finite conductivity. At $|\varepsilon| \to \infty$, diagonal elements of the matrix (8.68) tend to $(1 + R_{\mathrm{VP}})^2 \cos^2\theta$; $(1 + R_{\mathrm{VP}})^2(1 - R_{\mathrm{HP}})$, respectively.

The proper values of the matrix S would equal to its diagonal elements. Coordinates of the proper polarization basis (see Fig. 1.4) will, apparently, equal to $\alpha = \beta = 0$.

To identify output signals of the antenna, it is necessary to integrate all angles and according to Eq. (8.71), taking into account all radiated patterns of the antenna θ and φ the distance to it:

$$U_{ij} = \frac{k^2}{\pi}\left(\frac{\varepsilon - 1}{\varepsilon}\right)^2 \int\limits_{S_0} \frac{F(x, y)s_{ij}}{R_0^2} \xi(x, y) \exp(-2ikR_0)\mathrm{d}S_0. \tag{8.73}$$

8.4.4 The Scattering Matrix for Model 3

The task of reflection on a complicated surface (large-scale and small-scale irregularities) may be solved with a combination of two methods: Kirchhoff's method and the perturbation method. The target fields are determined in the form of the sum of random reflected fields on the large-scale surface and perturbation fields of small-scale unevenness.

Decomposing the boundary conditions by degrees $\chi(x, y)$, $\nabla\chi(x, y)$ into the row and confining to the first approximation of the perturbation theory, the task of scattering on the studied structure would reduce to Model 1. After several transformations, we shall get the following for the scattering matrix:

$$S = \left(\frac{2a_2 n_z \varepsilon}{a_2 n_z \varepsilon + q}\right)^2 \begin{pmatrix} \frac{2q\eta^2 n_z}{a_2 n_z + q} - \frac{\left[1 + a_2 n_z (q - a_2 n_z)\right]^2}{n_z\left(a_2^2 n_z^2 - q^2\right)} & -\frac{2a_1 \eta q n_z}{a_2 n_z + q} \\ -\frac{2a_1 \eta q n_z}{a_2 n_z + q} & \frac{2q a_1^2 n_z}{a_2 n_z + q} - \frac{\left[1 + a_2 n_z (q - a_2 n_z)\right]^2}{n_z\left(a_2^2 n_z^2 - q^2\right)} \end{pmatrix} \tag{8.74}$$

where a_1; a_2; η; n_z is determined with Eq. (8.66), the parameter $q = \sqrt{\varepsilon - \cos^2\theta}$.

Invariants of this matrix can be determined with Eqs. (6.27), (6.28), and (6.32).

At that, the Fresnel coefficient on horizontal polarization should be used instead of parameters ε and q.

$$\begin{cases} q = \frac{1 - R_{\mathrm{HP}}}{1 + R_{\mathrm{HP}}} \cos\theta \\ \varepsilon = 1 - 4\frac{R_{\mathrm{HP}}}{(1 + R_{\mathrm{HP}})} \cos^2\theta \end{cases}. \tag{8.75}$$

The equations relating to the considered model are too awkward, though they are not so difficult to be obtained. Calculations revealed that, first of all, proper values can be established with the Fresnel coefficient of reflection on the horizontal polarization, and, secondly, they depend on the value of the complex dielectric constant too much. The coordinates of the proper polarization basis conversely depend only on the roughness characteristics. Due to awkwardness of the obtained correlations, they are not represented in this text. Nevertheless, in respect of the real ε, it is possible to get equations for the proper values of the matrix S in rather compact form:

$$\begin{cases} \lambda_1 = \frac{2a_2^2\varepsilon^2(1+a_2n_z)^2(a_2n_z-q)}{a_1q\eta(a_2n_z\varepsilon-q)^2} \\ \lambda_2 = \left(\frac{a_1}{\eta}+\frac{\eta}{a_1}\right)\left(\frac{2a_2n_z\varepsilon}{a_2n_z\varepsilon-q}\right)^2+\lambda_1 \end{cases}. \tag{8.76}$$

To identify output signals of the antenna, it is necessary to integrate all angles θ and φ according to Eq. (8.74), taking into account all radiated patterns of the antenna and the distance to it:

$$U_{ij} = \frac{k^2}{\pi}\left(\frac{\varepsilon-1}{\varepsilon}\right)^2 \int\limits_{S_0} \frac{F(x, y)s_{ij}}{R_0^2}\xi_2(x, y)\exp(-2ikR_0)\mathrm{d}S_0, \tag{8.77}$$

where $\xi_2(x, y)$ are small-scale irregularities.

As it can be seen from Eq. (8.74), depolarized components of the filed may be generated during backscattering, only given a non-zero two-dimensional gradient of a large-scale structure of the surface.

The full scattering field from the surface for Model 3 can be determined as the sum of fields from the large-scale surface and a small-scale surface located on the large-scale one.

8.4.5 The Scattering Matrix for Model 4

This model includes such random uneven surfaces the geometrical profile of which represents rather plane faces or knobs of almost derived from (modeling of houses, cliffs, etc.), as well as all surfaces with a breaking profile (forests, vegetation, etc.).

One of variants to present uneven surface is Beckmann's model consisting of plane elements (faces) with random incline. Each face reflects a ray falling onto it with the Fresnel coefficient of reflection acting only in the local mirror direction, which is correct under $l \gg \lambda$. Partial waves reflected in one direction (faces with the same incline) are summed up by phases, generating a resultant wave in the same direction. Since structural profiles described with Model 4 are not common, it is impossible to develop a general scattering matrix. In this case, it would be more reasonable not to calculate the field scattered with the actual surface, but to construct a model of the required surface by the experimental study.

Though, let us illustrate Model 4 with some simple example.

Let us consider reflection from the structure in the form of square-topped projection. In this case, it is obvious that the ray falling onto the surface at an angle θ would fall on the vertical wall after reflection, where it would be reflected for the second time and become parallel to the falling ray. In this case, the total coefficient of reflection will be the product of Fresnel coefficients of reflection on angles θ and $(90^\circ - \theta)$.

8.4.6 Statistical Characteristics of the Scattering Parameters of Uneven Surfaces

The field scattered on statistically uneven surfaces are random, that is, why they can be described only statistically. In general, the random nature of the fields is functionally connected with statistical irregularities of the geometrical shape $\xi(x, y)$ and the complex dielectric constant $\varepsilon(x, y)$ of the surface.

The reflected field consists of non-coherent superposition of random partial fields scattered by local areas of the surface. According to the central limit theorem, the density of distribution of random characteristics of the scattered field may be considered correct, which could be proved experimentally. Consequently, full characteristics of the scattered field will include the mean field and dispersions.

To make mathematical description of statistical characteristics of the field, the following information on statistical characteristics of the surface $z = \xi(x, y; t)$ is required:

- for single-structure profile: total density of distribution of the height probs and their derivatives in two points of the surface located with the interval r $W_6\left(\xi, \xi_r; \frac{\partial \xi}{\partial x}, \frac{\partial \xi}{\partial y}, \frac{\partial \xi_r}{\partial x} \frac{\partial \xi_r}{\partial y}\right)$, the space–time correlation coefficient will be denoted as $\rho_\xi(r, \tau)$;
- for multi-structure profile: total density of distribution of probabilities $\prod_{k=1}^{n} W_6\left(\xi_k, \xi_{kr}; \frac{\partial \xi_k}{\partial x}, \frac{\partial \xi_k}{\partial y}, \frac{\partial \xi_{kr}}{\partial x} \frac{\partial \xi_{kr}}{\partial y}\right)$ or $W_{6n}(\xi_1, \xi_{1r}, \ldots, \xi_n; , \xi_{nr}; \frac{\partial \xi_1}{\partial x}, \frac{\partial \xi_1}{\partial y}, \ldots, \frac{\partial \xi_{nr}}{\partial x} \frac{\partial \xi_{nr}}{\partial y})$ for independent and dependent structures, respectively. Let us denote the space–time correlation coefficients as $\rho_{\xi_k}(r, \tau)$ and intercorrelation coefficients as $\rho_{\xi_k \xi_i}(r, \tau)$. If the surface is motionless, then $\rho_\xi(r, \tau) \equiv \rho_\xi(r)$.

Further on, we shall study the case of quasi-vertical sounding, which is generally rather characteristic for radiolocation. This allows exclusion of the shielding effect.

The mean field may be interpreted as a constant of the random field, coherent to the radiated field, given the radar and the surface are fixed.

With the help of Eqs. (8.70), (8.73), and (8.77), it is possible to write down a generalized equation for the field components by assemblies of surfaces under its different models:

$$U_{ij} = \int_{S_0} \frac{F(x, y)}{R_0^2} \exp(-2ikR_0) T_{ij}^N(\theta, \varphi; \varepsilon) \mathrm{d}S_0, \tag{8.78}$$

where N is the number of the model; $T_{ij}^{N}(\theta, \varphi; \varepsilon)$ are complex coefficients of reflection by the mean field of the partial wave, determined with the equations:

– for Model 1 (full field):

$$T_{ij}^{I} = \int \int\limits_{-\infty}^{\infty} \int s_{ij}\left(\frac{\partial \xi}{\partial x}, \frac{\partial \xi}{\partial y}; \theta, \varphi; \varepsilon\right) \exp(2ik_z\xi) \cdot W_3\left(\xi, \frac{\partial \xi}{\partial x}, \frac{\partial \xi}{\partial y}\right) \mathrm{d}\xi \cdot \mathrm{d}\frac{\partial \xi}{\partial x} \cdot \mathrm{d}\frac{\partial \xi}{\partial y}, \tag{8.79}$$

where s_{ij} is determined with Eq. (8.66).

– for Model 2 (perturbation field):

$$T_{ij}^{II} = \frac{k^2}{\pi}\left(\frac{\varepsilon - 1}{\varepsilon}\right)^2 s_{ij}(\theta, \varepsilon) \int\limits_{-\infty}^{\infty} \xi \cdot W(\xi)\mathrm{d}\xi, \tag{8.80}$$

where s_{ij} is determined with Eq. (8.71).

– for Model 3 (perturbation field):

$$T_{ij}^{III} = \frac{k^2}{\pi}\left(\frac{\varepsilon - 1}{\varepsilon}\right)^2 \int \int\limits_{-\infty}^{\infty} \int s_{ij}\left(\frac{\partial \xi_1}{\partial x}, \frac{\partial \xi_1}{\partial y}; \theta, \varphi; \varepsilon\right)$$
$$\times W_2\left(\frac{\partial \xi_1}{\partial x}, \frac{\partial \xi_1}{\partial y}\right) \xi_2 W(\xi_2)\mathrm{d}\xi_2 \cdot \mathrm{d}\frac{\partial \xi_1}{\partial x} \cdot \mathrm{d}\frac{\partial \xi_1}{\partial y} \tag{8.81}$$

where s_{ij} is determined with Eq. (8.74).

Thus, to determine the complex coefficient of reflection it is necessary to know definite laws of distribution of peaks and their gradients.

Experimental studies of distribution of peaks and inclines of the disturbed water surface and plowed fields reveal that they comply with the Gaussian statistics. The Gaussian distribution also applies to many land surfaces with different reliefs.

As it was already mentioned above, the mean field from the surface of Model 2 is the field of even surface $z = 0$, while the mean field of Model 3 is the mean field of a large-scale irregularity $z = \xi_1(x, y)$.

Thus, it remains to determine the mean field from the surface stated by Model 1.

Having done all necessary transformations (taking into account the above said), we got the following assessment of the complex coefficient of reflection by the mean field for the statistically isotropic surface with the finite conductivity:

$$\begin{cases} T_{\mathrm{VP(HP)}}^{I} = \left|R_{\mathrm{VP(HP)}}\right| \exp\left\{-2\left(\frac{2\pi\sigma\cos\theta}{\lambda}\right)^2\right\} \exp\left\{i\psi_{\mathrm{VP(HP)}}\right\} \cos\theta \\ T_{\mathrm{VP-HP}}^{I} = T_{\mathrm{HP-VP}}^{I} = 0 \end{cases}, \tag{8.82}$$

where $R_{\text{VP(HP)}}$ are the Fresnel coefficients of reflection determined with Eqs. (8.8) and (8.9); σ is a mean-squared height of irregularities.

Phases of the reflection coefficients are determined with the following equations:

$$\begin{cases} \psi_{\text{VP}} = \pi - \arctan \frac{2\text{Im}\sqrt{\varepsilon}\cos\theta}{|\varepsilon|\cos^2\theta - 1} \\ \psi_{\text{HP}} = \pi + \arctan \frac{2\text{Im}\sqrt{\varepsilon}\cos\theta}{|\varepsilon| - \cos^2\theta} \end{cases}. \tag{8.83}$$

As it is seen from Eq. (8.82), the mean field of orthogonal polarization is absent. Substituting Eq. (8.82) into Eq. (8.78) and making integration, we shall get:

$$U_{\text{VP(HP)}} = \sqrt{P_{\text{VP(HP)}}} F(0,0) \frac{\lambda}{2H} T_{\text{VP(HP)}}\left(\frac{\sigma}{\lambda}, \theta, |R_{\text{VP,HP}}|\right) \exp\left\{-i\left(\frac{4\pi H}{\lambda} - \psi_{\text{VP(HP)}}\right)\right\} \tag{8.84}$$

It is clear that the ranges of the mean fields depend both on statistical characteristics of the surface and on its electrical characteristics. Mean fields are almost absent in case of too rough surfaces $\left((2\pi\sigma/\lambda)^2 \gg 1\right)$. The phase of the mean field is proportional to the height of the mean level of the surface and depends on the phase of the coefficient of reflection on the corresponding polarization, i.e., from electrical characteristics of the surface.

Let us find out second moments of the reflected field now.

In general case (an elliptically polarized radiated field and the finite conductivity of the reflected surface), the matrix of central moments of the second kind appears to be a Hermitian matrix of non-coherent components of the complex random field consisting of dispersions and correlation moments:

$$M = \begin{pmatrix} \overline{|E_{\text{VV}}|^2} & \overline{E_{\text{VV}}E_{\text{VH}}^*} & \vdots & \overline{E_{\text{VV}}E_{\text{HV}}^*} & \overline{E_{\text{VV}}E_{\text{HH}}^*} \\ \overline{E_{\text{VV}}^*E_{\text{VH}}} & \overline{|E_{\text{VH}}|^2} & \vdots & \overline{E_{\text{VH}}E_{\text{HV}}^*} & \overline{E_{\text{VH}}E_{\text{HH}}^*} \\ \cdots & \cdots & \vdots & \cdots & \cdots \\ \overline{E_{\text{VV}}^*E_{\text{HV}}} & \overline{E_{\text{VH}}^*E_{\text{HV}}} & \vdots & \overline{|E_{\text{HV}}|^2} & \overline{E_{\text{HV}}E_{\text{HH}}^*} \\ \overline{E_{\text{VV}}^*E_{\text{HH}}} & \overline{E_{\text{VH}}^*E_{\text{HH}}} & \vdots & \overline{E_{\text{HV}}E_{\text{HH}}} & \overline{|E_{\text{HH}}|^2} \end{pmatrix},$$

$$= \begin{pmatrix} M_{\text{VV}} & \vdots & M_{\text{VH}} \\ \cdots & \cdots & \cdots \\ M_{\text{HV}} & \vdots & M_{\text{HH}} \end{pmatrix} \tag{8.85}$$

where suffix numbers denote the type of polarization of the radiated and the received radio waves.

Matrices M_{VV}, M_{HH} correspond to linearly polarized falling waves.

If the transmitter power for vertical and horizontal components of the falling field is denoted with $P_{\text{VP(HP)}}$, then the phase shift between these components would be denoted with δ, and then

$$\frac{E_{\mathrm{VH}}}{E_{\mathrm{HV}}} = \sqrt{\frac{P_{\mathrm{VP}}}{P_{\mathrm{HP}}}} = m\mathrm{e}^{-\mathrm{i}\delta}, \tag{8.86}$$

where $m = \sqrt{P_{\mathrm{VP}}/P_{\mathrm{HP}}}$.

Consequently, the matrix (8.85) will be fully determined, if its six components are known: dispersions $\overline{|E_{\mathrm{HH}}|^2}$, $\overline{|E_{\mathrm{VV}}|^2}$, $\overline{|E_{\mathrm{VH}}|^2}$ and complex product moments $\overline{E_{\mathrm{VV}}E^*_{\mathrm{VH}}}$, $\overline{E_{\mathrm{VV}}E^*_{\mathrm{HH}}}$, $\overline{E_{\mathrm{VH}}E^*_{\mathrm{HH}}}$. The remain 10 moments can be determined with the help of Eq. (3.15). In compliance with Eqs. (8.70), (8.73), and (8.77) for the field components of different models, it can be concluded that expressions for all elements of the matrix (8.85) differ from each other by the multiplier P_{ij}, subintegral functions s_{ij} and the dependence from the height of irregularities $\xi(x, y)$.

Let us introduce a generalized entry format for all moments of the field under different surface models:

$$\overline{E_i E_j^*} = P_{ij} \int\limits_{S_0}\int\limits_{S_{0r}} \frac{F(x, y)F(x_r, y_r)}{R_0^2 R_{0r}^2} \cdot \exp\{-2ik(R - R_{0r})\} Q_{ij}^N \mathrm{d}S_0 \mathrm{d}S_{0r} - \overline{E}\,\overline{E}_j^*. \tag{8.87}$$

In Eq. (8.87), suffixes (i, j) vary from 1 to 4, i.e., $i, j = \overline{1, 4}$. Meanwhile, the value $i = 1$ corresponds to $i =$ VV, $i = 2$ corresponds to $i =$ VH, $i = 3$ corresponds to $i =$ HV, and $i = 4$ corresponds to $i =$ HH.

For Model 1:

$$\begin{aligned} Q^I_{i,j} = \int\int\int\int\limits_{-\infty}^{\infty}\int\int s_i(\xi'_x, \xi'_y; \theta, \varphi; \varepsilon) s_j^*(\xi'_x, \xi'_y; \theta, \varphi; \varepsilon^*) \cdot \exp\{-2ik_z(\xi_r - \xi)\} \\ \times W_6(\xi, \xi_r, \xi'_x, \xi'_y, \xi'_{xr}, \xi'_{yr}) \mathrm{d}\xi, \mathrm{d}\xi_r, \mathrm{d}\xi'_x, \mathrm{d}\xi'_y, \mathrm{d}\xi'_{xr}, \mathrm{d}\xi'_{yr} \end{aligned}, \tag{8.88}$$

where $s_{i,j}$ is determined with Eq. (8.66).

– for Model 2 (moments of the perturbation field):

$$Q^{II}_{ij} = \frac{k^4}{\pi^2}\left|\frac{\varepsilon - 1}{\varepsilon}\right|^2 s_i(\theta, \varepsilon) s_j^*(\theta, \varepsilon^*) \cdot \int\limits_{-\infty}^{\infty}\int\limits_{-\infty}^{\infty} \xi \cdot \xi_r \cdot W_2(\xi, \xi_r) \mathrm{d}\xi \cdot \mathrm{d}\xi_r, \tag{8.89}$$

where $s_{i,j}$ is determined with Eq. (8.71).

– for Model 3 (moments of the perturbation field):

$$Q_{ij}^{III} = \frac{k^4}{\pi^2}\left|\frac{\varepsilon-1}{\varepsilon}\right|^4 \int \int_{-\infty}^{\infty} \int \int s_i\left(\xi'_{x1}, \xi'_{y1}; \theta, \varphi; \varepsilon\right) s_j\left(\xi'_{x1}, \xi'_{y1}; \theta, \varphi; \varepsilon^*\right)$$
$$\times W_4\left(\xi''_{x1}, \xi'_{y1}, \xi'_{x1r}, \xi'_{y1r}\right) \mathrm{d}\xi'_{x1}, \mathrm{d}\xi'_{y1}, \mathrm{d}\xi'_{x1r}, \mathrm{d}\xi'_{y1r} \int_{-\infty}^{\infty} \int_{-\infty}^{\infty} \xi_{2r}\xi_2 W_2(\xi_2, \xi_{2r}) \mathrm{d}\xi_2 \mathrm{d}\xi_{2r} \tag{8.90}$$

Coefficients P_{ij} are summarized in the table corresponding to the matrix (8.85):

$$P_{ij} = \begin{pmatrix} P_{\mathrm{VP}} & P_{\mathrm{VP}} & \sqrt{P_{\mathrm{VP}} P_{\mathrm{HP}}}\mathrm{e}^{-\mathrm{i}\delta} & \sqrt{P_{\mathrm{VP}} P_{\mathrm{HP}}}\mathrm{e}^{-\mathrm{i}\delta} \\ P_{\mathrm{VP}} & P_{\mathrm{VP}} & \sqrt{P_{\mathrm{VP}} P_{\mathrm{HP}}}\mathrm{e}^{-\mathrm{i}\delta} & \sqrt{P_{\mathrm{VP}} P_{\mathrm{HP}}}\mathrm{e}^{-\mathrm{i}\delta} \\ \sqrt{P_{\mathrm{VP}} P_{\mathrm{HP}}}\mathrm{e}^{\mathrm{i}\delta} & \sqrt{P_{\mathrm{VP}} P_{\mathrm{HP}}}\mathrm{e}^{\mathrm{i}\delta} & P_{\mathrm{HP}} & P_{\mathrm{HP}} \\ \sqrt{P_{\mathrm{VP}} P_{\mathrm{HP}}}\mathrm{e}^{\mathrm{i}\delta} & \sqrt{P_{\mathrm{VP}} P_{\mathrm{HP}}}\mathrm{e}^{\mathrm{i}\delta} & P_{\mathrm{HP}} & P_{\mathrm{HP}} \end{pmatrix}. \tag{8.91}$$

Further solution is possible if the type of the statistical dependence is specified. Let us consider several examples.

The case of extremely rough surface $\left((\sigma k_z)^2 \gg 1\right)$ is the most typical for *Model 1* during its practical application within the wave range shorter than the VHF band. At that, the main value of the integral (8.88) is within the range $r = \sqrt{(\Delta x)^2 + (\Delta y)^2} \ll \rho_0$, where is the space radius of correlation of the large-scale surface. In this case, the height ξ and its derivatives ξ'_x, ξ'_y may be considered to be dependent. Consequently,

$$W_6\left(\xi, \xi_r, \xi'_x, \xi'_y, \xi'_{xr}, \xi'_{yr}\right)\Big|_{r \ll \rho_0} \approx W_2(\xi, \xi_r) W_4\left(\xi'_x, \xi'_y, \xi'_{xr}, \xi'_{yr}\right). \tag{8.92}$$

Further reductions would lead to the following correlation:

$$Q_{ij}^{I} = \exp\left\{-0.5(\sigma k_z)^2[1 - \rho(\Delta x, \Delta y)]\right\} \cdot \int \int_{-\infty}^{\infty} \int \int s_i s_j^* W_4\left(\xi'_x, \xi'_y, \xi'_{xr}, \xi'_{yr}\right) \mathrm{d}\xi'_x \mathrm{d}\xi'_y \mathrm{d}\xi'_{xr} \mathrm{d}\xi'_{yr}, \tag{8.93}$$

where $(\Delta x, \Delta y)$ is the space coefficient of the surface correlation.

Usually, it is supposed for isotropic surfaces:

$$\rho(\Delta x, \Delta y) = \exp\left\{-\frac{(\Delta x)^2 + (\Delta y)^2}{\rho_0^2}\right\}, \tag{8.94}$$

where ρ_0^2 is the squared radius of correlation.

The correlation function considers for anisotropic surfaces that the surface has an average interval with the frequency Ω in the direction of waves (*OX* axis). There is no such interval along the wave. A typical representative of the considered surface is the sea "swell waves."

In this case, it is considered:

$$\rho(\Delta x, \Delta y) = \exp\left\{-\left[\left(\frac{\Delta x}{\rho_{0x}}\right)^2 + \left(\frac{\Delta y}{\rho_{0y}}\right)^2\right]\right\} \cos(\Omega \Delta x). \tag{8.95}$$

Further calculations are too awkward and are not stated here. The main conclusion that may be made from the latest correlations is that the relation of desired correlations and the complex dielectric constant can once again be expressed with Fresnel coefficients of reflection. At that, these coefficients are either included in subintegral functions, or present (their function) some multiplier against the function correlation containing only geometric parameters and their statistics. This may also be proved with Eqs. (8.66), (8.71), and (8.74).

The analysis performed on different models of surfaces revealed that correlation of the scattering parameters with the complex dielectric constant may be realized only with the help of Fresnel coefficients of reflection.

8.4.7 Statistical Characteristics of Electrical Parameters of Radar Targets

As it was already mentioned above, besides roughness of surfaces, changes in physical and chemical properties of surfaces, i.e., their dielectric constant, also contribute considerably to fluctuations of the reflection parameters of the radio wave reflected from such surface.

There is a lot of literature dedicated to development of physically substantiated statistical models of orthogonal components of partially polarized waves. Within the framework of the above considerations, this means that statistical models of underlying coverings were essentially developed. To develop such models on the basis of known models of partially polarized radio waves, it is required to know corresponding analytical dependencies.

To develop these dependencies, it is convenient to use equations concerning circular polarization:

$$\begin{cases} |q| = \frac{\sqrt[4]{(\varepsilon - \sin^2\theta) + (60\sigma\lambda)^2}}{\sin\theta\ \tan\theta} \\ \delta = 0.5\arctan\frac{60\sigma\lambda}{\varepsilon - \sin^2\theta} \end{cases}, \tag{8.96}$$

where $q = |q|e^{i\delta}$ is the polarization coefficient on the circular polarization, i.e., correlation of orthogonal components in the channels of right-hand and left-hand circles of polarization.

Applying common rules for transformation of the density of probabilities distribution, we shall get:

$$W(\varepsilon,\sigma) = W(|q(\varepsilon,\sigma)|,\delta(\varepsilon,\sigma))\left|\frac{\partial(|q|,\delta)}{\partial(\varepsilon,\sigma)}\right|. \tag{8.97}$$

The density of distribution of parameter probabilities for reflected radio wave $W(|q|,\delta)$, included in Eq. (8.97), is considered to be established.

To determine one-dimensional densities of probabilities $W(\varepsilon)$ and $W(\delta)$, it is necessary to integrate Eq. (8.97) by excluding variables. Though, in general case, the stated task reveals serious mathematical difficulties, and no results of integration of expression can be obtained in a closed cycle.

Let us consider a special case to illustrate the above said.

Let us determine the density of distribution of the probability of the dielectric constant for the case when the medium surface possesses small conductivity, which is often in the practice.

Let us suppose also that deterministic components of orthogonally polarized components of the partially polarized wave are absent, while the joint density of the probability $W(|q|, \delta)$ may be described with the equation, according to [1]:

$$W(|q|, \delta) = \frac{qh^2\left(1 - R^2\right)}{\pi\left(1 - 2qhR\,\cos\delta + q^2h^2\right)^2}, \tag{8.98}$$

where R is the coefficient of correlation of orthogonal components, $h = \sigma_R/\sigma_L$ is the correlation of mean-squared values of signals in orthogonal channels.

In this case, the required density of distribution of probabilities will be of the following appearance:

$$W(\varepsilon) = \frac{h^2\left(1 - R^2\right)\sin^2\theta\,\tan^2\theta}{8\sqrt{\varepsilon - \sin^2\theta}\left[\sin^2\theta\,\tan^2\theta - 2hR\,\sin\theta\,\tan\theta\sqrt{\varepsilon - \sin^2\theta} + h^2\left(\varepsilon - \sin^2\theta\right)\right]^2}. \tag{8.99}$$

Similarly, the desired correlations may be obtained under other statistical models of orthogonal components.

8.5 KLL-Sphere and Its Properties

One of the main results obtained from clauses (8.3)–(8.4) is the proof of the fact that relative power measurements in orthogonal channels of the receiver may lead to establishment of the value of the complex dielectric constant for rather wide range of underlying surfaces (first of all, for rather smooth surfaces, when no cross-components appear during reflection of radio waves of linear polarization). Regarding this aim, it is necessary to calculate the correlation of voltage ranges in the stated channels and to determine the phase differential of these voltages. This makes it possible to determine the main characteristic for the method proposed in clause (8.1)—the polarization correlation *f*. The module of these parameters equals to the voltage in the channel receiving the vertical component of the reflected radio wave and the same voltage in the channel receiving the horizontal component. The parameter argument *f* is the mentioned phase differential. In this case, the identity of directional properties of antennas for horizontal and vertical polarizations, as well as the identity of orthogonal channels of the receiver, is undoubtedly required.

As it was shown in clause (8.1), the complex dielectric constant can be determined with the help of the correlation (8.24), which allows presenting a real and an imaginary component of the complex dielectric constant in the following form:

$$\begin{cases} \mathrm{Re}\,\varepsilon = \left[1 + 4|f|\frac{(1+|f|^2)\cos\psi - 2|f|}{(1-2|f|\cos\psi + |f|^2)}\sin^2\theta\right]\tan^2\theta \\ \mathrm{Im}\,\varepsilon = 4|f|\frac{(1-|f|^2)\sin\psi}{(1-2|f|\cos\psi+|f|^2)}\sin^2\theta\ \tan^2\theta \end{cases}. \tag{8.100}$$

If waves of circular polarization are used, then instead of Eq. (8.24) we shall get:

$$\varepsilon = \left[1 + q^2\tan^2\theta\right]\sin^2\theta, \tag{8.101}$$

where $q = |q|e^{i\delta}$ is the polarized correlation in the circular polarization, i.e., correlation of orthogonal components in the channel with the right-hand and the left-hand circles of polarization.

Rather compact equations for the real component of the dielectric constant and the medium conductivity σ can be obtained from Eq. (8.101):

$$\begin{cases} \mathrm{Re}\,\varepsilon = \left[1 + q^2\cos 2\delta\tan^2\theta\right]\sin^2\theta \\ \sigma = \frac{1}{60\lambda}q^2\sin 2\delta\tan^2\theta\ \sin^2\theta \end{cases}. \tag{8.102}$$

Let us note that the complex dielectric constant can be expressed directly with the polarization ellipse parameter (r is the ellipticity ratio, and β is the polarization orientation) of the reflected radio wave:

$$\begin{cases} \varepsilon = \left[1 + \left(\frac{1-r}{1+r}\right)^2 e^{4i\beta}\tan^2\theta\right]\sin^2\theta \\ \mathrm{Re}\,\varepsilon = \left[1 + \left(\frac{1-r}{1+r}\right)^2\cos 4\beta\tan^2\theta\right]\sin^2\theta \\ \sigma = \frac{1}{60\lambda}\left(\frac{1-r}{1+r}\right)^2\sin 4\beta\tan^2\theta\ \sin^2\theta \end{cases}. \tag{8.103}$$

The stated correlations reveal that their application requires successive radiating of radio waves of the same voltage at horizontal and vertical polarization, meanwhile measuring the value of the voltage in the corresponding receiver channel, followed by calculations of the parameter f with Eq. (8.24).

Though, it can be performed in some other way. If a linearly polarized (the polarization angle 45°) radio wave (LP-45) is radiated, then we shall get the same situation. In this case, cophasal radio waves of the same power will be radiated at the same time in two orthogonal channels. As it was mentioned above, no cross-components appear for the considered surface class during reflection, and signals in the orthogonal receiver channels may be used for determination of the polarization ration f, which in this case would completely describe the polarization ellipse of the reflected wave. *Thus, the parameter f characterizes polarization of the reflected radio wave, provided that the surface is radiated with the LP-45 wave.*

Fig. 8.8 KLL-sphere

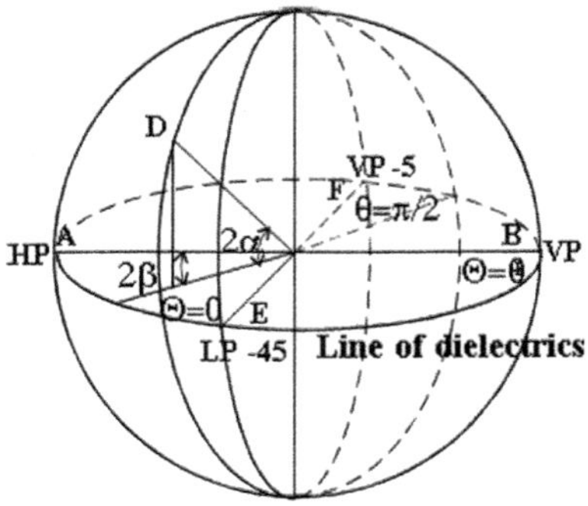

Consequently, each value f would comply with some polarization, which, as it is known, is reflected one-to-one in the corresponding point of the Poincare sphere. From the other hand, the parameter f uniquely determines the complex dielectric constant, i.e., the type of the surface. This means that *any point on the sphere can be one-to-one correlated with each type of the underlying surface*. Thus, a sphere can be developed where each point would determine one or another type of surface. Let us name such sphere a KLL-sphere (see Fig. 8.8).

The KLL-sphere reveals new opportunities to classify underlying surfaces. It is clear that the main classification parameter is the type of polarization of the reflected wave, provided that the studied surface is radiated with LP-45 polarization radio wave.

To establish quantitative ratios, let us state equations aligning the parameter f with the coordinates of points at the KLL-sphere representing the corresponding polarization. Let us use geographic coordinates for this reason.

The longitude of some arbitrary point D on the KLL-sphere would be denoted as 2β, and the latitude would be denoted as 2α (see Fig. 8.8).

Having these designations, we shall get the following correlations:

$$\left.\begin{aligned} 2\beta &= \arctan \frac{2|f|\cos\psi}{1-|f|^2} \\ \alpha &= \arctan \sqrt{\frac{\tan^2\beta - 2|f|\tan\beta\cdot\cos\psi + |f|^2}{1+2|f|\tan\beta\cdot\cos\psi + |f|^2\tan^2\beta}} \end{aligned}\right\}. \tag{8.104}$$

The designation $\psi = \arg f$ was introduced into Eq. (8.104).

The angle θ would be of significance under this approach, as it is seen from Eq. (8.104). Let us consider briefly its influence.

If the view angle is $\theta = 0$, then horizontal and vertical polarizations for any type of surface would not differ. That is why the parameter $f = 1$. This value of f complies with LP-45. In this case, all types of surfaces would reflect in the one and only point (point E in Fig. 8.8). Surfaces do not differ in this case. As per the second limit case when $\theta = \pi/2$, the parameter f would equal to -1 for any type of surface. In this case, all surfaces would also reflect in the same point (point F in Fig. 8.8). The point F is diametrically opposite to the point E. It complies with linear polarization

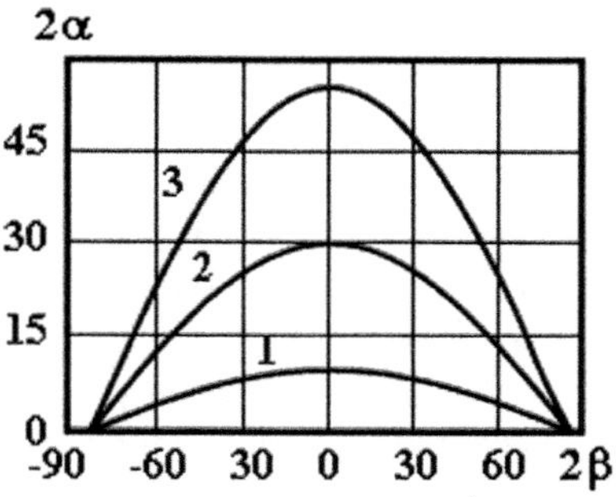

Fig. 8.9 Dependence of the geographical coordinates $(2\alpha, 2\beta)$ points on the KLL-sphere from different types of terrain

LP-135. If the view angle is changed, the point corresponding to different types of surfaces will begin tracing their own way, each starting in the point E and finishing in the point F.

Let us consider one more specific case. It would be assumed that the studied surface is an ideal dielectric (ε is a real number, $\psi = 0$). In this case, f would also be a real number that is why $\alpha = 0$. This means that changing of the view angle θ would result in the point representing the type of reflected radio wave polarization moving along the equator of the KLL-sphere. If the view angle equals to Brewster's angle, this point reaches the point B (see Fig. 8.8), i.e., the point corresponding to the vertical polarization. Thus, dielectrics would be reflected on the equator of the KLL-sphere. This allows naming the equator a line of dielectrics.

To illustrate the above stated, dependencies of the coordinates $(2\beta, 2\alpha)$ of the points on the KLL-sphere, representing different types of land surfaces, are shown in Fig. 8.9. As it can be seen on the figure, some curved lines trace on the KLL-sphere from the point corresponding to the coordinates $2\alpha = 0$, $2\beta = 90°$ to the point with the coordinates $2\alpha = 0$, $2\beta = -90°$ (negative angles 2β would comply with angles from 360° to 270° upon changing of the angle β) under changing of the view angle for each type of the underlying surface.

The surfaces were selected so that the relation of the real component of the complex dielectric constant to its imaginary component would be: (a) considerably less than 1; (b) considerably more than 1; and (c) of the value with the same order.

The following surfaces of this type were selected:

- an agricultural plant field (corn), characterized by $\varepsilon = 3 - i \cdot 15$ (curve 3 in Fig. 8.9);
- sea water (NaCl concentration is 2 g/mol*1) characterized by $\varepsilon = 40 - i \cdot 40$ (curve 2 in Fig. 8.9);
- watered sand (relative moisture content 12%), characterized by $\varepsilon = 8 - i \cdot 1$ (curve 1 in Fig. 8.9).

The highlighted points correspond to points with one and the same viewpoint θ.

It is rather evidenced from the figures that, when selecting a necessary view angle θ, a maximum difference of coordinate points reflecting corresponding surface types can be provided on the KLL-sphere. This may lead directly to increase in the efficiency of solving tasks of differentiation of underlying surfaces. Besides, the possibility to determine the complex dielectric constant on the basis of relative measurements combined with improvement of quality of underlying surfaces differentiating allows considerable increase in the efficiency of solving inverses.

Thus, the KLL-sphere allows rather extraordinary modeling of underlying surfaces. The essence of such modeling implies that *each type of smooth underlying surface viewed from the angle* θ *may comply with a certain type of polarization of the reflected wave*. This type of polarization corresponds with the polarization of reflected radio wave occurring as a result of reflection from the underlying surface of LP-45 polarization radio wave. As it was mentioned above, the said means that each type of underlying surface may be one-to-one matched with some point on the unit radius sphere (KLL-sphere). Any changes to physical and chemical properties of underlying surfaces would move this point through the KLL-sphere, testifying of changes in the stated properties.

It is not necessary to use LP-45 for development of the KLL-sphere; waves of other established polarization can be used. In this case, the polarization ratio should be adjusted, taking into account the available difference of ranges and phases of orthogonal components of the radiated radio wave. In this regard, the most interesting would be circular polarization. Application of the circular polarization allows avoiding of inaccuracies in installation of the antenna in reference to the ground surface, which is a certain challenge for LP-45 antennas. In order to apply Eqs. (8.104), it is required to change the difference of signal phases in orthogonal channels of the receiver by 90°.

Several models of rough surfaces were proposed in clause (8.4.1). Let us show that the complex dielectric constant may also be determined for some of them by means of corresponding relative measurements. This would allow developing of images of such surfaces on the KLL-sphere.

Let us begin with Model 1. For this purpose, we shall proceed from the equations stated in clause (8.4.2).

It is clear that the correlation of the voltage in orthogonal channels of the receiver would define the parameter f within the proper polarization basis. This offers scopes for determination of the complex dielectric constant of underlying surfaces that may be described with Model 1 and development of their images on the KLL-sphere.

The obtained correlations establish the principal scheme for realization of the stated possibilities. The first step is to select the type of polarization of the radiated wave that would eliminate (actually minimize) the cross-component in the reflected wave. Selection may be performed by application of the mode of complete polarization scanning. The second step involves selection of such type of polarization for the radiated radio wave that its orthogonal components in the proper polarization basis of the matrix S would be cophasal and of the same voltage. The third step is measurement of correlation of the difference in phases and ranges of voltage in orthogonal channels of the receiver. The last step would allow establishment of the required correlation f and the further determination of the complex dielectric constant and development of the KLL-sphere.

Knowledge of the polarization correlation allows determination of correlations of statistical parameters η/a. Following Eq. (8.67), it would be easy to find out the required correlation:

$$\left(\frac{\eta}{a}\right)^2 = \tan^2\zeta = \frac{U_{11}/U_{22} + f}{1 + fU_{11}/U_{22}}, \tag{8.105}$$

where U_{11} and U_{22} are the voltages in orthogonal channels, measured in the polarized basis (HP-VP).

Let us proceed to Model 2, and for this purpose, we shall address the equations stated in clause (8.4.3).

We shall show that relative measurements performed in orthogonal channels for this class of surfaces would provide the opportunity to determine the complex dielectric constant, as well as to develop a corresponding image on the KLL-sphere.

Correlation of voltages in the orthogonal channels of the receiver will be determined with the following equation, as it can be seen from Eq. (8.71):

$$\frac{U_{22}}{U_{11}} = 1 + \frac{(1 - R_{HP})\sin^2\theta}{L(\theta, \varepsilon)}. \tag{8.106}$$

After a number of rather awkward algebraic transformations, correlation of voltage in orthogonal channels which is connected with the polarization ratio f and the view angle θ may be established with the following equation:

$$\frac{U_{22}}{U_{11}} = \frac{1 + f^2 + 2f\sin^2\theta}{2\cos^2\theta}. \tag{8.107}$$

The correlation (8.107) solves the set task. It would not be difficult to get a direct equation for the polarized ratio:

$$f = -\sin^2\theta \pm \sqrt{\sin^4\theta + 2\frac{U_{22}}{U_{11}}\cos^2\theta - 1}. \tag{8.108}$$

Thus, the method for calculation of the complex dielectric constant for the surface described with Model 2, based on relative measurements of power in orthogonal channels of the receiver, remains in effect.

An essential reason to use this method is the same as used during the studies of smooth surfaces. A LP-45 radio wave or a wave with circular polarization with corresponding adjustment of phase correlations may be used as a sounding wave.

Let us proceed to Model 3.

Unfortunately, it is impossible to extract an explicit expression for the polarization ratio f from Eq. (8.74). Nonetheless, if it is supposed that there is some a priori knowledge on statistical characteristics of underlying surfaces, then in this case it would be possible to establish the complex dielectric constant, relying upon relative measurements. The correlation U_{11}/U_{22} reduces to the equation in the parameter

Table 8.1 Values of the module of the complex dielectric constant for some types of surfaces

Surface	$\overline{s_{11}}$	$\overline{s_{22}}$	$\psi°$	Target value $\|\varepsilon\|$	Value $\|\varepsilon\|$ [source]
Multi-year ice	0.0356	0.0689	7	2.7	3.0 [2]
One-year ice	0.0156	0.0277	5	3.5	3.4 [2]
Water surface	0.0084	0.0097	30	35	40–70 [3]
Field	0.283	0.235	10	14	12 [3]

q. Solving of this equation gives an opportunity to express the complex dielectric constant with the stated equation and statistical characteristics of the surface. A definite reduction may result in the following correlation obtained from relative measurements:

$$\frac{U_{11} - U_{22}}{U_{12}} = \frac{a_1^2 - \eta^2}{a_1 \eta} = \frac{a_1}{\eta} - \frac{\eta}{a_1}. \tag{8.109}$$

As per the underlying surfaces described with Model 4 (the surface has different complicated geometrical shapes), in this case it is obviously impossible to denote any definite value of the complex dielectric constant. Nevertheless, if we perform calculations of the complex dielectric constant formally using Eqs. (8.24), (8.75), or (8.108), we can talk, though conventionally, on some complex dielectric constant (pseudo-permittivity). This approach may be justified by the fact that it allows such structures as referred to Model 4 class to align some point on the KLL-sphere. Therefore, even the objects with a complicated geometrical structure may be included in the classification system with the help of the KLL-sphere.

The values of the module of the complex dielectric constant calculated with the stated equations for some types of surfaces, based on the results of experimental studies and the wavelength $\lambda = 21$ cm, are given in Table 8.1. The values of the complex dielectric constant from literature sources are also stated in the table.

Conditions of the measurements were as follows: multi-year ice, $\theta = 45°$; one-year ice $\theta = 45°$; water surface, wind rate 10–15 m/c, $\theta = 70°$; and grassy field, $\theta = 70°$.

For illustrative purposes, Fig. 8.10 shows correlation of the dielectric constant of the medium with the low conductivity (almost dielectric) and the phasor module $|f|$ under different view angles θ.

Figure 8.11 shows correlations of the dielectric constant of the same medium and the ellipticity ratio of the given electromagnetic wave K and also the polarization ratio q, during application of the circular polarization waves.

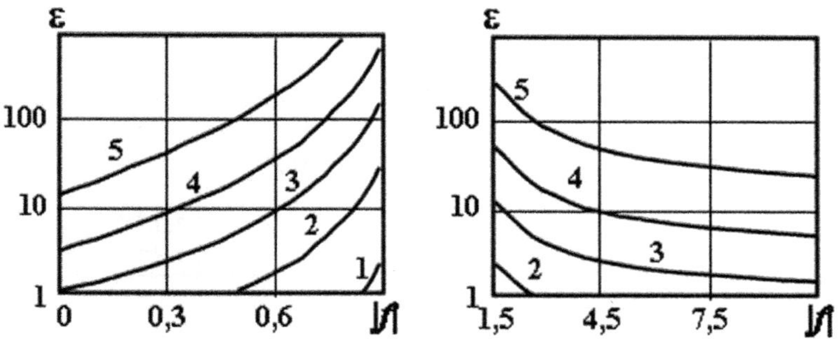

Fig. 8.10 Dependence of the permittivity ε on the phasor modulus 1—$\theta = 15°$; 2—$\theta = 30°$; 3—$\theta = 45°$; 4—$\theta = 60°$; 5—$\theta = 75°$

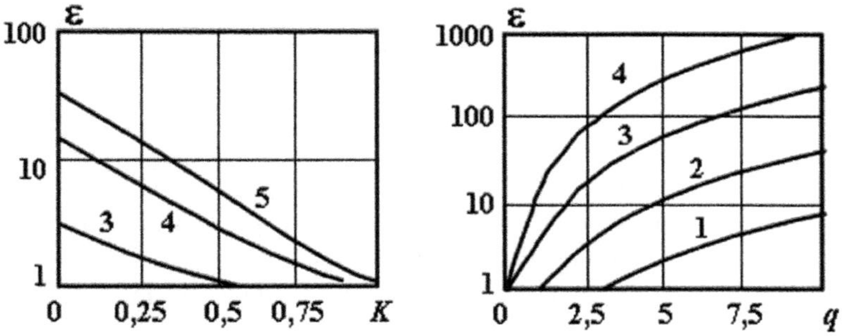

Fig. 8.11 Dependence of the permittivity ε on the polarization parameters q, K. 1—$\theta = 30°$; 2—$\theta = 45°$; 3—$\theta = 60°$; 4—$\theta = 75°$; 5—$\theta = 80°$

8.6 Determination of Electrophysical Characteristics of Observation Objects with Incomplete Information on Spatiotemporal State of Radiolocation Signals

Determination of the phase differential for signals in orthogonal channels of the receiver is an independent, technically rather complicated task. As per measurements of the signal range in orthogonal channels of the receiver, they can be performed with a rather high accuracy.

In the same time, there are no specific technical difficulties in measurements of the range under different sounding angles.

As a matter of practice, researches often have only some set of signal ranges in orthogonal channels of the receiver under a large number of view angles at their disposal. Any information of phase differential of these signals is not available.

It appears that even in this case, it is possible to determine the complex dielectric constant of underlying surfaces, having modified the proposed amplitude-phase differential method and basing only on measurements of correlation of signal voltage in orthogonal channels of the receiver. To implement this opportunity to the full, all

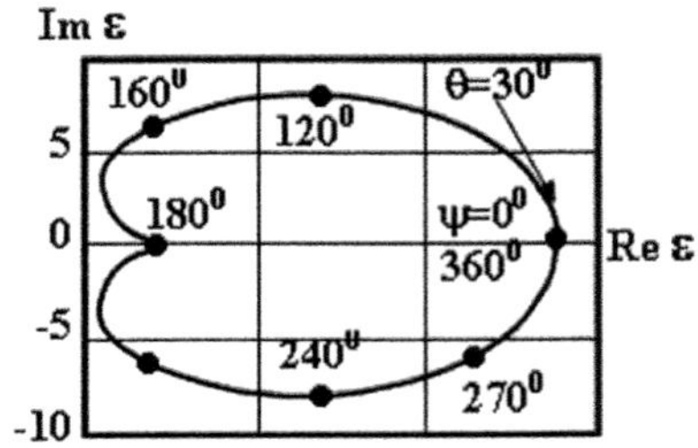

Fig. 8.12 Closed curve, corresponding to the value $\theta = 10°$

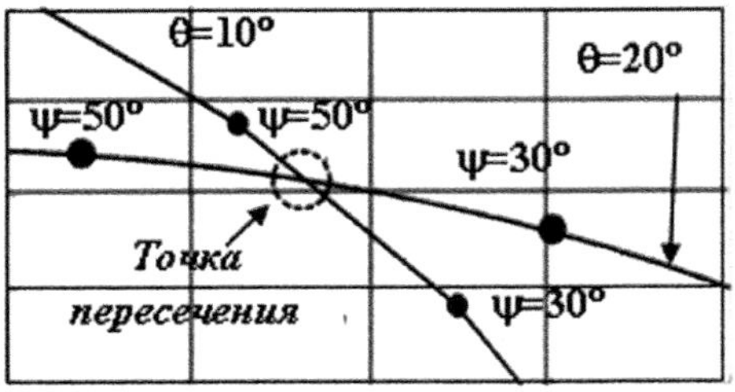

Fig. 8.13 Graphical method of determining the real and imaginary part of the complex permittivity

theoretical calculations stated for the range amplitude-phase differential method of the complex dielectric constant determination would fit.

Similarly to the amplitude-phase differential method, let us use the known Eq. (1.24) to determine the complex dielectric constant.

Let us look at the coordinate plane (see Fig. 2.6), where the vertical axis contains values of imaginary component of the complex dielectric constant Im ε (conductivity), and the horizontal axis contains values of the real component of the complex dielectric constant Re ε. A point on the stated coordinate plane (Im ε, Re ε) would correspond to any value of the argument ψ. When changing the argument ψ from zero to 360°, we obtain a closed curve, each point of which determines the value of the real and the imaginary component of the complex dielectric constant under a certain value of the argument ψ. The figure shows a closed curve which corresponds to the view angle $\theta = 10°$ (see Fig. 8.12). Similar curves may be obtained for all values of the view angle. The point corresponding to the real value of the complex dielectric constant would be the point of intersection of the stated curves traced for different values of the view angle (see Fig. 8.13). The coordinates of this point may also be obtained by solving the system of corresponding equations.

Thus, even non-availability of data on measurements of the phase differential in orthogonal channels of the receiver is not an obstacle to determine the complex dielectric constant of underlying surfaces subsequent to the results of relative measurements.

To illustrate the proposed method, we shall further give the results of determination of the complex dielectric constant obtained by corresponding processing of experimental information.

The experimental information comprises measurements of the specific scattering cross section on horizontal and vertical polarizations for several types of surfaces

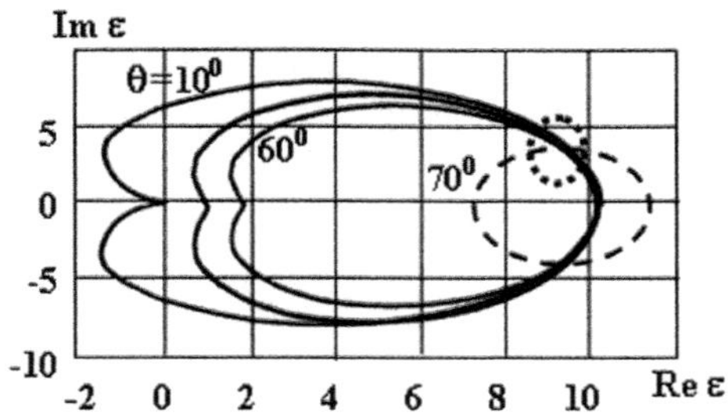

Fig. 8.14 Nomogram for the determination Im ε and Re ε green grass (X-band)

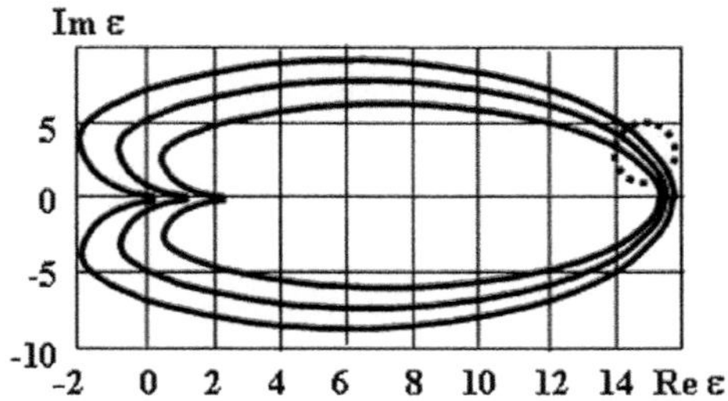

Fig. 8.15 Nomogram for determining Im ε and Re ε of the green grass (L-range)

within the frequency ranges 1–2 GHz (L), 2–4 GHz (S), 4–8 GHz (C), 8–12 GHz (X), given changing of the view angle within the range from zero to 90°. Nomographic charts were developed for each studied surface to determine and for each view angle θ, under the change of $\psi = 0\ldots360^{\circ}$, on the basis of the calculated Eq. (2.15). The cross-sectional area was established for intersected curves; this area complied with actual values of the real and the imaginary components of the complex dielectric constant of the studied surface. The obtained values of the complex dielectric constant were compared with the values from literature sources.

The following types of surfaces were selected by way of example.

1. Green grass with the height of 80 cm. Corresponding nomographic charts to determine Im ε and Re ε are given in Figs. 8.14 and 8.15:
   Figure 8.14 (X-range) shows that the real value is $\varepsilon = 10 - i \cdot 4$, which aligns rather well with that stated in the work [4], i.e., $\varepsilon = 9.2 - i \cdot 4$;
   Figure 8.15 (L-range) shows that the real value is $\varepsilon = 14 - i \cdot 6$, which aligns rather well with that stated in the work [4], i.e., $\varepsilon = 13.6 - i \cdot 5$.
2. Dry snow on the ground, snow depth is 58 cm, water equivalent is 13 cm, and snow temperature is −6 °C. Corresponding nomographic charts to determine Im ε and Re ε are given in Figs. 8.16 and 8.17:
   Figure 8.16 (X-range) shows that the real value is $\varepsilon = 1.2 - i \cdot 2 \times 10^{-4}$, which aligns rather well with that stated in literature, i.e., $\varepsilon = 1.2 - i \cdot 1.8 \times 10^{-4}$;
   Figure 8.17 (L-range) shows that the real value is $\varepsilon = 2.2 - i \cdot 10^{-4}$, which aligns rather well with that stated in literature, i.e., $\varepsilon = 2.1 - i \cdot 10^{-4}$.
3. Snow and rain mixed, snow depth is 48 cm, water equivalent is 10.5 cm, and relative humidity is 2%. A corresponding nomographic chart to determine Im ε and Re ε is given in Fig. 8.18 (X-range); as it is clear from the figure, the real

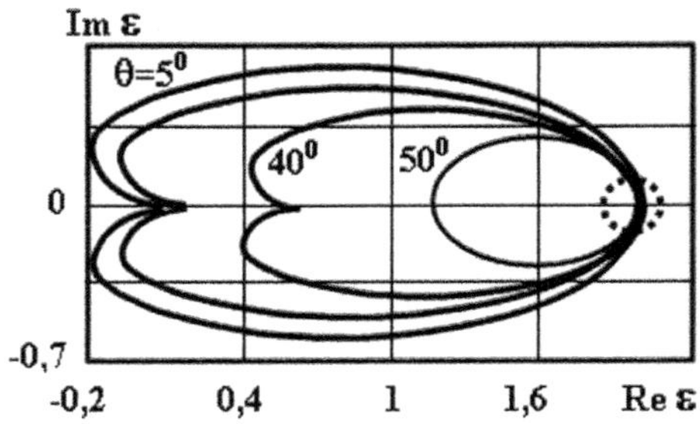

Fig. 8.16 Nomogram for determining Im ε and Re ε of the dry snow (X-range)

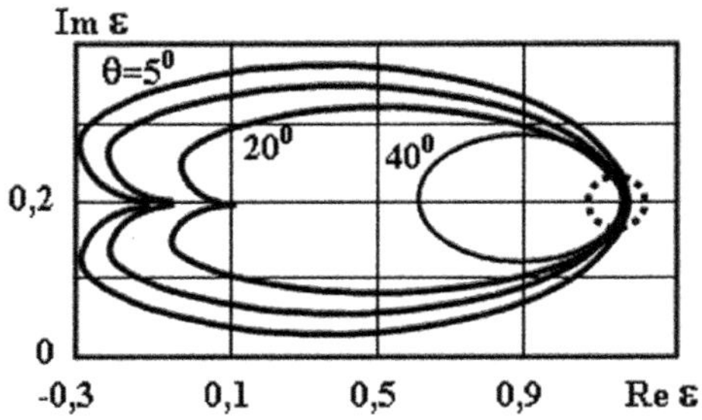

Fig. 8.17 Nomogram for determination Im ε and Re ε of the dry snow (L-band)

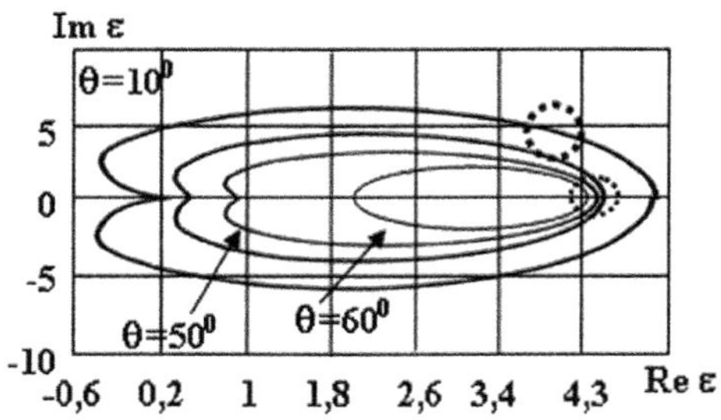

Fig. 8.18 Nomogram for determining Im ε and Re ε of the wet snow (X-range)

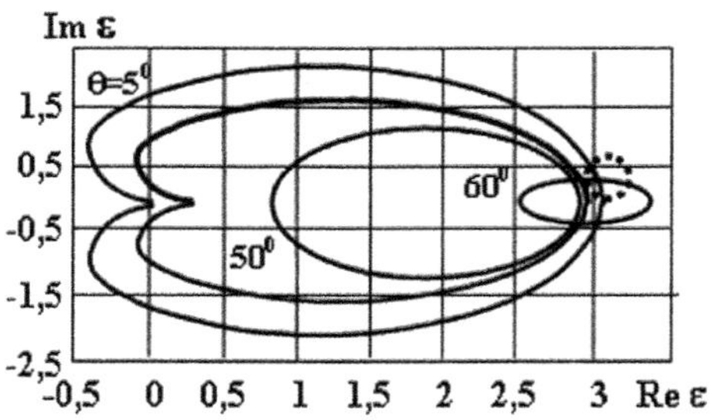

Fig. 8.19 Nomogram for determining Im ε and Re ε and of no vegetation field (X-range)

value is $\varepsilon = 4.5 - i \cdot 0.5$, which aligns rather well with that stated in literature, i.e., $\varepsilon = 4.8 - i$.

4. Greenless filed, mean-squared height of the surface is 2.8 cm, and absolute humidity is 0.36 g/cm^3. Corresponding nomographic charts to determine Im ε and Re ε are given in Figs. 8.19 and 8.20:

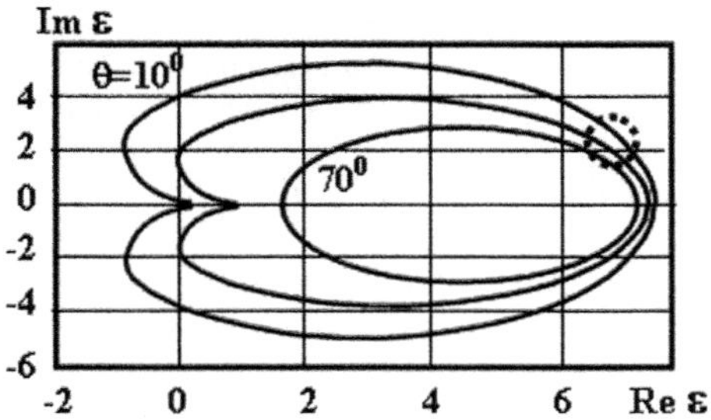

Fig. 8.20 Nomogram for determining Im ε and Re ε of no vegetation field (C-range)

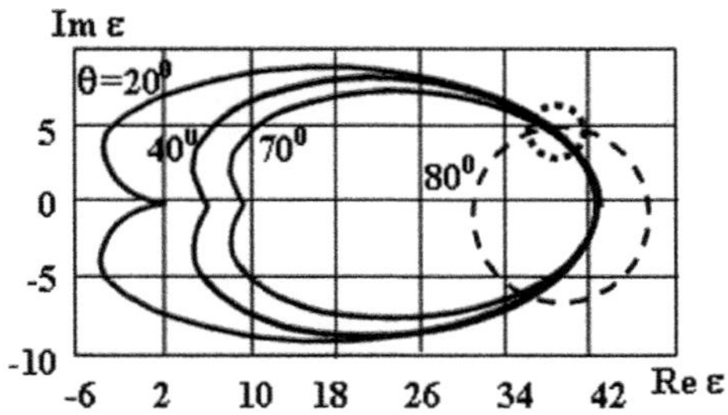

Fig. 8.21 Nomogram for determining Im ε and Re ε of the trees (X-range)

Figure 8.19 (X-range) shows that the real value is $\varepsilon = 3 - i \cdot 10^{-2}$, which aligns rather well with the literature data, i.e., $\varepsilon = 3.5 - i \cdot 10^{-2}$;
Figure 8.20 (C-range) shows that the real value is $\varepsilon = 8 - i \cdot 2$, which aligns rather well with the literature data, i.e., $\varepsilon = 8 - i \cdot 2.5$. This is in good agreement with the literature data.

5. Trees (poplar) with leaves, volumetric moisture content is 0.2. A corresponding nomographic chart to determine Im ε and Re ε is given in figure (X-range); as it is clear from Fig. 8.21, the real value is $\varepsilon = 40 - i \cdot 8$, which aligns rather well with that stated in the work, i.e., $\varepsilon = 41 - i \cdot 7$.

Thus, the proposed amplitude-deference method for determination of the complex dielectric constant, which is a modification for the amplitude-phase-difference method, provides for additional opportunities to classify a wide range of objects for remote sounding, by their electrophysical properties using only the results of relative measurements of signal correlations in orthogonal channels of the receiver performed under different view angles. To apply this method, it would be enough to use one impulse of the radiated wave for each view angle.

References

1. Antyufeev VI, Bykov VN, Grichanyuk AM, Krayushkin VA, Gakhov RP (2009) Application of the principles of radiometry in the correlation-extreme navigation systems of aircraft. Fizmatlit, Moscow

2. Beckmann P, Spizzichino A (1963) The scattering of electromagnetic waves from rough surfaces. Pergamon Press, Oxford-London-New-York-Paris
3. Bakulev PA (2007) Radar systems. Radio Engineering, Moscow
4. Kozlov AI, Logvin AI, Sarychev VA (2007) Polarization of radio waves. Book 2. Radar polarimetry. Radio Engineering, Moscow

Part III
Radiopolarimetry of Autonomous Navigation Systems

Chapter 9
Radiolocation in Radio Polarimetry Navigation Systems

One of the focus areas for radar monitoring tasks is to obtain non-coordinate target information. Such focus area implies tasks of target discrimination and evaluation of its geometric characteristics. Both tasks involve availability of information on scattering properties of radar targets due to their types. In particular, the solution of these tasks is based on the use of information on statistical characteristics of radar signal reflected from targets. Besides, it is necessary to be aware of the functional connection between parameters of scattered RF emission of radar targets and their physical, chemical, and electrical parameters. This task was solved in Chap. 1.

The problem of radar target discrimination, which is sometimes called the task of radar contrast enhancement, implies two targets; one of them is taken as a clutter. In other words, a radar signal reflected from the clutter target is a clutter signal.

Polarization characteristics of received radar signals are used to solve the task of enhancing discrimination of monitoring targets. Scattering matrix elements and anisotropy factors are considered as such characteristics.

When solving the discrimination task, they deal with low-contrast targets, i.e., targets that have the same scattering properties. But polarization change for two targets with the reflection of the same irradiating electromagnetic wave provides the basis for the solution of this task.

One of the prospective methods of enhancing the efficiency of airborne radar systems is to include in their structure devices using polarization effects in the radar channel.

A distinctive feature of technogenically hazardous and hard-to-reach areas is targets with geometrical discontinuities. Parameters of such discontinuities can be determined by means of airborne radar systems.

Roughness degree is considered as a geometric characteristic of monitoring targets in motion.

In the general case, to divide factors related to spatiotemporal state of received radio signals into those depending on geophysical characteristics and geometric configuration of visual targets seems to be impossible. In terms of practice, it is important not only to measure a reflected signal, but also to determine the structure and the state of different visual targets through it.

Kozlov A. I. et al., *Introduction to the Theory of Radiopolarimetric Navigation Systems*, Springer Aerospace Technology, https://doi.org/10.1007/978-981-13-8395-3_9

When spatiotemporal signals are processed in airborne radar systems, there are two fundamental tasks to be solved: The direct task is to determine characteristics of the reflected radar signal due to the kind, type, and state of monitoring targets, and the inverse task is to determine the kind, type, and state of monitoring targets due to the radar system output signal. When monitoring natural environments and technogenic targets located in technogenically hazardous and hard-to-reach areas, solution of the inverse task classified as an ill-conditioned problem is of principal interest; in this case, the final purpose of monitoring is to determine geophysical and geometric characteristics of targets.

Let us consider specific aspects of solving the inverse monitoring task. The final purpose of remote monitoring methods is to determine geographical parameters (ε, σ, tan δ), geometry (shape, dimensions, roughness degree), temperature, humidity, salinity, wind intensity, and other characteristics. This information on the structure and physical state of targets is determined due to parameters of the reflected radar signal.

In the general case, a wave reflected from any radar target is partially polarized. If partially polarized waves are received without regard to their polarization, a part of signal power can be lost.

The dielectric constant and conductivity can be found from polarization measurements in bistatic radio detection and ranging. If a transmitter antenna irradiates the studied smooth surface at a certain angle θ and the reflected signal is received in an inversed manner, the required parameters can be found from the Fresnel Formula (2.3):

$$\begin{aligned}
\operatorname{Re}\dot{\varepsilon} &= \left[1 + \frac{\left(1-p^2\right)^2 - 4p^2\cos^2\delta}{\left(1+p^2-2p\sin^2\delta\right)^2}\mathrm{tg}^2\theta\right]\sin^2\theta,\\
\sigma &= \frac{p\left(p^2-1\right)\cos\delta}{15\lambda\left(p^2-2p\sin\delta+1\right)^2}\mathrm{tg}^2\theta\,\sin^2\theta,
\end{aligned}$$

where p is the ratio of reflection factor moduli in vertical and horizontal polarization; δ is the difference of the arguments R_{VP} and R_{HP}.

If the measurements are conducted in a circularly polarized basis, the required values (2.3) can be found as per the following formulae:

$$\begin{aligned}
\operatorname{Re}\dot{\varepsilon} &= \left(1 + q^2\cos 2\delta_k \mathrm{tg}^2\theta\right)\sin^2\theta,\\
\sigma &= \frac{1}{60\lambda}q^2\sin 2\delta_k \mathrm{tg}^2\theta\,\sin^2\theta,
\end{aligned}$$

where q and δ_k are analogs of the values p and δ in the circular basis.

For autocorrelation function, if we apply the surface height represented by the Gaussian curve $e^{-\xi^2/L^2}$, where L is the so-called correlation interval, for the dependence of the surface scattering coefficient from the observation angle θ, the following formula can be used

$$\rho(\theta) = \frac{\rho(0)\theta}{\sin\,\theta}\mathrm{e}^{-\frac{L^2}{2\delta^2}\mathrm{tg}^2\theta}, \tag{9.1}$$

where δ is effective standard height deviations.

In the general case, parameters $L,\ \delta,\ \rho(0)$ depend on the polarization type of an irradiating wave.

Using Formula (9.1), you can easily receive the expression for anisotropy degree of rough surface (in this case, it is assumed that the emerging cross-polarized component is much lower than the basic one):

$$q = \left|\mathrm{th}\left[\frac{1}{4}\left(\frac{L_1^2}{\delta_1^2} - \frac{L_2^2}{\delta_2^2}\right)\mathrm{tg}^2\theta + \frac{1}{2}\ln\frac{\rho_1(0)}{\rho_2(0)}\right]\right|, \tag{9.2}$$

where the indices "1" and "2" refer to horizontal and vertical polarization, respectively.

The ratio obtained allows to represent the required difference of the geometric parameters $\left(\frac{L_1^2}{\delta_1^2} - \frac{L_2^2}{\delta_2^2}\right)$ via the measured values q and $\frac{\rho_1(0)}{\rho_2(0)}$:

$$\left(\frac{L_1^2}{2\delta_1^2} - \frac{L_2^2}{2\delta_2^2}\right) = \left(2\mathrm{Arth}\,q + \ln\frac{\rho_1(0)}{\rho_2(0)}\right)\mathrm{ctg}^2\theta. \tag{9.3}$$

If autocorrelation function differs from the Gaussian curve, certainly, (9.3) ratio estimates will not correspond to the true ones; nevertheless, the order of the value $\left(\frac{L_1^2}{\delta_1^2} - \frac{L_2^2}{\delta_2^2}\right)$ is reliably determined.

Chapter 10
Scattering Matrix as a Tool to Display Information on Visual Targets

Previously, we considered the impact of the reflectance profile on the scattering matrix elements of the reflecting surface. As was shown, the impact of the geometric characteristics comes down to a certain multiplier depending on the geometry of an irradiated area.

When studying the reflection process from an uneven surface, the situation is completely different. Generally, polarization effects cannot be divided into those caused by electrodynamic characteristics of the studied medium and those produced by its geometrical structure.

Further analysis will be performed using the following assumptions.

A radar station having an antenna with the radiation pattern $F(\theta, \varphi)$ is located at the height H above the statistically uneven two-dimensional rough surface $z = \xi(x, y)$ having dielectric constant $\varepsilon(x, y)$. The antenna produces a spherical electromagnetic wave of the length λ which irradiates a particular area on a certain uneven surface. We have to construct a scattering matrix for such area. As there are innumerable types of uneven surfaces, a problem inevitably arises how to construct a finite number of electrodynamic and statistical models comprising a large number of real surface types and kinds.

There is a vast literature on electrodynamic and statistical modeling of uneven and rough surfaces which review can compose an entire book. For further consideration, the models seem to be more suitable, that is why we will follow those works to describe uneven surfaces in accordance therewith using the following four models (Fig. 10.1).

Model 1 The surface is large-scale $(\rho_0/\lambda \gg 1)$, smooth $(\rho_{cor}/\lambda \gg 1)$, plain $\left((\nabla\xi_1)^2 < 1\right)$, with arbitrary roughness σ/λ, flat on the average $\left(\overline{\xi}_1 = 0\right)$, ρ_0, ρ_{cor} correlation and surface curvature radii; σ is the mean-square roughness height.

This model can be used to describe unfurrowed fields, hilly areas without significant plant cover, large waves of water surface, etc. (Fig. 10.1a).

Model 2 The surface is small scale $(\rho_0/\lambda \le 1)$, plain $(\nabla\xi_2)^2 \ll 1$, slightly rough $(\sigma/\lambda \ll 1)$, flat on the average $\left(\overline{\xi}_2 = 0\right)$.

Kozlov A. I. et al., *Introduction to the Theory of Radiopolarimetric Navigation Systems*, Springer Aerospace Technology, https://doi.org/10.1007/978-981-13-8395-3_10

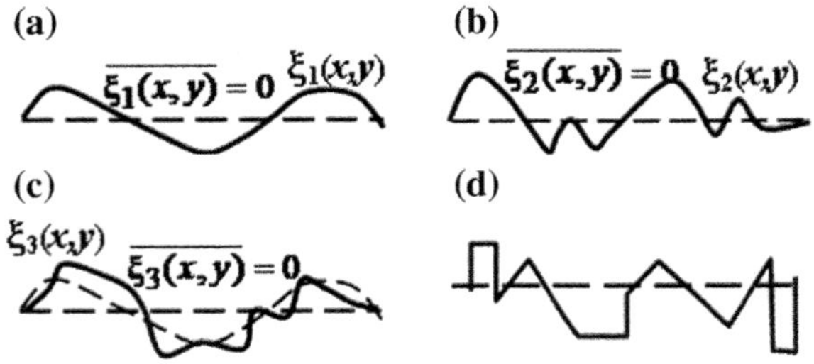

Fig. 10.1 Models of statistically uneven terrain: **a** model 1; **b** model 2; **c** model 3; **d** model

This model can be used to describe concrete, asphalt, sand, and gravel coverings, slightly rippled calm water surface, flat plains with low vegetation in centimeter and longer wave bands, etc. (Fig. 10.1b).

Model 3 The surface is complex: complete or discrete covering by small-scale irregularities over large-scale irregularities ($\xi_3(x, y) = \xi_1(x, y) + \xi_2(x, y)$); in this case, $\overline{\xi}_3 = \xi_1$.

This model can be used to describe a rolling water surface, deserts with large barchan dunes and sand ripples thereon, furrowed hilly surface, etc. (Fig. 10.1c).

Model 4 The surface has different geometric shapes.

This model can be used to describe settlements, mountains, forest areas, etc., in centimeter and decimeter wave band; these can be represented as a random composition of incoherent independent scatters (Fig. 10.1d).

Scattering matrix for model 1 Pursuant to conditions imposed on model 1, the following can be assumed $L \gg \rho_0,\ L \gg \sigma,\ H \gg \sigma$.

The problem is solved as follows. The incident and reflected spherical waves are decomposed into plane waves. Further in the Kirchhoff approximation, we have to solve a diffraction problem of a partial plane wave on the respective local surface areas s_i characterized by the following random values: the height $\xi(x, y)$, the gradient $\nabla\xi(x, y) = \mathbf{i}\frac{\partial\xi}{\partial x} + \mathbf{j}\frac{\partial\xi}{\partial y}$, and local Fresnel complex reflection factors. The field at the receiving point shall be found on the basis of the vector form of the Green's theorem (Stratton-Chu formula) with integration over illuminated surface areas [1].

As a result of the calculations, we obtain the following expressions for the scattering matrix:

$$\mathbf{S} = \begin{pmatrix} \frac{a_2 n_z^2}{1-a_2 n_z^2}\left(\eta^2 R_{\mathrm{VP}} - a_1^2 R_{\mathrm{HP}}\right) & \frac{a_1 a_2 n_z^2 \eta}{1-a_2 n_z^2}(R_{\mathrm{HP}} + R_{\mathrm{VP}}) \\ \frac{a_1 a_2 n_z^2 \eta}{1-a_2 n_z^2}(R_{\mathrm{HP}} + R_{\mathrm{VP}}) & \frac{a_2 n_z^2}{1-a_2 n_z^2}\left(a_1^2 R_{\mathrm{VP}} - \eta^2 R_{\mathrm{HP}}\right) \end{pmatrix}, \tag{10.1}$$

where $a_1 = \sin\theta - \gamma\cos\theta;\ a_2 = \cos\theta + \gamma\sin\theta;$

$$\eta = \frac{\mathrm{d}\xi}{\mathrm{d}x}\sin\phi - \frac{\mathrm{d}\xi}{\mathrm{d}y}\cos\phi;\quad \gamma = \frac{\mathrm{d}\xi}{\mathrm{d}x}\cos\phi + \frac{\mathrm{d}\xi}{\mathrm{d}y}\sin\phi;\quad n_z = \left[1 + \left(\frac{\mathrm{d}\xi}{\mathrm{d}x}\right)^2 + \left(\frac{\mathrm{d}\xi}{\mathrm{d}y}\right)^2\right]^{-1/2}.$$

To find S-matrix eigenvalues, let us multiply and divide each of its elements by the multiplier $\sqrt{a_1^2+\eta^2}$. In this case, we can assume that $a_1/\sqrt{a_1^2+\eta^2}=\cos\zeta$; $\eta/\sqrt{a_1^2+\eta^2}=\sin\zeta$. It allows to represent the scattering matrix (10.1) as

$$\mathbf{S}=-\frac{a_2n_z^2\sqrt{a_1^2+\eta^2}}{\left(1-a_2n_z^2\right)}\begin{bmatrix}R_{\mathrm{HP}}\cos^2\zeta-R_{\mathrm{VP}}\sin^2\zeta & -(R_{\mathrm{HP}}+R_{\mathrm{VP}})\sin\zeta\cos\zeta\\ -(R_{\mathrm{HP}}+R_{\mathrm{VP}})\sin\zeta\cos\zeta & R_{\mathrm{HP}}\sin^2\zeta-R_{\mathrm{VP}}\cos^2\zeta\end{bmatrix}. \tag{10.2}$$

Thus, we can write down matrix eigenvalue expressions (10.1) right away:

$$\begin{aligned}\lambda_1&=-\frac{a_2n_z^2}{1-a_2n_z^2}\sqrt{a_1^2+\eta^2}R_{HP},\\ \lambda_2&=\frac{a_2n_z^2}{1-a_2n_z^2}\sqrt{a_1^2+\eta^2}R_{VP}.\end{aligned} \tag{10.3}$$

The relations received enable to obtain both matrix invariants (10.1):

$$\begin{aligned}A=\sigma_\Sigma^2&=\left(\frac{a_2n_z^2}{1-a_2n_z^2}\sqrt{a_1^2+\eta^2}\right)^2\left(|R_{HP}|^2+|R_{VP}|^2\right),\\ B=|\det S|&=\left(\frac{a_2n_z^2}{1-a_2n_z^2}\sqrt{a_1^2+\eta^2}\right)^2|R_{HP}R_{VP}|.\end{aligned} \tag{10.4}$$

Proper base coordinates can be directly calculated as per:

$$\cos 2\beta=\cos 2\zeta=\left|\left(\eta^2-a_1^2\right)/\left(\eta^2+a_1^2\right)\right|.$$

As seen from the given formula and Fig. 10.1, one of the position coordinates of the proper polarization basis on the Poincaré sphere does not depend on complex dielectric constant and is only defined by roughness characteristics.

As for the second geographic coordinate: α, the similar calculations lead to relations showing that this coordinate also does not depend on the complex dielectric constant and is only defined by roughness characteristics. (Due to extraordinary awkwardness of the obtained relations, these are not included in this text.)

Horizontal and vertical polarizations do not represent the eigenpolarization of the surfaces under review. This results in the emergence of a cross constituent when horizontally and vertically polarized radio waves are reflected from such surfaces.

To determine the antenna output signals, an integration is needed in (10.1) overall the angles θ and φ with regard to the antenna directivity diagram and the distance to it:

$$U_{ij} = \frac{1}{(2\pi)^2} \int\limits_{S_{\mathrm{rad}}} \mathrm{d}s \int \int\limits_{-\infty}^{\infty} \int \int F(x, y) s_{ij}\left(\xi'_x, \xi'_y; \theta; \varphi; \varepsilon\right) n_z$$
$$\times \exp\left\{-i\left[\left(k_x + k'\right)x + \left(k_y + k'_y\right)y + \left(k_z + k'_z\right)(H - \xi(x, y))\right]\right\}$$
$$\times \left(\mathrm{d}k_x \mathrm{d}k_y / k_z\right)\left(\mathrm{d}k'_x \mathrm{d}k'_y / k'_z\right). \quad (10.5)$$

Thus, (10.5) shows that each constituent of the complex spherical field at the reception point represents a superposition of plane waves reflected from statistically uneven surface at different angles θ and φ.

Scattering matrix for model 2 Let a small-scale surface $z = \chi(x, y)$ be an interface for two media with dielectric constants 1 and ε.

For the surface $z = 0$, if we assume that respective electric and magnetic vectors have values $\mathbf{E}_0$ and $\mathbf{H}_0$, small-scale roughnesses cause slight changes of these vectors.

There are continuous tangential constituents of electrical and magnetic vectors on the surface $z = \chi\ (x, y)$. Deviation of the surface $z = \chi\ (x, y)$ from the plane $z = 0$ may be considered as a slight correction. Decomposing interface conditions in the raw by degrees $\chi\ (x, y)$, $\nabla\chi\ (x, y)$ and confining ourselves to the first approximation of the perturbation theory, after a number of transformations, we obtain the following expression for scattering matrix elements:

$$\mathbf{S} = \begin{bmatrix} (1 + R_{\mathrm{VP}})^2 L(\theta, \varepsilon) & 0 \\ 0 & (1 + R_{\mathrm{VP}})^2\left[(1 - R_{\mathrm{HP}})\sin^2\theta + L(\theta, \varepsilon)\right] \end{bmatrix}, \quad (10.6)$$

where $L(\theta, \varepsilon) = \frac{\left(\sin^2\theta + \sqrt{\varepsilon - \sin^2\theta}\cos\theta\right)^2}{\varepsilon - 1}$.

Invariants of this matrix are:

$$A = \sigma_\Sigma^2 = \left|\frac{(1 - R_{\mathrm{HP}})^2}{4R_{\mathrm{HP}}\cos^2 2\theta}\right|^2 + \left|\frac{R_{\mathrm{HP}}(1 - R_{\mathrm{HP}})\sin^2 2\theta - (1 - R_{\mathrm{HP}}\cos 2\theta)^2}{4R_{\mathrm{HP}}\cos^2\theta}\right|^2,$$
$$B = |\det S| = \left|\frac{(1 - R_{\mathrm{HP}})^2}{4R_{\mathrm{HP}}\cos^2 2\theta}\frac{R_{\mathrm{HP}}(1 - R_{\mathrm{HP}})\sin^2 2\theta - (1 - R_{\mathrm{HP}}\cos 2\theta)^2}{4R_{\mathrm{HP}}\cos^2\theta}\right|. \quad (10.7)$$

As seen in (10.6), in this case horizontal and vertical polarizations are the eigenpolarization of the surfaces under review. It explains the absence of a cross constituent when horizontally and vertically polarized radio waves are reflected from such surfaces.

This result shows that in the first approximation of the perturbation theory field, depolarization is not revealed even from the finite conductivity surface. At $|\varepsilon| \to \infty$, diagonal matrix elements (10.3) tend to $(1 + R_{\mathrm{VP}})^2\cos^2(\theta)$ and $(1 + R_{\mathrm{VP}})^2(1 - R_{\mathrm{HP}})$, respectively.

S-matrix eigenvalues will equal its diagonal elements. Coordinates of the proper polarization base, obviously, will be as follows: $\alpha = \beta = 0$.

To determine the antenna output signals, an integration is needed in (10.6) overall the angles θ and φ with regard to the antenna directivity diagram and the distance to it:

$$U_{ij} = \frac{k^2}{\pi}\left(\frac{\varepsilon - 1}{\varepsilon}\right)^2 \int\limits_{S_0} \frac{F(x, y)s_{ij}}{R_0^2} \xi(x, y)\exp(-2ikR_0)\mathrm{d}S_0. \tag{10.8}$$

Scattering matrix for model 3 Complex scattering problem (large-scale and small-scale irregularities) can be solved by a combination of two methods: Kirchhoff method and perturbation method. Required fields are determined as the sum of randomly reflected fields on the large-scale surface and perturbation fields from small-scale irregularities.

When decomposing interface conditions in the raw by degrees $\chi\ (x, y)$, $\nabla\chi\ (x, y)$ and confining ourselves to the first approximation of the perturbation theory, the scattering problem on the studied structure comes down to model 1. Having made a number of transformations for scattering matrix elements, we will obtain

$$\mathbf{S} = \left(\frac{2a_2n_z\varepsilon}{a_2n_z\varepsilon + q}\right)^2 \begin{bmatrix} \frac{2q\eta^2n_z}{a_2n_z+q} - \frac{[1+a_2n_z(q-a_2n_z)]^2}{n_z\left(a_2^2n_z^2-q^2\right)} & -\frac{2a_1\eta qn_z}{a_2n_z+q} \\ -\frac{2a_1\eta qn_z}{a_2n_z+q} & \frac{2qa_1^2n_z}{a_2n_z+q} - \frac{[1+a_2n_z(q-a_2n_z)]^2}{n_z\left(a_2^2n_z^2-q^2\right)} \end{bmatrix}, \tag{10.9}$$

where a_1, a_2, η, n_z are determined by the formula (10.1); $q = \sqrt{\varepsilon - \cos^2\theta}$—parameter.

Invariants of this matrix can be found via the Fresnel reflection factor for the horizontal polarization:

$$\begin{cases} q = \frac{1-R_{\mathrm{HP}}}{1+R_{\mathrm{HP}}}\cos\theta, \\ \varepsilon = 1 - 4\frac{R_{\mathrm{HP}}}{(1+R_{\mathrm{HP}})}\cos^2\theta. \end{cases} \tag{10.10}$$

Formulae related to the model under consideration are rather lengthy, although their derivation does not present any principal difficulties. The calculation shows that, firstly, eigenvalues can be expressed via the Fresnel reflection factor for the horizontal polarization, and secondly, to the great extent, they depend on the complex dielectric constant. On the contrary, coordinates of the proper polarization base depend on roughness characteristics only. (Due to extraordinary cumbersomeness of the obtained relations, these are also not included in this text.) Nevertheless, for the real ε, expressions for S-matrix eigenvalues can be obtained in a rather compact form:

$$\lambda_1 = \frac{2a_2^2\varepsilon^2(1 + a_2n_z)^2(a_2n_z - q)}{a_1q\eta(a_2n_z\varepsilon - q)^2},$$

$$\lambda_2 = \left(\frac{a_1}{\eta} + \frac{\eta}{a_1}\right)\left(\frac{2a_2 n_z \varepsilon}{a_2 n_z \varepsilon - q}\right)^2 + \lambda_1. \tag{10.11}$$

To determine antenna output signals, an integration is needed in the formula (10.9) overall the angles θ and φ with regard to the antenna directivity diagram and the distance to it:

$$U_{ij} = \frac{k^2}{\pi}\left(\frac{\varepsilon - 1}{\varepsilon}\right)^2 \int\limits_{S_0} \frac{F(x, y)s_{ij}}{R_0^2} \xi_2(x, y) \exp(-2ikR_0)\mathrm{d}S_0, \tag{10.12}$$

where $\xi_2(x, y)$ indicates small-scale irregularities.

The relation (10.9) implies that backscattered depolarized field constituents can only appear in case of nonzero two-dimensional gradient of a large-scale surface structure.

The complete field scattered from a surface given by model 3 is defined as the sum of fields from the large-scale surface to the small-scale surface located over the first one.

Scattering matrix for model 4 This model includes such random uneven surfaces which geometric profile represents rather flat faces or almost arbitrarily shaped convexities (modeling of houses, rocks, etc.) as well as all discontinuous surfaces (forests, vegetations, etc.).

Beckmann's model consisting of arbitrarily inclined flat elements (faces) is a representation of the uneven surface. Each face reflects the incident ray with a Fresnel reflection factor only in the local inverse direction that is true when $l \gg \lambda$. Partial waves reflected in the same direction (equally inclined faces) are summed by phases and form one resulting wave in the same direction. As profiles of the structures described by model 4 are not general, in this case, general scattering matrix cannot be constructed. In these circumstances, there is no sense to calculate the field scattered by a real surface, but the model of the required surface shall be synthesized through experimental study of the scattered field. Nevertheless, let us illustrate model 4 using a simple example.

Reference

1. Kramer HJ (1996) Observation of the earth and its environment. Survey of mission and sensors. Springer, Berlin, New-York

Chapter 11
Enhancing Radar Station Functionalities to Delineate Linear Dimensions of Extended Visual Targets for Incoherent Scattering of Electromagnetic Waves

Complexity of the problem related to radar target identification and (as a particular case) determination of its configuration and respective geometrical dimensions is primarily connected with the fact that a target detected by surveillance radar which dimensions are smaller than that of the respective resolution element is interpreted as *a point target* (a point that according to Euclid has no length, width, or height). In this regard, a conclusion about its linear dimensions is made that the target width L does not exceed a product of the target range R by the width of antenna directivity diagram $\Delta\theta_{\Gamma\Pi}$ in the horizontal plane, i.e., $0 < L \leq R * \Delta\theta_{\Gamma\Pi}$ its height H is the similar product for the vertical plane $0 < H \leq R * \Delta\theta_{\text{B}\Pi}$. For the target depth D (longitudinal dimension), the estimating inequality involves the light velocity c and the sounding impulse duration τ, namely $0 < D \leq c\tau/2$. If we are within the framework of the Rayleigh criterion, unfortunately, no conclusion can be made concerning configuration and dimensions of the target.

In this case, it should be specially emphasized that such situation also occurs when the target represents a number of spaced points called highlight*s*. At the same time, as numerous experimental data show, in a very large number of cases, highlights inhering both in natural and anthropogenic radar targets tend to be polarization sensitive, i.e., changing their reflectance profiles due to polarization type of the radiating electromagnetic wave. It means that, if changed, the source of the reflected electromagnetic wave received at the time of surveillance (i.e., the phase center of a locally spherical wave at the receiving point) will move in space along a certain space curve. Such "source" is called a radar phase center. With full polarization scanning, this curve will represent a closed line which configuration and dimensions are related to spatial arrangement geometry of the highlights mentioned above. It is reasonable to call such curve *a spatial polarization circuit* (SPC) of the radar target, and the construction procedure shall be called *polarization contouring* of this target.

Spatial movement of the phase center requires retuning of the receiving antenna to the maximum received signal, and accuracy of such procedure no longer depends on the width of the directivity diagram $\Delta\theta_{\Gamma\Pi}$ and $\Delta\theta_{\text{B}\Pi}$, but on the peak sharpness thereof, i.e., of its steepness toward the primary maximum. Therefore, if we associate dimensions of the spatial polarization contour with the real target dimensions, in

Kozlov A. I. et al., *Introduction to the Theory of Radiopolarimetric Navigation Systems*, Springer Aerospace Technology, https://doi.org/10.1007/978-981-13-8395-3_11

fact, we will be able to estimate its geometric dimensions or, at least, to reduce the aforesaid inequalities $0 < L \leq R * \Delta\theta_{\Gamma\Pi}$ and $0 < L \leq R * \Delta\theta_{\text{В}\Pi}$. It certainly requires analysis suggested below.

In the general case, a wave reflected from the radar target can be represented as the sum of so-called coherent and incoherent constituents.

Let the target be a certain structure containing a set of N highlights (HL) which spatial coordinates will be designated as (x_i, y_i). Let the S_i matrix of the form $S_i = \begin{pmatrix} s_{11i} & s_{12i} \\ s_{12i} & s_{22i} \end{pmatrix}$ be the i HL scattering matrix.

If a target is irradiated on a certain polarization and a reflected wave is also received on the same polarization, the power of the radio wave received at the receiving antenna output will be proportional $|s_{11i}|^2$, and at the same time, for incoherent reception, the aggregate power will represent a sum of signal strengths P_i corresponding to each HL, that is $P_\Sigma = \sum_{i=1}^{N} P_i = \sum_{i=1}^{N} |s_{11i}|^2$. As for the phase center position of the reflected wave, it will coincide with the gravity center of geometrical figures formed by HL with regard to their "weight" amounting to P_i. Thus, spatial coordinates of the phase center can be calculated as per the formulae

$$X_{\text{фц}} = \sum_{i=1}^{N} x_i P_i \bigg/ \sum_{i=1}^{N} P_i = \sum_{i=1}^{N} x_i |s_{11i}|^2 \bigg/ \sum_{i=1}^{N} |s_{11i}|^2, \tag{11.1}$$

$$Y_{\text{фц}} = \sum_{i=1}^{N} y_i P_i \bigg/ \sum_{i=1}^{N} P_i = \sum_{i=1}^{N} y_i |s_{11i}|^2 \bigg/ \sum_{i=1}^{N} |s_{11i}|^2. \tag{11.2}$$

When the polarization type of a wave irradiating the radar target changes, i.e., when polarization basis changes, scattering matrices S_i undergo congruent transformation via unitary matrix depending on four parameters only two of which are of interest for further consideration—$0 \leq \gamma \leq \pi$ and $0 \leq \varphi \leq 2\pi$ that determine a particular type of polarization observed.

Due to the aforesaid, an expression for the new s-matrix element to be designated as q_{11} that determines the reflected wave power in the case when radiation and reception occur on the same polarization, instead of $|s_{11i}|^2$, will be:

$$|q_{11}|^2 = \left|s_{11}e^{-2i\varphi}\cos^2\gamma + s_{22}e^{2i\varphi}\sin^2\gamma + s_{12}\sin 2\gamma\right|^2. \tag{11.3}$$

Formulae (11.16)—(11.18) solve this problem for incoherent scattering.

As an illustration, let us consider some examples based on the given formulae.

Figure 11.1 shows a spatial polarization circuit (phase center position) of the radar target that has four HL representing similar dipoles differently oriented in space and cruciformly arranged. As seen, in this case, SPC is a circle which radius is approximately 0.65 of the distance between the dipoles.

Ad hoc SPC transformation can be seen in Fig. 11.2 constructed for the cases when dipole reflectivities are different. As seen, the initial circle deforms. The effect of the parameter φ leads to vertical pattern movement. Nevertheless, again overall contour SPC dimensions are retained.

The given figures referred to the case of so-called degenerates which are the dipoles under review.

Figure 11.3 shows a SPC for the target having four HL which scattering matrix elements were given randomly via random number generator. HL are located at the top of the cross. Pattern transition corresponds to parameter change φ. As seen, even for these exotic SPC types, the relation of its dimensions to geometric dimensions of the radar target is rather clear again.

Let us consider one more illustration of obtained relations through a target that has three HL. As a base target, we will use three similar dipoles located at the vertices of an equilateral triangle; in this case, the first dipole is located vertically at the bottom left vertex, the second dipole is located horizontally at the bottom right vertex, and at the upper vertex, the third dipole is inclined at the angle $\beta = 45°$. The calculation results for such target are depicted in Fig. 11.4, which shows the evolution underwent by SPC through all polarization types (the change of γ and φ by all possible values).

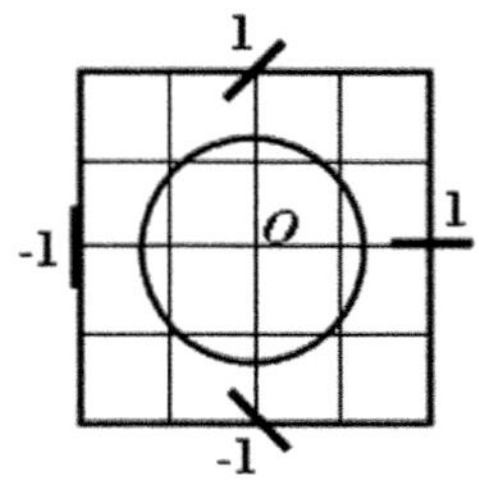

Fig. 11.1 SPC of radar target with four shining points

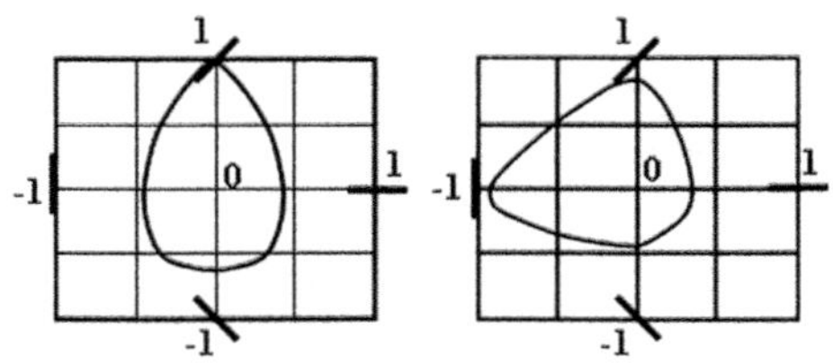

Fig. 11.2 Transformation of SPC

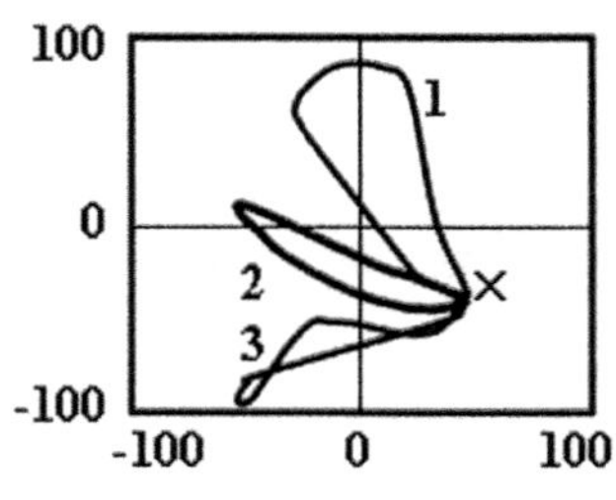

Fig. 11.3 SPC of radar target for the case of arbitrary elements of the loss matrix: (1) $\varphi = 0$; (2) $\varphi = \pi/4$; (3) $\varphi = \pi/2$

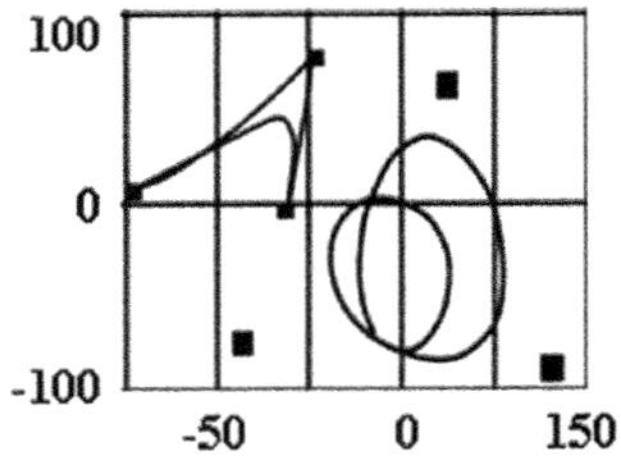

Fig. 11.4 SPC of the three-point target with arbitrary scattering matrices

It is important to note that in all cases, the maximum SPC size was close to the real target size.

The group (Fig. 11.5) describes the polarization impact of a radiated wave upon the target SPC which HL (three dipoles still located at the vertices of an equilateral triangle) have different reflectivities. As seen, the relation between the SPC dimensions and the real dimensions of the sounded target is obvious again.

Thus, the presented relations and the given illustrations rather convincingly and clearly highlight the principal possibility (at least in case of incoherent scattering) for estimating geometric dimensions of a certain class of radar targets even when their linear dimensions are smaller than those of the resolution element.

Let us consider polarization contouring for an independent reflector system. As the first step, let us discuss the following rather typical situation for radio detection and ranging.

Suppose there is a radar target representing an aggregate large number of statistically independent elementary reflectors (in this case, the target shall mean all reflectors in the RS resolution element). In this case, the amplitude of the total reflected signal resulting from incoherent addition of signals reflected from each elementary reflector obeys the Rayleigh law, and respectively, the reflected signal strength is subject to the exponential law. Therefore, the same law is adhered to by the total RCS of such system which is a sum of local RCS of each elementary reflector.

Such system of independent reflectors well describes earth and sea surfaces and large target surfaces.

When following such target, the phase center position of the reflected radio wave can be found from Formulae (11.1) and (11.2). If we take into consideration independence of elementary reflector coordinates and their RCS, for the mean values of the required phase center coordinates, a rather obvious result will be obtained:

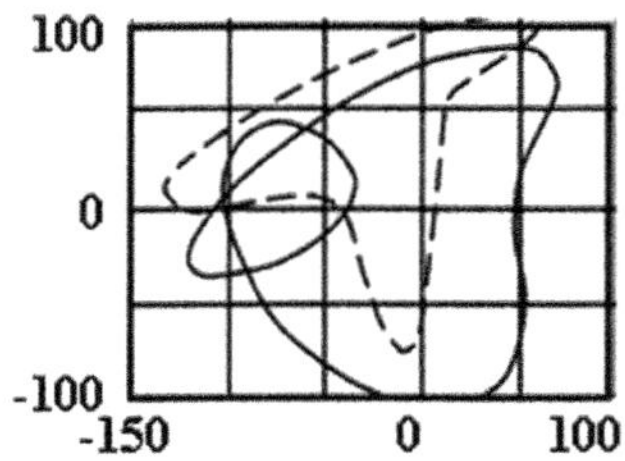

Fig. 11.5 SPC of the four-point target with arbitrary scattering matrices

$$\overline{X}^{n}_{\text{фц}} = \overline{\left(\frac{\sum_{i=1}^{N} x_i P_i}{\sum_{i=1}^{N} P_i}\right)} = \sum_{i=1}^{N} \overline{\frac{x_i P_i}{\sum_{j=1}^{N} P_j}} = \sum_{i=1}^{N} \overline{x}_i \overline{\left(\frac{P_i}{\sum_{j=1}^{N} P_j}\right)} = 0, \tag{11.4}$$

$$\overline{Y}^{n}_{\text{фц}} = \overline{\left(\frac{\sum_{i=1}^{N} y_i P_i}{\sum_{i=1}^{N} P_i}\right)} = \sum_{i=1}^{N} \overline{\frac{y_i P_i}{\sum_{j=1}^{N} P_j}} = \sum_{i=1}^{N} \overline{y}_i \overline{\left(\frac{P_i}{\sum_{j=1}^{N} P_j}\right)} = 0 \tag{11.5}$$

As for the mean square deviation of the phase center position from its mean value, with regard to the mentioned statistic independence of the random values x_i and P_i, we will have:

$$\sigma^2_{X\,\text{фц}} = \overline{\frac{\sum_{i=1}^{N} x_i P_i \sum_{j=1}^{N} x_j P_j}{\sum_{i=1}^{N} P_i \sum_{j=1}^{N} P_j}} = \overline{\frac{\sum_{i=1}^{N} \sum_{j=1}^{N} x_i x_j P_i P_j}{\left(\sum_{i=1}^{N} P_i\right)^2}} = \sum_{i=1}^{N} \sum_{j=1}^{N} \overline{x_i x_j}\, \overline{\frac{P_i P_j}{\left(\sum_{i=1}^{N} P_i\right)^2}} = \sum_{i=1}^{N} \overline{x_i^2}\, \overline{\frac{P_i^2}{\left(\sum_{i=1}^{N} P_i\right)^2}}. \tag{11.6}$$

Similarly, we have:

$$\sigma^2_{Y\,\text{фц}} = \overline{\frac{\sum_{i=1}^{N} y_i P_i \sum_{j=1}^{N} y_j P_j}{\sum_{i=1}^{N} P_i \sum_{j=1}^{N} P_j}} = \overline{\frac{\sum_{i=1}^{N} \sum_{j=1}^{N} y_i y_j P_i P_j}{\left(\sum_{i=1}^{N} P_i\right)^2}} = \sum_{i=1}^{N} \sum_{j=1}^{N} \overline{y_i y_j}\, \overline{\frac{P_i P_j}{\left(\sum_{i=1}^{N} P_i\right)^2}} = \sum_{i=1}^{N} \overline{y_i^2}\, \overline{\frac{P_i^2}{\left(\sum_{i=1}^{N} P_i\right)^2}}. \tag{11.7}$$

For the independent reflector system, it is natural to assume the uniform distribution law for x_i and y_i on the interval of their possible values, e.g., on the interval $[-L, L]$. In this case, the mean values of the squared random values x_i and y_i will equal:

$$\overline{x_i^2} = \overline{y_i^2} = L^2/3. \tag{11.8}$$

Having plugged the found expressions in (11.4) and (11.5), we get:

$$\sigma^2_{X\,\text{фц}} = \sigma^2_{Y\,\text{фц}} = \frac{L^2}{3} \overline{\sum_{i=1}^{N} P_i^2 \Big/ \left(\sum_{i=1}^{N} P_i\right)^2}. \tag{11.9}$$

From elementary mathematics, it is known that the minimum value of the relation $\sum_{i=1}^{N} Z_i^2 / \left(\sum_{i=1}^{N} Z_i\right)^2$ equals $1/N$ when all Z_i equal each other, and its maximum value equals 1 when all Z_i, except one, equal 0.

Given that for the situation under review the function of Z_i is exercised by the RCS of the elementary reflector P_i which are rather close to each other by value, let us approximately assume that the relation value for Formula (11.9) is close to 1/2. Thus, we have:

$$\sigma^2_{X\,\text{фц}} = \sigma^2_{Y\,\text{фц}} \approx L^2/6 = 0.17L^2. \tag{11.10}$$

In this case, the mean square deviation of the phase center position from the geometric center of independent reflector system will be as follows:

$$\sigma_{X\,\text{фц}} = \sigma_{Y\,\text{фц}} \approx 0.4L. \tag{11.11}$$

To check the correctness of the obtained relations, the system under review was modeled. We chose 100 elementary reflectors ($N = 100$), randomly located on the area of $2L \times 2L$. Each reflector was assigned a certain RCS (P_i) whose values were distributed according to the exponential law with the mean value $\lambda = 1$. Further, the phase center position was calculated. Such modeling was performed 100 times for different N and λ. Then the phase center position was calculated as per the Formulae (11.1) and (11.2). Following the results obtained, mean and mean square values of the phase center position were determined. For illustrative purposes, Fig. 11.6 shows the phase center position for certain particular cases.

If polarization of the radiated wave changes, as it was previously said, the phase center position changes. It can be stated that, for the reviewed independent reflector system, no qualitative change occurs in the phase center position. The mean values of its coordinates will still equal 0, and their mean square values shall remain within 0.4L.

Figure 11.7 illustrates the polarization type impact on the phase center position for the independent reflector system.

When averaging phase center coordinates, for several modeling options, the following results will be obtained from 100 realizations. As seen in Fig. 11.8, mean position coordinates of the phase center do not exceed the area limited by the expression $L^2/3N$. In this case, the mean value is virtually at the origin of coordinates.

Let us consider the situation when on the surface studied above there is a reflector whose contribution to the total RCS is significantly higher than that of each of the remaining reflectors.

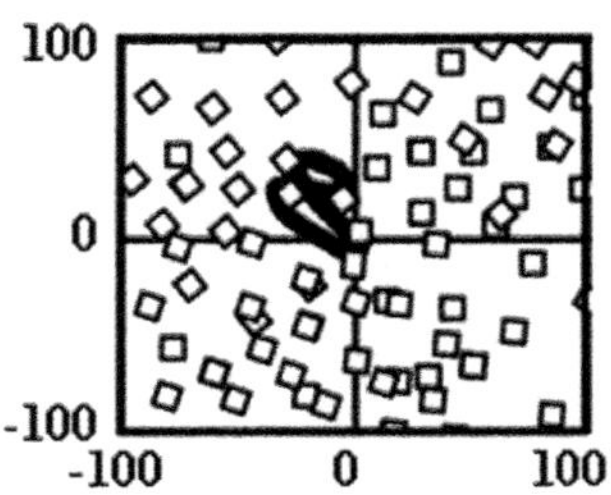

Fig. 11.6 The location of phase center when building models

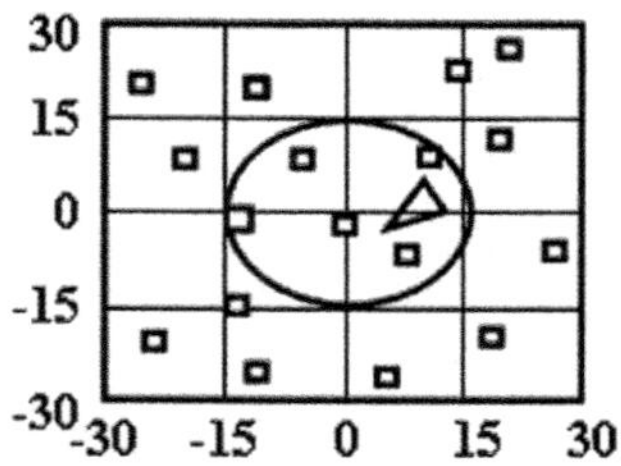

Fig. 11.7 Influence of polarization on location of phase center mean value

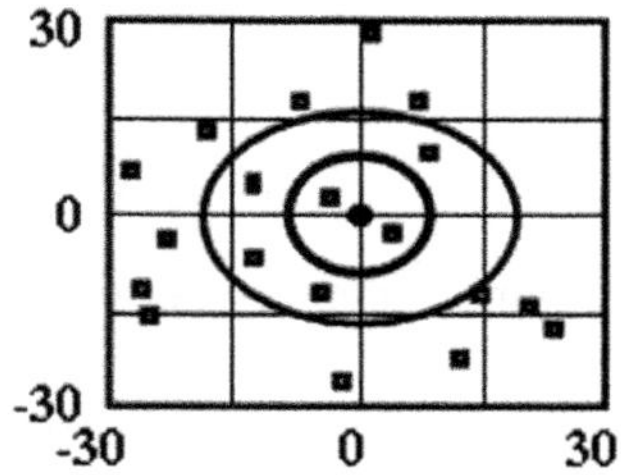

Fig. 11.8 Determining mean location of phase center

We will obtain an ad hoc expression for the RCS distribution density supposing that the strength of a signal reflected from the highlight, i.e., the corresponding RCS, is also distributed pursuant to the exponential law as this point signal can be also regarded as a sum of signals reflected from some independent elementary reflectors that form a structure called a "highlight".

So, the strength of the reflected signal P is composed of the two summands: $P_{\text{П}}$, which is the strength of a signal reflected from a surface structure formed by an independent reflector system, and, which is the strength of a highlight signal, i.e.,

$$P = P_{\text{п}} + P_{\text{бт}}. \tag{11.12}$$

Each of the two summands in (11.12) is distributed according to the exponential law:

$$W_1(P_{\text{п}}) = \lambda_{\text{п}} \mathrm{e}^{-\lambda_{\text{п}} P_{\text{п}}}, \; W_2(P_{\text{бт}}) = \lambda_{\text{бт}} \mathrm{e}^{-\lambda_{\text{бт}} P_{\text{бт}}}. \tag{11.13}$$

As is known, the distribution parameters $\lambda_{\text{П}}$ and $\lambda_{\text{бт}}$ are values inverse to the mean values of the random variables $P_{\text{П}}$ and $P_{\text{бт}}$.

Based on relations (11.13), statistic independence of the summands $P_{\text{П}}$ and $P_{\text{бт}}$, we will obtain the distribution density for the total RCS.

For this purpose, let us introduce two new variables:

$$P = P_{\text{п}} + P_{\text{бт}}, U = P_{\text{п}} - P_{\text{бт}}. \tag{11.14}$$

For such transition, the Jacobian module can be found easily; it equals 1/2.

Due to the aforesaid, the required distribution density of the random value P will be as follows:

$$W(P) = \lambda_{\text{п}}\lambda_{\text{бт}}\,\mathrm{e}^{-(\lambda_{\text{п}}+\lambda_{\text{бт}})\frac{P}{2}} \int_{-P}^{P} \mathrm{e}^{-(\lambda_{\text{п}}-\lambda_{\text{бт}})\frac{U}{2}}\,\mathrm{d}U = \frac{\lambda_{\text{п}}\lambda_{\text{бт}}}{\lambda_{\text{п}}-\lambda_{\text{бт}}}\left(\mathrm{e}^{-\lambda_{\text{бт}}P} - \mathrm{e}^{-\lambda_{\text{п}}P}\right). \tag{11.15}$$

Let us proceed to find statistic characteristics of the phase center position for the case under review.

First of all, we will obtain a relation for the phase center displacement. For this purpose, we will represent the Formula (11.1) as follows:

$$X_{\text{фц}}^{\text{бт}} = \frac{\sum_{i=1}^{N+1} x_i P_i}{\sum_{i=1}^{N+1} P_i} = \frac{x_{\text{бт}} P_{\text{бт}} + \sum_{i=1}^{N} x_i P_i}{P_{\text{бт}} + \sum_{i=1}^{N} P_i}. \tag{11.16}$$

Phase center displacement can be determined from the relation

$$\Delta X = X_{\text{фц}}^{\text{бт}} - X_{\text{фц}}^{n} = \frac{x_{\text{бт}} P_{\text{бт}} + \sum_{i=1}^{N} x_i P_i}{P_{\text{бт}} + \sum_{i=1}^{N} P_i} - \frac{\sum_{i=1}^{N} x_i P_i}{\sum_{i=1}^{N} P_i} = \frac{x_{\text{бт}} - X_{\text{фц}}^{n}}{1 + \frac{P_{\text{п}}}{P_{\text{бт}}}}. \tag{11.17}$$

The same expression for displacement along the Y-axis will be as follows:

$$\Delta Y = Y_{\text{фц}}^{\text{бт}} - Y_{\text{фц}}^{n} = \left(y_{\text{бт}} - Y_{\text{фц}}^{n}\right) \Big/ \left(1 + \frac{P_{\text{п}}}{P_{\text{бт}}}\right). \tag{11.18}$$

Chapter 12
Enhancing Radar Station Functionalities to Delineate and Estimate Linear Dimensions of Extended Visual Targets for Coherent Scattering of Electromagnetic Waves

Phase center position of the radar target for coherent scattering

Let N HL be at a certain area of the flat surface S radiated by an electromagnetic wave emitted by the antenna A (Fig. 12.1) located at the point with coordinates (x, y, z) (the area dimensions are determined by the width of the antenna A directivity diagram). In this case, each ith HL will be an electromagnetic field source which intensity I_i and initial phase δ_i will be determined by the scattering matrix S_i of such reflector.

Let electromagnetic waves reflected from HL are received via the same antenna A on the same polarization where the electromagnetic wave is emitted. In this case, the field E_Σ at the antenna input may be written as:

$$E_\Sigma = \mathrm{e}^{i\omega t} \sum_{i=1}^{N} I_i \mathrm{e}^{-i(2kR_i+\delta_i)} = I_\Sigma \mathrm{e}^{-i\psi_\Sigma} \mathrm{e}^{i\omega t}, \tag{12.1}$$

where I_Σ is the intensity; ψ_Σ is the phase of the resultant wave.

An obvious expression for the angle sine and cosine ψ_Σ results from Eq. (12.1):

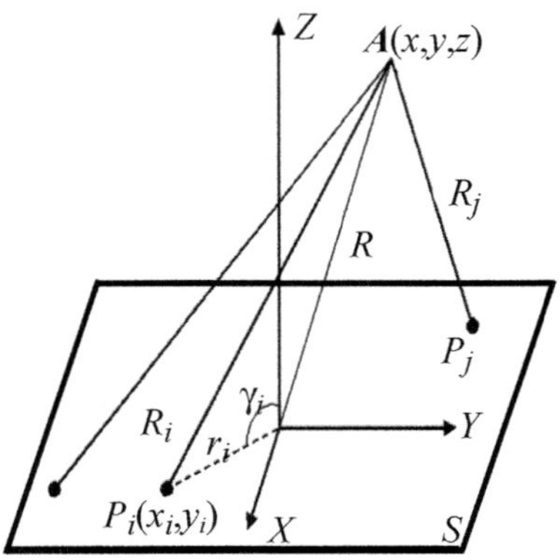

Fig. 12.1 Illustration of the problem of electromagnetic reflection waves from a radar object

Kozlov A. I. et al., *Introduction to the Theory of Radiopolarimetric Navigation Systems*, Springer Aerospace Technology, https://doi.org/10.1007/978-981-13-8395-3_12

$$\sin\psi_{\Sigma} = \frac{1}{I_{\Sigma}}\sum_{i=1}^{N} I_i \sin(2kR_i + \delta_i), \quad \cos\psi_{\Sigma} = \frac{1}{I_{\Sigma}}\sum_{i=1}^{N} I_i \cos(2kR_i + \delta_i). \quad (12.2)$$

An expression for this angle tangent does not require to know the intensity of the resultant wave.

$$\mathrm{tg}\psi_{\Sigma} = \left(\sum_{i=1}^{N} I_i \sin(2kR_i + \delta_i)\right) \Big/ \left(\sum_{i=1}^{N} I_i \cos(2kR_i + \delta_i)\right). \quad (12.3)$$

To determine the apparent reflection source position, a normal to the same phase surface, i.e., to the surface $\psi_{\Sigma} = \text{const}$, shall be constructed.

As follows from relation (12.3),

$$\psi_{\Sigma} = \mathrm{arctg}\left(\left(\sum_{i=1}^{N} I_i \sin(2kR_i + \delta_i)\right) \Big/ \left(\sum_{i=1}^{N} I_i \cos(2kR_i + \delta_i)\right)\right). \quad (12.4)$$

The required normal direction will coincide with that of the gradient to the equiphase surface.

Direct calculations grad ψ_{Σ} result in the following relations:

$$\frac{\partial\psi_{\Sigma}}{\partial x} = k\frac{\sum_{i=1}^{N} I_i \frac{\partial R_i}{\partial x}\cos(2kR_i + \delta_i)\sum_{j=1}^{N} I_j \cos\left(2kR_j + \delta_j\right) + \sum_{i=1}^{N} I_i \frac{\partial R_i}{\partial x}\sin(2kR_i + \delta_i)\sum_{j=1}^{N} I_j \sin\left(2kR_j + \delta_j\right)}{\left[\sum_{i=1}^{N} I_i \cos(2kR_i + \delta_i)\right]^2 + \left[\sum_{i=1}^{N} I_i \sin(2kR_i + \delta_i)\right]^2}. \quad (12.5)$$

Transformation (12.5) leads to the expression

$$\frac{\partial\psi_{\Sigma}}{\partial x} = k\frac{\sum_{i=1}^{N}\sum_{j=1}^{N} I_i I_j \frac{\partial R_i}{\partial x}\cos\left[2k\left(R_i - R_j\right) + \left(\delta_i - \delta_j\right)\right]}{\sum_{i=1}^{N}\sum_{j=1}^{N} I_i I_j \cos\left[2k\left(R_i - R_j\right) + \left(\delta_i - \delta_j\right)\right]}. \quad (12.6)$$

The same relations will be true for the y-component of the required gradient:

$$\frac{\partial\psi_{\Sigma}}{\partial y} = k\frac{\sum_{i=1}^{N}\sum_{j=1}^{N} I_i I_j \frac{\partial R_i}{\partial y}\cos\left[2k\left(R_i - R_j\right) + \left(\delta_i - \delta_j\right)\right]}{\sum_{i=1}^{N}\sum_{j=1}^{N} I_i I_j \cos\left[2k\left(R_i - R_j\right) + \left(\delta_i - \delta_j\right)\right]}, \quad (12.7)$$

and for the z-component:

$$\frac{\partial\psi_{\Sigma}}{\partial z} = k\frac{\sum_{i=1}^{N}\sum_{j=1}^{N} I_i I_j \frac{\partial R_i}{\partial z}\cos\left[2k\left(R_i - R_j\right) + \left(\delta_i - \delta_j\right)\right]}{\sum_{i=1}^{N}\sum_{j=1}^{N} I_i I_j \cos\left[2k\left(R_i - R_j\right) + \left(\delta_i - \delta_j\right)\right]}. \quad (12.8)$$

An obvious equation arises from Fig. 12.1:

$$R_i = \sqrt{(x - x_i)^2 + (y - y_i)^2 + z^2}, \tag{12.9}$$

whence we obtain:

$$\frac{\partial R_i}{\partial x} = \frac{x - x_i}{R_i}, \quad \frac{\partial R_i}{\partial y} = \frac{y - y_i}{R_i}, \quad \frac{\partial R_i}{\partial z} = \frac{z}{R_i}. \tag{12.10}$$

Relations (12.7)–(12.10) allow us to write down the equation of the normal to the equiphase surface at the antenna A point. For definiteness, let us assign to this point the coordinates x_A, y_A, z_A.

The required normal equation will be as follows:

$$\begin{aligned} &\frac{x - x_A}{\sum_{i=1}^{N}\sum_{j=1}^{N} I_i I_j \frac{x_A - x_i}{R_i}\cos\left[k\left(R_i - R_j\right) + \left(\delta_i - \delta_j\right)\right]} \\ &= \frac{y - y_A}{\sum_{i=1}^{N}\sum_{j=1}^{N} I_i I_j \frac{y_A - y_i}{R_i}\cos\left[k\left(R_i - R_j\right) + \left(\delta_i - \delta_j\right)\right]} \\ &= \frac{z - z_A}{\sum_{i=1}^{N}\sum_{j=1}^{N} I_i I_j \frac{z_A}{R_i}\cos\left[k\left(R_i - R_j\right) + \left(\delta_i - \delta_j\right)\right]}. \end{aligned} \tag{12.11}$$

Coordinates of the normal intersection point with the surface S, i.e., coordinates of the apparent reflection point, can be obtained from (12.11) if the latter assumes that $z = 0$. Thus, the phase center coordinates (x_0, y_0) will be as follows:

$$x_0 = x_A - \frac{\sum_{i=1}^{N}\sum_{j=1}^{N}(x_A - x_i)\frac{I_i I_j}{R_i}\cos\left[2k\left(R_i - R_j\right) + \left(\delta_i - \delta_j\right)\right]}{\sum_{i=1}^{N}\sum_{j=1}^{N}\frac{I_i I_j}{R_i}\cos\left[2k\left(R_i - R_j\right) + \left(\delta_i - \delta_j\right)\right]}, \tag{12.12}$$

$$y_0 = y_A - \frac{\sum_{i=1}^{N}\sum_{j=1}^{N}(y_A - y_i)\frac{I_i I_j}{R_i}\cos\left[2k\left(R_i - R_j\right) + \left(\delta_i - \delta_j\right)\right]}{\sum_{i=1}^{N}\sum_{j=1}^{N}\frac{I_i I_j}{R_i}\cos\left[2k\left(R_i - R_j\right) + \left(\delta_i - \delta_j\right)\right]}. \tag{12.13}$$

In (12.12) and (12.13), R_i and R_j mean the distances from the antenna to the ith and jth highlights, respectively, i.e.,

$$R_{i,j} = \sqrt{\left(x_A - x_{i,j}\right)^2 + \left(y_A - y_{i,j}\right)^2 + z_A^2} \tag{12.14}$$

As the observation is performed from the distance R_0 which considerably exceeds the linear dimensions of the radiated area of the surface S ($R_0 \gg r_{i,j}$), in this case, with an accuracy to the second order of smallness, we will have:

$$R_{i,j} = R_0\left[1 - \frac{r_{i,j}}{R_0}\cos\gamma_{i,j} + \frac{1}{2}\left(\frac{r_{i,j}}{R_0}\right)^2\sin^2\gamma_{i,j}\right], \tag{12.15}$$

where the angle γ_i is shown in Fig. 12.1.

Due to relation (12.14) and to $r_{i,j} = \sqrt{x_{i,j}^2 + y_{i,j}^2}$, $R_0 = \sqrt{x_A^2 + y_A^2 + z_A^2}$, we find from the triangle P_iOA as per the cosine theorem:

$$r_{i,j}\cos\gamma_{i,j} = \frac{x_A x_{i,j} + y_A y_{ij}}{\sqrt{x_A^2 + y_A^2 + z_A^2}}. \tag{12.16}$$

Equation (12.16) allows us to obtain the following relation:

$$r_{i,j}^2\sin^2\gamma_{i,j} = \frac{\left(x_A y_{i,j} - y_A x_{ij}\right)^2 + \left(x_{i,j}^2 + y_{i,j}^2\right)z_A^2}{x_A^2 + y_A^2 + z_A^2}. \tag{12.17}$$

The obtained formulae simplify if the origin of coordinates in the problem under review is brought into coincidence with the point corresponding to the projection of the antenna A location on the surface S, i.e., $x_A = y_A = 0$. In this case, relation (12.16) will become zero, and instead of Eq. (12.17) we will have:

$$r_{i,j}^2\sin^2\gamma_{i,j} = x_{i,j}^2 + y_{i,j}^2. \tag{12.18}$$

Accordingly, instead of (12.15) we will obtain the formula:

$$R_{i,j} = z_A + \frac{x_{i,j}^2 + y_{i,j}^2}{2z_A} \tag{12.19}$$

and therefore

$$R_i - R_j = \frac{x_i^2 - x_j^2 + y_i^2 - y_j^2}{2z_A}. \tag{12.20}$$

Let us notice that the limit values $x_{i,j}$ and $y_{i,j}$ are determined by the width of the antenna directivity diagram $\Delta\theta$. Their estimate can be obtained from the following assumptions: The value $\Delta\theta$ is determined by the antenna size—d_A, i.e., $\Delta\theta \sim \lambda/\mathrm{d}_A$, and the diameter of the radiated surface area will be $z_A\Delta\theta \sim z_A\lambda/\mathrm{d}_A$. It is this quantity that determines the limit values $x_{i,j}$ and $y_{i,j}$. Having plugged their values in Eq. (12.19), we will obtain $R_{i,j} = z_A\left(1 + \lambda^2/\mathrm{d}_A^2\right)$, that, within the framework of the considered approximation, gives all reasons to calculate $R_{i,j} \cong z_A$ with rather high accuracy in the denominators of the Formulae (12.12) and (12.13).

Thus, for the x-coordinate of the phase center, we will get:

$$x_0 = \frac{\sum_{i=1}^{N}\sum_{j=1}^{N} x_i I_i I_j \cos\left[2\pi\frac{x_i^2 - x_j^2 + y_i^2 - y_j^2}{z_A\lambda} + \left(\delta_i - \delta_j\right)\right]}{\sum_{i=1}^{N}\sum_{j=1}^{N} I_i I_j \cos\left[2\pi\frac{x_i^2 - x_j^2 + y_i^2 - y_j^2}{z_A\lambda} + \left(\delta_i - \delta_j\right)\right]}. \tag{12.21}$$

The same is for the y-coordinate:

$$y_0 = \frac{\sum_{i=1}^{N}\sum_{j=1}^{N} y_i I_i I_j \cos\left[2\pi\frac{x_i^2 - x_j^2 + y_i^2 - y_j^2}{z_A\lambda} + (\delta_i - \delta_j)\right]}{\sum_{i=1}^{N}\sum_{j=1}^{N} I_i I_j \cos\left[2\pi\frac{x_i^2 - x_j^2 + y_i^2 - y_j^2}{z_A\lambda} + (\delta_i - \delta_j)\right]}. \tag{12.22}$$

Let us transform the Formulae (12.21) and (12.22). First of all, let us select in their numerators and denominators the summands with the same indices where $i = j$. In this case, we will obtain:

$$x_0 = \frac{\sum_{i=1}^{N} x_i I_i^2 + \sum_{i=1}^{N} x_i I_i \sum_{j=1,\, j\neq i}^{N} I_j \cos\Phi_{i,j}}{\sum_{i=1}^{N} I_i^2 + \sum_{i=1}^{N} I_i \sum_{j=1,\, j\neq i}^{N} I_j \cos\Phi_{ij}}, \tag{12.23}$$

$$y_0 = \frac{\sum_{i=1}^{N} y_i I_i^2 + \sum_{i=1}^{N} y_i I_i \sum_{j=1,\, j\neq i}^{N} I_j \cos\Phi_{i,j}}{\sum_{i=1}^{N} I_i^2 + \sum_{i=1}^{N} I_i \sum_{j=1,\, j\neq i}^{N} I_j \cos\Phi_{ij}}, \tag{12.24}$$

where the phase designation was introduced:

$$\Phi_{i,j} = 2\pi\frac{x_i^2 - x_j^2 + y_i^2 - y_j^2}{z_A\lambda} + (\delta_i - \delta_j). \tag{12.25}$$

Squared intensities mentioned in (12.24) and (12.25) represent the HL return signal strengths. For these values, let us introduce the designations $I_i^2 = P_i$. In this case, the total strength of signals reflected from all HL will be as follows:

$$\sum_{i=1}^{N} I_i^2 = \sum_{i=1}^{N} P_i = P_{\Sigma 1}. \tag{12.26}$$

With regard to the designations introduced, the phase center coordinates will be written as follows:

$$x_0 = \frac{\sum_{i=1}^{N} x_i P_i + \sum_{i=1}^{N} x_i \sqrt{P_i} \sum_{j=1, j\neq i}^{N} \sqrt{P_j} \cos\Phi_{i,j}}{\sum_{i=1}^{N} P_i + \sum_{i=1}^{N} \sqrt{P_i} \sum_{j=1, j\neq i}^{N} \sqrt{P_j} \cos\Phi_{i,j}}, \tag{12.27}$$

$$y_0 = \frac{\sum_{i=1}^{N} y_i P_i + \sum_{i=1}^{N} y_i \sqrt{P_i} \sum_{j=1, j\neq i}^{N} \sqrt{P_j} \cos\Phi_{i,j}}{\sum_{i=1}^{N} P_i + \sum_{i=1}^{N} \sqrt{P_i} \sum_{j=1, j\neq i}^{N} \sqrt{P_j} \cos\Phi_{i,j}}. \tag{12.28}$$

As the next steps in the study of these capabilities, it is necessary to determine the classes of radar targets for which the proposed approach can be quite effective as well as to try to quantitatively link the SPC dimensions to the real target ones that will open the way for their specific identification.

Let us estimate the phase center position of a combined radar target (the case of incoherent and coherent reflection). Let us determine the sum wave amplitude E_Σ at the antenna A input. For this purpose, individual waves shall be added together.

To avoid a rather cumbersome procedure related to a large number of various trigonometric transformations, we will approach this problem in a slightly different way.

First of all, we will find the sum wave power at the receiving point:

$$P_{\Sigma} = |E_{\Sigma}|^2 = \left(e^{i\omega t}\sum_{i=1}^{N} I_i e^{2ikR_i} e^{i\delta_i}\right)\left(e^{-i\omega t}\sum_{j=1}^{N} I_j e^{-2ikR_j} e^{-i\delta_j}\right)$$
$$= \sum_{i=1}^{N}\sum_{j=1}^{N} I_i I_j e^{2ik(R_i - R_j)} e^{i(\delta_i - \delta_j)}. \tag{12.29}$$

Since P_{Σ} is a real number, we can immediately write down:

$$P_{\Sigma} = \sum_{i=1}^{N}\sum_{j=1}^{N} I_i I_j \cos\left[2ik(R_i - R_j) + (\delta_i - \delta_j)\right]. \tag{12.30}$$

With regard to the above transformations for the case when the observation is performed from the distance R_0 which considerably exceeds the linear dimensions of the radiated area of the surface ($R_0 \gg r_{i,j}$), Formula (12.30) is transformed as follows:

$$P_{\Sigma} = \sum_{i=1}^{N}\sum_{j=1}^{N} I_i I_j \cos \Phi_{ij}. \tag{12.31}$$

Thus, the input signal amplitude is easily found from Eq. (12.31):

$$|E_{\Sigma}| = \sqrt{\sum_{i=1}^{N}\sum_{j=1}^{N} I_i I_j \cos \Phi_{ij}}. \tag{12.32}$$

Let us emphasize one more aspect. Formula (12.31) can be interpreted as the sum of some "local" powers. It can be regarded (that follows from (12.23) and (12.24)) as the sum of strengths of signals reflected from highlights $P_{\Sigma 1} = \sum_{i=1}^{N} I_i^2$ and the sum of "interference" powers resulting from mutual wave interference—$P_{\Sigma 2} = \sum_{i=1}^{N} \sum_{j=1,\, j \neq i}^{N} I_i I_j \cos \Phi_{i,j}$.

The expression (12.31) exactly coincides with the denominators in the Formulae (12.21)–(12.22) that determine the phase center coordinates.

The numerators of these formulae represent weighted sums of "local" powers of "local" waves mentioned above where the HL coordinates serve as the "weighter."

If we compare the relations obtained for the phase center coordinates with the incoherent scattering case, the analogy is perfect. In both cases, it deals with the same mechanism. In the case of incoherent addition, the coordinate determination procedure only relates to the power $P_{\Sigma 1}$; in the case of coherent addition, it relates to the resultant power $(P_{\Sigma 1} + P_{\Sigma 2})$.

Let us consider the case when on the surface there is a set of N HL incoherently scattering the incident wave and M HL scattering it coherently. The task is to determine the phase center coordinates for this case. Since, as explained above, the indicated coordinates can be found via weighted averaging of each HL coordinates by power, this task can be approached as follows.

A set of incoherent reflectors determines its phase center position characterized by the coordinates $\left(x_{\text{фц}}^{\text{нек}}, y_{\text{фц}}^{\text{нек}}\right)$. The total input signal strength is a sum of HL return signals: $P_{\Sigma 1}^{\text{нек}} = \sum_{i=1}^{N} I_{(\text{нек})i}^{2}$. Thus, $P_{\Sigma}^{\text{нек}} = P_{\Sigma 1}^{\text{нек}}$.

A set of coherent reflectors determines its phase center position characterized by the coordinates $\left(x_{\text{фц}}^{\text{ког}}, y_{\text{фц}}^{\text{ког}}\right)$, which values are determined by Formulae (12.27) and (12.29). In this case, as already mentioned above, the total strength of the input signal represents the sum of HL return signal strengths $\left(P_{\Sigma 1}^{\text{ког}} = \sum_{i=1}^{M} I_{(\text{ког})i}^{2}\right)$ and the "interference" powers resulting from mutual wave interference— $P_{\Sigma 2}^{\text{ког}} = \sum_{i=1}^{M} \sum_{j=1, j \neq i}^{M} I_{(\text{ког})i} I_{(\text{ког})j} \cos\Phi_{i,j}$, i.e. $P_{\Sigma}^{\text{ког}} = P_{\Sigma 1}^{\text{ког}} + P_{\Sigma 2}^{\text{ког}}$.

The above formulae and the above arguments enable to write down the phase center coordinates for the case under review:

$$X_{\text{фц}} = \frac{x_{\text{фц}}^{\text{нек}} P_{\Sigma}^{\text{нек}} + x_{\text{фц}}^{\text{ког}} P_{\Sigma}^{\text{ког}}}{P_{\Sigma}^{\text{нек}} + P_{\Sigma}^{\text{ког}}}, \tag{12.33}$$

$$Y_{\text{фц}} = \frac{y_{\text{фц}}^{\text{нек}} P_{\Sigma}^{\text{нек}} + y_{\text{фц}}^{\text{ког}} P_{\Sigma}^{\text{ког}}}{P_{\Sigma}^{\text{нек}} + P_{\Sigma}^{\text{ког}}}. \tag{12.34}$$

Having plugged appropriate expressions in (12.33) and (12.34), we will obtain:

$$X_{\text{фц}} = \frac{\sum_{i=1}^{N} x_{i}^{\text{нек}} I_{(\text{нек})i}^{2} + \sum_{n=1}^{M} \sum_{m=1}^{M} x_{n}^{\text{ког}} I_{(\text{ког})n} I_{(\text{ког})m}}{\sum_{i=1}^{N} I_{(\text{нек})i}^{2} + \sum_{n=1}^{M} \sum_{m=1}^{M} I_{(\text{ког})n} I_{(\text{ког})m}}, \tag{12.35}$$

$$Y_{\text{фц}} = \frac{\sum_{i=1}^{N} y_{i}^{\text{нек}} I_{(\text{нек})i}^{2} + \sum_{n=1}^{M} \sum_{m=1}^{M} y_{n}^{\text{ког}} I_{(\text{ког})n} I_{(\text{ког})m}}{\sum_{i=1}^{N} I_{(\text{нек})i}^{2} + \sum_{n=1}^{M} \sum_{m=1}^{M} I_{(\text{ког})n} I_{(\text{ког})m}}. \tag{12.36}$$

Reasoning similarly, account can be taken of the signal reflected from a statistically homogeneous surface (background signal) where there are no highlights. If the strength of such input signal is designated as $P_{\text{фс}}$, instead of (12.8) and (12.9), we will have:

$$X_{\text{фц}} = \frac{\sum_{i=1}^{N} x_i^{\text{нек}} I_{(\text{нек})i}^2 + \sum_{n=1}^{M} \sum_{m=1}^{M} x_n^{\text{ког}} I_{(\text{ког})n} I_{(\text{ког})m}}{\sum_{i=1}^{N} I_{(\text{нек})i}^2 + \sum_{n=1}^{M} \sum_{m=1}^{M} I_{(\text{ког})n} I_{(\text{ког})m} + \frac{1}{2} P_{\text{фс}}}, \quad (12.37)$$

$$Y_{\text{фц}} = \frac{\sum_{i=1}^{N} y_i^{\text{нек}} I_{(\text{нек})i}^2 + \sum_{n=1}^{M} \sum_{m=1}^{M} y_n^{\text{ког}} I_{(\text{ког})n} I_{(\text{ког})m}}{\sum_{i=1}^{N} I_{(\text{нек})i}^2 + \sum_{n=1}^{M} \sum_{m=1}^{M} I_{(\text{ког})n} I_{(\text{ког})m} + \frac{1}{2} P_{\text{фс}}}. \quad (12.38)$$

These relations solve the required task.

By *point-to-point radar target,* we shall mean a target that can be modeled with two highlights. Let us put the target along the *OX* axis ($y = 0$).

In this case, (12.16) will be as follows:

$$x_0 = \frac{x_1 P_1 + x_2 P_2 + (x_1 + x_2)\sqrt{P_1 P_2} \cos \Phi_{12}}{P_1 + P_2 + 2\sqrt{P_1 P_2} \cos \Phi_{12}}. \quad (12.39)$$

Let us illustrate the relation obtained. For this purpose, we will introduce the dimensionless parameter $\xi^2 = P_2/P_1$. For definiteness, we will consider that $x_1 \geq x_2$. Let us also introduce the linear target dimension: (*X*-dimension) $\mathrm{d}_x = x_1 - x_2$. Let us rewrite relation (12.36) using the introduced designations:

$$x_0 = \frac{x_1\left(1 + \xi^2 + 2\xi \cos \Phi_{12}\right) - \mathrm{d}_x \xi(\xi + \cos \Phi_{12})}{1 + \xi^2 + 2\xi \cos \Phi_{12}} = x_1 - \mathrm{d}_x \frac{\xi(\xi + \cos \Phi_{12})}{1 + \xi^2 + 2\xi \cos \Phi_{12}}. \quad (12.40)$$

Relation (12.40) allows us to introduce one more rather natural dimensionless parameter μ_x that shows relative displacement of the phase center of the point-to-point target toward the HL 1 position which coordinate has a greater value (the right reflector):

$$\mu_x = \frac{x_1 - x_0}{\mathrm{d}_x} = \frac{\xi(\xi + \cos \Phi_{12})}{1 + \xi^2 + 2\xi \cos \Phi_{12}}. \quad (12.41)$$

The introduced parameter can be called a relative bearing error.

If the parameter $\mu_x < 0$, the phase center is to the right of the right HL, if $0 \leq \mu_x \leq 1$, the phase center is between the right and the left HL, and if $\mu_x > 1$, the phase center is to the left of the left reflector.

Considering the parameters ξ and Φ_{12} independent of one another, it is easily seen that the parameter μ_x reaches its extreme values when $\cos \Phi_{12} = \pm 1$ that allows us to write down the following equations for them:

If $\xi > 1$, then

$$\mu_{\max} = \frac{\xi}{\xi - 1}, \quad \mu_{\min} = \frac{\xi}{1 + \xi}, \quad (12.42)$$

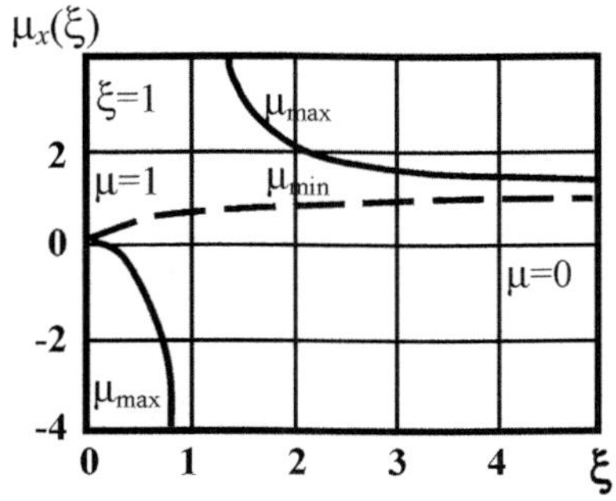

Fig. 12.2 Dependence of extreme values of parameter μ_x on relation if signals powers ξ, reflected from two shining points

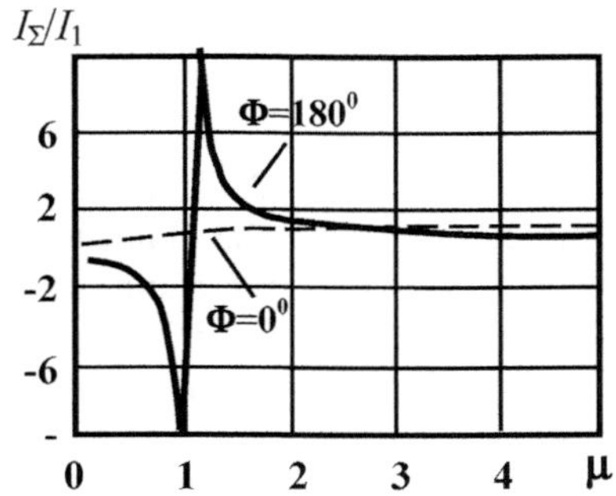

Fig. 12.3 Relation between the normalized amplitude of reflected signal and the relative bearing error μ at different phase shifts

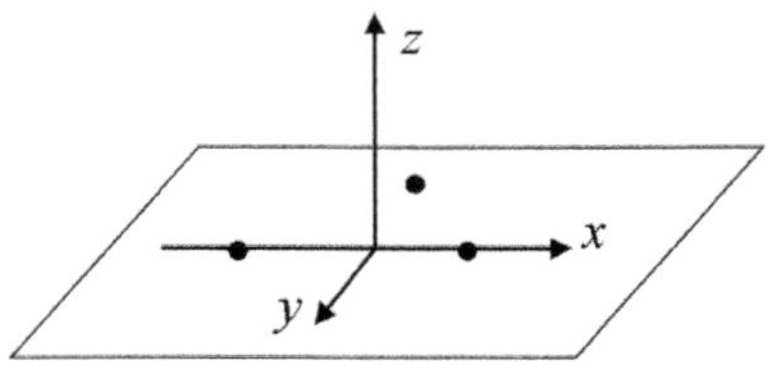

Fig. 12.4 Model of three-point object

if $\xi < 1$, then

$$\mu \text{max} = \frac{\xi}{1+\xi}, \quad \mu \text{min} = \frac{\xi}{\xi - 1}. \tag{12.43}$$

When $\xi = 1$, one of the extreme values equals 1/2 and the second value becomes $+\propto$ or $-\propto$. Nevertheless, it is worth noting, and it is extremely important, that in the latter case the resultant signal amplitude becomes zero.

Figure 12.2 shows charts depicting the dependences (12.42) and (12.43).

Figure 12.3 illustrates the relation between normalized amplitude of the reflected signal I_Σ/I_1 and relative bearing error μ at different phase shifts Φ_{12}. From the given figure, it is rather clear that a large bearing error occurs at small resultant amplitude.

By *three-point radar target,* we shall mean a target that can be modeled with three HL. Let us orient the axis *OX* ($y = 0$) to involve two of the three highlights and put the origin of coordinates in the middle of the appropriate segment (Fig. 12.4).

In this case, the Formulae (12.16) and (12.17) will be as follows:

$$x_0 = \frac{x_1 P_1 + x_2 P_2 + (x_1 + x_2)\sqrt{P_1 P_2}\cos\Phi_{12} + \Rightarrow}{P_1 + P_2 + 2\sqrt{P_1 P_2}\cos\Phi_{12} + \Rightarrow}$$
$$\rightarrow \frac{\Rightarrow + x_3 P_3 + (x_1 + x_3)\sqrt{P_1 P_3}\cos\Phi_{13} + (x_2 + x_3)\sqrt{P_2 P_3}\cos\Phi_{23}}{\Rightarrow P_3 + 2\sqrt{P_1 P_3}\cos\Phi_{13} + 2\sqrt{P_2 P_3}\cos\Phi_{23}}, \tag{12.44}$$

$$y_0 = \frac{y_1 P_1 + y_2 P_2 + (y_1 + y_2)\sqrt{P_1 P_2}\cos\Phi_{12} + \Rightarrow}{P_1 + P_2 + 2\sqrt{P_1 P_2}\cos\Phi_{12} + \Rightarrow}$$
$$\rightarrow \frac{\Rightarrow + y_3 P_3 + (y_1 + y_3)\sqrt{P_1 P_3}\cos\Phi_{13} + (y_2 + y_3)\sqrt{P_2 P_3}\cos\Phi_{23}}{\Rightarrow P_3 + 2\sqrt{P_1 P_3}\cos\Phi_{13} + 2\sqrt{P_2 P_3}\cos\Phi_{23}}. \tag{12.45}$$

As seen from (12.44) and (12.45), the emergence of the third HL led to the emergence of additional summands caused by interference of the new third wave with the two old waves in the numerator and denominator of the formulae determining the phase center position for the point-to-point target.

Elementary analysis shows that extreme values of the phase center coordinate, firstly, are reached when $\Phi_{ij} = n\pi$, i.e., when $\cos\Phi_{ij} = \pm 1$, and secondly, that is why the both phase center coordinates reach their extreme values at the same time:

$$x_0^{\text{экстр}} = \frac{x_1 P_1 + x_2 P_2 + x_3 P_3 \pm (x_1 + x_2)\sqrt{P_1 P_2} \pm (x_1 + x_3)\sqrt{P_1 P_3} \pm (x_2 + x_3)\sqrt{P_2 P_3}}{P_1 + P_2 + P_3 \pm 2\left(\sqrt{P_1 P_2} + \sqrt{P_1 P_3} + \sqrt{P_2 P_3}\right)}, \tag{12.46}$$

$$y_0^{\text{экстр}} = \frac{y_1 P_1 + y_2 P_2 + y_3 P_3 \pm (y_1 + y_2)\sqrt{P_1 P_2} \pm (y_1 + y_3)\sqrt{P_1 P_3} \pm (y_2 + y_3)\sqrt{P_2 P_3}}{P_1 + P_2 + P_3 \pm 2\left(\sqrt{P_1 P_2} + \sqrt{P_1 P_3} + \sqrt{P_2 P_3}\right)}. \tag{12.47}$$

As in the chosen coordinate system $x_1 = -x_2 = d_{12}$ and $y_1 = y_2 = 0$, (12.46) and (12.47) are somewhat reduced:

$$\rho_x^{\text{экстр}} = \frac{x_0^{\text{экстр}}}{x_3} = \frac{(\pm\alpha \pm \beta + 1) + \eta(\alpha - \beta)(\alpha + \beta \pm 1)}{(\alpha \pm \beta \pm 1)^2}, \tag{12.48}$$

$$\rho_y^{\text{экстр}} = \frac{y_0^{\text{экстр}}}{y_3} = \frac{\pm\alpha \pm \beta + 1}{(\alpha \pm \beta)^2 \pm 2(\alpha + \beta) + 1}, \tag{12.49}$$

where the parameters $\alpha^2 = P_1/P_3$, $\beta^2 = P_2/P_3$, $\eta = x_1/x_3 = d_{12}/x_3$.

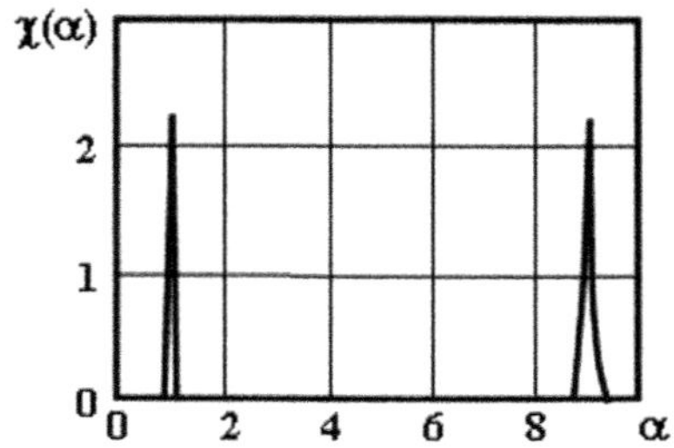

Fig. 12.5 "Background" signal in the case when the total the signal from the brilliant points is much less than the "background", $\varepsilon = 5, \gamma = 40, \beta = 4$

As the denominator in Formula (12.49) at $\beta' = \left(1 \pm \sqrt{\alpha}\right)^2$ becomes zero that entails an infinite value for $\rho_y^{\text{экстр}}$, it is reasonable to insert there an additional summand to take on the small given value ε at $\beta = \beta'$ and to be close to zero at $\beta \neq \beta'$. The physical basis of this procedure consists in the fact that there are always input signals reflected from different background objects, and they can also include self-noise of the receiving device recalculated to its input. It corresponds to the situation when these signals considerably exceed the resultant HL signal that occurs due to interference, as stated above.

As the mentioned additional summand, for example, the relation given below can be chosen to be interpreted as a background return signal (incoherent constituent):

$$\chi_\Phi^2 = P_\Phi / P_3 = \varepsilon\left\{ e^{-\gamma\left|\beta-\left(1+\sqrt{\alpha}\right)^2\right|} + e^{-\gamma\left|\beta-\left(1-\sqrt{\alpha}\right)^2\right|} \right\}. \qquad (12.50)$$

Here ε and γ are some given parameters determining the background signal and the degree of proposed approximation; Fig. 12.5 describes such dependence for $\varepsilon = 5, \gamma = 40$, and $\beta = 4$.

In view of the above, instead of Formula (12.50), we will write:

$$\rho_y^{\text{экстр}} = \frac{y_0^{\text{экстр}}}{y_3} = \frac{\pm\alpha \pm \beta + 1}{\left(\alpha \pm \beta\right)^2 \pm 2\left(\alpha + \beta\right) + 1 + \chi_\Phi^2}.$$

Figure 12.6 illustrates extreme deviation dependences of the x- and y-coordinates of the phase center for the equilateral triangle according to the changed strength of the signal reflected from the first highlight.

Thus, in this case, extreme values of the bearing error will be:

$$\eta_y = \frac{1 \pm (\alpha + \beta)}{1 + \alpha^2 + \beta^2 \pm 2(\alpha + \beta + \alpha\beta)}. \qquad (12.51)$$

To identify a quantitative relation between SPC and real target dimensions, let us model several target variants within the framework of methods described above.

This part of the work is divided into four stages:

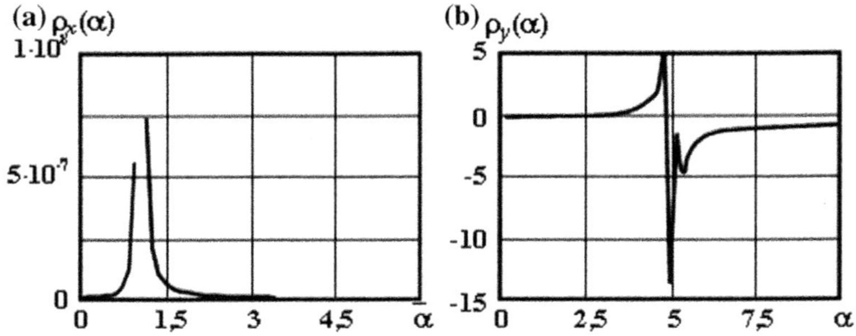

Fig. 12.6 Dependence of the limiting deviation of the x-coordinate (**a**) and y-coordinate (**b**) of the phase center for an equilateral triangle on the change in signal power reflected from the first shining point ($x_1 = 0.5, y_3 = 1, x_3 = 0, b = 1$)

1. The target is modeled by one HL.
2. The target is modeled by two HL with fixed coordinates.
3. The target is modeled by two HL which coordinates are arbitrarily defined.
4. The target is regarded as an aggregate of HL arbitrarily oriented in space.

For all the stages, two cases are considered:

1. HL targets and surface reflectors are modeled by dipoles.
2. HL targets and surface reflectors are modeled by arbitrary scattering matrices.

Let us determine the direction to a bright HL. Let us consider a plane of $2L \times 2L$. HL are scattered in the plane. Among these HL is the bright HL modeling the target. All HL position is randomly determined. The aggregate background brightness is much higher than that of the HL target. HL target brightness is A times higher than that of each background HL. HL is modeled by dipoles with the random inclination angle to the horizontal axis. The problem geometry and ad hoc SPC are given in Fig. 12.7.

As a result of modeling, it has become clear that, in this case, SPC has the form of a loop. The loop is directed to the bright HL modeling the target. SPC dependence on the HL target brightness consisted in more accurate direction to the bright HL when its brightness increased. The modeling results are given in Fig. 12.8.

Fig. 12.7 Geometry of the problem

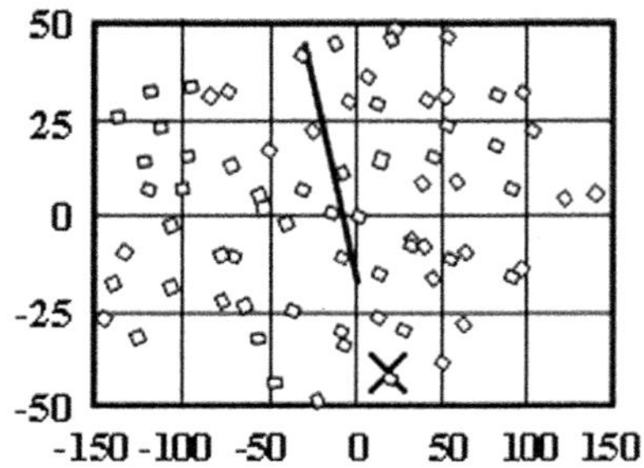

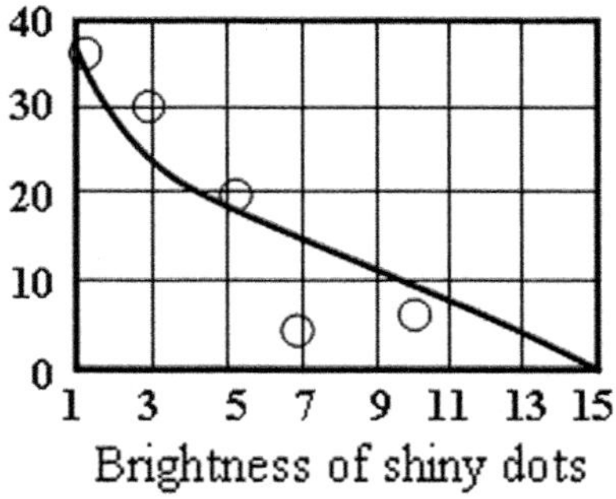

Fig. 12.8 Results of the modeling

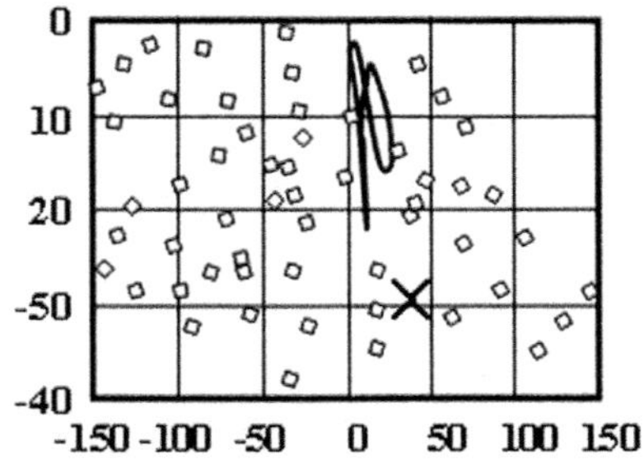

Fig. 12.9 Geometry of the problem and the result of the construction of a SPC

Figure 12.9 describes the mean error of direction to the bright HL due to the brightness A. The chart shows that the error does not exceed 5 degrees when the brightness reaches $A = 15$.

In the case of arbitrary SM, the problem geometry remains the same. HL are scattered in the plane $2L \times 2L$. Among these HL is the bright HL modeling the target. All HL position is randomly determined. The aggregate background brightness is much higher than that of the HL target. HL target brightness is A times higher than that of the HL background. The brightness of the HL background $A = 1$. HL are modeled by arbitrary SM. The problem geometry is given in Fig. 12.9.

In this case, SPC has a complex form. Let us define its maximum size and direction. As in the previous case, the maximum SPC size is directed to the HL target and depends on the brightness of this HL.

Following the modeling results, we constructed dependence of the mean error of direction to the target on the HL brightness. The dependence is given in Fig. 12.10.

The chart shows that the error of direction decreases when the HL brightness increases.

Let us consider the case of a point-to-point target described in Fig. 12.11. The target consists of two HL. HL are located at the points with coordinates $(-L, 0)$, $(L, 0)$. HL are modeled by dipoles oriented at the angle of 0 and $\pi/2$ to the horizontal axis. Thus, in this case, the i HL scattering matrix will be written as:

$$S_i = A_i \begin{pmatrix} \cos(\phi_i)^4 & \cos(\phi_i)^4 \sin(\phi_i)^4 \\ \cos(\phi_i)^4 \sin(\phi_i)^4 & \sin(\phi_i)^4 \end{pmatrix}, \tag{12.52}$$

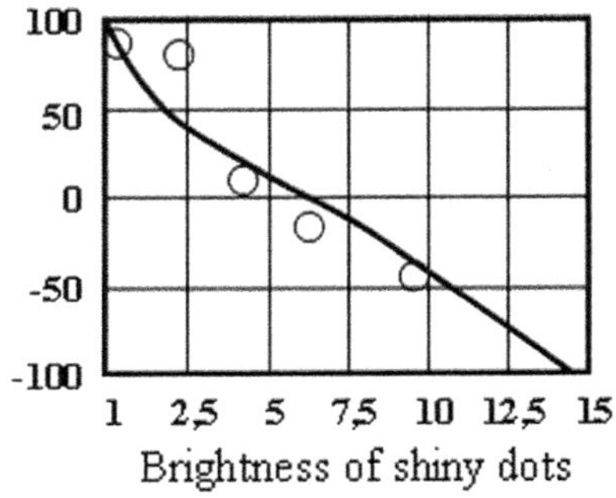

Fig. 12.10 Mean heading error

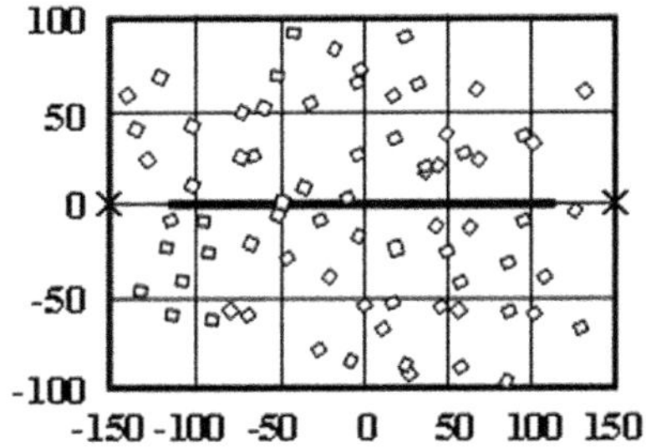

Fig. 12.11 Case of the two-point target

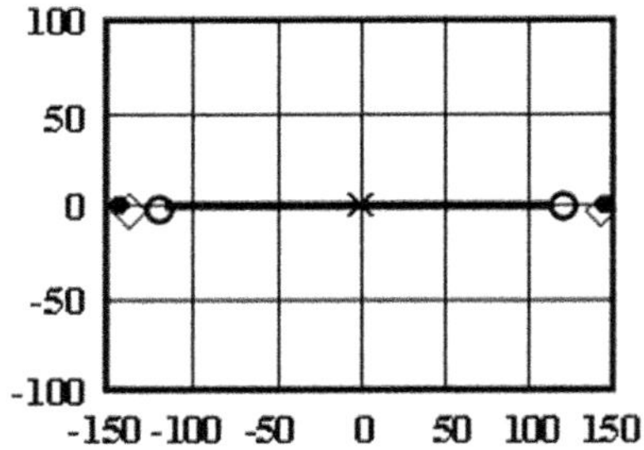

Fig. 12.12 SPC for target consisting of 2 shining points

where A_i is an i HL brightness, ϕ_i is an inclination angle to the horizontal axis.

HL modeled by dipoles with random rotation angles to the horizontal axis are scattered in the space $2L \times 2L$. These HL model a background. Their brightness $A = 1$. Although the target HL brightness is A times higher than that of each background HL, the total brightness of background HL is much higher than that of the target HL.

Plugging the scattering matrix expression for dipoles in (12.33), we will find a SPC expression for the target consisting of two fixed dipoles.

Based on the modeling and SPC construction results, for this case the following specific features were identified:

- SPC has the form of a narrow, strongly elongated loop,
- the loop is oriented along the line connecting bright HL while slightly deviated therefrom as shown in Fig. 12.12.

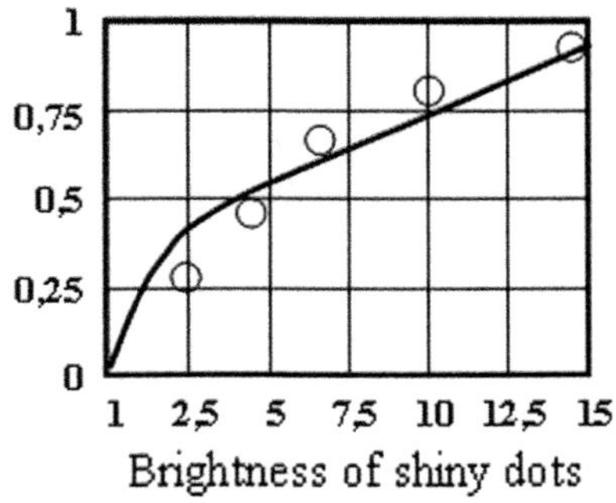

Fig. 12.13 Results of the modeling. ○ The maximum SPC size found

Based on the experiment results, the following feature was found that the SPC size and orientation strongly depend on the HL target brightness. When the target HL A equals 15, maximum SPC size virtually equals the real target size, and the SPC orientation is such that maximum size slightly deviates from the line connecting target HL. When the HL target brightness decreases, maximum SPC size also decreases and deviation from the target HL connecting line increases.

The next stage of study consists in comparison of SPC dimensions with the real target dimensions (the distance between HL). As SPC has the form of a strongly elongated loop, it is reasonable to take the maximum SPC size as the true target size and the maximum size orientation as the true target orientation in space.

Let us introduce the relation

$$k = \frac{m_0}{2L}, \tag{12.53}$$

where m_0 is the maximum SPC size, $2L$ is the true target size.

As a result of modeling, the chart shown in Fig. 12.13 was constructed.

The chart describes dependence of the coefficient k mean value on target HL brightness. The chart shows that we can speak of the described effect becoming apparent when the HL brightness is $A = 5$. When $A = 10$ and more, it is possible to determine the true target dimensions rather accurately. Thus, for example, when $A = 15$, $k = 0.9$.

Let us consider the case when arbitrary scattering matrices are used as both target and background HL. It means that values of scattering matrix elements are given by random number generator in the interval of $(0 \div 1)$ as per the uniform distribution law.

Similar to the above case, we will obtain expressions describing SPC for this case. We construct SPC as shown in Fig. 12.14.

Unlike the problem described above, SPC is not symmetric with respect to the origin of coordinates. The loop becomes significantly wider. Nevertheless, the SPC size dependence on the target HL brightness as well as its direction along the target HL connecting line remains. Thus, the assumptions described above are also applicable to this case.

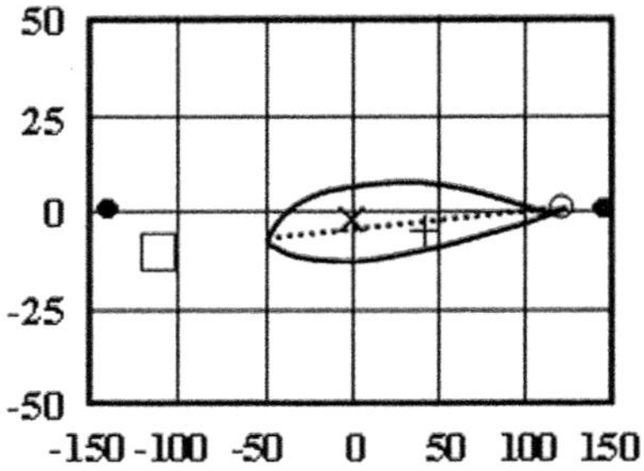

Fig. 12.14 Results of SPC modeling for arbitrary scattering matrices. □ The obtained coordinates BP with coefficient k; ● True BP coordinates; × Distance centers between bright BP; + The found centers of distances between the found BP

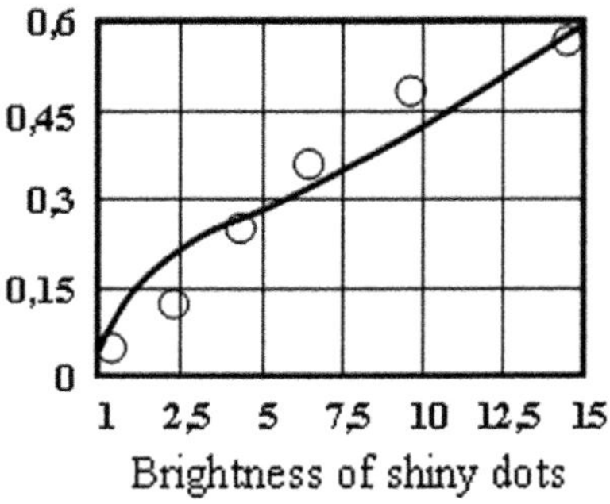

Fig. 12.15 Results of the modeling. ○ The maximum SPC size found

Let us take maximum SPC size as the true target size. Let us introduce the coefficient $k = \frac{m_0}{2L}$. As a result of modeling, the chart shown in Fig. 12.15 can be constructed.

Figure 12.6 describes dependence of the coefficient k mean value on target HL brightness. The general view of the chart is similar to that presented in Fig. 12.12. The difference is that the value which k tends to is limited by the angle 0.6, as compared to 0.9 in Fig. 12.13.

Let us consider a target consisting of 2 HL, as shown in Fig. 12.16; HL are randomly located in the space limited by the size of $2L \times 2$L. The target HL are modeled by dipoles. Background HL modeled by dipoles are scattered on the area $2L \times 2L$. All dipoles orient arbitrarily to the horizontal line. The background HL brightness $A = 1$, the target HL brightness equals A, but the total background brightness is much higher than that of the target HL. The SPC construction results are given in Fig. 12.17.

As seen from Fig. 12.17, the SPC also has the form of a loop oriented along the line connecting the target HL. Dependence of the maximum SPC size on the target HL brightness also remains. The modeling results are given in Fig. 12.18 that show dependence of the mean value of a relative target size error on the target HL brightness.

When modeling this problem, the SPC center coordinates were determined and the distance between the SPC center and the target HL distance center was found. As a result, the following dependence represented in Fig. 12.18 was obtained.

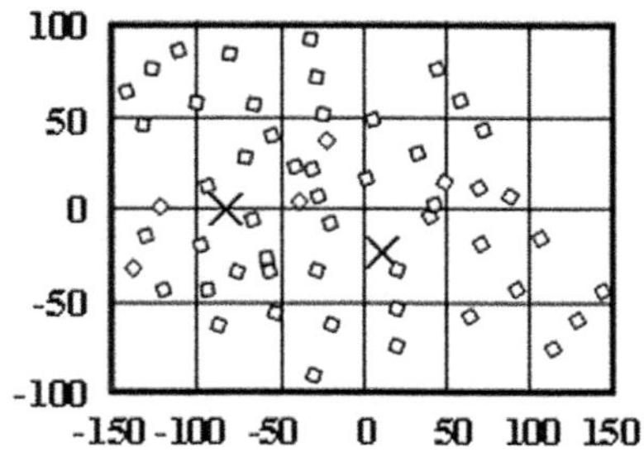

Fig. 12.16 Two-point target arbitrary oriented in space

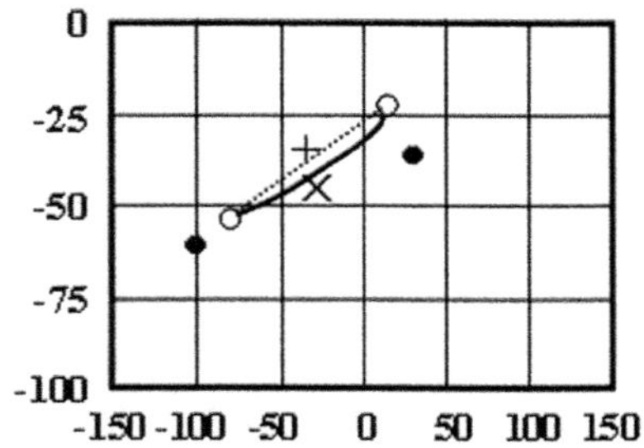

Fig. 12.17 Modeling SPC

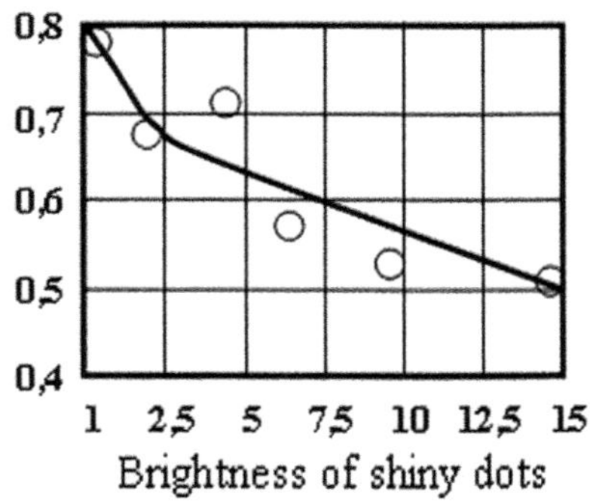

Fig. 12.18 Results of the modeling

Figure 12.19 presents dependence of the mean value of a target center error on the HL brightness. The chart shows that the curve is also sensible to the target HL brightness. It means that the distance between the maximum SPC size center and the target HL distance center decreases when the target HL brightness increases. Thus, SPC center location can be taken as the target center location.

Let us consider the case when both target and background HL are modeled by arbitrary scattering matrices. Let us consider a target consisting of 2 HL. HL are randomly located in the space limited by the size of $2L \times 2L$. HL modeling the background are scattered on the area $2L \times 2L$. These HL brightness $A = 1$, the target HL brightness equals A, but the total background brightness is much higher than that of the target HL as shown in Fig. 12.20. The SPC construction results are given in Fig. 12.21.

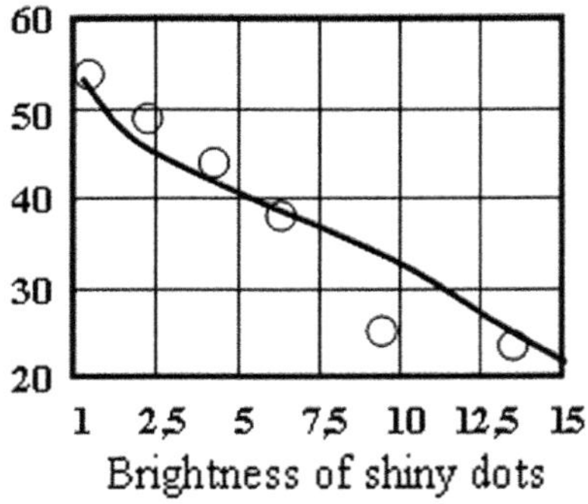

Fig. 12.19 Target center error

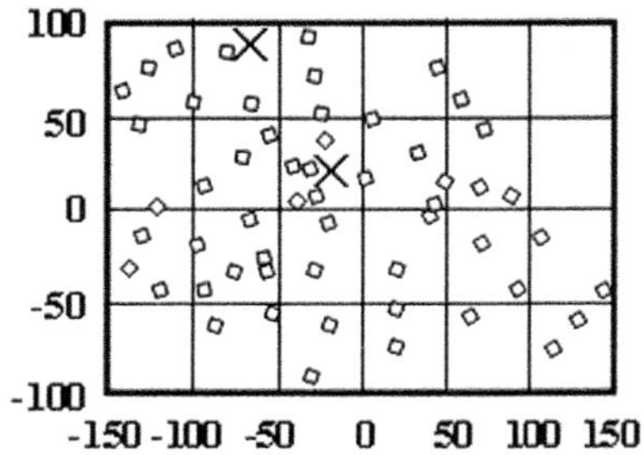

Fig. 12.20 Arbitrary two-point target

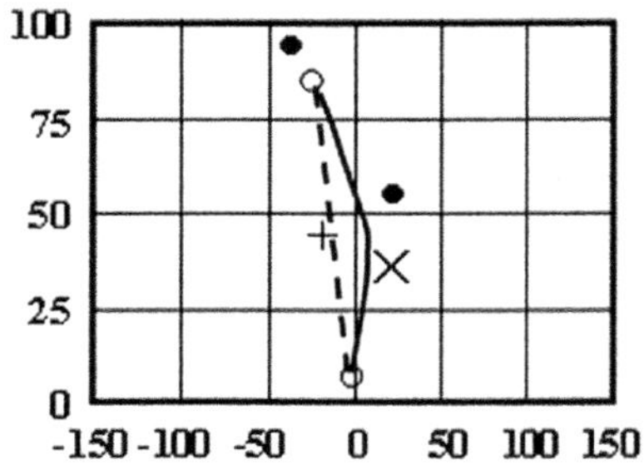

Fig. 12.21 Modeling SPC results

Unlike the previous cases, SPC has the form of a complex contour—a closed curve. As a result of modeling, it became clear that the maximum SPC size is also sensible to the target HL brightness. Thus, application of the accepted method to determine the real target size and its center position as per the maximum SPC size and its center position produced results demonstrated in Figs. 12.22 and 12.23.

The charts show that the general nature of curves is preserved, which means that determination accuracy of the true target size and center location increases with increasing the target HL brightness.

Let us consider a plane of $2L \times 2L$. HL are scattered in the plane. Among these HL are the bright HL modeling the target. All HL position is randomly determined. The aggregate background brightness is much higher than that of the HL target. Each HL target brightness is A times higher than that of the HL background. HL

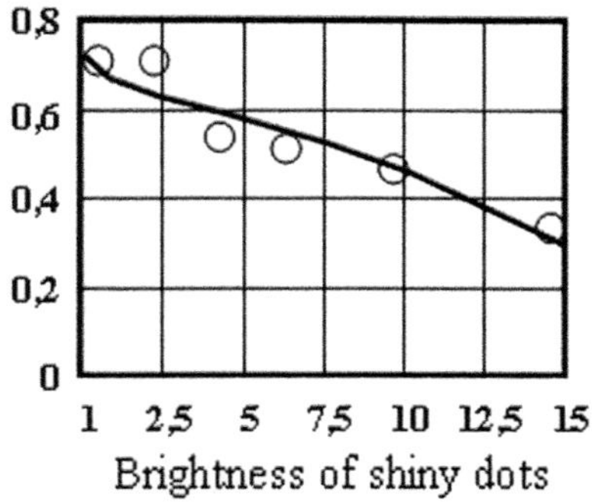

Fig. 12.22 Target size relative mean error

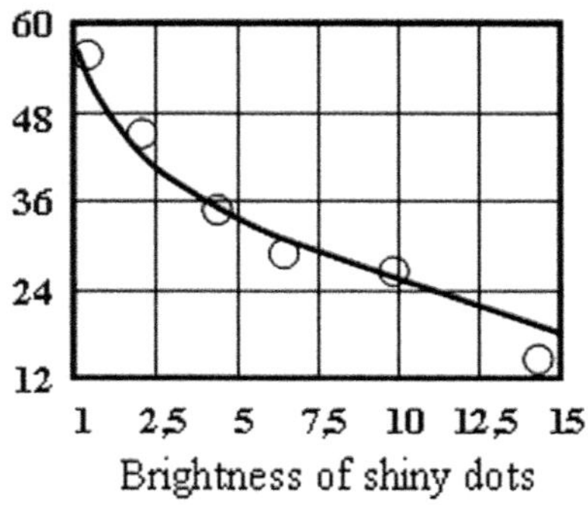

Fig. 12.23 Target center error

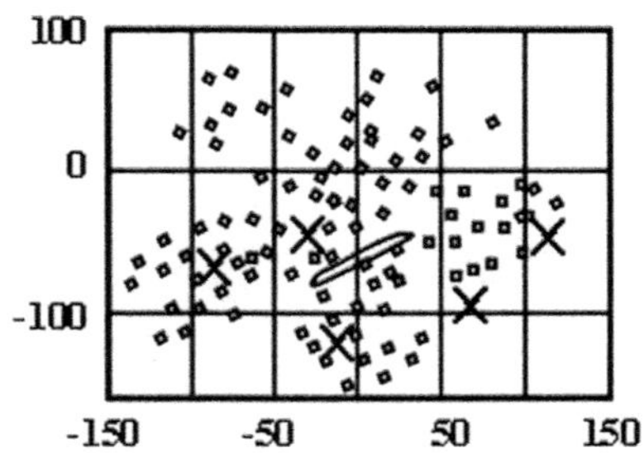

Fig. 12.24 Geometry of the problem

is modeled by dipoles with the random inclination angle to the horizontal axis as shown in Fig. 12.24. The constructed SPC is given in Fig. 12.25.

In this case, SPC has the form of a loop arbitrarily oriented in space. Let us take the maximum SPC size as the diameter of the area comprising the target HL and the maximum SPC size center as the center of this area as shown in Fig. 12.26.

In the process of modeling, the real size of the area (circle) comprising the target HL and its center was also determined. As a result of modeling, dependence of the SPC size on the target HL brightness is determined. Following the experiments, the charts presented in Figs. 12.27, 12.28 and 12.29 were constructed.

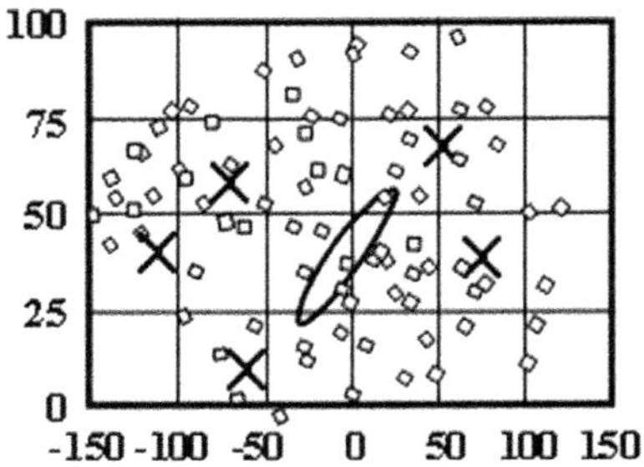

Fig. 12.25 SPC for scattering area of the shining points

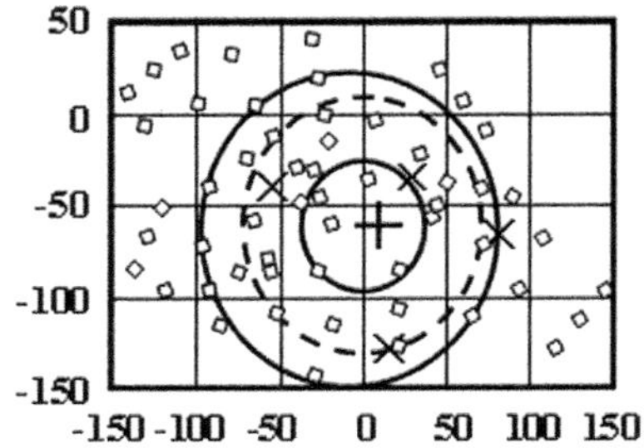

Fig. 12.26 Scattering areas for bright shining points

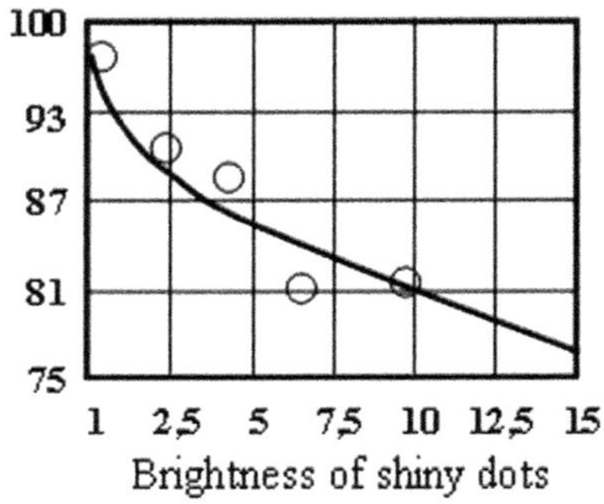

Fig. 12.27 Area size error

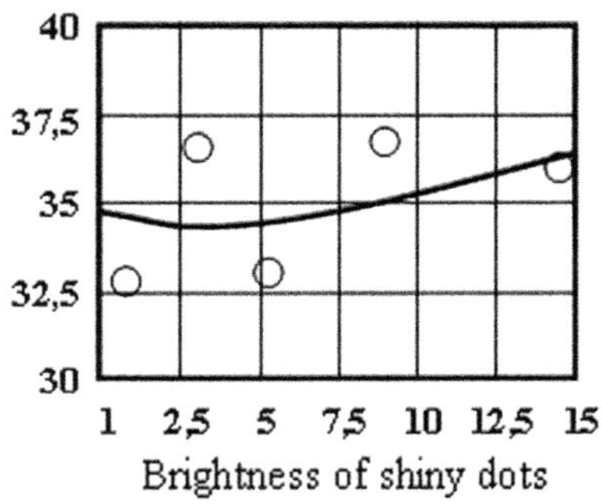

Fig. 12.28 Area size error taking into account K

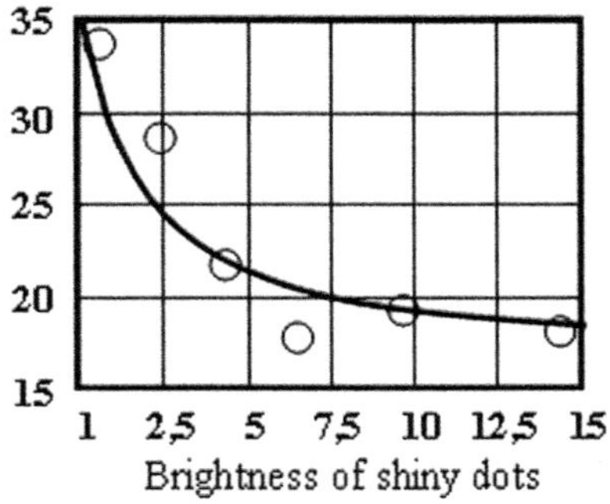

Fig. 12.29 Target center error

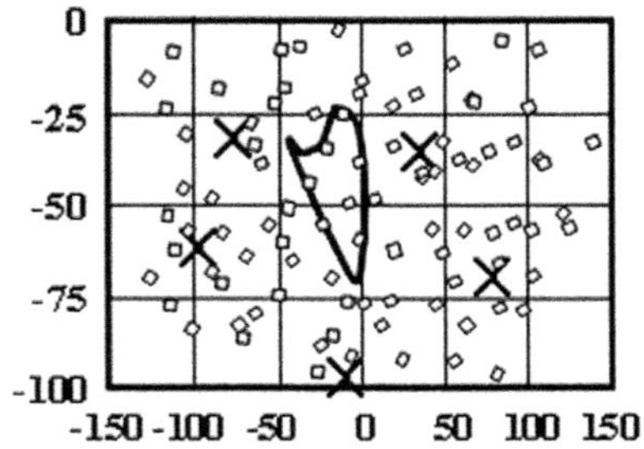

Fig. 12.30 Geometry of the problem. SPC for arbitrary scattering matrices

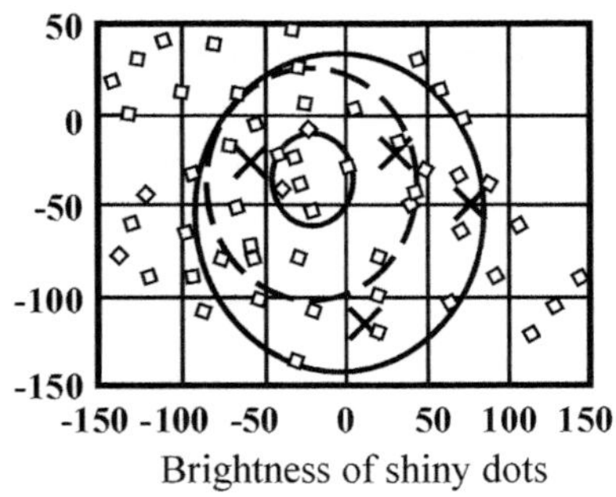

Fig. 12.31 Scattering areas for bright shining points

As seen in Figs. 12.26, 12.27, 12.28 and 12.29, the dependence of curves on the target HL brightness remains. When the brightness increases, the accuracy of parameter determination also increases.

Let us consider a plane of $2L \times 2L$. HL are scattered in the plane. Among these HL are the bright HL modeling the target. All HL position is randomly determined. The aggregate background brightness is much higher than that of the HL target. Each HL target brightness is A times higher than that of the HL background. The problem geometry and result of SPC construction are given in Fig. 12.30.

As seen in Fig. 12.31, SPC has a complex structure. Let us determine the maximum SPC size and take it as a size of the area comprising bright target HL. Let us take the maximum SPC size center as the center of the true area comprising the target HL as shown in Fig. 12.31.

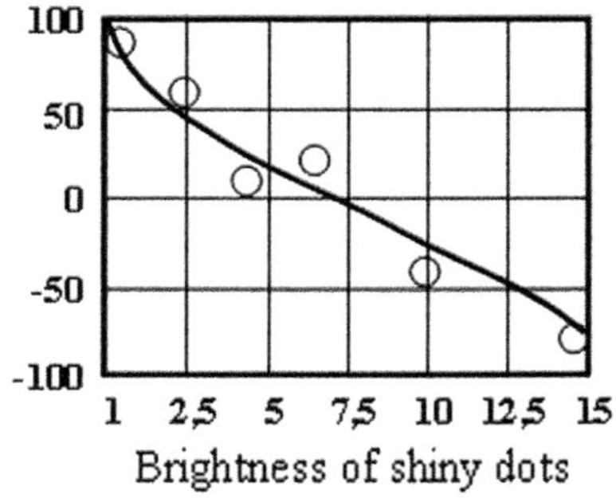

Fig. 12.32 Mean area size error

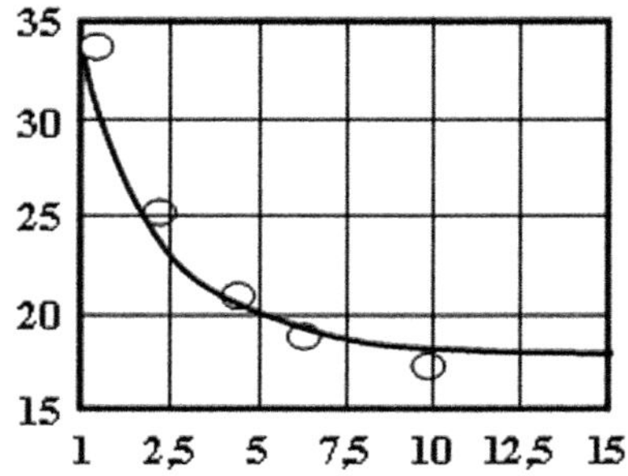

Fig. 12.33 Mean area size error taking into account *K*

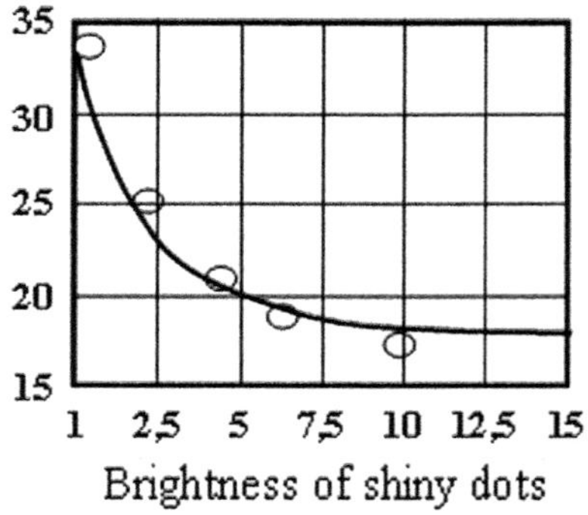

Fig. 12.34 Mean area center error

As previously, here remains the dependence of the mean area size and location errors on the target HL brightness. These dependences are presented in Figs. 12.32, 12.33 and 12.34.

The charts show that the general nature of curves is similar to the curves obtained for HL modeled by dipoles. When the target HL brightness increases, determination accuracy of the area center limiting bright HL also increases and the area size error decreases.

Bibliography

Part 1

1. Kozlov AI, Logvin AI, Sarychev VA (2008) Polarization of radio waves. Book 3. Radiopolarimetry of structurally complex signals. Radio engineering, Moscow
2. Beckmann P, Spizzichhino A (1963) The scattering of electromagnetic waves from rough surfaces. Pergamon Press, Oxford-London-New-York-Paris
3. Beloglazov IN, Kazarin SN, Kosyanchuk VV (2012) Information processing in iconic navigation, guidance and remote sensing systems. Fizmatlit, Moscow
4. Billingsley BJ (2013) Low-angle radar land clutter: measurements and empirical models. William Andrew Publishing, Norwick, NY
5. Bochkarev VV, Kravtsov VF, Sarychev VA, Kryzhanovskii GA, Kuzmin BI, Lipin AV, Lobachev EN, Sukhikh NN (2003) The concept and systems of CNS/ATM in civil aviation. Akademkniga ICC, Moscow
6. Bogorodsky VV, Kanareikin DB, Kozlov AI (1981) Polarization of scattered and intrinsic radio emission of terrestrial coverings. Gidrometeoizdat, Leningrad
7. Bogoslovsky AV (ed) (2012) Detection, recognition and definition of parameters of images of targets. Radio Engineering, Moscow
8. Bogoslovsky AV (ed) (2013) Processing of multidimensional signals. In: 2 books. Radio Engineering, Moscow
9. Branets VN (2009) Lectures on the theory of free-of-charge inertial navigation control systems. MIPT, Moscow
10. Branets VN, Sevastyanov NN, Fedulov RV (2013) Lectures on the theory of systems of orientation, motion control and navigation. TSU, Tomsk
11. Branets VN, Shmyglevsky IP (1992) Introduction to the theory of freeform inertial navigation systems. Science, Moscow

Part 2

12. Bogorodsky VV, Kanareikin DB, Kozlov AI (1981) Polarization of scattered and intrinsic radio emission of terrestrial coverings. Gidrometeoizdat, Leningrad
13. Branets VN, Shmyglevsky IP (1992) Introduction to the theory of freeform inertial navigation systems. Science, Moscow

Introduction to the Theory of Radiopolarimetric Navigation Systems,
Springer Aerospace Technology, https://doi.org/10.1007/978-981-13-8395-3

14. Branets VN, Shmyglevsky IP (1986) The use of quaternions in solid orientation problems. Science, Moscow
15. Kanareikin DB, Polyansky VA, Potekhin VA (1966) Polarization of radar signals. Moscow Sov. Radio
16. Kanareikin DB, Potekhin VA, Shishkin IF (1968) Marine polarimetry. Shipbuilding, Leningrad
17. Khizhnyak NA (1991) Modern problems of non-stationary macroscopic electrodynamics. NGO "Test Radio", Kharkov
18. Khudyakov GI (2009) Statistical theory of radio engineering systems. Academy, Moscow
19. Knott EF et al (1985) Radar cross section. Artech House, Dedham, MA
20. Kozlov AI, Lighart LP, Logvin AI (2001) Mathematical and physical modeling on microwave scattering and polarimetric remote sensing. Kluwer Academic Publishers, London

Part 3

21. Levanon N, Mozeson E (2004) Radar signals
22. Lvov LA (2003) Radar visibility of aircraft. RFNC-VNIITF, Snezhinsk
23. Nikolayev AI (ed) (2016) Radar systems. MSTU named after N.E. Bauman, Moscow
24. Proskurin VI, Yagolnikov SV, Shevchuk VI (2017) Radar surveillance. Methods, models, algorithms. Radio Engineering, Moscow
25. Richards MA, Scheer JA, Holm WA (eds) (2010) Principles of modern radar: basic principles. IET, Edison. SciTech Publishing, New York
26. Richards M (2015) Fundamentals of radar signal processing. McGraw-Hill, New York
27. Ostrovityanov RV, Basalov FA (1982) Statistical theory of the radar of extended targets. Radio and Communication, Moscow
28. Sharkov EA (2014) Radiothermal remote sensing of the Earth: physical basis, vol 2. IKI RAS, Moscow
29. Sherman SM, Barton DK (2011) Monopulse principles and techniques. Artech House
30. Shirman YD (2002) Computer simulation of aerial target radar scattering, recognition, detection, and tracking. Artech House Books, London
31. Shirman YD (ed) (2007) Handbook. radioelectronic systems. Fundamentals of construction and theory. Radio Engineering, Moscow
32. Skolnik M (ed) (2015) Handbook of radar. In: Two volumes. Technosphere, Moscow
33. Sosulin YG (1993) Theoretical basis of radar and radio navigation. Radio and Communication, Moscow
34. Sosulin YG, Kostrov VV, Parshin YN (2014) Evaluation-correlation-compensating signal processing and interference compensation. Radio Engineering, Moscow
35. Stager YA (1986) Scattering of radio waves on bodies of complex shape. Radio and Communication, Moscow
36. Tatarinov VN, Potekhin VA (1974) Questions of the classical theory of the partial coherence of a random electromagnetic field. TSU, Tomsk
37. Tatarinov VN, Tatarinov SV, Lighart LP. Introduction to the modern theory of polarized radar signals. Book 1. Tomsk
38. Tuchkov LT (ed) (1985) Radar characteristics of aircraft. Radio and Communication, Moscow
39. Verba VS, Tatarsky BG (ed) (2006) Complexes with unmanned aerial vehicles book 1-2. Radiotekhnika, Moscow

40. Lokhin VM, Zakharov VN (2001) Intelligent control systems: concepts, definitions, principles of construction. Fizmatlit, Moscow
41. White YI (ed) (2011) Radar systems for special and civil purposes 2010–2012. Radio Engineering, Moscow
42. Yagol'nikov SV (ed) (2013) Methods for studying the radar characteristics of objects. Radiotekhnika, Moscow

Printed by Printforce, the Netherlands